Biology and Ecology
of
Bluefin Tuna

Biology and Ecology
of
Bluefin Tuna

Editors

Takashi Kitagawa

Atmosphere & Ocean Research Institute
The University of Tokyo,
Kashiwa, Chiba
Japan

Shingo Kimura

Atmosphere & Ocean Research Institute
The University of Tokyo
Kashiwa, Chiba
Japan

CRC Press
Taylor & Francis Group
Boca Raton London New York

CRC Press is an imprint of the
Taylor & Francis Group, an **informa** business

A SCIENCE PUBLISHERS BOOK

Cover Illustrations:

· Background photograph reproduced by kind courtesy of Itaru Ohta (author of Chapter 6).

· Top left photograph reproduced by kind courtesy of Ko Fujioka (author of Chapter 5).

· Other three photographs reproduced by kind courtesy of the editor, Takashi Kitagawa.

CRC Press
Taylor & Francis Group
6000 Broken Sound Parkway NW, Suite 300
Boca Raton, FL 33487-2742

First issued in paperback 2020

© 2016 by Taylor & Francis Group, LLC
CRC Press is an imprint of Taylor & Francis Group, an Informa business

No claim to original U.S. Government works

ISBN-13: 978-1-4987-2487-6 (hbk)
ISBN-13: 978-0-367-73799-3 (pbk)

Visit the Taylor & Francis Web site at
http://www.taylorandfrancis.com

and the CRC Press Web site at
http://www.crcpress.com

Preface

Tuna are fascinating fishes that have been studied by fisheries scientists and fish ecologists for many years because of their high commercial value around the world. They are especially important in countries such as Japan, where tuna catches ranked the highest in the world in 2007, accounting for 14% (248,000 tons) of the world's total tuna catch. In addition, Japan is the largest tuna consumer in the world and is supplied with 473,000 tons of tuna (total amount of Japan's catches and imports; Fishery Agency 2009). Although Pacific (*Thunnus orientalis*), Atlantic (*T. thynnus*), and southern bluefin tuna (*T. maccoyii*) contribute relatively less in terms of total catch weight of the principal market tunas, their individual value is high because they are used for sashimi, which is a raw fish delicacy in Japan and increasingly in other countries (Majkowski 2007). Still fresh in our memories is the fact that a Pacific bluefin tuna caught off northeastern Japan fetched a record 155.4 million yen or about US\$ 1.76 million in the first auction in January 2013 at Tokyo's Tsukiji fish market. The price for the 222-kg Pacific bluefin tuna beat the previous year's record of 56.49 million yen (Kitagawa 2013).

Bluefin tuna numbers have decreased by 80% or more since 1970 as a result of overfishing (Dalton 2005). In 2010, at the 15th Conference of the Convention on International Trade in Endangered Species of Wild Fauna and Flora (CITES) in Doha, Qatar, a proposal to list Atlantic bluefin tuna in Appendix I of CITES presented by Monaco was discussed and put to a vote. Although the proposed ban was voted down, the stock management of the bluefin tuna species is being further strengthened. Therefore, for proper stock management of the species, more detailed biological information is needed for bluefin tuna, such as about their ecology, distribution, and movements.

As represented by Sharp and Dizon (1978), Hochachka and Mommsen (1991), and Block and Stevens (2001), from whom we learned much, studies on tuna were intensively conducted from the latter part of the previous century to the beginning of this century. However, thanks to the development of various kinds of technologies of microelectronics, microchemistry, molecular genetic science, and computer science including mathematical modeling, studies on the biology and ecology of tuna have progressed greatly after 2000. On the other hand, a book about tuna has not been published since Block and Stevens (2001), and the previous books only examined 'tuna' in general, that is, fishes of the genus *Thunnus* and/or Scombridae. This book, therefore, focuses on the latest information on the biology and ecology of the three bluefin tuna species that are the Pacific, Atlantic, and southern bluefin tuna.

In the book, the phylogeny of the three bluefin tuna species is described in the first part, then basic ecological information such as early life history, age and growth, food habits, feeding strategy and predators are covered in the second part. Information related to migratory ecology, and important biological aspects of each of the three species, such as metabolism and energetics, swimming performance, schooling, visual physiology, and reproductive physiology are included in the third and fourth parts, respectively. In the last part, new research insights about a few kinds of mathematical models about bluefin tuna ecology and a technique for measuring swimming behavior of bluefin tuna are introduced. All the chapters of the book have been contributed by active scientists engaged in bluefin tuna research.

We sincerely hope that this book will contribute to a better understanding of the biology and ecology of bluefin tuna, and that undergraduate and graduate students who read this book will be encouraged to become bluefin tuna scientists who can contribute to further understanding of the biology and ecology of bluefin tuna.

Lastly, we would like to thank all the contributors to the book and Charles J. Farwell, Monterey Bay Aquarium, USA and Michael J. Miller, Nihon University, Japan, for their advice and encouragement to make this book. We are grateful for the editorial assistance provided by Itsumi Nakamura, Atmosphere and Ocean Research Institute, The University of Tokyo, Japan.

Takashi Kitagawa
Shingo Kimura

References

Block, B.A. and E.D. Stevens. 2001. Tuna: Physiology, Ecology, and Evolution. Academic Press, 468 pp.

Dalton, R. 2005. Satellite tags give fresh angle on tuna quota. Nature 434: 1056–1057.

Fishery Agency. 2009. Fisheries of Japan (FY2009), Fisheries policy for FY2010 (White Paper on Fisheries), 30 pp.

Hochachka, P.W. and P.W. Mommsen. 1991. Phylogenetic and Biochemical Perspectives. Elsevier, 361 pp.

Kitagawa, T. 2013. Behavioral ecology and thermal physiology of immature Pacific bluefin tuna (*Thunnus orientalis*). pp. 152–178. *In*: H. Ueda and K. Tsukamoto (eds.). Physiology and Ecology of Fish Migration. CRC Press, Boca Raton, FL.

Majkowski, J. 2007. Global fishery resources of tuna and tuna-like species. FAO Fish. Tech. Pap. 483.

Sharp, G.D. and A.E. Dizon. 1978. The Physiological Ecology of Tunas. Academic Press, 485 pp.

Contents

IV. Behavioral or Physiological Aspects

V. New Insights: Mathematical model about Bluefin Tuna Ecology and Measuring Method for Swimming Behaviour of Bluefin Tuna

I. Phylogeny of Bluefin Tuna

CHAPTER 1

Phylogeny of Bluefin Tuna Species

Nobuaki Suzuki[1],* and *Seinen Chow*[2]

Phylogenetic Conflicts Among *Thunnus* Species: Morphology versus Molecular, Mitochondrial DNA versus Nuclear DNA

During the past half century, taxonomy, systematics and phylogenetic relationships among *Thunnus* tuna species have been investigated in depth by several eminent fish scientists. Before introducing the present view of phylogenetic relationships of *Thunnus* tuna species, we shall look at the history of tuna's systematics.

The systematic grouping and the phylogenetic implications were in complete harmony in the earlier decades but subsequently scientists made different postulates on the systematics of tuna based on detailed examinations for morphological characteristics. The comprehensive systematic studies of *Thunnus* were conducted from the 1960's and voluminous descriptions regarding measurements, meristics, coloration, osteological traits, viscera, vascular system and olfactory organ were made (Iwai et al. 1965; Nakamura 1965; Gibbs and Collette 1967). Comparing the 18 diagnostic characteristics, Gibbs and Collette (1967) recognized two groups of species in this genus; one the 'bluefin tuna group' consisted of *T. alalunga* (albacore), *T. thynnus* (northern bluefin tuna) and *T. maccoyii* (southern bluefin tuna) which were similar to each other in 14–16 diagnostic characters and the other the 'yellowfin tuna group' including *T. albacares* (yellowfin tuna), *T. atlanticus* (blackfin tuna) and *T. tonggol* (longtail tuna) which shared 15–16 characters. Gibbs and Collette also pointed out an intermediate state of diagnostic characters of *T. obesus* (bigeye tuna) between the two groups. These interpretations agreed with the intra-generic relationships presented by Iwai et al. (1965) and Nakamura (1965). Subsequently, Collette (1978) focused on internal morphology regarding adaptations for endothermy and considered that absence/presence of central heat exchanger within the bluefin tuna/yellowfin tuna

---
[1] National Research Institute of Far Seas Fisheries, Fisheries Research Agency, 5-7-1 Orido, Shimizu, Shizuoka 424-8633, Japan.
[2] National Research Institute of Fisheries Science, Fisheries Research Agency, 2-12-4 Fukuura, Kanazawa, Yokohama, Kanagawa 236-8648, Japan.
* Email: suzunobu@affrc.go.jp

groups, respectively, is a significant key to distinguish the two groups, and consequently interpreted the intermediate species of *T. obesus* as a member of the bluefin tuna group due to absence of this characteristic. Moreover, Collette (1978) hypothesized that the species of the bluefin tuna group have evolved from an ancestral tropical tuna and adapted into cooler waters due to the highly-developed lateral heat exchangers. However, there should be some concerns that the phylogenetic history was discussed based on adaptive characters.

Since 1980's, cladistic approach have been propagated to infer phylogenetic relationships of a wide variety of fishes. For scombroid fishes including the tunas, mackerels, billfishes and the relatives, several researches based on the approach have been conducted (Collette et al. 1984; Johnson 1986). Great accumulation in anatomical morphology of this group of fishes was strongly expected to find a trustable answer for the phylogenetic relationships. However, a robust phylogeny of scombroids was not able to be reconstructed and remained to be controversial because of conflicting and instability among resultant phylogenetic hypotheses (Carpenter et al. 1995). Since well-known morphology of scombroids could not even resolve the phylogenetic relationships among higher taxa than the genus level, it was certainly impossible to elucidate the interspecific relationships of the genus *Thunnus* in those days.

Next generation researches in tuna's phylogeny were brought by tremendous progress in biochemistry and molecular biology during the last decade of 20th century. Analytical techniques for allozymes (Elliott and Ward 1995) and mitochondrial DNA (mtDNA) (Bartlett and Davidson 1991; Block et al. 1993; Finnerty and Block 1995; Chow and Kishino 1995; Alvarado Bremer et al. 1997) were introduced to investigate the genetic relationships between *Thunnus* species and then genetic data different from morphological characters were collected. Based on the allozyme polymorphism analysis, Elliott and Ward (1995) reported genetic similarity among the five Pacific *Thunnus* species; Pacific northern bluefin (*T. t. orientalis*), southern bluefin, albacore, bigeye and yellowfin tunas, and implied the most divergent status of albacore. Finnerty and Block (1995) represented the novel phylogenetic hypothesis among scombroids including five tuna species (Atlantic northern bluefin, southern bluefin, albacore, bigeye and yellowfin tunas) inferred from partial nucleotide sequences of the mitochondrial cytochrome *b* gene, which indicated the sister relationship between albacore and the other monophyletic tunas and thus rejected monophyly of the bluefin tuna group previously supposed based on the morphology.

After the earlier molecular works, in order to illustrate the whole picture of *Thunnus* phylogeny and to conclude the taxonomy and systematics, Chow and Kishino (1995) and Alvarado Bremer et al. (1997) conducted comprehensive analyses including all of the seven species as well as the two northern bluefin subspecies. First, Chow and Kishino (1995) examined partial nucleotide sequences of the mitochondrial cytochrome *b* and ATPase genes in addition to Restriction Fragment Length Polymorphism (RFLP) of the internal transcribed spacer 1 (ITS1) in the nuclear rRNA gene family of all *Thunnus* species and then illustrated the comprehensive phylogeny. On the topology of the ATPase gene phylogeny, Pacific northern bluefin tuna (*T. t. orientalis*) was placed distant from the Atlantic subspecies (*T. t. thynnus*) but closely-related to albacore. Since the Atlantic northern bluefin and southern bluefin tunas were found to have mtDNA sequences very similar to species of yellowfin tuna group and not so similar to albacore and bigeye tunas which were morphologically assigned to the bluefin tuna group, rejection for monophyly of the bluefin tuna group was corroborated as proposed by the earlier molecular results. However, Chow and Kishino (1995) also reported an important finding that no differentiation in nuclear genome was observed between the Atlantic and Pacific northern bluefin tunas, in which the both subspecies were observed to share the same RFLP profiles for ITS1 generated by eight restriction enzymes. Here, we can see a more complicated conflict between mitochondrial and nuclear results as well as that between morphology and molecular data. Chow and Kishino (1995) concluded that albacore was the earliest offshoot, followed by bigeye tuna in this genus, which is inconsistent with the phylogenetic relationships between these tuna species inferred from morphology, and suggested possible hybridization that the mtDNA from albacore has been incorporated into the Pacific population of northern bluefin tuna and has extensively displaced the original mtDNA. Subsequently Alvarado Bremer et al. (1997) examined the rapidly evolving control region of mtDNA and presented the phylogenetic relationships almost consistent with Chow and Kishino (1995) except that bigeye tuna was identified as the sister species of the yellowfin tuna group. The authors concluded that, since the Atlantic and Pacific northern bluefin tunas were more divergent from each other than the average distance separating most species-pairs within the genus, a re-examination of their status as the Pacific subspecies of *T. thunnus* was warranted. After taking into account both molecular data introduced above and morphological differences (shape of the dorsal wall of the body cavity in large specimens and numbers of gill rakers), Collette (1999) and Collette et al. (2001) advocated separation of the northern bluefin tuna into the Atlantic species, *T. thynnus*, and the Pacific species, *T. orientalis* (Table 1.1).

Recently in order to elucidate the enigma regarding conflict between mitochondrial and nuclear phylogenies and to conclude phylogenetic relationships of *Thunnus* species, Chow et al. (2006) investigated the tuna's ITS1 nucleotide sequences with considering intraspecific variation of all of *Thunnus* species and with comparing the result to the updated phylogeny from mtDNA sequences. On both the ITS1 and the mtDNA phylogenies, the yellowfin tuna group was inferred as monophyletic, and southern bluefin and bigeye tunas showed a closer affinity to this tropical tuna group than to the northern bluefin tunas and albacore at least on the ITS1 phylogeny (Fig. 1.1). Therefore,

Table 1.1. Synopsis of current classification of *Thunnus* species.

Genus *Thunnus* South, 1845	
(Subgenus *Thunnus* South, 1845)	
Thunnus alalunga (Bonnaterre, 1788)	Albacore
Thunnus maccoyii (Castlenau, 1872)	Southern bluefin tuna
Thunnus obesus (Lowe, 1839)	Bigeye tuna
Thunnus thynnus (Linnaeus, 1758)	Atlantic bluefin tuna
Thunnus orientalis (Temminck and Schlegel, 1844)	Pacific bluefin tuna
(Subgenus *Neothunnus* Kishinouye, 1923)	
Thunnus albacares (Bonnaterre, 1788)	Yellowfin tuna
Thunnus atlanticus (Lesson, 1831)	Blackfin tuna
Thunnus tonggol (Bleeker, 1851)	Longtail tuna

This classification follows Collette (1999) and Collette et al. (2001).

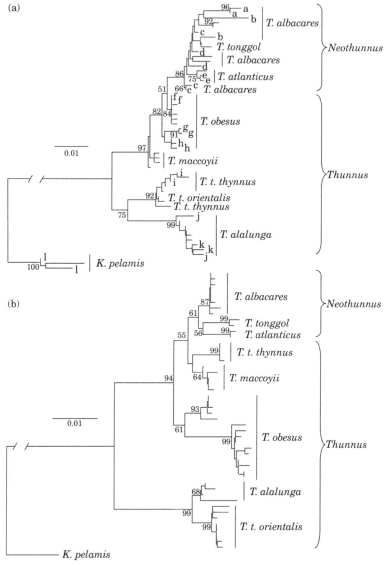

Figure 1.1. Neighbor-joining phylogenetic trees based on (a) nuclear rRNA ITS1 and (b) mtDNA AT sequence data (Chow et al. 2006). Bootstrap values >50% (out of 1000 replecates) are shown at the nodes.

although the relationships within the yellowfin group were still uncertain, the group should be existent based on similarities either from morphological and ecological characteristics or from nuclear and mitochondrial DNAs. On the other hand, the other morphological grouping, the bluefin tuna group, appeared to be questionable because the ITS1 and mtDNA topologies showing intermittent speciation of the bluefin tuna group members contradicted morphological subdivision of the *Thunnus* species. The resultant ITS1 phylogeny also provided remarkable similarity among the sequences

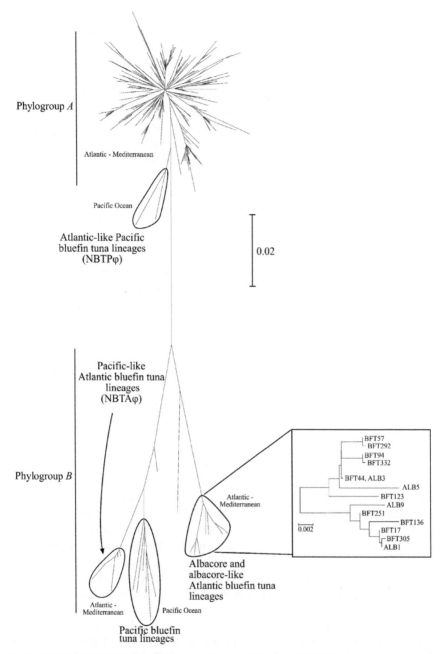

Figure 1.2. Unrooted neighbor-joining trees showing the relationships of mtDNA control region sequences for 334 Atlantic bluefin tuna haplotypes and for reference sequences of seven Pacific bluefin tuna and four albacore. Inset: subtree with albacore and albacore-like Atlantic bluefin tuna lineages. Alvarado Bremer et al. (2005) was modified.

from the Atlantic and Pacific northern bluefin tunas (Fig. 1.1a), which corresponded to the range of intraspecific variation, like as the identical RFLP profiles proposed by Chow and Kishino (1995). While the sister relationships among northern bluefin tunas and albacore were illustrated on the ITS1 phylogeny, the Pacific northern bluefin tuna alone was closely related with albacore on the mitochondrial phylogeny (Fig. 1.1b). The clustering the Pacific northern bluefin tuna with albacore in the mtDNA phylogeny may not be a simple consequence of the mitochondrial introgression from albacore to Pacific bluefin tuna proposed by Chow and Kishino (1995), but suggested an alternative mitochondrial introgression between the ancestral lineage of northern bluefin tuna and a species in the stem gene pool which subsequently led to the southern bluefin tuna, bigeye tuna and tropical tunas lineages. Consequently Chow et al. (2006) concluded the specific status of the two northern bluefin tunas would remain unresolved until more data from the nuclear genome become available.

As described above, a large amount of research focused on phylogenetic relationships and systematics of *Thunnus* tuna species have been performed during the last half century. Morphological, ecological and molecular data have also been accumulated continuously. The phylogeny that we can see, however, has been vague yet and some questions remain unresolved. Regarding the bluefin tuna species, it is sure that the most important issue is of phylogenetic relationship between Atlantic and Pacific northern bluefin tunas. Next we will focus further on inter- and intra-specific phylogeny of the two northern bluefin tunas.

Relationships Among Atlantic and Pacific Northern Bluefin Tunas

Alvarado Bremer et al. (2005) carried out comparative phylogeographic and historical demographic analyses for both Atlantic bluefin tuna and swordfish (*Xiphias gladius*) to clarify the complex phylogenetic signals in the North Atlantic-Mediterranean region. In the analysis for Atlantic bluefin tuna, nucleotide sequences of the mtDNA control region from the 607 individuals were investigated with those from Pacific bluefin tuna and albacore. The resultant mtDNA phylogeny obviously indicated reciprocal and monophyletic sister clusters of either Atlantic bluefin tuna or Pacific bluefin tuna in the Pacific and in the Atlantic, respectively. Definitely the Atlantic-like mtDNA lineage among Pacific bluefin tuna was reported by Chow and Kishino (1995) and Takeyama et al. (2001), while the Pacific-like mtDNA lineage among Atlantic bluefin tuna was found by Alvarado Bremer et al. (1999). These results should be interpreted as either Atlantic or Pacific bluefin tunas consisting of polyphyletic origins of mtDNA. The nucleotide sequence divergence between the Pacific bluefin tuna mtDNA and the Pacific-like Atlantic bluefin tuna mtDNA was estimated to be approximately 4.7% in the mtDNA control region, which was almost consistent with the reciprocal comparison between the Atlantic bluefin tuna mtDNA and the Atlantic-like Pacific bluefin tuna mtDNA (4.8%). Alvarado Bremer et al. (2005) attempted to propose a possible scenario to explain the reciprocal relationships between the Atlantic and the Pacific bluefin tunas, in which the common ancestors were sympatric until being separated by the rise of the Isthmus of Panama, about 3.0–3.5 million years ago and then gene flow between them have been prevented for a long time. According to the scenario, however, mutation rate for the mtDNA control region was estimated between 1.3–1.6% substitutions per million years, which would be too slow to compare the previous estimates for other teleosts. Therefore the authors explored an alternative calibration for the mtDNA control region in teleosts and adopted the rate of 4.9–5.7% substitutions per million years according to Donaldson and Wilson (1999) and Tringali et al. (1999). Using this calibration, cladogenesis splitting the Atlantic bluefin tuna mtDNA and the Atlantic-like Pacific bluefin tuna mtDNA was estimated to occur just prior to the glacial events about 980–840 thousand years ago, followed by population size expansion in the Atlantic Ocean during a warm period that occurred approximately 450–390 thousand years ago. It should be a likely estimation at present that induced from molecular phylogeny, historical demography and geographic information regarding the Atlantic bluefin tuna and its relative lineages.

When we consider phylogenetic relationships between Atlantic and Pacific bluefin tunas, one more confusing issue emerges from their evolutionary histories. That is possible introgressions between the northern bluefin tunas and albacore. First Chow and Kishino (1995) described closer relationships between Pacific bluefin tuna and albacore mtDNAs than that of Atlantic bluefin tuna in contrast with no differentiation regarding electropholetic patterns of nuclear ITS1 RFLP between Atlantic and Pacific bluefin tunas. Since Atlantic bluefin and southern bluefin tunas were also found to have similar mtDNA sequences, Chow and Kishino (1995) proposed the hypothesis that albacore mtDNA has been introgressed into the Pacific bluefin tuna and that three bluefin tuna species should be closely-related to each other. Subsequently Chow et al. (2006) presented an alternative hypothesis regarding albacore mtDNA introgression because of clustering the two northern bluefin tunas with albacore in ITS1 sequence phylogeny, which has been uncertain yet but possibly occurred between the ancestral lineage of northern bluefin tunas and a species in the stem gene pool which subsequently led to the southern bluefin tuna, bigeye tuna and tropical tunas lineages. In Atlantic bluefin tuna, however, a different case of close relationships between Atlantic bluefin tuna and albacore mtDNAs was found by Alvarado Bremer et al. (2005). About 3.3% of Atlantic bluefin tuna mtDNA sequences were interspersed among the albacore mtDNA cluster in the phylogeny and also found to include one shared sequence. This implies a recent hybridization event between Atlantic bluefin tuna and albacore and may support an introgression hypothesis causing confusing phylogenetic relationships among bluefin tuna species as well as albacore.

In contrast to the confusing relationships in mtDNA lineages among Pacific and Atlantic bluefin tunas together with albacore, clear geographic population subdivision among Atlantic bluefin tuna was detected using nuclear DNA markers. Albaina et al. (2013) discovered a total of 616 Single Nucleotide Polymorphisms (SNPs) from 54 PCR amplicons of nuclear DNA among 35 fish samples of albacore. Of all the SNPs, 128 SNPs were attempted in cross-species genotyping for Atlantic bluefin tuna, of which 17 SNPs corresponding to 15 markers (13 SNPs and two haplotype blocks) were validated for PCR amplification, data quality and Mendelian population genetic assumptions. Despite a small number of markers, these 15 markers succeeded to detect significant overall differentiation among 107 individuals from three geographic samples in the northwestern Atlantic, northeastern Atlantic (Bay of Biscay) and Mediterranean

(Balearic Sea) (P < 0.05). Individual clustering analysis illustrated that the three samples most likely represented two genetically distinguishable populations, that is, northwestern Atlantic or northeastern Atlantic plus Mediterranean (Fig. 1.3), that corresponded to the two major spawning grounds of Atlantic bluefin tuna. In contrast that the young-of-the-year samples from northwestern Atlantic and Mediterranean should be represented for each spawning group, the northeastern Atlantic sample composed mixed age classes from juveniles to adults foraging in the Bay of Biscay. Therefore the SNP analysis indicated the Mediterranean origins of foraging individuals in the Bay of Biscay.

The bluefin tuna phylogeny and their evolutionary history with the close relatives have remained to be ambiguous as described above. However, new effective approaches such as the SNP markers by Albaina et al. (2013) are appearing in fish researches. Recently, genomic approach is becoming common not only in medical sciences for human health but also in natural sciences for wild organisms. Next, some brand-new genomic studies targeted at *Thunnus* tuna species are introduced. A variety of genomic approaches will be forthcoming powerful tools for improving population genetics, phylogenetic systematics and evolutionary ecology, and may provide breakthroughs to disentangle confusing relationships among bluefin tuna species.

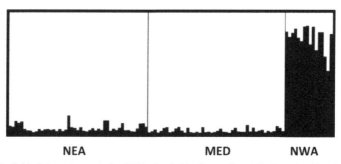

NEA   **MED**   **NWA**

Figure 1.3. Individual clustering analysis of 107 Atlantic bluefin tuna for 17 single nucleotide polymorphisms (SNPs) on 15 independent DNA fragments. Each vertical bar represents an individual, and sampling locations are separated by vertical black lines. The color proportions of each bar correspond to the individual's estimated membership fractions to each of the clusters (cluster membership coefficient) (Albaina et al. 2013).

An Epoch in Tuna Genomics

In the last decade, nucleotide sequencing capability has been dramatically progressing due to spread of multi-capillary sequencers or to the appearance of Next-Generation Sequencers (NGS), which necessitate rapid development of bioinformatic technology to analyze a huge amount of sequence reads generated from NGS. This situation allows us to perform much larger-scale analyses beyond our earlier expectations.

The evolutionary origin of the family Scombridae including *Thunnus* tuna species has been a longtime issue despite long-term arguments based on morphological and molecular data (Collette et al. 1984; Johnson 1986; Block et al. 1993; Orrell et al. 2006). As a new approach to address the issue, Miya et al. (2013) collected more than 10,000 sequences of percomorph fishes from international DNA database and bioinformatically sorted into nine homologous protein-coding (six mitochondrial and three nuclear) genes, which finally comprised 5,367 species across 1,558 genera and 215 families (87% of total percomorph family diversity). A whole result from each phylogeny of the individual genes suggested that 15 perciform families formed the least monophyletic group that contains all core members of the classical scombroid families (Gempylidae, Trichiuridae, Scombridae). According to the bioinformatic result, Miya et al. (2013) sampled a total of 124 complete mitochondrial genomic sequences, which consisted of 56 species from those 15 families (in group) and 68 out group species including representative pelagic percomorphs, with generating original sequences from 37 in group species and from 17 out group species. Large data sets including more than 10,000 mitochondrial nucleotides from 12 protein-coding genes, 22 tRNA genes and 2 rRNA genes were produced and then analyzed by phylogenetic maximum likelihood estimation. The result from the mitogenomic analysis supported a monophyly of the 15 families noticed above with 100% bootstrap probability and the family Scombridae also indicated to form a robust monophyletic group with strong statistical support (Fig. 1.4). Additionally Miya et al. (2013) proposed the time-calibrated phylogeny and the ecological mapping of specific habitat depth on the topology, consequently suggesting a possible origin of scombrids from a deep-water ancestor and subsequent radiation following the Cretaceous-Paleogene mass extinction including large predatory epipelagic fishes. In their study, a new approach that combined an unprecedented scale of both partial and complete mitochondrial genomic data with bioinformatic technology was able to illustrate the phylogenetic framework and the evolutionary scenario regarding scombrids clearly. Although phylogenetic relationships among and within tuna species were not focused in this study, large-scale mitogenomic analyses with many more samples covering a geographic range of each species might be able to disentangle mitochondrial phylogenetic uncertainty of *Thunnus*.

Another innovative work was brought by an application of NGS. Nakamura et al. (2013) reported the complete genome sequences of a male of Pacific bluefin tuna caught as a young-of-the-year juvenile off the Pacific coast of Japan and reared for about three years, in which genetic and evolutionary basis of optic adaptation of tuna was elucidated. A whole-genome shotgun sequencing and subsequent bioinformatic assembling provided a total of 740.3 Mb sequences consisting of 192,169 contigs (>500 bp) and 16,802 scaffolds (> 2 kb), corresponding to 92.5% of the estimated genome size (~800 Mb). Additional NGS sequencing was carried out for total cDNA from another adult female reared at age five, and mapping the cDNA sequences as well as Expressed Sequence Tags (ESTs) from the Atlantic bluefin tuna on the genome of Pacific bluefin tuna predicted more than 26,000 protein-coding gene candidates. Pairwise comparison with known genome sequences from six teleosts (cod, medaka, zebrafish, stickleback, greenpuffer, fugu) found out 6,170 conserved genes in the Pacific bluefin tuna genome. Among the genes, 10 visual pigment genes were identified and compared with fish opsin sequences deposited in the DNA databases. Based on the multiple alignments, Nakamura et al. (2013) found amino acid substitutions at spectral tuning sites of rhodopsin and four green-sensitive opsin genes (Table 1.2) which involved light sensitivity and may be tuned to blue light. It is likely that these substitutions contribute to spectral tuning to the offshore environment, for better

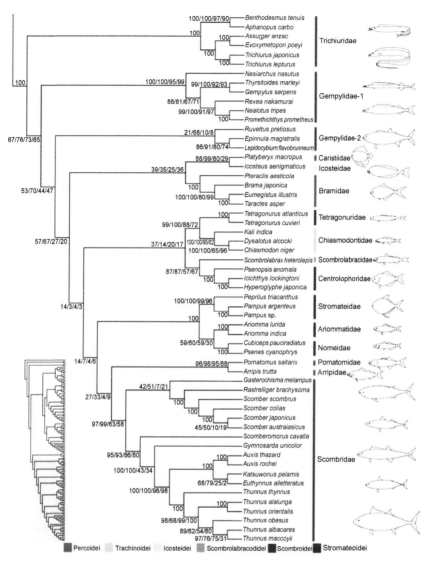

Figure 1.4. The best-scoring maximum likelihood tree based on whole mitogenome sequences. Bootstrap values >50% (out of 1000 replecates) are shown at the nodes. Miya et al. (2013) was modified.

Table 1.2. Representative amino acid sites involved in the light sensitivity of rhodopsin in comparison with seven teleosts.

Tuning site	zebrafish	cod	medaka	greenpuffer	fugu	stickleback	tuna
83	D	N	D	N	D	D	D
122	E	E	E	E	E	E	Q
261	F	F	F	F	F	F	F
292	A	A	A	A	A	A	A

cognition for bluish contrasts, and consequently to effective detection of prey in the bluish ocean. Phylogenetic analysis illustrated that the amino acid substitution of rhodopsin was widely conserved among the suborder Scombroidei containing the family Scombridae (Fig. 1.5), suggesting that the molecular evolutionary change

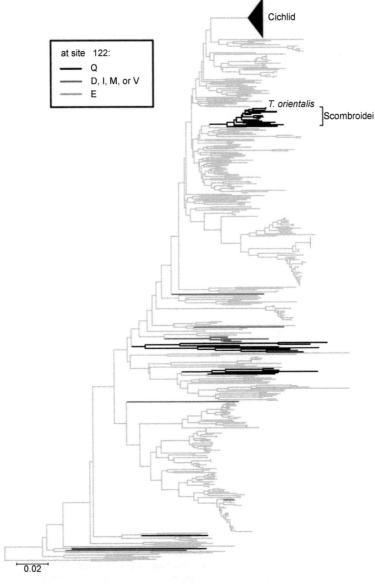

Figure 1.5. A neighbor-joining tree of fish rhodopsin sequences. A total of 557 nucleotide sequences including that for tuna (accession no. BAG14281) were collected from the GenBank database. The species that have glutamine (Q) at site 122 are colored in blue. For site 122, the species with aspartic acid (D), isoleucine (I), methionine (M), or valine (V) are colored in red, and those with glutamic acid (E) are colored in gray. Nakamura et al. (2013) was modified.

occurred in an ancestor of this lineage. Phylogenetic analysis also suggested that gene conversions have occurred in each of the blue- and green-sensitive gene loci in a short period. Those genetic recombinations may have facilitated adaptation to the offshore environment. Further genomic information from a wide range of pelagic fishes including tunas will provide significant clues for comprehensive understanding of biodiversity and adaptation of *Thunnus*.

In this chapter we followed the progress of researches about taxonomy, systematics and phylogenetic relationships for *Thunnus* species mainly focusing on bluefin tunas during the last half century, recognizing that many issues remain such as conflicts among morphology, mitochondrial DNA and nuclear genome. And finally we reached two large-scale genomic researches introduced earlier. Genome-wide sequence analysis was performed only in the Pacific bluefin tuna in the genus *Thunnus*, but the up-to-date genomic data provided a new interesting and detailed insight for tuna evolution and adaptation. As noted by Albaina et al. (2013), the NGS technology will undoubtedly realize a wide variety of genome-wide examinations in non-model organisms in the near future. Using NGS approaches targeted at total genomes as well as mitochondrial genomes, further applications for more samples of *Thunnus* species are largely expected to disentangle the phylogenetic uncertainty.

References

Albaina, A., M. Iriondo, I. Velado, U. Laconcha, I. Zarraonaindia, H. Arrizabalaga, M.A. Pardo, M. Lutcavage, W.S. Grant and A. Estonba. 2013. Single nucleotide polymorphism discovery in albacore and Atlantic bluefin tuna provides insights into worldwide population structure. Anim. Genet. 44: 678–692.

Alvarado Bremer, J.R., I. Naseri and B. Ely. 1997. Orthodox and unorthodox phylogenetic relationships among tunas revealed by the nucleotide sequence analysis of the mitochondrial DNA control region. J. Fish Biol. 50: 540–554.

Alvarado Bremer, J.R., I. Naseri and B. Ely. 1999. Heterogeneity of northern bluefin tuna populations. ICCAT Coll. Vol. Sci. Pap. 49: 127–129.

Alvarado Bremer, J.R., J. Viñas, J. Mejuto, B. Ely and C. Pla. 2005. Comparative phylogeography of Atlantic bluefin tuna and swordfish: the combined effects of vicariance, secondary contact, introgression, and population expansion on the regional phylogenies of two highly migratory pelagic fishes. Mol. Phylogenet. Evol. 36: 169–87.

Bartlett, S.E. and W.S. Davidson. 1991. Identification of *Thunnus* tuna species by the polymerase chain reaction and direct sequence analysis of their mitochondrial cytochrome b genes. Can. J. Fish. Aquat. Sci. 48: 309–317.

Block, B.A., J.R. Finnerty, A.F.R. Stewart and J. Kidd. 1993. Evolution of endothermy in fish: mapping physiological traits on a molecular phylogeny. Science 260: 210–214.

Carpenter, K.E, B.B. Collette and J.L. Russo. 1995. Unstable and stable classifications of Scombroid fishes. Bull. Mar. Sci. 56: 379–405.

Chow, S. and H. Kishino. 1995. Phylogenetic relationships between tuna species of the genus *Thunnus* (Scombridae: Teleostei): inconsistent implications from morphology, nuclear and mitochondrial Genomes. J. Mol. Evol. 41: 741–748.

Chow, S., T. Nakagawa, N. Suzuki, H. Takeyama and T. Matsunaga. 2006. Phylogenetic relationships among *Thunnus* Species inferred from rDNA ITS1 Sequence. J. Fish Biol. 68(supple. A): 24–35.

Collette, B.B. 1978. Adaptation and systematics of the mackerels and tunas. pp. 7–39. In: G.D. Sharp and A.E. Dizon (eds.). The Physiological Ecology of Tunas. Academic Press, New York.

Collette, B.B. 1999. Mackerels, molecules, and morphology. pp. 149–164. *In*: Proc. 5th Indo-PaciWc Fish. Conf., Nouméa, 1997, Soc. Fr. Icthyol.

Collette, B.B., T. Pothoff, W.J. Richards, S. Ueyanagi, J.L. Russo and Y. Nishikawa. 1984. Scombroidei: Development and relationships. pp. 591–620. *In*: H.G. Moser, W.J. Richards, D.M. Cohen, M.P. Fahay,

A.W. Kendall, Jr. and S.L. Richardson (eds.). Ontogeny and Systematics of Fishes. Spec. Publ. 1, Am. Soc. Ichthyol. Herpetol. Allen Press, Lawrence.

Collette, B.B., C. Reeb and B.A. Block. 2001. Systematics of the tunas and mackerels (Scombridae). pp. 1–33. In: B.A. Block and E.D. Stevens (eds.). Tuna: Physiology, Ecology, and Evolution. Fish Physiology Series, vol. 19. Academic Press, San Diego.

Donaldson, K.A. and R.R. Wilson, Jr. 1999. Amphi-panamic geminates of snook (Percoidei: Centropomidae) provide a calibration of the divergence rate in the mitochondrial DNA control region of fishes. Mol. Phylogenet. Evol. 13: 208–213.

Elliott, N.G. and R.D. Ward. 1995. Genetic relationships of eight species of Pacific tunas (Teleostei: Scombridae) inferred from allozyme analysis. Mar. Freshwater Res. 46: 1021–1032.

Finnerty, J.R. and B.A. Block. 1995. Evolution of cytochrome b in the Scombroidei (Teleostei): molecular insights into billfish (Istiophoridae and Xiphiidae) relationships. Fish. Bull. 93: 78–96.

Gibbs, R.H. and B.B. Collette. 1967. Comparative anatomy and systematics of the tunas, genus *Thunnus*. Fish. Bull. 66: 65–130.

Iwai, T., I. Nakamura and K. Matsubara. 1965. Taxonomic study of the tunas. Bull. Misaki Mar. Biol. Inst. Kyoto Univ. [Special Report] 2: 1–51.

Johnson, G.D. 1986. Scombroid phylogeny: an alternative hypothesis. Bull. Mar. Sci. 39: 1–41.

Miya, M., M. Friedman, T.P. Satoh, H. Takeshima, T. Sado, W. Iwasaki, Y. Yamanoue, M. Nakatani, K. Mabuchi, J.G. Inoue, J.Y. Poulsen, T. Fukunaga, Y. Sato and M. Nishida. 2013. Evolutionary origin of the Scombridae (tunas and mackerels): members of a Paleogene adaptive radiation with 14 other pelagic fish families. PloS One 8: e73535.

Nakamura, I. 1965. Relationships of fishes referable to the subfamily Thunninae on the basis of the axial skeleton. Bull. Misaki Mar. Biol. Inst. Kyoto Univ. 8: 7–38.

Nakamura, Y., K. Mori, K. Saitoh, K. Oshima, M. Mekuchi, T. Sugaya, Y. Shigenobu, N. Ojima, S. Muta, A. Fujiwara, M. Yasuike, I. Oohara, H. Hirakawa, V. Sur Chowdhury, T. Kobayashi, K. Nakajima, M. Sano, T. Wada, K. Tashiro, K. Ikeo, M. Hattori, S. Kuhara, T. Gojobori and K. Inouye. 2013. Evolutionary changes of multiple visual pigment genes in the complete genome of Pacific bluefin tuna. Proc. Natl. Acad. Sci. USA 110: 11061–11066.

Orrell, T.M., B.B. Collette and G.D. Johnson. 2006. Molecular data support separate scombroid and xiphioid clades. Bull. Mar. Sci. 79: 505–519.

Tringali, M.D., T.M. Bert, S. Seyoum, E. Bermingham and D. Bartolacci. 1999. Molecular phylogenetics and ecological diversification of the transisthmian fish genus *Centropomus* (Perciformes: Centropomidae). Mol. Phylogenet. Evol. 13: 193–207.

II. Basic Ecology of Bluefin Tuna

CHAPTER 2

Early Life History

Yosuke Tanaka[1],* and *Nobuaki Suzuki*[2]

Introduction

Many studies concerning the early life history of Pacific bluefin tuna *Thunnus orientalis* have been conducted since the 1950s in Japan. Generally, studies of the early life history of marine fishes are composed of two approaches; field surveys and rearing experiments. At the beginning, the studies conducted for the early life history of Pacific bluefin tuna were the former. From the 1950s to the 1980s, large scale surveys of field sampling of the larvae were conducted around the spawning grounds formed in the northwestern Pacific Ocean. A lot of information about the species identification and the distribution of the larvae were accumulated through these surveys. Thereafter, in the 2000s studies concerning the growth and development of larvae and survival mechanisms in relation to those have been conducted using various analysis methods.

On the other hand, studies on technique development for artificial mass culture have been promoted from the 1990s aimed at promoting aquaculture and stock enhancement in Japan. The most remarkable work in this process is the life completion of Pacific bluefin tuna under aquaculture conditions at Kinki University accomplished in 2002 (Sawada et al. 2005). In the processes of the technique development of mass cultures, various rearing experiments have revealed the ecological, physiological and behavioral traits of the early life stages of Pacific bluefin tuna, which are difficult to be researched based on field captured specimens.

Thus, recent studies on the early life history of Pacific bluefin tuna have been conducted by field surveys and rearing experiments. Since Pacific bluefin tuna is the only species to be reared throughout their whole life stages among other tuna species, we focus on the early life history of Pacific bluefin tuna. In this chapter, we review the eco-physiological and behavioral traits and the survival mechanisms in the early life history of Pacific bluefin tuna, revealed by both field surveys and rearing experiments.

[1] Research Center for Tuna Aquaculture, Seikai National Fisheries Research Institute, Fisheries Research Agency, 955 Hyou, Setouchi, Kagoshima, 894-2414, Japan.
[2] National Research Institute of Far Seas Fisheries, Fisheries Research Agency, 5-7-1 Orido, Shimizu, Shizuoka 424-8633, Japan.
* Email: yosuket@affrc.go.jp

Distribution of Larvae and Juveniles of Pacific Bluefin Tuna

Spatiotemporal extent of the larval distribution in the Pacific Ocean

Research surveys to analyze the geographical distribution of Pacific bluefin tuna larvae can be categorized into three generations, which have been carried out for over more than a half of century by Japanese scientists. The first generation of surveys (1956–1989) was summarized by Nishikawa et al. (1985) and Yonemori (1989), focused on understanding the spawning grounds and periods based on the extent of larval distribution in the northwestern Pacific Ocean. First, Ueyanagi (1969) threw light upon the geographical patterns of the larval distributions of the tuna species within the Indo-Pacific oceans by presenting charts where sampling sites and catch records of larvae of albacore, yellowfin, bigeye, skipjack, Pacific bluefin and southern bluefin tunas were given. Regarding the Pacific bluefin tuna, a relatively restricted larval distribution was found around the Kuroshio Current and its counter current regions from the south of Japan to the east of Taiwan (see Appendix Fig. 8 in Ueyanagi 1969). Next, Nishikawa et al. (1978) and Nishikawa et al. (1985) attempted to compile all of the data collected by both research vessels as well as local government vessels during the past large-scale sampling program, to calculate the Catch Per Unit Effort (CPUE) in terms of the number of larvae per 1,000 m^3, and to illustrate either annual or quarterly global distributions of larval densities of various scombrid fishes. Complete data analyzed in Nishikawa et al. (1985) was compiled from 63,017 tows of 1.4 m or 2.0 m diameter plankton net conducted from 1956 to 1981. The resultant chart succeeded to enable a more detailed distribution of Pacific bluefin tuna larvae than that by Ueyanagi (1969), identifying the putative spawning area ranging from the Bashi Strait to the southern coast of central Japan with the spawning season mainly from April to June. Additionally Nishikawa et al. (1985) referred to the occurrence of larvae from the Sea of Japan during early August in the text despite no plot being given on the chart, and then discussed the strong 'homing' nature of this species based on the limited spawning area compared to the Pacific-wide adult distribution. In order to clarify whether this species does in fact show 'homing' behavior, it is necessary to consider the spawning ecology of this species.

Following the earliest works that had provided basic information on spawning grounds of wild Pacific bluefin tuna, large-scale surveys called 'Marine Ranching Project' were conducted from 1980 to 1989 with a preliminary survey in 1979. Yonemori (1989) summarized the 10-year project, clearly indicating two geographically-isolated spawning grounds, the Nansei area and the Sea of Japan, as well as the spatiotemporal extent of the larval distribution (Fig. 2.1). Through the project, 928 surface trawl tows in surrounding waters of the Nansei Islands, 152 tows in the Pacific offshore waters of the Japanese main island, Honshu, and 138 tows in the Sea of Japan were carried out using 2 m diameter plankton net, consequently more than 10,000 individuals of Pacific bluefin tuna larvae were sampled and examined. Although the Nansei area had relatively higher CPUEs than those either in the Pacific offshore waters or in the Sea of Japan, sampling sites with high CPUEs varied among year to year fluctuations within the areas. Therefore the Nansei area should be considered as the main ground for spawning of the Pacific bluefin tuna but the spawning behavior

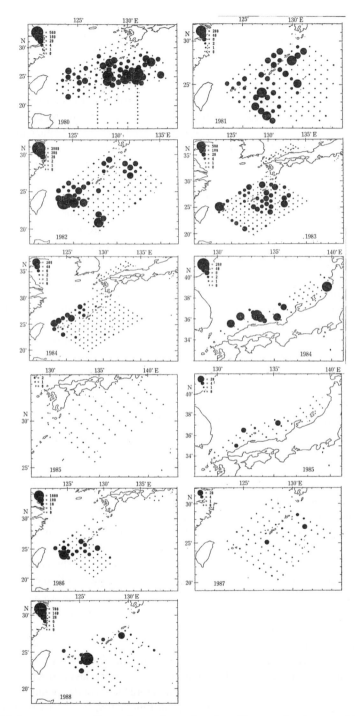

Figure 2.1. Survey area, sampling stations and annual catches of larvae and early stage of juveniles of Pacific bluefin tuna from 1980 to 1988 (Yonemori 1989).

may be adaptively changeable. The larvae occurred at 23.5 to 29.5°C of Sea Surface Temperature (SST), most frequently at 25 to 27°C (Yonemori 1989), and this SST can therefore be considered as the most appropriate condition at the main spawning ground. In addition, since the larvae tended to occur around the Yaeyama Islands, the western part of the Nansei area, and near the offshore edge of the Kuroshio Current including the counter current region, and not occurring on the continental shelf of the East China Sea, Yonemori (1989) concluded that the Pacific bluefin tuna might spawn more actively in the counter current region with oceanographic eddies than in the main stream region of the Kuroshio Current. Lower CPUEs observed in the Pacific offshore waters of Honshu and in the Sea of Japan, in contrast to the Nansei area, also provided evidence of spawning events in both areas. These findings succeeded to illustrate details of the spatiotemporal extent of the spawning grounds of Pacific bluefin tuna; middle May to late June in the Nansei area, late June to early July in the Pacific offshore waters of Honshu, and late July to mid August in the Sea of Japan. In addition, the spawning in the Sea of Japan was considered to occur every year based on larval catches in 1984 and 1985, alternatively the spawning in the Pacific offshore waters of Honshu appeared to be more restricted. As noted above, a whole picture of the distribution of the Pacific bluefin tuna larvae was drawn in the first generation of surveys, but environmental details providing the basis for better understand the larval distribution were still lacking.

Since there was a suspension of surveys following the Marine Ranching Project, we needed to wait subsequent progress until the beginning the second generation of surveys from 2004 to 2010. In the second period, two types of surveys were conducted focusing on high density larval distributions (patches) of the larvae; radio buoy tracking (2004–2008) and environmental adaptive sampling (2007–2010). These surveys were designed to develop new strategies for both larval sampling and oceanographic observations in waters of the major spawning ground, the Nansei area. Radio buoy tracking accompanying oceanographic observations by Conductivity Temperature Depth (CTD) profiling and Acoustic Doppler Current Profiler (ADCP) represented detailed distribution of larval patches in relation to mesoscale eddies successfully as well as the vertical and horizontal extents of the patch structure, larval growth and natural mortality (Satoh et al. 2008; Satoh 2010; Satoh et al. 2013). Satoh (2010) described patches entrained in mesoscale eddies (~100 to 500 km diameter), and therefore considered that the positional relationship between spawning events and mesoscale eddies was important for the recruitment process of the larvae (Fig. 2.2). Similarly adaptive sampling trials adjusting to oceanographic conditions in the Nansei area detected occurrence of larval patches around mesoscale eddies and succeeded to converge the survey area to monitor year-to-year fluctuations in amounts of larval catches irrespective of oceanographic conditions (Suzuki et al. 2014) (Fig. 2.3). Marked collections of Pacific bluefin tuna larvae were made at sampling sites located between 24° and 26° N along a 126° E line throughout the four-year surveys, which seemed to be influenced by mesoscale eddies based on sea surface level structure and local currents resulting from bottom topography. Therefore the second generation of surveys provided fragmentary information of relationships between the larval distribution and oceanographic condition but enabled development of the next scheme for further surveys focusing on fluctuations in biological and environmental factors.

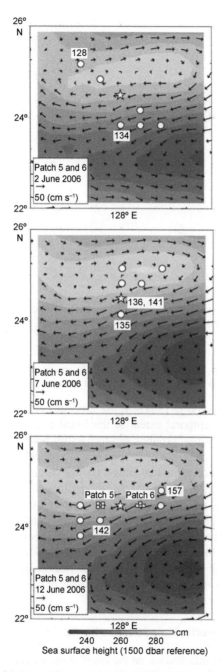

Figure 2.2. Sea surface height, sea surface currents and sampling stations within the Nansei area in June 2006. Open circles with/without numbers show sampling stations. Stars indicate stations at the center of the schematic observation. Patches 5 and 6 were detected at the stations 146 and 151, respectively. Modified from Satoh (2010).

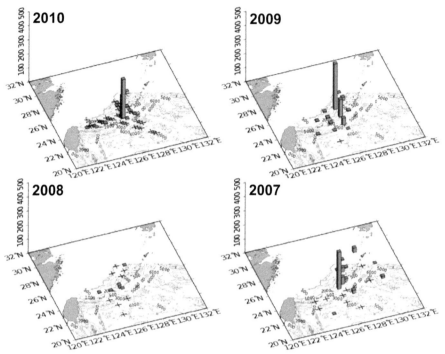

Figure 2.3. Horizontal plots of annual catches of Pacific bluefin larvae in the Nansei area from 2007 to 2010. Bars indicate numbers of larvae and '+' indicates no catch at a given station (Suzuki et al. 2014).

The most recent surveys (2011-), the third generation, have been continuing to evaluate the spatiotemporal extent of the larval distribution in Japanese waters including the two spawning grounds and to identify both spawning spots and their surrounding environments. Possible spawning spots, where the spawning events actually occurred, were estimated based on the catch of the larvae using back calculations on an oceanographic numerical model, though much less larval collections in the Sea of Japan may provide rough estimates of the spots in the area. The calculation also can provide possible trajectory of advection history and oceanographic conditions experienced by the larvae. This new approach that consists of biological sampling including the Pacific bluefin tuna larvae and other plankton species, oceanographic observations, and back calculations on the numerical model is anticipated to clarify the spawning ecology and larval survival processes of the Pacific bluefin tuna in the near future.

Analytical and theoretical approaches for the larval distribution

As noted above, basic information on the distribution and environmental conditions of the larvae have been accumulated over about a half of the century, but analytical and/or theoretical works to validate the favorable conditions for the larvae are strongly required for a better understanding larval distribution, spawning and reproductive ecology, and the early life history of the wild Pacific bluefin tuna. In the Atlantic congener, *T. thynnus*, many more studies to analyze field data using statistics and numerical modeling have been energetically performed for both spawning grounds, the Gulf of Mexico (Teo et al. 2007; Muhling et al. 2010; Lindo-Atichati et al. 2012; Muhling et al. 2013) and the Mediterranean Sea (Alemany et al. 2010; Mariani et al. 2010; Muhling et al. 2013; Rodriguez et al. 2013). In contrast, in the Pacific bluefin tuna larvae, few studies have been made except for the modeling approaches on larval and juvenile dispersal processes by Kitagawa et al. (2010) and to evaluate possible impacts of the global warming by Kimura (2010). Regarding the larval transporting processes and the evolutionary interpretation on having both restricted spawning area and season compared to other tropical tuna species, Kitagawa et al. (2010) presented clear evidence based on numerical simulations that reproduction of larvae around the Nansei area, not in the Pacific offshore waters of Honshu, should enable effective recruitment of the Pacific bluefin tuna larvae into the nursery area located in Japanese coastal waters, irrespective of annual fluctuations in oceanographic conditions (Fig. 2.4). According to Kitagawa et al. (2010), restricted spawning area and season of Pacific bluefin tuna could be explained as a consequence to maximize success of larval recruitment into optimal thermal conditions for better growth and survival with low energetic costs rather than as a preferable environment for spawning adults. This is considered to be strongly related to the evolutionary history of the Pacific bluefin tuna and seems to be largely consistent with a hypothetical scenario that the period from larval to very early juvenile stages is critical for density-dependent controls operating on the population dynamics. Theoretically possible hypotheses of evolutionary mechanisms that had generated restricted spawning grounds and limited larval distribution in bluefin tuna species have been proposed by Bakun (2006; 2013), based on strongly convergent environments around energetically forced ocean eddies and interactions between larvae and their predators. Further analytical as well as modeling studies based on methods such as described by Kitagawa et al. (2010) are likely to present some kind of empirical evidence for ecological and evolutionary hypotheses of recruitment processes of the Pacific bluefin tuna.

Distribution of juvenile Pacific bluefin tuna

Juvenile Pacific bluefin tuna (< 50 cm Fork Length, FL) are distributed around the coastal area of Japan and are caught by troll and/or purse seine fisheries in the Pacific Ocean and in the Sea of Japan, part of which are used as seedlings for aquaculture. Fisheries-based information on the juvenile distribution, therefore, has been sufficiently accumulated but scientific research surveys have been lacking with the fisheries due to the absence of appropriate sampling gears and skills. To capture sufficient numbers of tuna species juveniles, midwater trawl net with a large mouth opening capable of high speed towing was developed and adopted for research surveys targeting tuna juveniles (Tanabe and Niu 1998; Itoh et al. 1999; Mohri et al. 2005). The first record of a large amount of Pacific bluefin tuna juveniles was from the survey during June 11–14, 1997, around the Kuroshio Current region north of the Okinawa main Island in the Nansei area by using a large mouth midwater trawl net (total length: 86 m, expected mouth opening: 30 x 30 m) (Fig. 2.5, Itoh et al. 1999). Based on molecular species identification of mtDNA haplotypes, 164 juveniles of Pacific bluefin tuna (19.5–46.9 mm FL, 35.7 mm mean FL) were collected (Chow et al. 2003) and found to inhabit

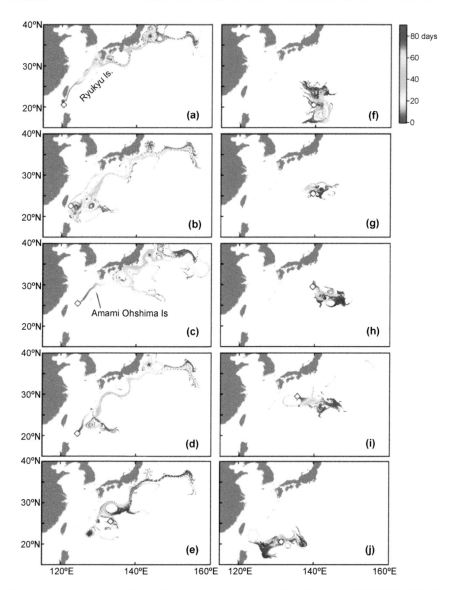

Figure 2.4. Representative particle trajectories from a release point (a) to the south (20.5°N, 120.5°E) and (b) east (22.5°N, 122.5°E) of Taiwan, (c) close to the Kuroshio (25.5°N, 125.5°E), (d) 20.5°N, 124.5°E and (e) 26.5°N, 129.5°E in waters on the eastern edge of the putative spawning area of bluefin tuna, and (f) 29.5°N, 130.5°E, (g) 25.5°N, 139.5°E, (h) 29.5°N, 139.5°E, (i) 29.5°N 135.5°E and (j) 20.5° N 131.5°E in waters outside of the putative spawning area. The diamond in each panel indicates the release point. The colors indicate the number of days after hatching (Kitagawa et al. 2010).

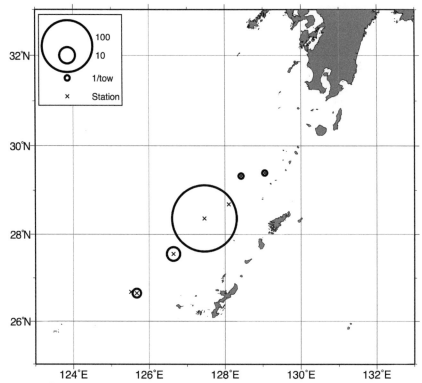

Figure 2.5. Horizontal distribution of juvenile Pacific bluefin tuna in the Nansei area in 1997. Modified from Itoh et al. (1999).

in waters within a range of 26.9–27.5°C of surface temperatures (Itoh et al. 1999). This finding was largely expected as the starting point to uncover early life history of the juvenile Pacific bluefin tuna including distribution, growth, feeding habitat, and recruitment process in the nursery area. Although a little success of such an amount of the juvenile catch has been achieved after that, trawl surveys have been continued around the the Nansei spawning ground to date.

Trawl surveys like those performed in the Nansei area were also carried out in the Sea of Japan. Mid-water trawls conducted from late August to late September in 1999 and 2004 succeeded to capture 100 and 86 individuals of Pacific bluefin tuna juveniles, respectively (Tanaka et al. 2007a). The juveniles appeared widely within the survey area in 1999 but restrictedly in the near-shore area in 2004 (Fig. 2.6), where mean ambient temperatures ranged from 23.4–25.9°C (mean from surface to 30 m deep), likely inhabiting at slightly lower temperatures than in the Nansei area. Since juveniles from the Sea of Japan (108–280 mm FL) were larger than those from the Nansei area (19.5–46.9 mm FL) sampled by Itoh et al. (1999), it was impractical to compare both results directly. Further surveys are expected to capture juveniles larger or smaller than 100 mm FL in the Nansei area or in the Sea of Japan, respectively. Accumulating knowledge from many more cases of Pacific bluefin juvenile catches would be crucial for illustrating distribution patterns and recruitment process of them.

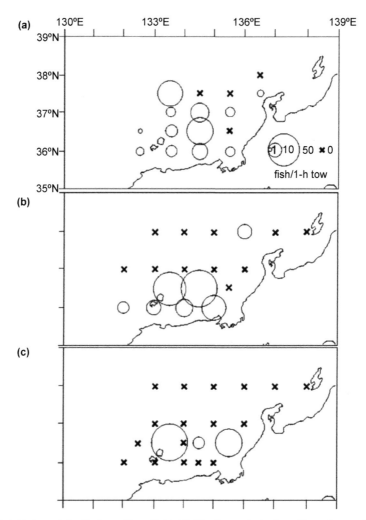

Figure 2.6. Horizontal distributions of juvenile Pacific bluefin tuna in the Sea of Japan from cruises in (a) 1999, (b) 2004 first cruise, and (c) 2004 second cruise (Tanaka et al. 2007b).

Eco-physiological Traits and Survival Strategy of Larval Pacific Bluefin Tuna

Growth and morphological development of larvae

Body length of newly hatched Pacific bluefin tuna larvae is 3 mm. Field-caught tuna larvae grow to 4.5 mm in Standard Length (SL) by seven days after hatching (DAH), 6 mm SL on 10 DAH and 7 mm SL on 14 DAH, estimated by otolith microstructure analysis (Tanaka et al. 2006). Morphological development of field-caught larvae has been reported in Tanaka et al. (2006). The shift from the preflexion to flexion phase is first noted at 4.0 mm SL on 7 DAH. Flexion larvae are observed up to 7.5 mm SL on 13 DAH. Postflexion larvae are first noted at the 6.0 mm SL on 10 DAH and all captured larvae are in the postflexion phase at 7.5 mm SL on 14 DAH. However, fish in the late-larval and early juvenile stages have been rarely captured in the field.

On the other hand, information of the growth and morphological traits of hatchery-reared larvae and juveniles have been accumulated since Pacific bluefin tuna were reared under aquaculture conditions throughout their complete life cycle (Sawada et al. 2005). Hatchery-reared tuna larvae grow to 5 mm in Total Length (TL) by 10 DAH, 9 mm TL on 20 DAH and 30 mm SL on 30 DAH (Miyashita et al. 2001). Particularly, Pacific bluefin tuna larvae show very high growth rates after 20 DAH, which corresponds to the onset of piscivory (Tanaka et al. 2007b; Tanaka et al. 2014). Morphological developments of the hatchery-reared larvae are as follows (Fig. 2.7).

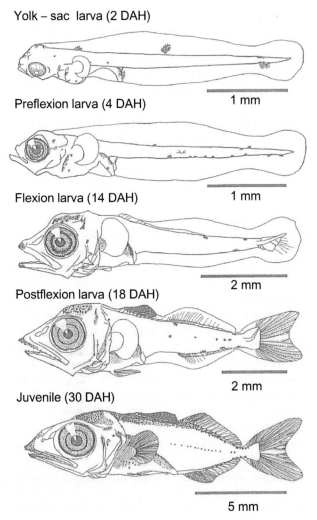

Figure 2.7. Morphological development of hatchery reared Pacific bluefin tuna (modified from Kaji et al. 1996).

The shift from the preflexion to flexion phase occurs at 5.0 mm SL on 10 DAH, from the flexion to postflexion phase at 7 mm SL on 14 DAH and from postflexion phase to juvenile at 16 mm SL on 23 DAH (Kaji et al. 1996; Tanaka et al. 2007b).

Precocious development of digestive system of Pacific bluefin tuna larvae

Generally, newly hatched larvae of most mass-spawning marine fish species are underdeveloped in terms with their digestive systems. The mouth and anus of the newly hatched larvae do not open and their digestive tract is a simple tube. Two transition periods in the process of development of the larval digestive system are observed. The former is the establishment of a primitive digestive system in the early larval stages, which enable to digest and absorb the exogenous nutrients. The latter is the establishment of an adult type digestive system which has a functional stomach with well-developed gastric glands and pyloric caeca occurring in the juvenile stage (Tanaka 1973).

The development of a digestive system of larval Pacific bluefin tuna is more precocious relative to other marine fishes. The gut opens from the mouth to anus on 2 DAH, differentiating the rudimentary stomach, intestine and rectum. The larvae at first feeding on 3 DAH have a primitive digestive system which enables the larva to feed on exogenous food although the gastric glands and blind sac is not differentiated (Kaji et al. 1996). By 12 DAH the stomach with a blind sac and gastric glands and the pyloric caecum become differentiated. At that time, an adult type digestive system is established (Kaji et al. 1996). Miyashita et al. (1998) also summarized the ontogenetic development of the digestive system. The pharyngeal teeth, mucous cells of oesophagus, blind sac and gastric gland differentiated in preflexion phase around 10 DAH. Pyloric caecum differentiated in the flexion phase on 15 DAH. Corresponding to the development of the digestive system, the activities of digestive enzymes increase. After the establishment of larval type digestive system, the activity of amylase rapidly increases. Thereafter, the activities of pepsin-like and trypsin-like enzymes quickly increase on 15 DAH immediately after the establishment of the adult type digestive system (Miyashita et al. 1998).

Thus, the larval and adult type digestive systems in the early life stages of Pacific bluefin tuna are established at first feeding on 3 DAH and on flexion phase on 12–15 DAH, respectively. The larval type digestive system of Pacific bluefin tuna is established in the typical periods of other marine fish species. However, the adult type digestive system establishes prior to the juvenile stage, which are earlier periods than other marine fishes. Pacific bluefin tuna shows rapid growth in the late larval to early juvenile stages after the onset of piscivory. The precocious development of the digestive system should be important characteristics for the rapid growth in the early life stages of Pacific bluefin tuna in relation to piscivory.

Ontogenetic changes in behavior in early life stages of Pacific bluefin tuna

The behavioral studies in early life stages of Pacific bluefin tuna have been conducted such as schooling and aggressive behavior using hatchery-reared fish in order to examine the problem of cannibalism that occurs in the mass culture process.

The swimming speed of tuna larvae is approximately 22.4 mm/s, and the speed does not change largely from the preflexion phase on 5 DAH to early juvenile on 21 DAH (Sabate et al. 2010). The swimming ability elevates with the increase of swimming speed after 25 DAH, reaching 500 mm/s on 55 DAH (Fukuda et al. 2010). The distance to the neighboring fish decreases, with the increase of the swimming speed. Swimming Separation Index (SSI) is often used as an index of schooling behavior (Nakayama et al. 2003). The value of SSI ranges between 0 and 2. When the value is 1.414, two fish show random swimming. The value decreases to 0 as the swimming of the two fish becomes aligned. The value of SSI of larval Pacific bluefin tuna is approximately 1.4 from 10 to 20 DAH (Sabate et al. 2010) (Fig. 2.8). Thereafter, the value of SSI is significantly less than 1.414 on 25 DAH after the transition to juvenile, which indicates the onset of schooling behavior. On 29 DAH, the value of SSI decreases to 0.5 and juvenile Pacific bluefin tuna shows clear schooling behavior. Pacific bluefin tuna develops schooling behavior later than other species such as anchovy *Engraulis mordax* (Hunter and Coyne 1982), Atlantic herring *Clupea harengus* (Gallego and Hearth 1994), yellowtail *Seriola quinqueradiata* (Sakakura and Tsukamoto 1999) and striped jack *Pseudocaranx dentex* (Masuda and Tsukamoto 1999), that develop schooling behavior in the late larval stage. The relatively late

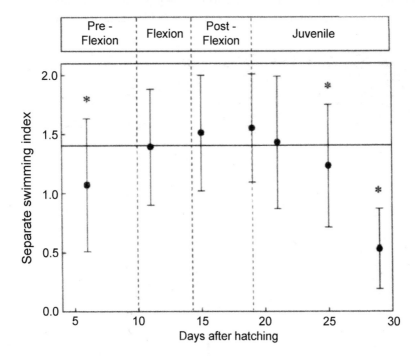

Figure 2.8. Changes in schooling behavior of Pacific bluefin tuna based on separation swimming index (SSI: Nakayama et al. 2003). Dots indicate the average SSI with standard deviation for every age group. Asterisks indicate significant differences between the mean and the expected SSI value for fish swimming at random direction and speed (t-test, p < 0.05) (modified from Sabate et al. 2010).

development of schooling behavior in Pacific bluefin tuna is presumed to reflect development of the central nerve system in the early juvenile stage, which is needed for schooling behavior (Sabate et al. 2010).

In the process of mass culture, cannibalism of larvae and juveniles is frequently observed after the onset of piscivory, which is one of the causes of the low survival rate in hatcheries (Sawada et al. 2005). Aggressive behavior to neighboring individuals which induces the cannibalism first occurs in the postflexion phase (Fig. 2.9). Thereafter, the aggressive behavior is continuously observed after the juvenile stage (Sabate et al. 2010). The onset of aggressive behavior corresponds to the timing of the onset of piscivory.

Pacific bluefin tuna larvae show piscivory with the development of aggressive behavior. The larvae which can feed on other fish larvae show high growth rates, consequently resulting high survival rates of the larvae in the field. The high growth rate can increase the swimming speed and rapidly develop the schooling behavior, which are advantageous to the foraging and escaping from predators. Thus, ontogenetic development of these behaviors should closely relate to the survival of larval and juvenile Pacific bluefin tuna in the field. Although these behavioral approaches are conducted using laboratory-reared fish, the results of behavioral studies will largely help the understanding of the mechanisms of transportation, survival process and estimation of the recruitment abundance in the early life stages of Pacific bluefin tuna in northwestern Pacific Ocean.

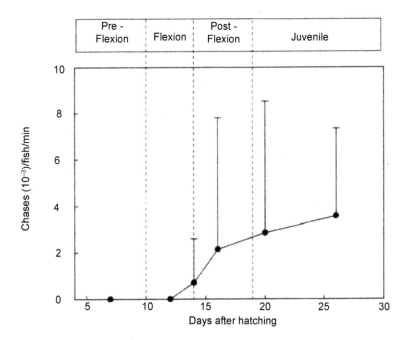

Figure 2.9. Mean number of chases per minute with standard deviation as an estimate of aggressive behavior in Pacific bluefin tuna (modified from Sabate et al. 2010).

Feeding habits

Pacific bluefin tuna larvae are visual day feeders and they do not feed at nighttime (Uotani et al. 1990). Uotani et al. (1990) examined the gut contents of 1939 Pacific bluefin tuna larvae (2.28–14.60 mm BL) collected at the main spawning ground for the present species in the northwestern Pacific Ocean. The main prey item of the larvae is copepods, accounting for 97.0% in number (Table 2.1). The larvae smaller than 5 mm BL mainly preyed on copepod nauplii smaller than 0.3 mm, while copepodites of genus *Corycaeus* were the main component for the larvae larger than 5 mm BL. Thus, a shift in feeding habit is observed at 5 mm BL, from small zooplankton (copepod nauplii) to larger zooplankton (copepods of *Corycaeus* spp.). The relative growth of the head length, mouth size and body depth of larval Pacific bluefin tuna changes at approximately 5 mm BL (Uotani et al. 1990; Miyashita et al. 2001). Furthermore, the pharyngeal teeth differentiate at this size (Miyashita et al. 1998). The shift in feeding habit at 5 mm BL is observed corresponding to these morphological changes, which could be due to the improvement in ability to catch and digest the prey.

Although Pacific bluefin tuna larvae tend to prey on larger zooplankton with fish growth, prey fish larvae were not found in their guts in the report of Uotani et al. (1990). On the other hand, laboratory-reared larvae show piscivory from sizes larger than 8 mm SL (Tanaka et al. 2014). However, the larvae ranging 10 to 20 mm SL have been rarely captured in the field and the ecological traits in relation to piscivory of field-captured larvae are still unknown. Since the piscivory could play an important role for the early growth, survival and consequent recruitment success of Pacific bluefin tuna, it is expected to elucidate the feeding habits in the late larval to early juvenile stages.

Table 2.1. Number and frequency of occurrence of forageorganisms found in the guts from 1939 larvae (Uotani et al.1990).

Food item	Number of organisms	Frequency of occurrence (%)
COPEPODA	1885	97
Paracalanus	62	3.2
Clausocalanus	217	11.2
Corycaeus	722	37.1
Temora	1	0.1
Oncaea	1	0.1
Copepoda nauplii	667	34.3
Copepoda eggs	57	2.9
unidentified Copepoda	158	8.1
OTHERS	61	3.3
Evadne	58	3
Mysidacea	1	0.1
Macrura nauplius	1	0.1
Fish egg	1	0.1
	1946	100.3

Survival mechanism in relation to growth and nutritional condition

Pacific bluefin tuna produce a huge number of small pelagic eggs. The newly hatched larvae are very vulnerable. These small larvae experience various environmental conditions through their ontogeny. Then they show heavy mortality due to the dispersion, predation, starvation and disease and so on, and a very few survived fish can recruit to the stock. The survival mechanisms in early life stages of mass-spawning fishes are concerned with the population dynamics. The elucidation of the survival mechanisms in early life stages is indispensable for optimal stock management. Here, we describe the survival mechanisms in the early life stages of Pacific bluefin tuna in the northwestern Pacific Ocean in relation to their growth and nutritional condition.

The otolith microstructure has daily increments which enable estimation of the daily age of fish by enumeration of the increments number. The increment width is generally proportional to the somatic growth (Campana and Neilson 1985). Because the growth history of each individual is recorded in the otolith (Degens et al. 1969; Dunkelberger et al. 1980; Watanabe et al. 1982; Mugiya 1987), growth histories can be back-calculated.

The otolith microstructure analysis of tuna larvae collected around the Ryukyu Islands in the northwestern Pacific Ocean revealed that marked growth and developmental variations of the larvae occur even at the same age (Tanaka et al. 2006). Juvenile tuna ranging from 15 to 30 cm TL are caught in Japan by troll fisheries in the coastal areas off Kochi and Nagasaki prefectures in July and August. These juveniles are considered to be survivors of the larval cohorts distributed around the Ryukyu Islands. In Tanaka et al. (2006), the growth histories of juveniles in their larval periods were compared to those of larvae at each age ranging from 6 to 13 DAH, estimated by otolith microstructure analysis. Figure 2.10 shows frequency distributions of logarithm otolith radius (ln OR) of larvae in the preflexion, flexion and postflexion phases and back-calculated for juvenile tuna at 10, 11, 12 and 13 DAH. Since SL was positively correlated with the ln OR during the larval stage, ln OR were used for the index of SL at each age. Arrows indicate smallest value of ln OR in juveniles and represent the smallest possible size of larvae for successful recruitment to juvenile stage, indicating that the larvae with smaller otolith than that value did not survive to juvenile stage. A large part of preflexion, and flexion larvae had smaller otoliths than those of the juveniles. On the other hand, the otolith radius of postflexion larvae was comparable to the juveniles. The results showed that surviving juveniles originated from the larvae with high growth and developmental rate at catch among the larvae with various growth and developmental rate distributed around the Ryukyu Islands.

Growth of larvae and juveniles is one of the most important factors for their survival. In several fish species, larvae with lower growth rates have been reported to have a higher mortality (Hovenkamp 1992; Meekan and Fortier 1996; Searcy and Sponaugle 2001; Takasuka et al. 2003; Takahashi and Watanabe 2004), because bigger larvae have a higher tolerance to starvation and a greater ability to escape from predators than smaller larvae (Anderson 1988; Miller et al. 1988; Bailey and Houde 1989). In Japanese anchovy, growth and developmental rate-dependent mortality occurred at 50 to 60 days. Growth selective mortality occurred at 41 to 80 days in Atlantic cod *Gadus morhua* (Meekan and Fortier 1996). In the case of Pacific bluefin

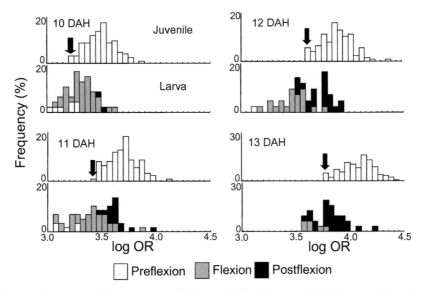

Figure 2.10. Frequency distributions of ln OR in juvenile bluefin tuna at 10, 11, 12 and 13 DAH (each upper panels) and larval bluefin tuna in the preflexion, flexion and postflexion phases at 10, 11, 12 and 13 DAH (each lower panels). ln OR indicates logarithm of otolith radius. Arrows indicate smallest value of ln OR in juveniles. Modified from Tanaka et al. (2006).

tuna, the growth rates within two weeks of hatching are concluded to be crucial for survival to recruitment, which occurs at a relatively earlier period compared to other neritic fish species.

Larval Pacific bluefin tuna shows a conspicuous patch-like distribution (Nishikawa 1985; Satoh 2006) similar to that reported for the southern bluefin tuna *Thunnus maccoyii* (Davis et al. 1990; Jenkins and Davis 1990; Jenkins et al. 1991). On the basis of the characteristics of the distribution of the present species, it is assumed that the survival of larval Pacific bluefin tuna in the patches could be important for their recruitment. Tanaka et al. (2008) assess the nutritional status of field-captured larvae from three cohorts collected in the northwestern Pacific Ocean using the RNA/ DNA ratio. RNA/DNA ratio has been proven as a useful and sensitive indicator of the nutritional condition and recent growth of larval and juvenile fishes in relation to protein synthesis, which has been widely applied to both laboratory-reared and field-caught fish (e.g., Clemmesen 1987). The quantity of DNA in an animal cell is believed to be normally stable but the quantity of RNA, primarily associated with ribosomes, is closely related to the rate of protein synthesis. The nutritional condition is associated with the food supply and feeding success of the fish and therefore, variability in the trophic environment is reflected in the nutritional condition.

The criteria of the RNA/DNA ratio representing the 'starving condition' is required in order to allow the assessment of nutritional status of field-caught fish. Tanaka et al. (2008) carried out rearing experiments without food (starvation experiment) using hatchery-reared larvae to determine these criteria before assessment of nutritional condition of field sampled tuna larvae. Firstly, it is found that the survival rate of the larvae one day after starved condition was less than 50% and all larvae died three to four days after the starved condition, suggesting that larval Pacific bluefin tuna have a very low tolerance for starvation. Secondly, RNA/DNA ratio of the larvae one day after the starved condition (one day starved larvae) rapidly decreased and the value was significantly lower than the values of fed larvae. Accordingly, the RNA/DNA ratios of the one day starved group as the criteria representing the crucial starved condition of larvae was used for the comparison with field-caught Pacific bluefin tuna larvae.

The nutritional status of three patches of field-caught tuna larvae collected in the northwestern Pacific Ocean in 2004 and 2005 was examined based on the value of the RNA/DNA ratio of the one day starved larvae. Field-caught larvae with a RNA/DNA ratio below the 95% confidence upper limit of one day starved larvae were considered as 'starving fish'. Percentages of the starving larvae in three patches were 15.15, 4.35 and 25.77%, respectively (Fig. 2.11). Furthermore, the patch with the highest ambient prey density showed lowest percentage of the starving larvae and vice versa, indicating a negative correlation with the ambient prey densities. Occurrence of starving larvae is variable depending on the fish species. Percentage of starving larvae in the field is estimated as 5.5% for Japanese flounder *Paralichthys olivaceus* (Gwak and Tanaka 2001), 0.0–4.8% for sardine *S. pilchardus* (Chicharo1998; Chicharo et al. 2003). In anchovy *Engraulis anchoita*, although the percentage of starving fish was found to be high (52.4%) only in the size class of 5–6 mm and the percentage was 0.0–1.75% in all other size classes (Clemmesen 1996). In larval Japanese sardine *Sardinops*

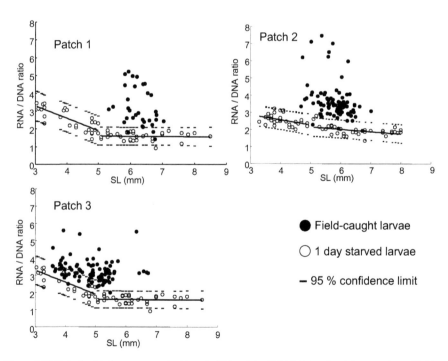

Figure 2.11. Relationship between standard length (SL) and the RNA/DNA ratio of three patches of field-caught larvae, compared to those of one day starved fish. Modified from Tanaka et al. (2008).

melanostictus at first feeding, the percentage of starving larvae is zero (Kimura et al. 1996). Thus, the percentage of starving larvae in Pacific bluefin tuna appears to be relatively higher than those of other species. Although predation is generally considered to be most important factor which controls larval mortality in marine fishes (Houde 1987; Bailey and Houde 1989), starvation itself as well as the delayed response of starvation weakened larvae to predatory attack could greatly contribute to mortality in the larval period of Pacific bluefin tuna.

As mentioned above, larval Pacific bluefin tuna shows a conspicuous patch-like distribution. In the studies of Satoh et al. (2008) and Satoh et al. (2013), they observed the growth and survival of larval Pacific bluefin tuna in the patch-like distribution in the north western Pacific Ocean using drifting reference buoy tracking the same cohorts. Seven high larval patches of Pacific bluefin tuna found from 2004 to 2008 were tracked each for 28–171 hours and the relationship between the daily growth and survival of the larvae investigated on the basis of ambient water temperature, stratification and prey density. These studies revealed that ambient water temperature and prey density influenced larval growth rate for short periods and larger and faster growing larvae could survive on a daily basis. In these tracking periods of the patches, the estimated daily percentage of instantaneous mortality were 90.9% at maximum (Satoh et al. 2008), which is higher than most estimates for other marine fish larvae (Houde and Zastrow 1993). Thus, Pacific bluefin tuna larvae experience short-term heavy mortality during a short period after hatching.

Generally, oceanic sub-tropical waters where larval Pacific bluefin tuna are distributed are oligotrophic and the productivity of zooplankton is low relative to more neritic waters. The survival mechanism in early life stages of Pacific bluefin tuna in such an environment are characterized by the following two traits. Firstly, larval growth in very short periods after hatching such as two weeks markedly influences their survival. Secondly, larval Pacific bluefin tuna have very low tolerance to starvation and the starvation itself could greatly contribute to their mortality. These traits in the survival mechanism explain the biological backgrounds of extremely high mortality rates in of the larvae and the species-specific survival mechanism for Pacific bluefin tuna which exist in oligotrophic, oceanic sub-tropical waters.

New Approaches for the Survival Strategy in Early Life History of Pacific Bluefin Tuna

Importance of piscivory for Pacific bluefin tuna larvae estimated using hatchery-reared fish

Scombrid fish larvae including Pacific bluefin tuna show piscivory from very early stages of their ontogeny relative to other fish species. In larval chub mackerel and Japanese Spanish mackerel, the importance of piscivory for their growth and survival has been pointed out. As mentioned earlier, precocious development of the digestive system of hatchery-reared Pacific bluefin tuna corresponds with the occurrence of piscivory. However, there are no reports of the feeding habits of Pacific bluefin tuna larvae in relation to the piscivory in the field because the field collection of Pacific bluefin tuna larvae larger from 10 to 20 mm SL is very rare mainly due to net avoidance.

On the other hand, the techniques of mass culture of Pacific bluefin tuna have been developing day by day in Japan and large numbers of samples can be obtained under various designs of rearing experiment. These samples are good materials for observations of piscivory, which is still unverified in the field. Here, we describe the potential importance of the piscivory for the larval growth observed using hatchery-reared Pacific bluefin tuna.

In general, feeding regimes for the mass culture of Pacific bluefin tuna are composed of four prey items, which are rotifers, *Brachionus plicatilis*, *Artemia* nauplii, yolk-sac larvae of other species and minced fish (Sawada et al. 2005; Tanaka et al. 2007b; Tanaka et al. 2010; Tanaka et al. 2014). These prey items are fed in the following order with larval growth; rotifers from the first feeding to 15 mm TL, *Artemia* nauplii from 5 to 20 mm TL, yolk-sac larvae from 7–8 to 30 mm TL and minced fish from 20 to 50 mm TL. The growth of Pacific bluefin tuna was found to be accelerated after the onset of piscivory (Kaji et al. 1996; Miyashita et al. 2001). Tanaka et al. (2007b) showed ontogenetic changes in RNA/DNA ratios of hatchery reared larval and juvenile Pacific bluefin tuna in order to elucidate the growth performance. The values of RNA/DNA ratio from 13 DAH to 19 DAH increased slightly from 4.0 to 7.0 (Fig. 2.12). After that, the ratio steeply increased to 20 on 23 DAH. In this study, yolk-sac larvae were fed from 19 DAH. Thus, the value of RNA/DNA ratio which is an indicator of protein synthesis activity rapidly increased after the first feeding of yolk-sac larvae. It is suggested that high activity of protein synthesis induced by the onset of piscivory supports the high somatic growth of late-larval to early juvenile stages. Such drastic

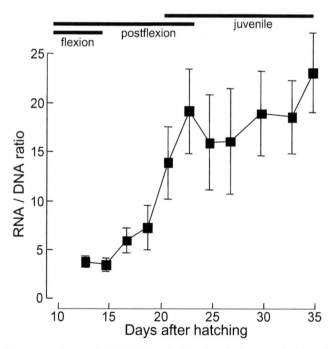

Figure 2.12. Ontogenetic changes in RNA/DNA ratio. Error bars indicate standard deviation. Modified from Tanaka et al. (2007b).

changes in RNA/DNA ratio are not found in the early ontogeny of Japanese flounder *Paralichthys olivaceus* and Japanese temperate bass *Lateolabrax japonicus* which are representative target species of aquaculture and stock enhancement with the technology related to the rearing techniques being well developed (Tanaka et al. 2007b). This also indicates that drastic changes in RNA/DNA ratio of tuna larvae could be a species/genus specific pattern reflected in their growth performance in relation to the piscivory.

In mass culture of Pacific bluefin tuna, heavy mortalities due to cannibalism during the late larval and juvenile stages occur (Sawada et al. 2005). The cannibalism is mainly caused by marked growth variations. As mentioned above, two kinds of prey, rotifers and yolk-sac larvae are simultaneously used as prey items in the mass culture tank when the size of Pacific bluefin tuna larvae ranges from 7–8 to 15 mm TL. Growth variations in the larvae conspicuously occur after the initial feeding of yolk-sac larvae. Tanaka et al. (2014) conducted prey switch experiments to examine the relationship between prey utilization and growth variations in larval Pacific bluefin tuna. A prey switch experiment was conducted under two different feeding regimes: a group that kept being fed on rotifers (rotifer fed group) and a group that were switched to feed on yolk-sac larvae of other species (fish fed group) from 15 DAH (6.87 mm BL). The fish fed group showed much higher growth than the rotifer fed group. The mean BL on nine days after the initial day was 17.21 ± 2.31 (SD) mm for the fish fed group and 8.67 ± 0.38 mm for the rotifer fed group. At that time, dry body weight of the fish fed group was approximately 50 times greater than that of the rotifer fed group (Fig. 2.13). These results suggest that the growth variation in tuna larvae can be strongly affected by individual differences in the availability to prey on yolk-sac larvae.

From the results of the prey switch experiment, the prey utilization is concluded to be a key factor for the growth variations in actual mass culture. Two kinds of prey,

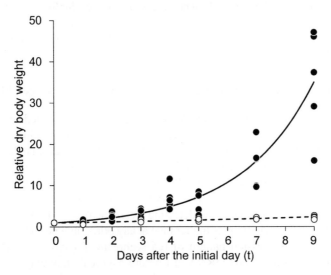

Figure 2.13. Relative dry body weight (RDBW) increase in the prey switch experiment. Closed and open circles indicate the fish fed and rotifer fed groups respectively (modified from Tanaka et al. 2014). RDBW indicates the relative increase of dry body weight of larvae. RDBW at the initial day of the experiment means 1.0.

rotifers and yolk-sac larvae, simultaneously exist in the mass culture tank when the size of Pacific bluefin tuna larvae ranges from 7–8 to 15 mm TL (Sawada et al. 2005; Tanaka et al. 2010). Tanaka et al. (2014) have also investigated the nitrogen stable isotope ratio ($\delta^{15}N$) of various sized hatchery-reared Pacific bluefin tuna larvae in actual mass culture tank with the existence of two prey items, rotifer and yolk-sac larvae. As stomach or gut content analysis shows only the instantaneous prey utilization of predators, other longer term methods are necessary to elucidate the cause of growth variations. Stable isotope analysis is a useful tool for investigating the food webs in terrestrial and aquatic ecosystems. Nitrogen stable isotope ratios have been used to estimate the food web through the prey–predator relationship. The $\delta^{15}N$ has been used to examine the trophic position of organisms in food webs (DeNiro and Epstein 1978; Owens 1987; Wada et al. 1987; Hobson and Welch 1992; Hesslein et al. 1993). Recently, stable isotope ratios have been applied to aquaculture and stock enhancement (Yokoyama et al. 2002; Tominaga et al. 2003). Since the value of $\delta^{15}N$ changes after the switch of prey (Fig. 2.14), prey utilization of various sized larvae in the mass culture tank after initial feeding of yolk-sac larvae were investigated using $\delta^{15}N$ (Tanaka et al. 2014). The values of $\delta^{15}N$ of the large larvae (fast growing larvae) were significantly higher than those of the small larvae (slow growing larvae) (Fig. 2.15). This indicates that the large larvae could utilize the yolk-sac larvae rapidly after the yolk-sac larvae feeding and small larvae depended more on rotifers as the main prey item relative to the large group. Thus, the individual difference in the utilization of the prey items is one of the factors inducing larval growth variations even in the same age.

Fast growing larvae can survive selectively in the field as mentioned earlier. Rearing experiment revealed that piscivory is a very important factor for fast growth of Pacific bluefin tuna larvae, which is advantageous for their survival in the field. In order to elucidate the survival mechanisms of Pacific bluefin tuna larvae in the piscivorous stage, the food availability such as yolk-sac larvae of other species as their prey item should be investigated in the area where Pacific bluefin tuna larvae and juveniles are distributed.

Advanced tools for physiological assessment of an early stage of juveniles

In recent studies, aside from RNA/DNA ratio and $\delta^{15}N$ ratio mentioned above, several new advanced tools to assess physiological status potentially regulating body growth and organic development have been introduced in early stage of juvenile Pacific bluefin tuna. Adachi et al. (2008) focused on the Growth Hormone (GH) and investigated the daily expression profile of GH mRNA in juvenile Pacific bluefin tuna under aquacultured conditions. Continuous observations from 10 am to 12 pm of the next day at 1 hour intervals succeeded in detecting a clear peak of GH mRNA expression level at 3–4 am of about 10 times higher than at any other time period, which clearly differed from the expression pattern of red seabream, *Pagrus major* (Fig. 2.16). The pulsed expression of GH mRNA just before daybreak may play a key role for achieving the extremely rapid somatic growth during the early juvenile stage through efficiently regulating secretion of the functional GH and subsequent assimilation of nutrients during the daytime.

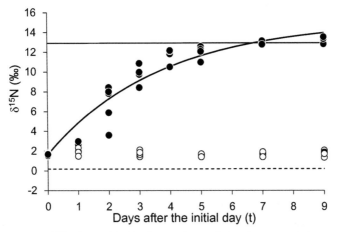

Figure 2.14. Values of δ¹⁵N for Pacific bluefin tuna larvae during the prey switch experiment with equations of the model of elapsed time. Closed and open circles indicate the fish fed and rotifer fed groups respectively. Dashed and solid lines indicate δ¹⁵N of rotifers and yolk-sac larvae respectively. Modified from Tanaka et al. (2014).

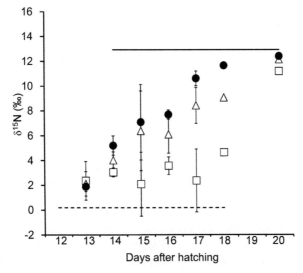

Figure 2.15. Changes in mean δ¹⁵N of the three size groups of tuna larvae collected in the mass culture tank (modified from Tanaka et al. 2014). Open squares, open triangles and open circles indicate the small, intermediate and large groups respectively. Error bars show the standard deviation. Dashed and solid lines indicate δ¹⁵N of rotifers and yolk-sac larvae respectively.

Contrastingly, Suda et al. (2012) analyzed quantitatively the expressions of the gastric ghrelin and pepsinogen 2 genes during early stage of somatic growth from the underyearling to yearling juveniles. Ghrelin has been shown to be produced predominantly in the stomach of various vertebrate species, stimulating the release of GH and other pituitary hormones, and therefore having a role in the regulation of feeding behavior, water intake and locomotor activity. The expression profiles of the

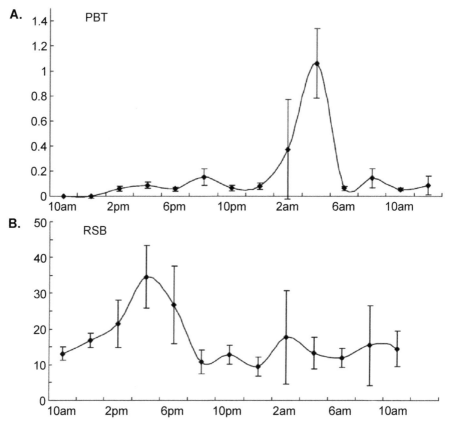

Figure 2.16. Daily GH mRNA expression patterns in the pituitary. (A) Pacific bluefin tuna. (B) Red seabream. The peaks at 1–2 and 3–4 am in Pacific bluefin tuna are significantly higher than those at other time periods by one-way ANOVA and post-hoc Bonferroni's multiple comparison tests ($p < 0.05$). Twilight ended at 6:06 pm and began at 5:43 am on the sampling day. Data are presented as means ± SE (Adachi et al. 2008).

gastric ghrelin and pepsinogen 2 genes based on real-time PCR assays showed no significant changes across the early growth stages, suggesting physiological completion of the digestive function at least in yearling juveniles with about 55 cm FL.

These gene-targeting expression analyses, as well as the large-scale exhaustive expression analyses using next generation sequencers, will make it possible to detail biological processes such as growth, survival and adaptation to the environment. Further exploration of the gene assemblage in the Pacific bluefin tuna genome is expected to rapidly progress the studies concerning the early life history of Pacific bluefin tuna.

Future Prospects of the Studies for Early Life History of Pacific Bluefin Tuna

The studies for early life history of tuna based on the field surveys have been conducted to elucidate the mechanisms of annual variations of recruitment and consequently to aim at optimizing stock management. On the other hand, studies for early life history of tuna based on rearing experiments have been conducted mainly to develop the rearing techniques of mass culture for the application to aquaculture and stock enhancement. These two types of studies based on different approaches have been carried out largely independently of each other. However, studies on the early life history in the future should be developed with close linkages of field surveys and rearing experiments because most of the captured Pacific bluefin tuna by fisheries are immature fish immediately after recruitment and young of the year of tuna which are used as fingerlings for aquaculture.

For instance, although tuna juveniles at onset of piscivory (1–2 cm TL) are rarely captured in the field, results of rearing experiments under various food designs concerning piscivory may enable the late larval—early juvenile tuna to be captured in the field and also can help to understand the role of piscivory in early life stages in relation to the annual variability of the recruitment to the stock. Conversely, if eco-physiology of the field captured larvae and juvenile under various environments can be elucidated, these results also can be used in the development of mass culture technique of Pacific bluefin tuna. Thus, integration of the field surveys and rearing experiments are necessary for contribution of the studies of early life history of Pacific bluefin tuna on the tuna fisheries.

References

Adachi, K., K. Kato, M. Yamamoto, K. Ishimaru, T. Kobayashi, O. Murata and H. Kumai. 2008. Pulsed expression of growth hormone mRNA in the pituitary of juvenile Pacific bluefin tuna under aquacultured conditions. Aquaculture 281: 158–161.

Alemany, F., L. Quintanilla, P. Velez-Belchí, A. García, D. Cortés, J.M. Rodríguez, M.L. Fernández de Puelles, C. González-Pola and J.L. López-Jurado. 2010. Characterization of the spawning habitat of Atlantic bluefin tuna and related species in the Balearic Sea (western Mediterranean). Prog. Oceanogr. 86: 21–38.

Anderson, J.T. 1988. A review of size dependent survival during pre-recruit stages of fishes in relation to recruitment. J. Northwest Atl. Fish. Sci. 8: 55–66.

Bailey, K.M. and E.D. Houde. 1989. Predation on eggs and larvae and the recruitment problem. Adv. Mar. Biol. 25: 1–83.

Bakun, A. 2006. Wasp-waist populations and marine ecosystem dynamics: navigating the "predator pit" topographies. Prog. Oceanogr. 68: 271–288.

Bakun, A. 2013. Ocean eddies, predator pits and bluefin tuna: implications of an inferred 'low risk-limited payoff' reproductive scheme of a (former) archetypical top predator. Fish and Fisheries 14: 424–438.

Campana, S.E. and J.D. Neilson. 1985. Microstructure of fish otoliths. Can. J. Fish. Aquat. Sci. 42: 1014–1032.

Chicharo, M.A. 1998. Nutritional condition and starvation in *Sardina pilchardus* (L.) larvae off southern Portugal compared with some environmental factors. J. Exp. Mar. Biol. Ecol. 225: 123–137.

Chicharo, M.A., E. Esteves, A.M.P. Santos, A. dos Santos, A. Peliz and P. Re. 2003. Are sardine larvae caught off northern Portugal in winter starving? An approach examining nutritional conditions. Mar. Ecol. Prog. Ser. 257: 303–309.

Chow, S., K. Nohara, T. Tanabe, T. Itoh, S. Tsuji, Y. Nishikawa, S. Uyeyanagi and K. Uchikawa. 2003. Genetic and morphological identification of larval and small juvenile tunas (Pisces: Scombridae) caught by a mid-water trawl in the western Pacific. Bull. Fish. Res. Agency 8: 1–14.

Clemmesen, C. 1987. Laboratory studies on RNA/DNA ratios on starved and fed herring (*Clupea harengus*) and turbot (*Scophthalmus maximus*) larvae. J. Cons. Int. Explor. Mer. 43: 122–128.

Clemmesen, C. 1996. Importance and limits of RNA/DNA ratios as a measure of nutritional condition in fish larvae. pp. 67–82. *In*: Y. Watanabe, Y. Yamashita and Y. Oozeki (eds.). Survival Strategies in Early Life Stages of Marine Resources. A. A. Balkema, Rotterdam.

Davis, T.L.O., G.P. Jenkins and J.W. Young. 1990. Patterns of horizontal distribution of the larvae of southern bluefin (*Thunnus maccoyii*) and other tuna in the Indian Ocean. J. Plankton Res. 12: 1295–1314.

Degens, E.T., W.G. Deuser and R.L. Haedrich. 1969. Molecular structure and composition of fish otoliths. Mar. Biol. 2: 105–113.

DeNiro, M.J. and S. Epstein. 1978. Influence of diet on the distribution of carbon isotopes in animals. Geochimica et Cosmochimica Acta 42: 495–506.

Dunkelberger, D.G., J.M. Dean and N. Watabe. 1980. The ultrastructure of the otolithic membrane and the otolith in juvenile mummichog, *Fundulus heteroclitus*. J. Morphol. 163: 367–377.

Fukuda, H., S. Torisawa, Y. Sawada and T. Takagi. 2010. Ontogenetic changes in schooling behaviour during larval and early juvenile stages of Pacific bluefin tuna *Thunnus orientalis*. J. Fish Biol. 76: 1841–1847.

Gallego, A. and M.R. Hearth. 1994. The development of schooling behaviour in Atlantic herring *Clupea harengus*. J. Fish Biol. 45: 569–588.

Gwak, W.S. and M. Tanaka. 2001. Developmental changes in RNA: DNA ratios of fed and starved laboratory-reared Japanese flounder larvae and juveniles, and its application to assessment of nutritional condition for wild fish. J. Fish Biol. 59: 902–915.

Hesslein, R.H., K.A. Hallard and P. Ramlal. 1993. Replacement of sulfur, carbon, and nitrogen in tissue of growing broad whitefish (*Coregonus nasus*) in response to a change in diet traced by $\delta^{34}S$, $\delta^{13}C$, and $\delta^{15}N$. Can. J. Fish. Aquat. Sci. 50: 2071–2076.

Hobson, K.A. and H.E. Welch. 1992. Determination of trophic relationships within a high Arctic marine food web using $\delta^{13}C$ and $\delta^{15}N$ analysis. Mar. Ecol. Prog. Ser. 84: 9–18.

Houde, E.D. 1987. Fish early life dynamics and recruitment variability. Am. Fish. Soc. Symp. 2: 17–29.

Houde, E.D. and C.E. Zastrow. 1993. Ecosystem- and taxon-specific dynamic and energetics properties of larval fish assemblages. Bull. Mar. Sci. 53: 290–335.

Hovenkamp, F. 1992. Growth-dependent mortality of larval plaice *Pleuronectes platessa* in the North Sea. Mar. Ecol. Prog. Ser. 58: 95–101.

Hunter, J.R. and K.M. Coyne. 1982. The onset of schooling in northern anchovy larvae, *Engraulis mordax*. Calif. Coop. Ocean. Fish. Investig. Rep. 23: 246–251.

Itoh, T., D. Inagake, T. Kaji, S. Tsuji, K. Tsuchiya, S. Chow, J. Mori, M. Moteki and T. Yoshimura. 1999. Survey for larvae and juveniles of northern bluefin tuna. Report of the first Research Cruise in 1997 by RV Shunyo Maru. National Research Institute of Far Seas Fisheries. 89 pp. (in Japanese).

Jenkins, G.P. and T.L.O. Davis. 1990. Age, growth rate, and growth trajectory determined from otolith microstructure of southern bluefin tuna *Thunnus maccoyii* larvae. Mar. Ecol. Prog. Ser. 63: 93–104.

Jenkins, G.P., J.W. Young and T.L.O. Davis. 1991. Density dependence of larval growth of a marine fish, the southern bluefin tuna, *Thunnus maccoyii*. Can. J. Fish. Aquat. Sci. 48: 1358–1363.

Kaji, T., M. Tanaka, Y. Takahashi, M. Oka and N. Ishibashi. 1996. Preliminary observations on development of Pacific bluefin tuna *Thunnus thynnus* (Scombridae) larvae reared in the laboratory, with special reference to the digestive system. Mar. Freshwater Res. 47: 261–269.

Kimura, R., C. Sato and K. Nakata. 1996. Nutritional condition of first-feeding larvae of sardine *Sardinops melanostictus*. pp. 105–113. *In*: Y. Watanabe, Y. Yamashita and Y. Oozeki (eds.). Survival Strategies in Early Life Stages of Marine Resources. A. A. Balkema, Rotterdam.

Kimura, S., Y. Kato, T. Kitagawa and N. Yamaoka. 2010. Impacts of environmental variability and global warming scenario on Pacific bluefin tuna (*Thunnus orientalis*) spawning grounds and recruitment habitat. Prog. Oceanogr. 86: 39–44.

Kitagawa, T., Y. Kato, M.J. Miller, Y. Sasai, H. Sasaki and S. Kimura. 2010. The restricted spawning area and season of Pacific bluefin tuna facilitate use of nursery areas: a modeling approach to larval and juvenile dispersal processes. J. Exp. Mar. Biol. Ecol. 393: 23–31.

Lindo-Atichati, D., F. Bringas, G. Goni, B. Muhling, F.E. Muller-Karger and S. Habtes. 2012. Varying mesoscale structures influence larval fish distribution in the northern Gulf of Mexico. Mar. Ecol. Prog. Ser. 463: 245–257.

Mariani, P., B.R. MacKenzie, D. Iudicone and A. Bozec. 2010. Modelling retention and dispersion mechanisms of bluefin tuna eggs and larvae in the northwest Mediterranean Sea. Prog. Oceanogr. 86: 45–58.

Masuda, R. and K. Tsukamoto. 1999. School formation and concurrent developmental changes in carangid fish with reference to dietary conditions. Environ. Biol. Fish 56: 243–252.

Meekan, M.G. and L. Fortier. 1996. Selection for fast growth during the larval life of Atlantic cod *Gadus morhua* on the Scotian shelf. Mar. Ecol. Prog. Ser. 137: 25–37.

Miller, T.J., L.B. Crowder, J.A. Rice and E.A. Marschall. 1988. Larval size and recruitment mechanisms in fishes: toward a conceptual framework. Can. J. Fish. Aquat. Sci. 45: 1657–1670.

Miyashita, S., K. Kato, Y. Sawada, O. Murata, Y. Ishitani, K. Shimizu, S. Yamamoto and H. Kumai. 1998. Development of digestive system and digestive enzyme activity of larval and juvenile bluefin tuna, *Thunnus thynnus*, reared in the laboratory. Suisan Zoushoku 46: 111–120 (in Japanese with English abstract).

Miyashita, S., Y. Sawada, T. Okada, O. Murata and H. Kumai. 2001. Morphological development and growth of laboratory-reared larval and juvenile *Thunnus thynnus* (Pisces: Scombridae). Fish. Bull. 99: 601–616.

Mohri, M., H. Yamada, Y. Tanaka and K. Hukada. 2005. Towing method for the most efficient estimates on the vertical distribution of bluefin tuna juveniles—study for towing method in the west coasts on the Sea of Japan by using surface and mid-water trawl net. Math. Phys. Fish. Sci. 3: 26–35.

Mugiya, Y. 1987. Phase difference between calcification and organic matrix formation in diurnal growth of otoliths in the rainbow trout, *Salmo gairdneri*. Fish. Bull. 85: 395–401.

Muhling, B.A., J.T. Lamkin and M.A. Roffer. 2010. Predicting the occurrence of Atlantic bluefin tuna (*Thunnus thynnus*) larvae in the northern Gulf of Mexico: building a classification model from archival data. Fish. Oceanogr. 19: 526–539.

Muhling, B.A., P. Reglero, L. Ciannelli, D. Alvarez-Berastegui, F. Alemany, J.T. Lamkin and M.A. Roffer. 2013. Comparison between environmental characteristics of larval bluefin tuna *Thunnus thynnus* habitat in the Gulf of Mexico and western Mediterranean Sea. Mar. Ecol. Prog. Ser. 486: 257–276.

Nakayama, S., R. Masuda, J. Shoji, T. Takeuchi and M. Tanaka. 2003. Effect of prey items on the development of schooling behavior in chub mackerel *Scomber japonicus* in the laboratory. Fish. Sci. 69: 670–676.

Nishikawa, Y. 1985. Identification for larvae of three species of genus *Thunnus* by melanophore patterns. Bull. Far Seas Fish. Res. Lab. 22: 119–129.

Nishikawa, Y., S. Kikawa, M. Honma and S. Ueyanagi. 1978. Distribution atlas of larval tunas, billfishes and related species—Results of larval surveys by R/V Shunyo Maru and Shoyo Maru, 1956–1976. Far Seas Fish. Res. Lab. S series 9: 1–99 (in Japanese with English abstract).

Nishikawa, Y., M. Honma, S. Ueyanagi and S. Kikawa. 1985. Average distribution of larvae of oceanic species of scombrid fishes, 1956–1981. Bull. Far. Seas. Fish. Res. Lab. 12: 1–99 (in Japanese with English abstract).

Owens, N.J.P. 1987. Natural variation in ^{15}N in the marine environment. Adv. Mar. Biol. 24: 389–451.

Rodriguez, J.M., I. Alvarez, J.L. Lopez-Jurado, A. Garcia, R. Balbin, D. Alvarez-Berastegui, A.P. Torres and F. Alemany. 2013. Environmental forcing and the larval fish community associated to the Atlantic bluefin tuna spawning habitat of the Balearic region (Western Mediterranean), in early summer 2005. Deep Sea Res. I 77: 11–22.

Sabate, F. de la S., Y. Sakakura, Y. Tanaka, K. Kumon, H. Nikaido, T. Eba, A. Nishi, S. Shiozawa, A. Hagiwara and S. Masuma. 2010. Onset and development of cannibalistic and schooling behavior in the early life stages of Pacific bluefin tuna *Thunnus orientalis*. Aquaculture 301: 16–21.

Sakakura, Y. and K. Tsukamoto. 1999. Ontogeny of aggressive behavior in schools of yellowtail, *Seriola quinqueradiata*. Environ. Biol. Fish 56: 231–242.

Satoh, K. 2006. Growth, distribution and advection transport of tuna larva in the northwestern Pacific Ocean. Nippon Suisan Gakkaishi 72: 939–940 (in Japanese).

Satoh, K. 2010. Horizontal and vertical distribution of larvae of Pacific bluefin tuna *Thunnus orientalis* in patches entrained in mesoscale eddies. Mar. Ecol. Prog. Ser. 404: 227–240.

Satoh, K., Y. Tanaka and M. Iwahashi. 2008. Variations in the instantaneous mortality rate between larval patches of Pacific bluefin tuna *Thunnus orientalis* in the northwestern Pacific Ocean. Fish. Res. 89: 248–256.

Satoh, K., Y. Tanaka, M. Masujima, M. Okazaki, Y. Kato, H. Shono and K. Suzuki. 2013. Relationship between the growth and survival of larval Pacific bluefin tuna, *Thunnus orientalis*. Mar. Biol. 160: 691–702.

Sawada, Y., T. Okada, M. Miyashita, O. Murata and H. Kumai. 2005. Completion of the Pacific bluefin tuna *Thunnus orientalis* (Temmincket Schlegel) life cycle. Aquaculture Research 36: 413–421.

Searcy, S.P. and S. Sponaugle. 2001. Selective mortality during the larval-juvenile transition in two coral reef fishes. Ecology 82: 2452–2470.

Suda, A., H. Kaiya, H. Nikaido, S. Shiozawa, K. Mishiro and H. Ando. 2012. Identification and gene expression analyses of ghrelin in the stomach of Pacific bluefin tuna (*Thunnus orientalis*). Gen. Comp. Endocrinol. 178: 89–97.

Suzuki, N., T. Tanabe, K. Nohara, W. Doi, H. Ashida, T. Kameda and Y. Aonuma. 2014. Annual fluctuation in Pacific bluefin tuna (*Thunnus orientalis*) larval catch from 2007 to 2010 in waters surrounding the Ryukyu Archipelago, Japan. Bull. Fish. Res. Agen. 38: 87–99.

Takahashi, M. and Y. Watanabe. 2004. Growth rate-dependent recruitment of Japanese anchovy *Engraulis japonicus* in the Kuroshio-Oyashio transitional waters. Mar. Ecol. Prog. Ser. 266: 227–238.

Takasuka, A., I. Aoki and I. Mitani. 2003. Evidence of growth-selective predation on larval Japanese anchovy *Engraulis japonicus* in Sagami Bay. Mar. Ecol. Prog. Ser. 252: 223–238.

Tanabe, T. and K. Niu. 1998. Sampling juvenile skipjack tuna, *Katsuwonus pelamis*, and other tunas, *Thunnus* spp., using midwater trawls in the tropical western Pacific. Fish. Bull. 96: 641–646.

Tanaka, M. 1973. Studies on the structure and function of the digestive system of teleost larvae. PhD thesis Kyoto University, 136 pp.

Tanaka, Y., K. Satoh, M. Iwahashi and H. Yamada. 2006. Growth-dependent recruitment of Pacific bluefin tuna *Thunnus orientalis* in the northwestern Pacific Ocean. Mar. Ecol. Prog. Ser. 319: 225–235.

Tanaka, Y., M. Mohri and H. Yamada. 2007a. Distribution, growth and hatch date of juvenile Pacific bluefin tuna *Thunnus orientalis* in the coastal area of the Sea of Japan. Fish. Sci. 73: 534–542.

Tanaka, Y., W.S. Gwak, M. Tanaka, Y. Sawada, T. Okada, S. Miyashita and H. Kumai. 2007b. Ontogenetic changes in RNA, DNA and protein contents of laboratory-reared Pacific bluefin tuna *Thunnus orientalis*. Fish. Sci. 73: 378–384.

Tanaka, Y., K. Satoh, H. Yamada, T. Takebe, H. Nikaido and S. Shiozawa. 2008. Assessment of the nutritional status of field-caught larval Pacific bluefin tuna by RNA/DNA ratio based on a starvation experiment of hatchery-reared fish. J. Exp. Mar. Biol. Ecol. 354: 56–64.

Tanaka, Y., H. Minami, Y. Ishihi, K. Kumon, T. Eba, A. Nishi, H. Nikaido and S. Shiozawa. 2010. Prey utilization by hatchery-reared Pacific bluefin tuna larvae in mass culture tank estimated using stable isotope analysis, with special reference to their growth variation. Aquaculture Science 58: 501–508.

Tanaka, Y., H. Minami, Y. Ishihi, K. Kumon, K. Higuchi, T. Eba, A. Nishi, H. Nikaido and S. Shiozawa. 2014. Relationship between prey utilization and growth variation in hatchery-reared Pacific bluefin tuna, *Thunnus orientalis* (Temminck et Schlegel), larvae estimated using nitrogen stable isotope analysis. Aquac. Res. 45: 537–545.

Teo, S.L.H., A.M. Boustany and B.A. Block. 2007. Oceanographic preferences of Atlantic bluefin tuna, *Thunnus thynnus*, on their Gulf of Mexico breeding grounds. Mar. Biol. 152: 1105–1119.

Tominaga, O., N. Uno and T. Seikai. 2003. Influence of diet shift from formulated feed to live mysids on the carbon and nitrogen stable isotope ratio ($\delta^{13}C$ and $\delta^{15}N$) in dorsal muscles of juvenile Japanese flounder, *Paralichthys olivaceus*. Aquaculture 218: 265–276.

Ueyanagi, S. 1969. Observations on the distribution of tuna larvae in the Indo-Pacific Ocean with emphasis on the delineation of the spawning areas of albacore, *Thunnus alalunga*. Bull. Far Seas Fish. Res. Lab. 2: 177–256 (in Japanese with English abstract).

Uotani, I., T. Saito, K. Hiranuma and Y. Nishikawa. 1990. Feeding habit of bluefin tuna *Thunnus thynnus* larvae in the western North Pacific Ocean. Nippon Suisan Gakkaishi 56: 713–717 (in Japanese with English abstract).

Yokoyama, H., J. Higano, K. Adachi, Y. Ishihi, Y. Yamada and P. Pichitkul. 2002. Evaluation of shrimp polyculture system in Thailand based on stable carbon and nitrogen isotope ratios. Fish. Sci. 68: 745–750.

Yonemori, T. 1989. Towards enhancing stocks of highly migratory pelagic fishes: stock enhancement of Pacific bluefin tuna. pp. 8–59. *In*: Agriculture, Forestry and Fisheries Research Council (eds.). Marine ranching. Koseisha-Koseikaku, Tokyo, Japan [in Japanese, the title was translated by authors].

Wada, E., M. Terazaki, Y. Kabaya and T. Nemoto. 1987. ^{15}N and ^{13}C abundances in the Antarctic Ocean with emphasis on the biogeochemical structure of the food web. Deep Sea Res. 34: 829–841.

Watanabe, N., K. Tanaka, J. Yamada and J.M. Dean. 1982. Scanning electron microscope observations of the organic matrix in the otolith of the teleost fish *Fundulus heteroclitus* (Linnaeus) and *Tilapia nilotica* (Linnaeus). J. Exp. Mar. Biol. Ecol. 58: 127–134.

CHAPTER 3

Age, Growth and Reproductive Biology of Bluefin Tunas

Tamaki Shimose[1], and *Jessica H. Farley[2]*

Introduction

Among the 15 tuna species (Scombridae, Thunnini) found worldwide, bluefin tunas grow to be the largest (Collette et al. 2001). The angling record for the heaviest tuna is 679 kg for an Atlantic bluefin tuna (*Thunnus thynnus*) caught off Nova Scotia in 1979 (Collette and Nauen 1983). The heaviest documented Pacific bluefin (*Thunnus orientalis*) and southern bluefin (*Thunnus maccoyii*) are 555 kg and 260 kg respectively (Foreman and Ishizuka 1990; Nakamura 1990). Given the large body size of these animals, the question arises: how fast do they grow, what is their life-span and at what size or age do they start to reproduce?

Information on age, growth and maturity for commercial fish species is important for assessing the status of stocks. Bluefin tuna stocks are routinely assessed using age-based models which require estimates of length-at-age and age-at-maturity. Despite the importance of this information, the population biology of bluefin tunas has not been studied well until relatively recently. While there have been many growth studies, validated estimates of age have only become available in recent years. Similarly, the spawning areas and spawning seasons of bluefin tunas have been studied for decades, yet valid estimates of length- or age-at-maturity are rare.

In this chapter, we briefly review the methods used to estimate the age, growth and maturity of bluefin tunas. We then review the existing literature on the estimation of these life-history parameters for the three bluefin species. Finally, these parameters are compared within the genus *Thunnus*.

---

[1] Research Center for Subtropical Fisheries, Seikai National, Fisheries Research Institute, Fisheries Research Agency, 148-446, Fukai-Ohta, Ishigaki, Okinawa 907-0451, Japan.

[2] CSIRO Oceans and Atmosphere Flagship, Hobart, Tasmania 7004, Australia.

* Email: shimose@affrc.go.jp

Age and Growth Estimation

The age and growth of teleost fish have been estimated using several methods, e.g., captive rearing, tag-recapture analysis, modal progression analysis, and age estimation using hard structures. The former two are methods that measure an individual's change in length over a specific period, while the latter two are methods that estimate length-at-age for a population. All four methods have been applied to bluefin tunas.

Captive rearing provides an absolute measure of growth for individuals as it is based directly on measurements of fish length (or weight) at multiple time points so that the change in length over time is known with few measurement errors. Growth in captive situations has been estimated for two bluefin tuna species (e.g., Kumai 1997; Farwell 2001; Tičina et al. 2007; Masuma 2009; Estess et al. 2014). However, captive rearing conditions may differ from wild conditions (e.g., food supply, temperature, swimming range), which might produce different growth rates to those from wild populations (Fig. 3.1).

Growth can be estimated for wild fish populations through tag-recapture programs as the length of each fish is measured when initially tagged and again when recaptured after a known period at liberty. As only a single growth estimate is obtained for each fish, the method requires a large number of fish to be tagged so that a sufficient number of recaptures can be obtained, at various time scales, to estimate growth for the entire life history. This method has been applied to Pacific bluefin tuna since the 1950's (Yukinawa and Yabuta 1967; Bayliff et al. 1991), and to Atlantic bluefin tuna and southern bluefin tuna since the 1960's (Kirkwood 1983; Turner et al. 1991). Since recapture rates are generally high for juveniles and decrease for older age classes due to mortality and/or tag-shedding, etc. (Hampton 1997), it is often difficult to estimate growth rate for the full life span of long-lived species such as bluefin tunas from tag-return data. Growth estimates can also be influenced by trauma and/or injury during tag-attachment. Tag-recapture experiments have also been used to validate otolith increment deposition rates in bluefin tunas (Foreman 1996; Clear et al. 2000).

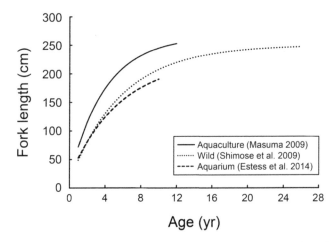

Figure 3.1. Comparison among three growth curves of Pacific bluefin tuna from different habitats.

Modal progression analysis is often applied to commercially harvested fishes and shellfishes, especially species that do not have hard structures for direct age estimation. Growth rates are estimated by following seasonal changes in modes apparent in length-frequency distribution of commercial landings. Modes are often clearly visible in small/young fish for species with relatively short spawning seasons, but can only be recognized if sufficient number of fish are measured on a regular (e.g., monthly) basis. Several methods have been developed to estimate growth parameters from length measurement data (e.g., MULTIFAN; Fournier et al. 1990), and length frequency analysis is useful when accompanied by direct ageing methods (Eveson et al. 2004; Wells et al. 2013).

Directly estimating the age of fish from hard structures is the most common and reliable method used, and is potentially applicable to all size and age classes (Campana 2001). The fundamental technique is based on counting periodic growth marks (increments) on structures such as caudal vertebrae, scales, sagittal otoliths or dorsal fin spines. However, direct validation of the periodicity of the marks being counted (i.e., daily or annual) is crucial for the success of this method (Wright et al. 2002). Direct age determination workshops have been held to discuss the ageing protocol and techniques for all three bluefin tuna species; southern bluefin tuna in 2002 (Anonymous 2002), Atlantic bluefin tuna in 2006 (Rodríguez-Marín et al. 2007), and Pacific bluefin tuna in 2013 (Shimose and Ishihara 2015).

Vertebrae and scales have been used to estimate the age of all bluefin species (e.g., Mather and Schuck 1960; Prince et al. 1985; Foreman 1996; Gunn et al. 2008). Vertebrae are relatively easy to collect and the 35th or 36th are often selected for ageing (Prince et al. 1985; Gunn et al. 2008). They are generally stained with alizarin and sometimes sectioned before being read, however, the growth increments become very closely spaced towards the edge of the structure limiting their use in older fish (Gunn et al. 2008). Historically, scales have been a preferred structure for age estimation because they can be collected without damaging the fish, crucial for fish with high market values, and can be read with almost no preparation. Scales, however, are also not a reliable structure to estimate the age of larger individuals (Rodríguez-Marín et al. 2007; Gunn et al. 2008).

Fin spines have been used to estimate the age of Atlantic bluefin (e.g., Cort 1991; Santamaria et al. 2009) but not Pacific or southern bluefin tuna. They are collected without damaging the fish, and are thinly sectioned near the base for viewing under a microscope. A potential problem of using this structure is that the core area is often vascularized and the early annuli are obscured. Methods to account for the loss of these early annuli have been proposed by some authors (Compeán-Jimenez and Bard 1983; Cort 1991; Rodríguez-Marín et al. 2012a). As multiple translucent/opaque zones can form annually in spines, a reader's experience is important for interpreting the increment pattern present (Compeán-Jimenez and Bard 1983; Cort 1991). Back-calculating length-at-age from hard structures has been applied to Atlantic bluefin using fin spines (Compeán-Jimenez and Bard 1983; Santamaria et al. 2009).

Otoliths are the most widely used hard structure for age estimation of teleost fishes (Secor et al. 1995). Of the three types of otoliths present in the inner ear of fish, the largest sagittal otolith is generally used for age estimation in tunas (Fig. 3.2). Otoliths can be difficult to locate and remove without experience; however, a method has been

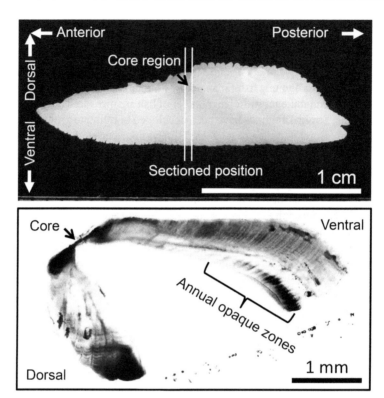

Figure 3.2. Photographs of whole (254 cm FL, female) and sectioned (243 cm FL, male) sagittal otoliths of Pacific bluefin tuna.

developed to 'drill' out the otoliths from the underside of the head without damaging the fish (Thorogood 1986a; Clear et al. 2000). Otoliths can be read whole for small tunas or sectioned for both small and large tunas. Both daily and annual increments are counted on sectioned otoliths, however, the disadvantage of estimating daily age is that it is time consuming and restricted to young age classes (Foreman 1996).

The accuracy (counts correctly representing the absolute individual age) and precision (consistency of multiple readings) of age estimates must to be assessed (Morales-Nin and Panfili 2002). Accuracy can be confirmed through various validation methods such as bomb radiocarbon studies, mark-recapture programs, and marginal increment analysis (Campana 2001), which have all been used for bluefin tunas (e.g., Kalish et al. 1996; Clear et al. 2000; Gunn et al. 2008; Neilson and Campana 2008). Standard methods to estimate the precision of increment counts such as calculating the Index of Average Percent Error (IAPE), Coefficient of Variation (CV), and age bias plots (Campana 2001) for multiple counts have also been applied to bluefin tunas. To assign fish to their correct age class or to calculate a decimal age, counts of annual increments from hard structures may need to be adjusted. These adjustments can be made by using an algorithm based on the date of capture and the assumed birth date and increment formation date (e.g., Eveson et al. 2004; Shimose et al. 2009). The method corrects for the time difference between sampling date and the individual's last birthday.

Growth of fish is generally described by the von Bertalanffy growth function (VBGF) (von Bertalanffy 1938):

$$L_t = L_\infty \times (1 - \exp(-k \times (t - t_0)))$$

where L_t is length at age t, L_∞ is mean asymptotic length, k is a growth coefficient, and t_0 is the theoretical age at length = 0. In other words, L_∞ is the estimated mean maximum body length of the population, while k indicates the growth rate of the fish; higher values indicate faster growth in early age classes and the slower growth in older age classes. The standard VBGF is most commonly used in bluefin tuna growth studies because of its simplicity and generality. There is evidence, however, that the standard VBGF may not adequately describe tuna growth and so a two stanza von Bertalanffy growth model has been developed for southern bluefin tuna (Hearn and Polacheck 2003). Composite growth models have also been developed that integrate age and growth data from several sources (e.g., Eveson et al. 2004; Restrepo et al. 2010). Fork Length (FL) is exclusively used for body length in bluefin tuna growth study.

Maturity Schedules

Precise determination of spawning seasons, length/age at maturity and fecundity rely on the correct interpretation of gonad development and maturity stage. Gonad index and maximum oocyte size have been used to assess ovary development in tuna, and occasionally to estimate approximate size at maturity (e.g., Yabe 1966; Davis 1995). However, histological analysis of ovaries is considered a more accurate method to determine reproductive state and maturity of fish (West 1990). A classification scheme, based on histological criteria, was first developed for tunas by Schaefer (1996; 1998) and later adapted for other tuna species (e.g., Farley and Davis 1998; Schaefer et al. 2005; Chen et al. 2010; Farley et al. 2013). It is recognized that sexual maturity generally occurs over a range of sizes/ages. While many studies report the size at first maturity as being equal to the smallest mature fish sampled, the more appropriate parameter useful in fisheries assessments is the length at 50% maturity (L_{50}) (also termed mean size at first maturity). This is the average size at which 50% of the individuals examined are sexually mature and is a required biological parameter used for modeling population dynamics of wild stocks. For many species, L_{50} is estimated for females only as it is assumed that male sperm production is much less limiting. This is important in the context of estimating the relationship between the size and/or age distribution of the population and the expected reproductive output of the population as a whole. It is the proportion of the population at each size or age class capable of reproduction that is of interest in a population dynamics context, not the extremes of the distribution (i.e., the earliest or latest age of first reproduction).

There are three main requirements for estimating size- and/or age-at-maturity for a fish population (Schaefer 2001):

 i. Precise criteria to identify mature and immature fish.
 ii. Unbiased sampling of ovaries from fish in the appropriate size range that includes both immature and mature females, and at the time of year when it is possible to distinguish between the two reproductive states.

iii. Fitting an appropriate statistical model to the maturity at length (or age) data to estimate the maturity schedule (or 'ogive'). The maturity ogive is a plot of the relationship between proportion mature against size or age. This estimated relationship can then be used to predict the proportion that is sexually mature at specific lengths and/or ages (e.g., length/age at 50% maturity).

Estimating a maturity ogive is complicated for species, such as bluefin tunas, where the mature fish migrate to relatively discrete areas to spawn, as this can produce a bias towards mature or immature (virgin) fish in sampling programs, depending on the area and time sampled.

Pacific Bluefin Tuna

Age and growth

Growth rates of Pacific bluefin tuna have been estimated using various methods. The earliest attempt was by Aikawa and Kato (1938) using vertebrae. Age was estimated up to 10 years, and estimated size at age was similar to the current otolith-based estimates (Shimose et al. 2009) for growth up to age five years. Subsequently, growth of Pacific bluefin tuna was estimated using scales with supporting information obtained from modal progression analysis of commercial catch data and by growth rates estimated from tag-recapture data (Yukinawa and Yabuta 1967). Age estimates from scales produced similar growth rates to vertebrae (Aikawa and Kato 1938), and was supported by modal progression analysis up to age two years (ca. 80 cm FL) and by tag-recapture analysis up to age seven years (ca. 180 cm FL; Yukinawa and Yabuta 1967). By accumulating tag-recapture data, new growth curves were estimated which showed similar growth rates to previous results for fish up to 153 cm FL (ca. five years old) (Bayliff et al. 1991).

Foreman (1996) examined fin spines, scales, vertebrae, and otoliths of Pacific bluefin tuna, and validated the daily deposition rate of increments in otoliths through an oxytetracycline mark tag-recapture experiment. The daily growth micro-increments were counted on acid-etched whole otoliths, up to ca. 2000 (5.5 years). Although the sample size was not large and length-at-age highly variable, the average length-at-age was not substantially different to previously estimated growth rates for juveniles. Recent counts of otolith micro-increments age estimates up to 420 days, and showed that growth during days 180-420 was slow, corresponding with the winter season (Itoh 2009). Predicted length at age 365 days was estimated at 57–60 cm FL from VBGF equation (Itoh 2009).

Shimose (2009) used sectioned otoliths to estimate the annual age of 806 Pacific bluefin tuna caught around Japan and Taiwan between 1992 and 2008. Fish ranged in length from 47 to 260 cm FL, encompassing almost the full size range currently harvested in the North Pacific Ocean (Shimose et al. 2009). The seasonality of otolith opaque zone formation was assessed by edge type analysis and validated as being deposited annually for larger individuals only (≥150 cm FL) (Shimose et al. 2009). Age estimates ranged from one to 26 years. These ages were adjusted into quarter year intervals, and the VBGF was fitted using nonlinear least square regression (Shimose et al. 2009). The estimated growth curve for fish up to ca. five years is similar to growth estimated by tag-recapture data (Bayliff et al. 1991), otolith micro-increment analysis (Foreman 1996), scale analysis (Yukinawa and Yabuta 1967), and vertebrae analysis (Aikawa and Kato 1938) (Fig. 3.3).

The von Bertalanffy growth parameters described by Shimose et al. (2009) were L_∞ = 249.6 cm, k = 0.173, t_0 = −0.254 years. Asymptotic length is reached at ~250 cm FL which is similar to the maximum size of Pacific bluefin tuna commonly caught by Japanese and Taiwanese fisheries. Predicted length at age one, five, 10, and 20 are 48.7, 149.0, 207.3, and 242.1 cm FL, respectively. Length at age one and five are not substantially different to other VBGF reported; 43–60 cm FL at one year, 143–145 cm at five years (Aikawa and Kato 1938; Yukinawa and Yabuta 1967; Itoh 2009), but length at ages >10 years had not been previously estimated. Estimated growth was rapid for the first 10 years and then gradually declined (Shimose et al. 2009).

Individual length-at-age for Pacific bluefin tuna is highly variable (Shimose et al. 2009). The reason for this phenomenon could be related to the long spawning season, which occurs from May to August (Itoh 2009), and individual variation in growth rates. Late hatched individuals (peak in early August) will have grown less than early hatched individuals (peak in mid-June) if caught at the same time. The actual age of 'age one year' fish includes those from 0.75 years (i.e., hatched in August and aged one year in May, the assumed birthday for all fish) to 1.00 years old. Individual somatic growth may also differ among individuals due to environmental factors or competition for food. Pacific bluefin tuna is a highly migratory species (Bayliff 1994) with a large geographic range so spatial variation in the abundance of prey or water temperatures may result in different growth rates. Inter-annual variation in prey and environmental factors may produce variation in length-at-age between fish spawned in different years. Variation in length-at-age may also be due to error in age estimation (Shimose and Ishihara 2015).

Sex specific growth of Pacific bluefin tuna is under investigation (T.S., unpubl. data), but mean fork length in older age groups (10–14 years) are reported to be 6.8 cm larger in males than females (Shimose et al. 2009), indicating differential growth by sex.

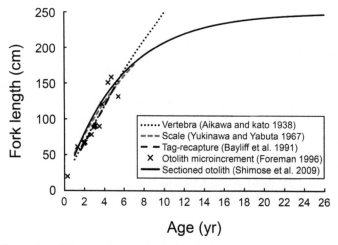

Figure 3.3. Comparison of five growth curves for Pacific bluefin tuna estimated by various methods.

Two length-weight relationships of Pacific bluefin tuna are shown here (Fig. 3.4):

$$W = 2.3058 \times 10^{-5} \times FL^{2.934}; \text{ range of } FL = 50-290 \text{ cm}$$

$$n = 1774, R^2 = 0.987 \text{ (Hsu et al. 2000)}$$

$$PW = 1.930 \times 10^{-5} \times FL^{2.981}$$

$$n = 49,234, R^2 = 0.985 \text{ (Itoh 2001)}$$

Where W is eviscerated (gutted) body weight in kg, PW is processed (gilled and gutted) body weight in kg, FL is Fork Length in cm. Body weights are 17–18 kg at 100 cm and 130–140 kg at 200 cm (Hsu et al. 2000; Itoh 2001). Body weights are highly variable for fish of the same length and also change seasonally (Itoh 2001).

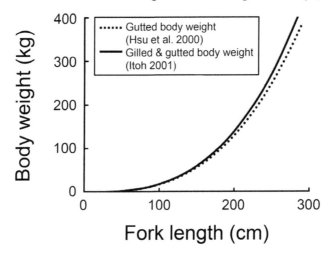

Figure 3.4. Length-weight relationships of Pacific bluefin tuna.

Maturation and spawning

There are two known spawning grounds for Pacific bluefin tuna. The main spawning ground is an extensive area in the southwestern North Pacific; around the Ryukyu Islands, from southern Japan to northeast of the Philippines (Yabe et al. 1966; Chen et al. 2006). Spawning in this area is estimated to start in late April and the activity decreases in late June as indicated by gonad index and larval occurrence (Yabe et al. 1966). Like all tunas, Pacific bluefin are multiple spawners with indeterminate annual fecundity (Schaefer 2001). A preliminary estimated mean spawning interval for females during May to June was 3.3 days (Chen et al. 2006). Estimated batch fecundity increased from around 10 million oocytes for a 190 cm FL fish to 25 million for a 240 cm fish (Chen et al. 2006). Almost all female Pacific bluefin tuna caught from this spawning ground, including the smallest individual (172 cm FL), are sexually mature

(Chen et al. 2006). Using the growth curve of Shimose (2009), the smallest female corresponded to an age of ca. six years.

Large adults are thought to spawn in the southern area of this spawning ground earlier in the season and the smaller adults in the northern area later in the season (Chen et al. 2006; Itoh 2006). Condition factor, which indicates fatness of the body, decreased during the spawning season (Chen et al. 2006). Females in spawning condition and larvae are caught in surface temperatures >24°C (Yabe et al. 1966; Tanaka 1999). Larval sampling surveys indicate that an occasional spawning may also occur southeast of Japan to ~150°E (Yabe et al. 1966; Nishikawa et al. 1985) and as far east as the Hawaiian Islands (Miller 1979), assuming larval identification was correct. The spawning season has also been estimated by back-calculating hatch date of juveniles caught in the Pacific; most spawning was estimated to have occurred from mid-May to late-June in 1992–1997 (Itoh 2009).

The second spawning ground for Pacific bluefin tuna is in the southern part of the Sea of Japan, confirmed by the occurrence of larvae (Okiyama 1974; Kitagawa et al. 1995) and evaluation of ovaries (Tanaka 1999; 2006). Adult Pacific bluefin tuna in spawning condition have been caught by purse seine in this region and the gonads have been examined histologically (Tanaka 1999; 2006). The smallest female in an active reproductive condition was 107 cm FL (Okochi 2010) corresponding to an age of three years (Shimose et al. 2009), much smaller than individuals observed in the southwestern North Pacific spawning ground (172 cm FL; Chen et al. 2006). In the Sea of Japan, 80% of age three females (<125 cm) and 90% of age four females (126–150 cm) are assumed to be mature, but these maturation rates vary slightly between years, probably due to environmental factors (Tanaka 2006). Length/age at 50% maturity has not yet been determined (Tanaka 2006). Spawning was confirmed from mid-June to August by the presence of postovulatory follicles in ovaries (Tanaka 2006; Okochi 2010). A mean spawning interval is estimated at 1.2 days (Okochi 2010) with batch fecundity ranged from 1 to 12 million oocytes for females ranging from 156–170 cm FL (Tanaka 1999). The spawning season has also been estimated by the back-calculation of hatch dates of juveniles caught in the region; spawning was estimated to have occurred from mid-July to late-August in 1992–1997 (Itoh 2009) and from late-June to late-July in 1999 and 2004 (Tanaka et al. 2007). Itoh (2009) back-calculated the hatch dates of juveniles to determine the overall spawning season (two spawning areas combined); estimated to be from mid-March to early-December.

The two spawning grounds for Pacific bluefin tuna are considered to be isolated and non-continuous. The maturity ogive currently used in the stock assessment of Pacific bluefin tuna is for a single stock, i.e., 20% mature at age three years, 50% at age four years, and 100% at age >5 years (Pacific Bluefin Tuna Working Group 2012). This was based on the work of Chen et al. (2006) and Tanaka (2006) based on the assumption that some individuals of age three and four years are not mature and not present on the spawning grounds during the spawning season. To improve knowledge on maturity schedules, continued investigations on the apparent differences in reproductive parameters from fish of the two spawning grounds are required.

Atlantic Bluefin Tuna

Age and growth

The growth of Atlantic bluefin tuna has been estimated using various methods (Rodríguez-Marín et al. 2007; Rooker et al. 2007). Unlike the other two bluefin tuna species, fin spines are often used to estimate the age of Atlantic bluefin, along with vertebrae and otoliths (Rodríguez-Marín et al. 2007). Three age determination workshops for Atlantic bluefin tuna have been held and the methodology and validation has been well documented (Hunt et al. 1978; Prince and Pulos 1983; Rodríguez-Marín et al. 2007). To review growth and longevity of Atlantic bluefin tuna, two stocks should be considered, i.e., western and eastern (including Mediterranean Sea) Atlantic.

For the western stock of Atlantic bluefin tuna, growth has been estimated by analysis of length frequency data, tag-recapture data, and age estimated using hard structures such as scales and vertebrae (e.g., Mather and Schuck 1960; Parrack and Phares 1979; Farber and Lee 1981; Turner and Restrepo 1994). Estimated growth rates using these methods were similar up to 10 years at ca. 196–220 cm FL. Turner and Restrepo (1994) estimated a growth curve based on tag-return data with a maximum estimated age of ca. 18 years, and this growth model was used in the stock assessment for western Atlantic bluefin until the 2000s.

Sectioned otoliths have also been used for annual age determination of western Atlantic bluefin tuna since the 1970's (Hurley and Iles 1983). Although not validated at the time, Hurley and Iles (1983) described sex specific growth parameters. Recently, bomb radiocarbon methods were used to validate the annual periodicity of otolith opaque zone formation in western Atlantic bluefin tuna (Neilson and Campana 2008) and a maximum age of over 30 years was obtained. Secor et al. (2009) also examined sectioned otoliths of 121 fish from the western stock, and estimated growth parameters for fish aged four–33 years. Predicted lengths at age from these three otolith studies were relatively similar; 199–224 cm FL at age 10, 251–268 cm at age 20, 256–280 cm at age 30. However, the growth curves from the latter two studies did not represent early growth well (age one–three years) due to a lack of small fish examined (Restrepo et al. 2010).

Restrepo et al. (2010) estimated VBGF parameters for the western stock using the age estimates for large fish (>4 years) from Neilson and Campana (2008) and Secor et al. (2009), and length frequency data for small fish (40–110 cm FL; one–three years). Predicted lengths at age were similar to the modal sizes for young ages, and the estimated L_∞ of 314.9 cm FL was consistent with the maximum sizes observed in the catch (Restrepo et al. 2010). Predicted mean lengths at age one, five, 10, 20, and 30 were 54.5, 132.4, 198.0, 266.9, and 295.2 cm FL, respectively (Fig. 3.5).

Sectioned spines and vertebrae have been commonly used for age determination of the eastern Atlantic (and Mediterranean) stock of Atlantic bluefin tuna (Rodríguez-Marín et al. 2007). The early studies tended to use vertebra to estimate age (e.g., Sella 1929; Hamre 1960; Rodriguez-Roda 1964; Farrugio 1980) and the structure is still currently used (Olafsdottir and Ingimundardottir 2003). Growth parameters were also estimated by analyzing length frequency data and a maximum age of 16 years was obtained (Farrugio 1980; Arena et al. 1980). However, spines have been the more

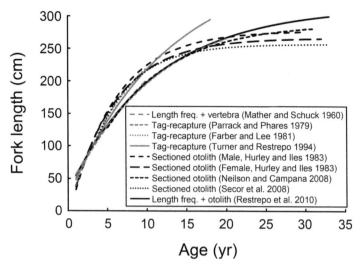

Figure 3.5. Comparison of nine selected growth curves for western stock of Atlantic bluefin tuna. Hurley and Iles (1983) includes sex specific growth curves.

frequently used structure for age determination of the Mediterranean population in recent years (e.g., Cort 1991; Megalofonou and de Metrio 2000; El-Kebir et al. 2002; Rodríguez-Marín et al. 2004; Corriero et al. 2005; Santamaria et al. 2009) because of the ease of processing. Validation of the ageing method using spines is currently being undertaken (Rodríguez-Marín et al. 2012b). A maximum age of 19 years has been estimated (Compeán-Jimenez and Bard 1983), but commonly only to 15–17 years (Cort 1991; Rodríguez-Marín et al. 2004; Santamaria et al. 2009). Most studies (Rodriguez-Roda 1964; Farrugio 1980; Compeán-Jimenez and Bard 1983; Cort 1991; Santamaria et al. 2009) produced similar growth curves for the eastern stock, and the growth curve by Cort (1991) is used in the current stock assessment. Predicted mean lengths at age one, five, 10, and 15 were at 53–66, 135–143, 203–216, and 247–258 cm FL, respectively (Fig. 3.6). Megalofonou (2006) estimated the age of 8.5–55.5 cm FL fish using counts of assumed daily growth increments in otoliths; mean age at 55 cm FL was 106 days on average.

Sex specific growth parameters were estimated for both western and eastern stocks of Atlantic bluefin tuna (Hurley and Iles 1983; Santamaria et al. 2009). Estimated L_∞ of VBGF was larger for males than for females in the western (female: 266 cm, male: 278 cm; Hurley and Iles 1983) and the eastern (female: 349 cm, male: 382 cm; Santamaria et al. 2009) stocks. Observed maximum ages were similar between sexes within the western (female: 32 years, male: 30 years; Hurley and Iles 1983) and the eastern (female: 14 years, male: 15 years; Santamaria et al. 2009) stocks. Sectioned otoliths have recently been used to estimate the annual age of the eastern stock (Secor et al. 2009), and it is likely that growth parameters and estimates of maximum age may change for that stock in the future.

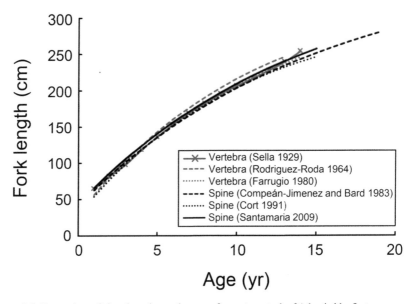

Figure 3.6. Comparison of six selected growth curves for eastern stock of Atlantic bluefin tuna.

There are many length-weight relationships for Atlantic bluefin tuna, and three examples are shown here (Fig. 3.7):

$RW = 3.00 \times 10^{-5} \times FL^{2.9192}$; range of FL = ca. 40–300 cm, western Atlantic

$n = 3578$, $R^2 = 0.997$ (Baglin 1980)

$RW = 7.2606 \times 10^{-5} \times FL^{2.7206}$; range of FL = 130–295 cm, eastern Atlantic

$n = 336$, $R^2 = 0.9618$ (Aguado-Giménez and García-García 2005)

$RW = 0.28 \times 10^{-5} \times FL^{3.34}$; range of FL = 114–256 cm, eastern Atlantic

$n = 363$, $R^2 = 0.955$ (Perçin and Akyol 2009)

Where RW is round (whole) body weight in kg, FL is fork length in cm.

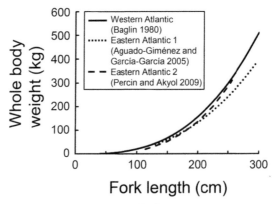

Figure 3.7. Length-weight relationships of Atlantic bluefin tuna.

Maturation and spawning

Although feeding grounds are mixed in the Atlantic Ocean, spawning grounds are thought to be separated for the two stocks of Atlantic bluefin tuna (e.g., Block et al. 2005; Rooker et al. 2008); the Gulf of Mexico (Richards 1976; Baglin 1982) where western Atlantic bluefin spawn, and the Mediterranean Sea (Corriero 2003; Karakulak et al. 2004; Heinisch et al. 2008) where eastern Atlantic bluefin spawn. However, some authors note the possibility that some proportion of the eastern stock of bluefin may spawn in the western spawning area (Nemerson et al. 2000; Goldstein et al. 2007).

In the Gulf of Mexico, spawning occurs from April to June with a peak spawning in May based on gonad analysis (Baglin 1982). Back calculation of hatch date from otolith micro-increments of larvae is also consistent with peak spawning in May (Brothers et al. 1983). The size range of spawning females is 190–282 cm FL (Baglin 1982), which corresponds to ages of ca. ≥9 years (Restrepo et al. 2010). Age at maturity for the western stock has been estimated based on the length frequency distribution of the commercial catch on the spawning ground (inside the Gulf of Mexico) assuming that all females were mature (Diaz and Turner 2007). Diaz (2011) estimated 50% maturity by this relatively crude method at 15.8 years, using size data from 15,585 fish collected in the late 1970s to early 1980s from the Japanese longline catch, assuming mortality = 0.14 and the growth equation of Restrepo et al. (2010). 'Fecundity', based on counts of fully yolked oocytes, was estimated at 13.6–57.6 (mean = 34.2) million oocytes for 205–269 cm FL females (Baglin 1982). Since tuna have indeterminate annual fecundity, this estimate is not useful and potential annual fecundity would be higher. Some larval collection surveys suggest that Atlantic bluefin tuna spawn near but outside of the Gulf of Mexico, north of Miami, Florida (McGowan and Richards 1989) and the western Caribbean (Muhling et al. 2011).

In the Mediterranean Sea, spawning varies both spatially and temporally (e.g., Corriero 2003; Karakulak et al. 2004; Heinisch et al. 2008). Spawning starts from mid-May in the eastern areas, from early-June in the central area, and from mid-June in the western area of the Mediterranean Sea. The spawning season lasts from 1 to 1.5 months in each area (Heinisch et al. 2008). Body size and gonadosomatic indices (GSI) were also found to be different among areas; smaller fish with higher GSI were found in the eastern Mediterranean, while larger fish with the lower GSI were found in the western Mediterranean (Heinisch et al. 2008). Length at 50% maturity is estimated at 103.6 cm FL (age three years from analysis of fin spines) and 100% maturity at 135 cm FL (age four to five years) (Corriero et al. 2005). ICCAT, however, assumes 50% maturity at four years (Diaz 2011; Schirripa 2011). These estimated length and age at 50% maturity are much lower than estimated for the western stock. A recent meeting to review biological parameters for Atlantic bluefin recommended that the maturity schedules currently used for both the eastern and western Atlantic stocks be revised (Anon. 2013).

The spawning interval of Atlantic bluefin tuna in the western Mediterranean Sea was estimated by Medina et al. (2007). Based on samples from purse seine and longline catch, mean spawning interval was estimated at 1.2 days and 3.1 days, respectively (Medina et al. 2007). This difference in spawning interval was caused by the fishing gear selectivity; purse seines usually catch fish gathering near surface and almost all fish sampled had evidence of spawning activity. Alternatively, longlines target deeper zone and only 35.6% of females caught were actively spawning. Mean value of batch fecundity for Atlantic bluefin tuna in the Mediterranean Sea was estimated at 6.5 million eggs for a 122.8 ± 83.7 kg (15–349 kg) female caught by purse seine, and only 216,000 eggs for a 201.5 ± 52.1 kg (15–375 kg) female caught by longline (Medina et al. 2007). Aranda et al. (2013) also estimated a mean spawning interval of 1.20 days for purse seine caught Atlantic bluefin in the western Mediterranean, and 1.12 days when only reproductive active females were analyzed. Using a stereological method, Aranda et al. (2013) estimated realized and potential batch fecundity of Atlantic bluefin tuna from counts of postovulatory follicles and mature stage (migratory nucleus or hydrated) oocytes respectively. Mean realized batch fecundity was estimated at 6.23 million (0.358 million in a 98 cm FL to 18.3 million in a 208 cm FL fish) and was less than the estimated mean potential batch fecundity of 8.80 million (1.17 million in a 98 cm fish to 27.7 million in a 199 cm fish).

Recent studies suggest that skipped spawning may occur in Atlantic bluefin tuna, i.e., sexually matured individuals do not spawn every year (Secor 2007). This is supported by two lines of evidence; 1) the presence of adult sized fish outside the known spawning ground during the spawning season (Lutcavage et al. 1999; Block et al. 2005), and 2) the presence of females in non-reproductive state on the spawning ground during the spawning season in the Mediterranean Sea (Zupa et al. 2009). Zupa et al. (2009) sampled three non-reproductive females during the spawning period in the Mediterranean Sea over a 10 year period, and given their large size (212–237 cm FL) suggested that they may be 'older skipping' fish, although further studies were needed.

Southern Bluefin Tuna

Age and growth

Age and growth estimation of southern bluefin tuna is the most advanced among all tuna species. Early estimates were based on the analysis of length frequency data (Serventy 1956; Hynd 1965; Shingu 1970; Kirkwood 1983). The preliminary estimated von Bertalanffy growth curve parameters all had relatively high L_∞ (>200 cm FL) and low k (<0.14) values, possibly as a result of low numbers of large fish included in the analyses (Hampton 1991). Seasonal variation in juvenile growth was first observed in the modal progression of length data where growth is fastest during the austral summer (Serventy 1956). Fournier et al. (1990) applied MULTIFAN to several length frequency data sets for southern bluefin tuna and estimated lower L_∞ and higher k values.

In the 1970s, growth parameters were estimated by analyzing tag-return data from a large scale tagging program undertaken in the previous decade (Shingu 1970; Lucas 1974; Murphy 1977; Hearn 1979). These studies obtained L_∞ values ranging between 171 and 187 cm FL, and k values between 0.15 and 0.19. Kirkwood (1983) proposed a method to combine length frequency and tagging-return data to estimate growth. Unfortunately, at that time very few fish had been at liberty for longer than one year and most were <150 cm FL, which may have led to an underestimation of age for large fish (Hearn 1986).

Tagging programs continued for southern bluefin tuna in the 1970s and 1980s and a large number of investigations of the data have been undertaken (e.g., Hampton 1991; Laslett et al. 2002; Hearn and Polacheck 2003; Eveson et al. 2004; Laslett et al. 2004a; Polacheck et al. 2004). Seasonal variation in growth was also evident in the tag data with juveniles growing at faster rates in late summer/early autumn and slowest in late winter/early spring (Caton 1991). In the early 1990s, it became clear that the traditional VBGF did not adequately describe southern bluefin tuna growth because of a suspected change in the pattern of growth during their transition from juveniles to sub adults (Anon. 1994). A two stanza von Bertalanffy growth model fitted the data better than the VBGF (Anon. 1994; Hearn and Polacheck 2003). The two-stage process allows for growth in each stage to follow a different VBGF curve, with a discontinuity in the growth rates at the transition point between the two. Hearn and Polacheck (2003) suggested that the change in growth pattern may be related to the transition from a tightly schooling fish that spends substantial time in near and surface shore waters to one that is found primarily in more offshore and deeper waters.

Leigh and Hearn (2000) analyzed length frequency data collected from southern Australia between 1963 and 1991 and found that the mean length of two- to four-year-old fish had increased over the 50-year history of the fishery. They suggested that the increase in growth was most likely the result of a decline in the population due to high fishing levels. The tag-return data from the 1980s and 1990s confirmed that there had been a substantial increase in juvenile growth rates between these two periods (Hearn and Polacheck 2003). Laslett et al. (2004b) developed a two-stage model to analyze length frequency data which included seasonal growth, and concluded that growth was faster in the 1980s compared to the 1960, and that growth in the 1970s was variable. Polacheck et al. (2004) used an integrated method developed by Eveson et al. (2004) that combined growth information from tagging, length frequency and direct age data from otoliths to show that the increase in growth occurred over four decades. They also showed that growth in the 1980s was substantially higher than in the 1960s, the 1970s was a transition period, but that growth continued to increase in the 1990s. They suggested that growth of fish up to about age four years was faster in the 1990s than in the 1980s. Eveson and Polacheck (2005) suggested that growth of juveniles in 2001–2002 was similar to the early 1990s, or slightly higher.

Estimating the age of southern bluefin tuna directly from hard structures began in the 1960s. Yukinawa (1970) examined scales and found that increments form annually in September/October (period of slow growth) in fish up to 130 cm FL (age seven years). After that size, he found that age estimates were not reliable. Thorogood (1987) was the first to count growth zones on whole otoliths to estimate annual age. He used marginal increment analysis to determine that the zones formed annually in June/July in young fish, and confirmed that juvenile growth was fastest in summer/autumn.

In the 1980s and 1990s, hard structures (i.e., sagittal otoliths, vertebrae and scales) were sampled from southern bluefin tuna caught by Australian, Japanese and Indonesian fishing vessels (Gunn et al. 1996). Techniques were developed to determine the daily age of larval and small southern bluefin tuna using light and scanning electron microscopy of micro-increments in whole, fractured or sectioned otoliths (Jenkins and Davis 1990; Itoh and Tsuji 1996; Rees et al. 1996). Jenkins and Davis (1990) validated the daily deposition rate of otoliths micro-increments in larval fish through marginal increment analysis and following the daily progression of increments on otoliths collected on six successive days from fish in the same school. Itoh and Tsuji (1996) and Rees et al. (1996) showed that southern bluefin tuna are one year old at around 50 cm FL, and are two years old at around 79 cm FL; both estimates were larger-at-age than previously estimated.

Gunn et al. (1996; 2008) presented a comprehensive ageing study comparing several hard structures including otoliths, scales and vertebrae and found that otoliths were the most accurate structures to estimate the age of southern bluefin tuna over its full length range. Scales were useful for age estimates up to four years and vertebrae up to 10 years, but analysis of otoliths showed that southern bluefin tuna can live up to at least 40 years. The ageing methods were validated using three methods. Firstly, bombradiocarbon chronometry determined the absolute age of southern bluefin tuna and confirmed that increments in otoliths are deposited annually throughout life (Kalish et al. 1996). The level of radiocarbon near the primordium of otoliths from large southern bluefin was measured using accelerator mass spectrometry, and compared to known changes in the levels of bomb-generated radiocarbon in the ocean (from nuclear testing in the 1950s and 1960s). This provided an approximate birth year (and absolute age) for the southern bluefin analyzed. The oldest fish using this method was 34 years caught in 1994. Secondly, a strontium chloride mark-recapture experiment was conducted in the early 1990s and by 1996 over 20,000 fish had been injected (Clear et al. 2000). Of these, 961 fish were recaptured and 616 otoliths sampled. Strontium marks were detected in 59 of 67 otoliths analyzed using a scanning electron microscope. The work showed that one increment was laid down per year at liberty for fish aged one–six years (Clear et al. 2000). In addition, the return of two tagged (but not chemically marked) fish after a long period at liberty showed that increments form annually in otoliths to at least the age of 13 years (Clear et al. 2000). Finally, marginal increment analysis of otoliths from two-year-old fish confirmed that the increments counted form annually during the austral winter (Gunn et al. 2008). In 2002, the techniques developed to prepare and read otoliths were transferred to participants at an international direct ageing workshop to standardize the interpretation of southern bluefin tuna otoliths (Anon. 2002). Subsequent work has confirmed that the age composition of the southern bluefin catch varies regionally (Farley et al. 2007; Shiao et al. 2008), juvenile growth rates increased in the 1980s and 1990s (Farley and Gunn 2007) and there is sexually dimorphic growth with males growing slightly faster than females and reaching larger length-at-age (Farley et al. 2007; Gunn et al. 2008; Lin and Tzeng 2010; Farley et al. 2014). Age is routinely estimated using otoliths for most southern bluefin tuna fisheries (CCSBT 2014).

As noted above, Eveson et al. (2004) developed an integrated method to model growth that used information on length-at-age from length-frequency, tagging, and otolith data from the 1960s to the early 1990s, and also included a seasonal growth component. Each source of data was useful for different life history stages of southern bluefin tuna. The direct age estimates included in the model were limited to those fish sampled in the southern oceans from October to April to remove estimates from fish caught during the time that the increments formed. The number of increments counted in each otolith was adjusted to assign fish to their correct age class (Eveson et al. 2004). The integrated approach provided more comprehensive estimates of growth (with associated variances for the cohorts represented in the data) especially given that the growth data available from a single source is limited and patchy. Eveson (2011) updated the growth analyses using additional tagging and direct age data from the late 1990s and 2000s. The new growth schedule was adopted by the Commission for the Conservation of Southern Bluefin Tuna (CCSBT 2011). In the 2000s, predicted length at age 10 was 156 cm FL, age 20 was 176 cm, and age 30 was 180 cm (Fig. 3.8).

Several length-weight relationships are available for southern bluefin tuna depending on the size of fish and season caught (Caton 1991; Anon. 1994) (Fig. 3.9).

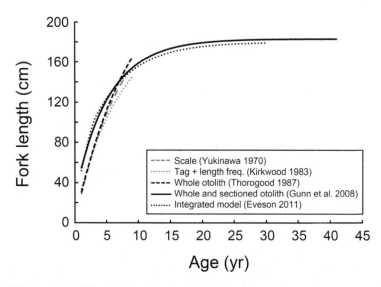

Figure 3.8. Comparison of five growth curves for southern bluefin tuna. Integrated model (Eveson 2011) used tag-recapture, length frequency, and otolith ageing data.

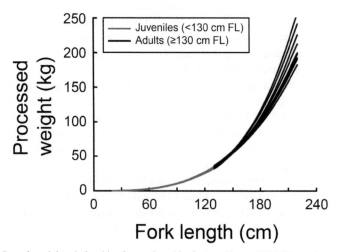

Figure 3.9. Length-weight relationships for southern bluefin tuna (Anon. 1994) illustrating regionally and seasonally different curves based on three juvenile and 10 adult curves.

As an example, the agreed relationships used by the CCSBT (Anon. 1994) for quarters 1 and 4 are:

$PW = 1.3545 \times 10^{-5} \times FL^{3.0214}$; FL < 130 cm (juveniles)

$PW = 7.3465 \times 10^{-6} \times FL^{3.1570}$; FL > 130 cm (adults)

Where PW is processed weight in kg and FL is fork length in cm. Processed weight (gilled and gutted) to whole weight is 1:1.15 (Caton 1995).

Maturation and spawning

Larval studies have shown that southern bluefin tuna spawn in a relatively large area in the north-east Indian Ocean between Indonesia and the north-west coast of Australia (Nishikawa et al. 1985). When mature, fish migrate to this spawning ground between September and April (Farley and Davis 1998). Initial studies reported that southern bluefin tuna mature at a length of around 130 cm FL based on gonad index data and the smallest lean (post-spawning) fish caught on the spawning ground (Shingu 1970; Warashina and Hisada 1970). Using length-at-age curves available at the time (Robins 1963), fish of this size were estimated to be eight years old. This estimate was essentially the minimum length that southern bluefin tuna can mature rather than the length at 50% maturity. Campbell (1994) used Warashina and Hisada's (1970) data to demonstrate that the length at 50% maturity (L_{50}) was 146 cm in the late 1960s, and then estimated that L_{50} had increased to 154 cm FL in the 1980s and 157 cm FL by the 1990s.

Thorogood (1986b) sampled ovaries from southern bluefin tuna caught around the coast of Australia and suggested that the minimum length-at-maturity was 110–125 cm FL (aged five–seven years). The study used partially yolked (yolk vesicle) oocytes or more advanced as the criteria to identify mature fish. Although this maturity classification has been used for other fish species (see Brown-Peterson 2011), in tunas the presence of fully yolked oocytes is the minimum needed to identify sexually mature fish (Schaefer 2001). It is unknown if the fish with partially yolked oocytes would have gone on to develop fully yolked oocytes and spawn in the following spawning season. Using validated ageing methods (Gunn et al. 2008) fish of 110–115 cm FL would now be considered ~four year olds.

In the early 1990s, a large-scale sampling program was established in the southern oceans to investigate the reproductive biology and maturity of southern bluefin tuna (Farley and Davis 1998). Ovaries were collected by scientific observers on board Japanese longline vessels fishing from South of Africa across the southern Indian Ocean to New Zealand, including a prespawning area well south of the spawning ground (35–45°S, 90–120°E). Using measurements of egg diameter and gonad index data for the fish caught in the prespawning area between August and December, Davis (1995) estimated L_{50} at between 152 and 162 cm FL. An estimated mean of 157 cm was recommended by Davis (1995) to use for stock assessment, which is consistent with that estimated by Campbell (1994). Using direct ageing data, Gunn et al. (2008)

estimate southern bluefin tuna of this size to be 10–12 years old. These estimates of L_{50} were considered preliminary given that it was not known if all southern bluefin tuna classed as mature in the pre-spawning area would have gone to the spawning ground to spawn (Davis et al. 2001).

In 1992, a program was established to monitor the catch and size composition of southern bluefin tuna landed by the Indonesian longline fishery on the spawning ground (Davis et al. 1995). The early data showed that most fish were aged 10–35 years (Gunn et al. 1996). As fish aged <10 years were almost absent, it was suggested that L_{50} was more likely to be 12–13 years rather than eight years as agreed by the Commission for the Conservation of Southern Bluefin Tuna (CCSBT) Scientific Committee (Anon. 1994; Gunn et al. 1996). Davis et al. (2001) estimated the L_{50} by comparing the length distribution of fish in the Indonesian catch (assumed to be an unbiased size distribution of the spawning population) with the length distribution from the Japanese longline catch in the southern oceans (assumed to be an unbiased size distribution of southern bluefin tuna off the spawning ground). They estimated the L_{50} of the population that had recruited to the spawning ground as between 158.4–163.1 cm for the six years of data examined. These estimates were similar to the earlier estimates of L_{50}, and they concluded that for stock assessment purposes, the age at 50% maturity in southern bluefin tuna is around 11–12 years old.

There was concern, however, that length data from the Indonesian fishery were not representative of the spawning population, given that southern bluefin tuna caught by Indonesia were generally larger than those caught historically by Japan. Based on the Indonesian catch data, Davis and Farley (2001) found that southern bluefin tuna appear to segregate by size and depth on the spawning ground and that the difference in the size distribution of the catch between fisheries was most likely due to different fishing depths: Indonesia generally used shallow-set longliners as it was targeting yellowfin tuna (*Thunnus albacares*) whereas Japan had traditionally used deep-set longlines to target bigeye tuna (*Thunnus obesus*) when fishing on the spawning ground. It was suggested that the larger fish would make a greater contribution to egg production than smaller fish, and may be more representative of the actual spawning population than those caught in deep longline sets.

Through the Indonesian catch monitoring program, samples of ovaries were also collected in 1992–1995 and 1999–2002 for reproductive work. Histological analysis confirmed that southern bluefin tuna have asynchronous oocytes development and indeterminate annual fecundity, and only sexually mature fish are caught on the spawning ground (Farley and Davis 1998). Although the main spawning period is between September and April, individuals do not spawn for that entire time and return to the southern oceans to feed when spawning is complete. The length of time that individual fish spawn is currently unknown. When in prime spawning condition, southern bluefin tuna spawn on average every 1.1 days and have an average batch fecundity of 6.5 million oocytes, but the relationship with length was variable (Farley et al., in press). Tagging data suggests that southern bluefin tuna may not spawn every year (Evans et al. 2012).

Comparison among *Thunnus* Species

Validated annual age determination studies using sectioned otolith have now been conducted on several *Thunnus* species, and estimated maximum ages have generally increased for these species compared to previous studies using other hard structures. Here we summarize annual age and growth studies using sectioned otolith to estimate the age of seven *Thunnus* species (Table 3.1). Although the maximum age of blackfin tuna *Thunnus atlanticus* has been estimated as 2.7 years from counts of micro-increments in sectioned otoliths (Doray et al. 2004), maximum ages may increase if annuli were present in otoliths, especially for larger fish. It appears that the growth rates of Atlantic bluefin tuna may differ between the western and the eastern (and Mediterranean Sea) populations. However, the age of most fish from the eastern population has been estimated from fin spines (e.g., Cort 1991; Santamaria et al. 2009) and only a small sample size has been estimated by sectioned otoliths (Secor et al. 2009). Therefore, these eastern fish were not included in the current comparison, but validated estimates of length-at-age of this species/population are expected to be available in the future.

Table 3.1. Parameters of von Bertalanffy growth function (VBGF) and ranges of estimated ages from analysis of sectioned otolith for seven *Thunnus* species.

Species	Location	Authors (year)	Sex	VBGF parameters L_∞	k	t_0	Age
Atlantic bluefin tuna	W Atlantic	Hurley and Iles (1983)	Female	266.4	0.17	0.106	1–32
Atlantic bluefin tuna	W Atlantic	Hurley and Iles (1983)	Male	277.8	0.169	0.254	1–30
Atlantic bluefin tuna	W Atlantic	Neilson and Campana (2008)	Combine	289	0.116	−0.06	5–31
Atlantic bluefin tuna	W Atlantic	Secor et al. (2009)	Combine	257	0.200	0.830	4–33
Atlantic bluefin tuna*	W Atlantic	Restrepo et al. (2010)	Combine	314.9	0.089	−1.13	1–33
Pacific bluefin tuna*	N Pacific	Shimose et al. (2009)	Combine	249.6	0.173	−0.254	1–26
Southern bluefin tuna*	Southern Oc.	Gunn et al. (2008)	Combine	183.2	0.185	−0.923	1–41
Bigeye tuna	SW Pacific	Farley et al. (2006)	Combine	169.1	0.238	−1.706	1–15
Bigeye tuna*	E Indian	Farley et al. (2006)	Combine	178.4	0.176	−2.500	1–16
Yellowfin tuna*	W Indian	Shih et al. (2014)	Combine	166.9	0.209	−2.663	1–11
Albacore	N Pacific	Chen et al. (2012)	Female	103.5	0.340	−0.53	1–10
Albacore	N Pacific	Chen et al. (2012)	Male	114.0	0.253	−1.01	1–14
Albacore	S Pacific	Williams et al. (2012)**	Combine	104.5	0.40	−0.49	1–14
Albacore*	N Pacific	Wells et al. (2013)	Combine	124.1	0.164	−2.239	1–15
Longtail tuna*	SW Pacific	Griffiths et al. (2010)**	Combine	99.7	0.230	−1.50	1–18

*Used for comparison among species because of large coverage of age (Fig. 3.12)
**Authors used other growth models (logistic or Schnute-Richards) rather than VBGF

It is clear that estimates of L_∞ are larger for the three bluefin tuna species than the other four *Thunnus* species given their larger maximum body sizes. Conversely, growth coefficients k are relatively low for bluefin tunas, as L_∞ and k are negatively correlated (Fig. 3.10). This negative correlation between the two parameters is well known for other fishes (e.g., Manooch 1987; Morales-Nin 1994; Nanami and Takegaki 2005). Estimated maximum ages are high for bluefin tunas exceeding 25 years in all three species, and over 40 years for southern bluefin tuna. No record of >20 years has been obtained for the other four *Thunnus* species examined. L_∞ and the maximum age are positively correlated, and the larger species have the greatest longevity (Fig. 3.11).

Growth and longevity are similar in Atlantic bluefin tuna (western Atlantic population) and Pacific bluefin tuna, which is not surprising since they are closely related (Collette et al. 2001). Estimated maximum age and L_∞ are higher for Atlantic bluefin than Pacific bluefin, which is consistent with the largest size of fish commonly caught (Collette and Nauen 1983; Foreman and Ishizuka 1990). Large sizes (>240 cm FL) of Pacific bluefin tuna were more abundant in the period of 1993–1996 than in 1997–2004 around Taiwan (Hsu et al. 2000; Chen et al. 2006). If these larger individuals could be included for VBGF estimation, parameters may be closer to the values of Atlantic bluefin tuna. It is also possible that the maximum age of Atlantic bluefin may exceed 33 years if the otoliths of large fish, such as the International Game Fish Association record fish (at 679 kg, 384 cm), could have been examined. Growth parameters of southern bluefin tuna, bigeye tuna, and yellowfin tuna are quite similar, but maximum ages of southern bluefin tuna are much higher than other two species. Their similarities and differences may be explained by their distribution range and migration (water temperature), reproductive strategies and/or feeding habits in future studies.

Some reproductive characteristics are similar between Pacific and Atlantic bluefin tunas. Both species have two known spawning grounds thought not to be connected; one is located in subtropical waters around 20–30°N (southwestern North Pacific and the Gulf of Mexico) and another in temperate waters around 30–40°N (Sea of Japan and the Mediterranean Sea). However, the spawning grounds for Pacific bluefin

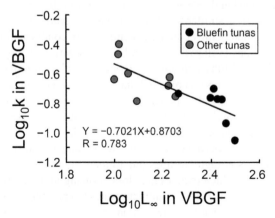

Figure 3.10. Relationship between log transformed VBGF parameters L_∞ and k for 15 sources of seven *Thunnus* species. Data used in this figure are shown in Table 3.1.

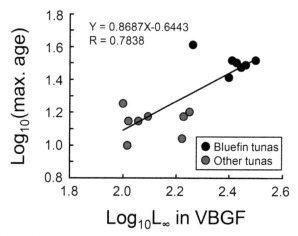

Figure 3.11. Relationship between log transformed VBGF parameter L_∞ and observed maximum age for 15 sources of seven *Thunnus* species. Data used in this figure are shown in Table 3.1.

tuna are almost adjacent to each other separated only by the continental shelf of the northeastern East China Sea, while the spawning grounds for Atlantic bluefin tuna are located on opposite sides of the Atlantic Ocean. Although maturity schedules are not well estimated, size/age at maturity for fish caught in the two spawning areas appear to be different for both species. In the northerly spawning grounds (30–40°N, Sea of Japan and Mediterranean Sea), females start to spawn at ca. three years and 100% are mature at >5 years. In the more southerly spawning grounds (20–30°N southwestern North Pacific and Gulf of Mexico), only larger/older mature fish are caught. According to the theory of a tropical origin for tunas (Boyce et al. 2008), in the Northern Hemisphere, southern (subtropical) spawning grounds are believed to be original, and northern (temperate) spawning grounds are thought to have been established secondarily. The latitudinal range of the southern bluefin spawning ground (10–20°S) is more similar to southern spawning grounds for Pacific and Atlantic bluefin tunas (20–30°N).

Spawning grounds and seasons are restricted for all three bluefin tuna species. This spawning pattern is similar to albacore (*Thunnus alalunga*) and categorized as 'migratory and spatiotemporally confined', compared to 'confluent throughout tropical and subtropical regions (yellowfin tuna and bigeye tuna)' and 'regionally confined and protracted (longtail tuna *Thunnus tonggol* and blackfin tuna)' (Schaefer 2001). This spawning pattern means that bluefin tunas migrate to feeding grounds in the non-spawning season and to specific spawning grounds during the spawning season. All tunas (Thunnini) are thought to be of tropical origin; spawning is restricted to warm waters to increase egg and larval developmental rates, hence increasing the chance of survival (Boyce et al. 2008). Bluefin tunas are 'warm blooded' and can migrate to cool water feeding grounds at high latitudes due to their highly evolved counter-current heat exchange system, which reduces heat loss allowing individuals to elevate body temperature as much as 20°C above ambient water temperature (Carey and Lawson 1973; Sharp 1979). This enables bluefin tuna species to migrate to and utilize rich feeding grounds at higher latitudes which may enable them to grow to larger sizes than other tuna species.

Sexual dimorphism in body size is found in most *Thunnus* species, with males reaching larger sizes than females. A dominance of males in the larger length classes for yellowfin tuna is thought to be caused by differential natural mortality (due to higher cost of reproduction in females) rather than differential growth or vulnerability to capture (Schaefer 1996; 2001). Unfortunately, sex specific mortality has not been estimated for any *Thunnus* species to date. Observed maximum ages are similar between the sexes for many *Thunnus* species (female vs. male = 32 vs. 30 years for Atlantic bluefin tuna, Hurley and Iles 1983; 38 vs. 41 years for southern bluefin tuna, Farley et al. 2007; 10 vs. 14 years for North Pacific albacore, Chen et al. 2012; 14 vs. 14 years for South Pacific albacore, Williams et al. 2012). Studies have also found sexual dimorphism in growth for other *Thunnus*, i.e., males are larger-at-age than females in older age classes for southern bluefin tuna (Gunn et al. 2008) and Pacific bluefin tuna (Shimose et al. 2009) or differential growth parameters for Atlantic bluefin tuna (Hurley and Iles 1983), bigeye tuna (Farley et al. 2006), and albacore (Chen et al. 2012; Williams et al. 2012). It is possible that sexual difference in the maximum body size observed in *Thunnus* species may also be caused by differential growth by sex. However, sexual differences in mortality and vulnerability to capture should be examined in the future. Sexual differences in growth may be due to a difference in reproductive cost, i.e., much more energy use for reproduction in females than males (Schaefer 1996) leading to slower growth rates in females after maturity.

Summary and Conclusion

The age and growth of bluefin tuna have been estimated by various methods over the past century. Most methods produced similar growth rates for the youngest age classes of each of the three bluefin tuna species. Sectioned otoliths, however, appear to be the only structure applicable for age determination of very old/large fish, and can provide growth rates for the full life history of each species. According to the results, growth rates of bluefin tunas are relatively rapid among the genus *Thunnus* (Fig. 3.12). Growth is fairly rapid during the first 10–15 years of life after which growth slows. Bluefin tunas can live for a further 15–25 years, depending on the species, with little change in length. The accuracy of age estimates has improved through validation work including large-scale mark-recapture experiments and bomb-radiocarbon analysis. Integrated growth models using length-at-age data from multiple sources have improved estimates of growth of some bluefin tunas and continuous research is needed for the other species. Regional differences and decadal changes in growth rates and age structures are reported for southern bluefin tuna, and need to be investigated for Pacific and Atlantic bluefin tunas in the future.

Estimating total reproductive output throughout the whole life period of bluefin tunas is difficult, requiring estimates of maturity-at-length/age, spawning duration, spawning interval, batch fecundity, the occurrence and frequency of skipped spawning for all age classes. Some of these parameters may be estimated through bio-logging studies, while others are appropriately estimated using gonad samples along with histological techniques and appropriate criteria. It is hoped that updated reproductive

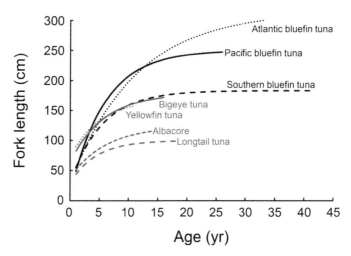

Figure 3.12. Comparison of the growth curves for seven *Thunnus* species, estimated from counts of annuli in sectioned otolith. Parameters of VBGF used in this figure are shown in Table 3.1.

parameters will be available in the future to improve the assessment of bluefin tuna stocks. General comparison of all life history traits among the eight *Thunnus* species will help to understand the evolutionary process within the genus *Thunnus* in terms of their life history and ecology.

Acknowledgments

The authors thank N.P. Clear (CSIRO, Australia), C. Farwell (Monterey Bay Aquarium), Y. Okochi (National Research Institute of Far Seas Fisheries), K. Schaefer (Inter-American Tropical Tuna Commission), and T. Tanabe (Seikai National Fisheries Research Institute) for reviewing and improving this manuscript.

References

Aguado-Giménez, F. and B. García-García. 2005. Changes in some morphometric relationships in Atlantic bluefin tuna (*Thunnus thynnus thynnus* Linnaeus, 1758) as a result of fattening process. Aquacul. 249: 303–309.

Aikawa, H. and M. Kato. 1938. Age determination of fish. I. Nippon Suisan Gakkaishi 7: 79–88 (in Japanese).

Anonymous. 1994. Report of the Southern Bluefin Tuna Trilateral Workshop. Hobart, 17 January–4 February 1994. CSIRO Marine Research.

Anonymous. 2002. A manual for age determination of southern bluefin tuna, *Thunnus maccoyii*. Otolith sampling, preparation and interpretation. The Direct Age Estimation Workshop of the CCSBT 11–14 June, 2002, Queenscliff, Australia, 1–36 pp. http://www.ccsbt.org/docs/pdf/about the commission/age determination manual.pdf.

Anonymous. 2013. Report of the 2013 bluefin meeting on biological parameters review. http:/ www.iccat. es/Documents/Meetings/Docs/2013-BFT_BIO_ENG.pdf.

Aranda, G., A. Medina, A. Santos, F.J. Abascal and T. Galaz. 2013. Evaluation of Atlantic bluefin tuna reproductive potential in the western Mediterranean Sea. J. Sea Res. 76: 154–160.

Arena, P., A. Cefali and F. Munao. 1980. Analysis of the age, weight, length and growth of *Thunnus thynnus* (L.) captured in Sicilian Seas. Mem. Biol. Mar. Ocean. 10: 119–134.

Baglin, R.E., Jr. 1980. Length–weight relationships of western Atlantic bluefin tuna, *Thunnus thynnus*. Fish. Bull. 77: 995–1000.

Baglin, R.E., Jr. 1982. Reproductive biology of western Atlantic bluefin tuna. Fish. Bull. 80: 121–134.

Bayliff, W.H. 1994. A review of the biology and fisheries for northern bluefin tuna, *Thunnus thynnus*, in the Pacific Ocean. FAO Fish. Tech. Pap. 336: 244–295.

Bayliff, W.H., Y. Ishizuka and R.B. Deriso. 1991. Growth, movement, and attrition of northern bluefin tuna, *Thunnus thynnus*, in the Pacific Ocean, as determined by tagging. IATTC Bull. 20: 1–94.

Block, B.A., S.L.H. Teo, A. Walli, A. Boustany, M.J.W. Stokesbury, C.J. Farwell, K.C. Weng, H. Dewar and T.D. Williams. 2005. Electronic tagging and population structure of Atlantic bluefin tuna. Nature 434: 1121–1127.

Boyce, D.G., D.P. Tittensor and B. Worm. 2008. Effects of temperature on global patterns of tuna and billfish richness. Mar. Ecol. Prog. Ser. 355: 267–276.

Brothers, E.B., E.D. Prince and D.W. Lee. 1983. Age and growth of young-of-the-year bluefin tuna, *Thunnus thynnus*, from otolith microstructure. NOAA Tech. Rep. NMFS 8: 49–59.

Brown-Peterson, N.J., D.M. Wyanski, F. Saborido-Rey, B.J. Macewicz and S.K. Lowerre-Barbieri. 2011. A standardized terminology for describing reproductive development in fishes. Mar. Coast. Fish.: Dyn. Mgmt. Ecosys. Sci. 3: 52–70.

Campana, S.E. 2001. Accuracy, precision and quality control in age determination, including a review of the use and abuse of age validation methods. J. Fish Biol. 59: 197–242.

Campbell, R. 1994. SBT length at maturity. Report of SBT Trilateral Workshop. Hobart, 17 January–4 February 1994. CSIRO Marine Research. Appendix 6, 26 pp.

Carey, F.G. and K.D. Lawson. 1973. Temperature regulation in free-swimming bluefin tuna. Comp. Biochem. Physiol. 44A: 375–392.

Caton, A.E. 1991. Review of aspects of southern bluefin tuna biology, population and fisheries. pp. 181–357. *In*: R.B. Deriso and W.H. Bayliff (eds.). World Meeting on Stock Assessment of Bluefin Tuna: Strengths and Weaknesses. Inter-American Tropical Tuna Commission, La Jolla, California, Special Report 7.

CCSBT. 2011. Report of the sixteenth meeting of the Scientific Committee, Commission for the Conservation of Southern Bluefin Tuna, 19–28 July, Bali, Indonesia.

CCSBT. 2014. Report of the Nineteenth meeting of the Scientific Committee, Commission for the Conservation of Southern Bluefin Tuna, 6 September, Auckland, New Zealand.

Chen, K.S., P. Crone and C.C. Hsu. 2006. Reproductive biology of female Pacific bluefin tuna *Thunnus orientalis* from south-western North Pacific Ocean. Fish. Sci. 72: 985–994.

Chen, K.S., P.R. Crone and C.C. Hsu. 2010. Reproductive biology of albacore *Thunnus alalunga*. J. Fish Biol. 77: 119–136.

Chen, K.S., T. Shimose, T. Tanabe, C.Y. Chen and C.C. Hsu. 2012. Age and growth of albacore *Thunnus alalunga* in the North Pacific Ocean. J. Fish Biol. 80: 2328–2344.

Clear, N.P., J.S. Gunn and A.J. Rees. 2000. Direct validation of annual increments in the otoliths of juvenile southern bluefin tuna, *Thunnus maccoyii*, by means of a large-scale mark-recapture experiment with strontium chloride. Fish. Bull. 98: 25–40.

Collette, B.B. and C.E. Nauen. 1983. FAO species catalogue. Scombrids of the world. An annotated and illustrated catalogue of tunas, mackerels, bonitos and related species known to date. FAO Fish. Synop. (125) 2: 137.

Collette, B.B., C. Reeb and B.A. Block. 2001. Systematics of the tunas and mackerels (Scombridae). pp. 1–33. *In*: B.A. Block and E.D. Stevens (eds.). Fish Physiology Vol. 19, Tuna: Physiology, Ecology, and Evolution. Academic Press, London.

Compeán-Jimenez, G. and F.X. Bard. 1983. Growth increments on dorsal spines of eastern Atlantic bluefin tuna *Thunnus thynnus*, and their possible relation to migration patterns. NOAA Tech. Rep. NMFS 8: 77–86.

Corriero, A., S. Desantis, M. Deflorio, F. Acone, C.R. Bridges, J.M. de la Serna, P. Megalofonou and G. De Metrio. 2003. Histological investigation on the ovarian cycle of the bluefin tuna in the western and central Mediterranean. J. Fish Biol. 63: 108–119.

Corriero, A., S. Karakulak, N. Santamaria, M. Deflorio, D. Spedicato, P. Addis, S. Desantis, F. Cirillo, A. Fenech-Farrugia, R. Vassallo-Agius, J.M. de la Serna, Y. Oray, A. Cau and G. De Metrio. 2005. Size and age at sexual maturity of female bluefin tuna (*Thunnus thynnus* L. 1758) from the Mediterranean Sea. J. Appl. Ichthyol. 21: 483–486.

Cort, J.L. 1991. Age and growth of the bluefin tuna, *Thunnus thynnus* (L.), of the northeast Atlantic. ICCAT Coll. Vol. Sci. Pap. 35: 213–230.

Davis, T., J. Farley and J. Gunn. 2001. Size and Age at 50% Maturity in SBT: An integrated view from published information and new data from the spawning ground. CCSBT Scientific Meeting; 28–31 August 2001, Tokyo, Japan. CCSBT/SC/0108/16.

Davis, T.L.O. 1995. Size at first maturity of southern bluefin tuna. Commission for the Conservation of Southern Bluefin Tuna Scientific Meeting, 11–19 July 1995, Shimizu, Japan Rep. CCSBT/SC/95/9, Far Seas Fish. Res. Lab. Shimizu. 8 pp.

Davis, T.L.O. and J.H. Farley. 2001. Size partitioning by depth of southern bluefin tuna (*Thunnus maccoyii*) on the spawning ground. Fish. Bull. 99: 381–386.

Davis, T.L.O., S. Bahar and J.H. Farley. 1995. Southern bluefin tuna in the Indonesian longline fishery: historical development, composition, season, some biological parameters, landing estimation and catch statistics for 1993. Indonesian Fish. Res. J. 1: 68–86.

Diaz, G.A. 2011. A revision of western Atlantic bluefin tuna age of maturity derived from size samples collected by the Japanese longline fleet in the Gulf of Mexico (1975–1980). Coll. Vol. Sci. Pap. ICCAT 66: 1216–1226.

Diaz, G.A. and S.C. Turner. 2007. Size frequency distribution analysis, age composition, and maturity of western bluefin tuna in the Gulf of Mexico from the U.S. (1981–2005) and Japanese (1975–1981) longline fleets. ICCAT Col. Vol. Sci. Pap. 60: 1160–1170.

Doray, M., B. Stéquert and M. Taquet. 2004. Age and growth of blackfin tuna (*Thunnus atlanticus*) caught under moored fish aggregating devices, around Martinique Island. Aquat. Living Resour. 17: 13–18.

El-Kebir, N.K., C. Rodriguez-Cabello and M.Y.O. Tawil. 2002. Age estimation of bluefin tuna (*Thunnus thynnus* L.) caught by traps in Libyan waters based on spine reading. ICCAT Col. Vol. Sci. Pap. 54: 641–648.

Estess, E.E., D.M. Coffey, T. Shimose, A.C. Seitz, L. Rodriguez, A. Norton, B. Block and C. Farwell. 2014. Bioenergetics of captive Pacific bluefin tuna (*Thunnus orientalis*). Aquacul. 434: 137–144.

Evans, K., T.A. Patterson, H. Reid and S.J. Harley. 2012. Reproductive schedules in southern bluefin tuna: are current assumptions appropriate? PLoS ONE 7(4): e34550. doi:10.1371/journal.pone.0034550.

Eveson, J.P., G.M. Laslett and T. Polacheck. 2004. An integrated model for growth incorporating tag–recapture, length–frequency, and direct aging data. Can. J. Fish. Aquat. Sci. 61: 292–306.

Eveson, J.P. and T. Polacheck. 2005. Updated estimates of tag reporting rates for the 1990's tagging experiments. CCSBT-MPTM/0502/05.

Eveson, P. 2011. Updated growth estimates for the 1990s and 2000s, and new age-length cut-points for the operating model and management procedures. Working Paper CCSBT-ESC/1107/9, 16th meeting of the Extended Scientific Committee, Commission for the Conservation of Southern Bluefin Tuna, 19–28 July, Bali, Indonesia.

Farber, M.I. and D.W. Lee. 1981. Ageing western Atlantic bluefin tuna, *Thunnus thynnus*, using tagging data, caudal vertebrae and otoliths. ICCAT Col. Vol. Sci. Pap. 15: 288–301.

Farley, J.H., N.P. Clear, B. Leroy, T.L.O. Davis and G. McPherson. 2006. Age, growth and preliminary estimates of maturity of bigeye tuna, *Thunnus obesus*, in the Australian region. Mar. Freshw. Res. 57: 713–724.

Farley, J.H., J.P. Eveson, T.L.O. Davis, R. Andamari, C.H. Proctor, B. Nugraha and C.R. Davies. 2014. Demographic structure, sex ratio and growth rates of southern bluefin tuna (*Thunnus maccoyii*) on the spawning ground. PLoS ONE 9(5): e96392. doi:10.1371/journal.pone.0096392.

Farley, J.H., T.L.O. Davis, M. Bravington, R. Andamari and C.R. Davies. in press. Spawning dynamics and size related trends in reproductive parameters of southern bluefin tuna, *Thunnus maccoyii*. PLoS ONE.

Farley, J.H. and T.L.O. Davis. 1998. Reproductive dynamics of southern bluefin tuna, *Thunnus maccoyii*. Fish. Bull. 96: 223–236.

Farley, J.H., T.L.O. Davis, J.S. Gunn, N.P. Clear and A.L. Preece. 2007. Demographic patterns of southern bluefin tuna, *Thunnus maccoyii*, as inferred from direct age data. Fish. Res. 83: 151–161.

Farley, J.H. and J.S. Gunn. 2007. Historical changes in juvenile southern bluefin tuna, *Thunnus maccoyii*, growth rates based on otolith back-calculation. J. Fish Biol. 71: 852–867.

Farley, J.H., A.J. Williams, S.P. Hoyle, C.R. Davies and S.J. Nicol. 2013. Reproductive dynamics and potential annual fecundity of South Pacific albacore tuna (*Thunnus alalunga*). PLoS ONE 8(4): e60577. doi:10.1371/journal.pone.0060577.

Farrugio, H. 1980. Age et croissance du thon rouge (*Thunnus thynnus*) dans la pecherie Francaise de surface en Mediterranee. Cybium 9: 45–59.

Farwell, C. 2001. Tunas in captivity. pp. 391–412. *In*: B.A. Block and E.D. Stevens (eds.). Fish Physiology Vol. 19, Tuna: Physiology, Ecology, and Evolution. Academic Press, London, UK.

Foreman, T. 1996. Estimates of age and growth, and an assessment of ageing techniques, for northern bluefin tuna, *Thunnus thynnus*, in the Pacific Ocean. IATTC Bull. 21: 75–123.

Foreman, T.J. and Y. Ishizuka. 1990. Giant bluefin tuna off southern California, with a new California size record. Calif. Fish Game 76: 181–186.

Fournier, D.A., J.R. Sibert, J. Majkowski and J. Hampton. 1990. MULTIFAN a likelihood-based method for estimating growth parameters and age composition from multiple length frequency data sets illustrated using data for southern bluefin tuna (*Thunnus maccoyii*). Can. J. Fish. Aquat. Sci. 47: 301–317.

Griffiths, S.P., G.C. Fry, F.J. Manson and D.C. Lou. 2009. Age and growth of longtail tuna (*Thunnus tonggol*) in tropical and temperate waters of the central Indo-Pacific. ICES J. Mar. Sci. 67: 125–134.

Goldstein, J., S. Heppell, A. Cooper, S. Brault and M. Lutcavage. 2007. Reproductive status and body condition of Atlantic bluefin tuna in the Gulf of Maine, 2000–2002. Mar. Biol. 151: 2063–2075.

Gunn, J.S., N.P. Clear, T.I. Carter, A.J. Rees, C.A. Stanley, J.H. Farley and J.M. Kalish. 2008. Age and growth in southern bluefin tuna, *Thunnus maccoyii* (Castelnau): direct estimation from otoliths, scales and vertebrae. Fish. Res. 92: 207–220.

Gunn, J.S., T.L.O. Davis, J.H. Farley, N.P. Clear and K. Haskard. 1996. Preliminary estimations of the age structure of the SBT spawning stock (including a revised estimate of age at first spawning). Commission for the Conservation of Southern Bluefin Tuna Scientific Meeting; 26 August–5 September 1996, CSIRO, Hobart, Australia. Rep. CCSBT/SC/96/10, 23 pp.

Hampton, J. 1991. Estimation of southern bluefin tuna *Thunnus maccoyii* growth parameters from tagging data, using von Bertalanffy models incorporating individual variation. Fish. Bull. 89: 577–590.

Hampton, J. 1997. Estimates of tag-reporting and tag-shedding rates in a large-scale tuna tagging experiment in the western tropical Pacific Ocean. Fish. Bull. 95: 68–79.

Hamre, J. 1960. Tuna investigation in Norwegian coastal waters 1954–1958. Annales biologiques. Conseil permanent International pour l'Exploration de la Mer. 15: 197–211.

Hearn, W.S. 1979. Growth of southern bluefin tuna. Paper presented at the Indo-Pacific Tuna and Billfish Stock Assessment Workshop, June 1979, Shimizu, Japan. Doc. Far Seas Fish. Res. Lab., SAWS/BP/14.

Hearn, W.S. 1986. Mathematical methods for evaluating marine fisheries. Ph.D. Thesis, The University of New South Wales, Kensington, NSW, Australia, 195 p.

Hearn, W.S. and T. Polacheck. 2003. Estimating long-term growth-rate changes of southern bluefin tuna (*Thunnus maccoyii*) from two periods of tag-return data. Fish. Bull. 101: 58–74.

Heinisch, G., A. Corriero, A. Medina, F.J. Abascal, J.-M. de la Serna, R. Vassallo-Agius, A.B. Ríos, A. García, F. de la Gándara, C. Fauvel, C.R. Bridges, C.C. Mylonas, S.F. Karakulak, I. Oray, G. De Metrio, H. Rosenfeld and H. Gordin. 2008. Spatial–temporal pattern of bluefin tuna (*Thunnus thynnus* L. 1758) gonad maturation across the Mediterranean Sea. Mar. Biol. 154: 623–630.

Hsu, C.C., H.C. Liu, C.L. Wu, S.T. Huang and H.K. Liao. 2000. New information on age composition and length–weight relationship of bluefin tuna, *Thunnus thynnus*, in the southwestern North Pacific. Fish. Sci. 66: 485–493.

Hunt, J.J., M.J.A. Butler, F.H. Berry, J.M. Mason and A. Wild. 1978. Proceedings of Atlantic bluefin tuna ageing workshop. Col. Vol. Sci. Pap. ICCAT 7: 332–348.

Hurley, P.C.F. and T.D. Iles. 1983. Age and growth estimation of Atlantic bluefin tuna, *Thunnus thynnus*, using otoliths. NOAA Tech. Rep. NMFS 8: 71–75.

Hynd, J.S. 1965. Southern bluefin tuna populations in south-west Australia. Aust. J. Mar. Freshw. Res. 16: 25–32.

Itoh, T. 2001. Estimation of total catch in weight and catch-at-age in number of bluefin tuna *Thunnus orientalis* in the whole Pacific Ocean. Bull. Nat. Res. Inst. Far Seas Fish. 38: 83–111 (in Japanese with English abstract).

Itoh, T. 2006. Sizes of adult bluefin tuna *Thunnus orientalis* in different areas of the western Pacific Ocean. Fish. Sci. 72: 53–62.

Itoh, T. 2009. Contributions of different spawning seasons to the stock of Pacific bluefin tuna *Thunnus orientalis* estimated from otolith daily increments and catch-at-length data of age-0 fish. Nippon Suisan Gakkaishi 75: 412–418 (in Japanese with English abstract).

Itoh, T. and S. Tsuji. 1996. Age and growth of juvenile southern bluefin tuna *Thunnus maccoyii* based on otolith microstructure. Fish. Sci. 62: 892–896.

Jenkins, G.P. and T.L.O. Davis. 1990. Age, growth rate, and growth trajectory determined from otolith microstructure of southern bluefin tuna *Thunnus maccoyii* larvae. Mar. Ecol. Prog. Ser. 63: 93–104.

Kalish, J.M., J.M. Johnston, J.S. Gunn and N.P. Clear. 1996. Use of the bomb radiocarbon chronometer to determine age of southern bluefin tuna *Thunnus maccoyii*. Mar. Ecol. Prog. Ser. 143: 1–8.

Karakulak, S., I. Oray, A. Corriero, M. Deflorio, N. Santamaria, S. Desantis and G. De Metrio. 2004. Evidence of a spawning area for the bluefin tuna (*Thunnus thynnus* L.) in the Eastern Mediterranean. J. Appl. Ichthyol. 20: 318–320.

Kirkwood, G.P. 1983. Estimation of von Bertalanffy growth curve parameters using both length increment and age-length data. Can. J. Fish. Aquat. Sci. 40: 1405–1411.

Kitagawa, Y., Y. Nishikawa, T. Kubota and M. Okiyama. 1995. Distribution of ichthyoplanktons in the Japan Sea during summer, 1984, with special reference to scombroid fishes. Bull. Jpn. Soc. Fish. Oceanogr. 59: 107–114.

Kumai, H. 1997. Present state of bluefin tuna aquaculture in Japan. Suisanzoshoku 45: 293–297.

Laslett, G.M., J.P. Eveson and T. Polacheck. 2002. A flexible maximum likelihood approach for fitting growth curves to tag–recapture data. Can. J. Fish. Aquat. Sci. 59: 976–986.

Laslett, G.M., J.P. Eveson and T. Polacheck. 2004a. Estimating the age at capture in capture-recapture studies of fish growth. Aust.-N.Z. J. Stat. 46: 59–66.

Laslett, G.M., J.P. Eveson and T. Polacheck. 2004b. Fitting growth models to length frequency data. ICES J. Mar. Sci. 61: 218–230.

Leigh, G.M. and W.S. Hearn. 2000. Changes in growth of juvenile southern bluefin tuna (*Thunnus maccoyii*): an analysis of length-frequency data from the Australian fishery. Mar. Freshw. Res. 51: 143–154.

Lin, Y.-T. and W.-N. Tzeng. 2010. Sexual dimorphism in the growth rate of southern bluefin tuna *Thunnus maccoyii* in the Indian Ocean J. Fish. Soc. Taiwan 37: 135–151.

Lucas, C. 1974. Working paper on southern bluefin tuna population dynamics ICCAT, SCRS/77/4. Coll. Vol. Sci. Pap. 111: 110–124.

Lutcavage, M., R. Brill, G. Skomal, B. Chase and P. Howey. 1999. Results of pop-up satellite tagging on spawning size class fish in the Gulf of Maine. Do North Atlantic bluefin tuna spawn in the mid-Atlantic? Can. J. Fish. Aquat. Sci. 56: 173–177.

Manooch, C.S. III. 1987. Age and growth of snappers and groupers. pp. 329–374. In: J.J. Polovina and S. Ralston (eds.). Tropical Snappers and Groupers: Biology and Fisheries Management. Westview Press, Colorado.

Masuma, S. 2009. Biology of Pacific bluefin tuna inferred from approaches in captivity. ICCAT Coll. Vol. Sci. Pap. 63: 207–229.

Mather, F.J. and H.A. Shuck. 1960. Growth of bluefin tuna of the western North Atlantic. Fish. Bull. 61: 39–52.

McGowan, M.F. and W.J. Richards. 1989. Bluefin tuna, *Thunnus thynnus*, larvae in the Gulf Stream off the southeastern United States: satellite and shipboard observations of their environment. Fish. Bull. 87: 615–631.

Megalofonou, P. 2006. Comparison of otolith growth and morphology with somatic growth and age in young-of-the-year bluefin tuna. J. Fish Biol. 68: 1867–1878.

Megalofonou, P. and G. De Metrio. 2000. Age estimation and annulus formation in dorsal spines of juvenile bluefin tuna, *Thunnus thynnus*, from the Mediterranean Sea. J. Mar. Biol. Assoc. UK. 80: 753–754.

Medina, A., F.J. Abascal, L. Aragón, G. Mourente, G. Aranda, T. Galaz, A. Belmonte, J.M. de la Serna and S. García. 2007. Influence of sampling gear in assessment of reproductive parameters for bluefin tuna in the western Mediterranean. Mar. Ecol. Prog. Ser. 337: 221–230.

Miller, J.M. 1979. Nearshore abundance of tuna (Pisces: Scombridae) larvae in the Hawaiian Islands. Bull. Mar. Sci. 29: 19–26.

Morales-Nin, N. 1994. Growth of demersal fish species of the Mexican Pacific Ocean. Mar. Biol. 121: 211–217.

Morales-Nin, N. and J. Panfili. 2002. Validation and verification methods—verification. pp. 138–142. In: J. Panfili, H. de Pontual, H. Troadec and P.J. Wright (eds.). Manual of Fish Sclerochronology. Vol. IV: A Brest: IFREMER.

Muhling, B.A., J.T. Lamkin, J.M. Quattro, R.H. Smith, M.A. Roberts, M.A. Roffer and K. Ramírez. 2011. Collection of larval bluefin tuna (*Thunnus Thynnus*) outside documented western Atlantic spawning grounds. Bull. Mar. Sci. 87: 687–694.

Murphy, G.I. 1977. A new understanding of southern bluefin tuna. Aust. Fish. 36: 2–6.

Nakamura, I. 1990. Scombridae. pp. 404–405. In: O. Gon and P.C. Heemstra (eds.). Fishes of the Southern Ocean. J.L.B. Smith Institute of Ichthyology, Grahamstown.

Nanami, A. and T. Takegaki. 2005. Age and growth of the mudskipper *Boleophthalmus pectinirostris* in Ariake Bay, Kyushu, Japan. Fish. Res. 74: 24–34.

Neilson, J.D. and S.E. Campana. 2008. A validated description of age and growth of western Atlantic bluefin tuna (*Thunnus thynnus*). Can. J. Fish. Aquat. Sci. 65: 1523–1527.

Nemerson, D., S. Berkeley and C. Safina. 2000. Spawning site fidelity in Atlantic bluefin tuna, *Thunnus thynnus*: the use of size-frequency analysis to test for the presence of migrant east Atlantic bluefin tuna on Gulf of Mexico spawning grounds. Fish. Bull. 98: 118–126.

Nishikawa, Y., M. Honma, S. Ueyanagi and S. Kikawa. 1985. Average distribution of larvae of oceanic species of scombrid fishes, 1956–1981. Far Seas Fish. Res. Lab. S. Ser. 12: 1–99.

Okiyama, M. 1974. Occurrence of the postlarvae of bluefin tuna, *Thunnus thynnus* (Linnaeus), in the Japan Sea. Bull. Japan Sea Reg. Fish. Res. Lab. 25: 89–97.

Okochi, Y. 2010. Spawning of Pacific bluefin tuna, *Thunnus orientalis*, around Japan. M.S. Thesis, Tokai University, Shizuoka, Japan.

Olafsdottir, D. and Th. Ingimundardottir. 2003. Age-size relationship for bluefin tuna (*Thunnus thynnus*) caught during feeding migrations to the northern N. Atlantic. ICCAT Col. Vol. Sci. Pap. 55: 1254–1260.

Pacific Bluefin Tuna Working Group. 2012. Stock assessment of Pacific bluefin tuna in 2012. International Scientific Committee for Tuna and Tuna—Like Species in the North Pacific Ocean.

Parrack, M.L. and P.L. Phares. 1979. Aspects of growth of Atlantic bluefin tuna determined from mark-recapture data. ICCAT Col. Vol. Sci. Pap. 8: 356–366.

Perçin, F. and O. Akyol. 2009. Length–weight and length–length relationships of the bluefin tuna, *Thunnus thynnus* L., in the Turkish part of the eastern Mediterranean Sea. J. Appl. Ichthyol. 25: 782–784.

Polacheck, T., J.P. Eveson and G.M. Laslett. 2004. Increase in growth rates of southern bluefin tuna (*Thunnus maccoyii*) over four decades: 1960 to 2000. Can. J. Fish. Aquat. Sci. 61: 307–322.

Prince, E.D., D.W. Lee and J.C. Javech. 1985. Internal zonations in sections of vertebrae from Atlantic bluefin tuna, *Thunnus thynnus*, and their potential use in age determination. Can. J. Fish. Aquat. Sci. 42: 938–946.

Prince, E.D. and L.M. Pulos (eds.). 1983. Proceedings of the International Workshop on Age Determination of Oceanic Pelagic Fishes: Tunas, Billfishes, and Sharks. NOAA Technical Report NMFS 8. 211 p.

Rees, A.J., J.S. Gunn and N.P. Clear. 1996. Age determination of juvenile southern bluefin tuna, *Thunnus maccoyii*, based on scanning electron microscopy of otolith microincrements (appendix). *In*: J.S. Gunn, N.P. Clear, A.J. Rees, C. Stanley, J.H. Farley and T. Carter (eds.). The Direct Estimation of Age in Southern Bluefin Tuna, Final report to FRDC, Report No. 1992/42.

Restrepo, V.R., G.A. Diaz, J.F. Walter, J.D. Neilson, S.E. Campana, D. Secor and R.L. Wingate. 2010. Updated estimate of the growth curve of Western Atlantic bluefin tuna. Aquat. Living Resour. 23: 335–342.

Richards, W.J. 1976. Spawning of bluefin tuna (*Thunnus thynnus*) in the Atlantic Ocean and adjacent seas. ICCAT Col. Vol. Sci. Pap. 5: 267–278.

Robins, J.P. 1963. Synopsis of biological data on bluefin tuna, *Thunnus thynnus maccoyii* (Castlenau) 1872. FAO Fish. Rep. 6: 562–587.

Rodríguez-Marín, E., N. Clear, J.L. Cort, P. Megalofonou, J.D. Neilson, M. Neves dos Santos, D. Olafsdottir, C. Rodriguez-Cabello, M. Ruiz and J. Valeiras. 2007. Report of the 2006 ICCAT workshop for bluefin tuna direct ageing (Instituto Español de Oceanografía, Santander, Spain, 3–7 April 2006). Col. Vol. Sci. Pap. ICCAT 60: 1349–1392.

Rodríguez-Marín, E., J. Landa, M. Ruiz, D. Godoy and C. Rodríguez-Cabello. 2004. Age estimation of adult bluefin tuna (*Thunnus thynnus*) from dorsal spine reading. ICCAT Coll. Vol. Sci. Pap. 56: 1168–1174.

Rodríguez-Marín, E., P.L. Luque, M. Ruiz, P. Quelle and J. Landa. 2012a. Protocol for sampling, preparing and age interpreting criteria of Atlantic bluefin tuna (*Thunnus thynnus*) first dorsal fin spine sections. ICCAT Coll. Vol. Sci. Pap. 68: 240–253.

Rodríguez-Marin, E., J. Neilson, P.L. Luque, S. Campana, M. Ruiz, D. Busawon, P. Quelle, J. Landa, D. Macías and J.M.O. de Urbina. 2012b. Blueage, a Canadian-Spanish joint research project. Validated age and growth analysis of Atlantic bluefin tuna (*Thunnus thynnus*). Coll. Vol. Sci. Pap. ICCAT 68: 254–260.

Rodríguez-Roda, J. 1964. Biología del atún, *Thunnus thynnus* (L.), de la costa sudatlántica de España. Invest. Pesq. 25: 33–146.

Rooker, J.R., D.H. Secor, G. De Metrio, R. Schloesser, B.A. Block and J.D. Neilson. 2008. Natal homing and connectivity in Atlantic bluefin tuna populations. Science 322: 742–744.

Rooker, J.R., J.R. Alvarado Bremer, B.A. Block, H. Dewar, G. de Metrio, A. Corriero, R.T. Kraus, E.D. Prince, E. Rodríguez-Marín and D.H. Secor. 2007. Life history and stock structure of Atlantic bluefin tuna (*Thunnus thynnus*). Rev. Fish. Sci. 15: 265–310.

Santamaria, N., G. Bello, A. Corriero, M. Deflorio, R. Vassallo-Agius, T. Bök and G. De Metrio. 2009. Age and growth of Atlantic bluefin tuna, *Thunnus thynnus* (Osteichthyes: Thunnidae), in the Mediterranean Sea. J. Appl. Ichthyol. 25: 38–45.

Schaefer, K.M. 1996. Spawning time, frequency, and batch fecundity of yellowfin tuna, *Thunnus albacares*, near Clipperton Atoll in the eastern Pacific Ocean. Fish. Bull. 94: 98–112.

Schaefer, K.M. 1998. Reproductive biology of yellowfin tuna (*Thunnus albacares*) in the eastern Pacific Ocean. IATTC Bull. 21: 489–528.

Schaefer, K.M. 2001. Reproductive biology of tunas. pp. 225–270. *In*: B.A. Block and E.D. Stevens (eds.). Fish Physiology Vol. 19, Tuna: Physiology, Ecology, and Evolution. Academic Press, London.

Schaefer, K.M., D.W. Fuller and N. Miyabe. 2005. Reproductive biology of bigeye tuna (*Thunnus obesus*) in the eastern and central Pacific Ocean. IATTC Bull. 23: 1–31.

Schirripa, M.J. 2011. A literature review of Atlantic bluefin tuna age at maturity. Coll. Vol. Sci. Pap. ICCAT 66: 898–914.

Secor, D.H. 2007. Do some Atlantic bluefin tuna skip spawning? Coll. Vol. Sci. Pap. ICCAT 60: 1141–1153.

Secor, D.H., J.M. Dean and S.E. Campana (eds.). 1995. Recent Developments in Fish Otolith Research. University of South Carolina Press, Columbia.

Secor, D.H., R.L. Wingate, J.D. Neilson, J.R. Rooker and S.E. Campana. 2009. Growth of Atlantic bluefin tuna: direct age estimates. Coll. Vol. Sci. Pap. ICCAT 64: 405–416.

Sella, M. 1929. Migrations and habitat of the tuna (*Thunnus thynnus* L.), studied by the method of the hooks, with observations on growth, on the operation of the fisheries, etc. (Translated from Italian by W.G. Van Campen). U.S. Department of the Interior Fish and Wildlife Service. Special Scientific Report: Fisheries No 76. 20 p.

Serventy, D.L. 1956. The southern bluefin tuna, *Thunnus thynnus maccoyii* (Castlenau), in Australian waters. Aust. J. Mar. Freshw. Res. 7: 1–43.

Sharp, G. 1978. Behavioural and physiological properties of tunas and their effects on vulnerability to fishing gear. pp. 397–450. *In*: G.D. Sharp and A.E. Dizon (eds.). The Physiological Ecology of Tunas. Academic Press, New York.

Shiao, J.-C., S.-K.Chang, Y.-T. Lin and W.-N. Tzeng. 2008. Size and age composition of southern bluefin tuna (*Thunnus maccoyii*) in the central Indian Ocean inferred from fisheries and otolith data. Zool. Stud. 47: 158–171.

Shih, C.-L., C.-C. Hsu and C.-Y. Chen. 2014. First attempt to age yellowfin tuna, *Thunnus albacares*, in the Indian Ocean, based on sectioned otoliths. Fish. Res. 149: 19–23.

Shimose and Ishihara. 2015. A manual for age determination of Pacific bluefin tuna *Thunnus orientalis*. Bull. Fish. Res. Agen. 40: 1–11.

Shimose, T., T. Tanabe, K.-S. Chen and C.-C. Hsu. 2009. Age determination and growth of Pacific bluefin tuna, *Thunnus orientalis*, off Japan and Taiwan. Fish. Res. 100: 134–139.

Shingu, C. 1970. Studies relevant to distribution and migration of southern bluefin tuna. Bull. Far Seas Fish. Lab. (Shimizu) 3: 57–114.

Tanaka, S. 1999. Spawning season of Pacific bluefin tuna estimated from histological examination of the ovary. ANRIFSF/IATTC Joint Workshop on Pacific Northern Bluefin Tuna, NRIFSF, Shimizu, Japan. ISC. 11 p.

Tanaka, S. 2006. Maturation of bluefin tuna in the Sea of Japan. ISC/06/PBF-WG/09. 6 p.

Tanaka, Y., M. Mohri and H. Yamada. 2007. Distribution, growth and hatch date of juvenile Pacific bluefin tuna *Thunnus orientalis* in the coastal area of the Sea of Japan. Fish. Sci. 73: 534–542.

Thorogood, J. 1986a. New technique for sampling otoliths of sashimi-grade scombrid fishes. Trans. Am. Fish. Soc. 115: 913–914.

Thorogood, J. 1986b. Aspects of the reproductive biology of the southern bluefin tuna (*Thunnus maccoyii*). Fish. Res. 4: 297–315.

Thorogood, J. 1987. Age and growth rate determination of southern bluefin tuna, *Thunnus maccoyii*, using otolith banding. J. Fish. Biol. 30: 7–14.

Tičina, V., I. Katavić and L. Grubišić. 2007. Growth indices of small northern bluefin tuna (*Thunnus thynnus* L.) in growth-out rearing cages. Aquacul. 269: 538–543.

Turner, S.C., V.R. Restrepo and A.M. Eklund. 1991. A review of the growth of Atlantic bluefin tuna, *Thunnus thynnus*. ICCAT Col. Vol. Sci. Pap. 35: 271–293.

Turner, S.C. and V.R. Restrepo. 1994. A review of the growth rate of west Atlantic bluefin tuna, *Thunnus thynnus*, estimated from marked and recaptured fish. ICCAT Col. Vol. Sci. Pap. 42: 170–172.

von Bertalanffy, L. 1938. A quantitative theory of organic growth (Inquiries on growth laws II). Human Biol. 10: 181–213.

Warashina, L. and K. Hisada. 1970. Spawning activity and discoloration of meat and loss of weight in the southern bluefin tuna. Bull. Far Seas Fish. Res. Lab. 3: 147–165.

Wells, R.J.D., S. Kohin, S.L.H. Teo, O.E. Snodgrass and K. Uosaki. 2013. Age and growth of North Pacific albacore (*Thunnus alalunga*): implications for stock assessment. Fish. Res. 147: 55–62.

West, G. 1990. Methods of assessing ovarian development in fishes: a review. Aust. J. Mar. Freshw. Res. 41: 199–222.

Williams, A.J., J.H. Farley, S.D. Hoyle, C.R. Davies and S.J. Nicol. 2012. Spatial and sex-specific variation in growth of albacore tuna (*Thunnus alalunga*) across the South Pacific Ocean. PLoS ONE 7: e39318.

Wright, P.J., J. Panfili, A. Folkvord, H. Mosegaard and F.J. Meunier. 2002. Validation and verification methods—direct validation. pp. 114–129. *In*: J. Panfili, H. de Pontual, H. Troadec and P.J. Wright (eds.). Manual of Fish Sclerochronology. Vol. IV: A Brest: IFREMER.

Yabe, H., S. Ueyanagi and H. Watanabe. 1966. Studies on the early life history of bluefin tuna *Thunnus thynnus* and on the larva of the southern bluefin tuna *T. maccoyii*. Rep. Nankai Reg. Fish. Res. Lab. 23: 95–129.

Yukinawa, M. 1970. Age and growth of southern bluefin tuna, *Thunnus maccoyii* (Castlenau) by use of scale. Bull. Far Seas Fish. Res. Lab. (Shimizu) 3: 229–257.

Yukinawa, M. and Y. Yabuta. 1967. Age and growth of the bluefin tuna *Thunnus thynnus* (Linnaeus), in the North Pacific Ocean. Rep. Nankai Reg. Fish. Res. Lab. 25: 1–18 (in Japanese).

Zupa, R., A. Corriero, M. Deflorio, N. Santamaria, D. Spedicato, C. Marano, M. Losurdo, C.R. Bridges and G. De Metrio. 2009. A histological investigation of the occurrence of non-reproductive female bluefin tuna *Thunnus thynnus* in the Mediterranean Sea. J. Fish Biol. 75: 1221–1229.

CHAPTER 4

Feeding Ecology of Bluefin Tunas

Tamaki Shimose[1],* and *R.J. David Wells*[2]

Introduction

Bluefin tunas are large pelagic fishes that primarily prey on fishes and squids (Collette and Nauen 1983). The feeding ecology of bluefin tunas has been studied in depth primarily due to their high value in commercial and recreational fisheries (Collette and Nauen 1983). Because stock biomass and growth rates of bluefin tunas are potentially influenced by prey abundance, having a good understanding of species interactions in oceanic ecosystems is important. Furthermore, fishing grounds of bluefin tuna may be formed in prey abundant areas, thus fishermen have a vested interest in preference of bluefin tuna prey. In contrast, abundance of commercially important fishes and squids could also be influenced by abundance of predatory tunas. In the aquaculture field, information of prey species in their natural condition is useful to develop equivalent food for captive individuals. Consequently, the feeding ecology of bluefin tunas is important throughout all ocean basins.

Feeding provides the energy source needed for fish growth and reproduction and thus the degree of feeding activity can influence growth rates and reproductive output. Bluefin tunas are capable of migrating over large distances (Fournier and Sibert 1990; Bayliff 1994; Fromentin and Powers 2005) during their relatively long life spans (>25 years; Farley et al. 2007; Shimose et al. 2009; Restrepo et al. 2010), and feeding habits are thought to be a function of life stage, oceanographic region, and/or seasonal variation. In this chapter, we aim to review available information on the feeding ecology of three bluefin tuna species and their trophic position in oceanic ecosystems.

[1] Research Center for Subtropical Fisheries, Seikai National, Fisheries Research Institute, Fisheries Research Agency, 148-446, Fukai-Ohta, Ishigaki, Okinawa 907-0451, Japan.
[2] Texas A&M University, Department of Marine Biology, 1001 Texas Clipper Rd., Galveston, Texas USA, 77553.
* Email: shimose@affrc.go.jp

Pacific Bluefin Tuna

Early works on diet

Some of the earliest studies examining the stomach contents of Pacific bluefin tuna (*Thunnus orientalis*) occurred in the western Pacific; however, size information was often limited or unavailable in the studies. Results of one of the first feeding studies by Kishinouye (1923) found sardine (*Sardinops melanostictus*), anchovy (*Engraulis japonicus*), flying fish (Exocoetidae), scad (Carangidae), sand-eel (*Ammodytes personatus*), squids (presumably *Todarodes pacificus*, *Loligo* spp., etc.), Pteropoda, *Pyrosoma*, crustacean larvae, and bottom fishes were recorded in the stomachs of Pacific bluefin tuna.

Yabe et al. (1953) examined stomach contents of 50 juvenile Pacific bluefin tuna collected by trolling off Aburatsu (southern Japan) from August–October of 1950. Anchovy (*E. japonicus*) were the most abundant prey; crustacean larvae were also noted to occur frequently. Watanabe (1960) examined stomach contents of 20 individual Pacific bluefin tuna caught by longline off Sanriku (Pacific side of northern Japan) in August and September of 1954 and squid were most abundant, in addition to flying fish (Exocoetidae), bigeye (Priacanthidae), and dolphinfish (Coryphaenidae). Although sample size was small, the dominant prey item changed to skipjack tuna (*Katsuwonus pelamis*) by late September. Yokota et al. (1961) performed a more comprehensive examination of stomach contents of 1277 young Pacific bluefin tuna caught by various fishing techniques in southern Japan from December–April during 1959–1961. Squid (*T. pacificus*), anchovy (*E. japonicus*), sand-eel (*Spratelloides gracilis*), and lanternfishes (Myctophidae) were most abundant, although mackerels (Scombridae) and jacks (Carangidae) were occasionally recorded.

Diet of larvae and juveniles

Uotani et al. (1990) investigated the feeding habits of larval Pacific bluefin tuna around the Nansei Islands spawning grounds from May–July of 1981 and 1982. Gut contents of 1939 larvae (2.28–14.60 mm Total Length, TL) were examined and the morphological development of the larvae was discussed. Copepods were the most abundant prey among all samples (97.0% in number); however, dominant prey changed with respect to length. Copepoda nauplii (<0.3 mm) was the most abundant prey item in smaller larvae (≤4.00 mm TL), *Corycaeus* copepod (>0.4 mm) was most abundant in medium sized larvae (4.01–7.00 mm TL), and a mix of other copepods (>0.7 mm) were most common in large larvae (>7 mm TL). A diet shift from Copepoda nauplii to *Corycaeus* was strongly correlated with an increase in mouth size of larvae. Feeding activity of larvae increased in day light hours relative to night and the feeding activity of larger larvae (>5 mm TL) exceeded that of smaller larvae.

Shimose et al. (2013) examined stomach contents of 437 age-0 Pacific bluefin tuna (20.3–59.4 cm Fork Length, FL) caught by trolling in the Tsushima Current region (Sea of Japan) and Kuroshio region (Pacific Ocean) between August and December of 2008. Although stomach contents differed between the two regions, ontogenetic diet shifts shared a common pattern. Small Pacific bluefin (20–25 cm FL) in the Tsushima

Current preyed upon small squid (juvenile *Enoploteuthis chunii*), and larger bluefin (25–35 cm FL) shifted their diet to mesopelagic fish (*Maurolicus japonicus*). In the Kuroshio region, small Pacific bluefin (20–25 cm FL) preyed upon small zooplankton (mostly crustacean larvae), while larger individuals (25–40 cm FL) shifted to epipelagic fishes (*Etrumeus teres, S. melanostictus, E. japonicus*). These findings suggest a diet switch to a piscivorous diet after reaching 25 cm FL where others have documented a concomitant increase in endothermy (Kubo et al. 2008), likely coinciding with enhanced diving and swimming abilities.

In the Eastern North Pacific feeding ground

Spawning of Pacific bluefin tuna occurs in the western North Pacific (Chen et al. 2006; Tanaka et al. 2007) with an unknown fraction migrating to the eastern North Pacific Ocean at approximately age-one, where they may remain until attaining sexual maturity (>3 years). While in the eastern North Pacific, bluefin tuna feed in the productive California Current during their juvenile to sub-adult stages (Bayliff 1994).

Blunt (1958; cited by Bell 1963) examined stomach contents of 168 Pacific bluefin tuna caught off California in 1957. Anchovy (*Engraulis mordax*) were the most abundant (70% in number), with jacksmelt (*Atherinopsis californiensis*), round herring (*Etrumeus acuminatus*) and squids also being found. Some coastal species, such as Pacific sand dab (*Citharichthys sordidus*), white croaker (*Genyonemus lineatus*), white seaperch (*Phanerodon furcatus*), and starfish were found in stomachs. Pinkas et al. (1971) examined stomach contents of 1073 Pacific bluefin tuna (531–1360 mm FL) caught by purse seines off California in 1968 and 1969, primarily from June to September. It was found that northern anchovy (*E. mordax*) was the most important prey among years and areas. Northern anchovy totaled 86.6% in total prey number, 80.0% in total prey volume, and was found in 72.0% of the total stomachs examined. Common squid (*Loligo opalescens*) and red swimming crab (*Pleuroncodes planipes*) were the most important species comprising the prey groups of mollusks and crustaceans, respectively, with their relative proportions dependent on the area of collection.

Diet of adults

Yamauchi (2011) examined stomach contents of 53 adult Pacific bluefin tuna (176–248 cm FL) caught by longlines throughout the southwestern North Pacific spawning ground from April–June of 2009 and 2010. Thaliaceans (Doliolida and Pyrosomida) and planktonic crustacean (e.g., Phronimidae and Platyscelidae) were dominant prey items; fishes (e.g., *Alepisaurus ferox*) and cephalopods were rarely found. Mean stomach content index (stomach content weight × 100/body weight) was quite low (female = 0.11, male = 0.17) and common prey species for Pacific bluefin tuna were thought to be scarce in this spawning area.

In the Sea of Japan, prey species of adult Pacific bluefin (115–260 cm FL) are thought to be relatively abundant compared to the southwestern North Pacific spawning region (S. Tanaka, personal communication). Dominant prey species in the Sea of

Japan consist of anchovy (*E. japonicus*) in June and common squid (*T. pacificus*) from July to August. This prey also dominates in the Tsugaru Strait area in northern Japan.

Relation to prey abundance

Pacific bluefin tuna are thought to be opportunistic feeders that feed on nekton or zooplankton available in the surrounding area. Therefore, the relative abundance of prey species in stomachs has the potential to reflect the abundance of the prey species in the surrounding environment. Prey abundance is also thought to be partly mediated by the abundance of the predator. Doi (1960) applied Volterra's equation to assess the relationship between Pacific bluefin tuna catch and common squid (*T. pacificus*) abundance in southern Japan from 1950–1955. Catch of common squid decreased while catches of Pacific bluefin increased, and thus a prey-predator relationship was thought to exist. Itoh (1961) also assessed the relationship between abundance of Pacific bluefin tuna and sardine (*S. melanostictus*) and found synchronized trends over a 100 year time frame (e.g., relatively abundant during 1920–1940, scarce during 1940–1950, increased after 1951).

Competitors, predators and parasites

Potential competitors of bluefin tunas may include other tunas (tribe Thunnini), Spanish mackerels (*Scomberomorus*), billfishes (Istiophoridae), and sharks (Carcharhinidae) similar in size (Yamanaka and staff 1963; Tiews 1963; Robins 1963; Bayliff 1994).

Toothed whales (Odontoceti), seals, sharks, billfishes, and their larger relatives (Scombridae) are thought to be potential predators of all three species of juvenile bluefin tuna (Yamanaka and staff 1963; Tiews 1963; Robins 1963). Yokota et al. (1961) reported that juvenile Pacific bluefin tuna (40–50 mm FL) were found from stomachs of skipjack tuna in the Tokara area (southern Japan) from April–June. Shimose et al. (2012) also found juvenile Pacific bluefin tuna (14–24 cm Standard Length, SL) from stomachs of blue marlin (*Makaira nigricans*) off the Pacific coast of Japan. Tunas have also been recorded as prey to orcas (e.g., *Orcinus orca*) (Yamanaka and staff 1963). As bluefin tuna increase in size and age, they reduce the risk of predation (Kishinouye 1923; Collette and Nauen 1983); however, longline hooked bluefin tunas are still preyed upon by sharks and toothed whales.

Munday et al. (2003) listed several parasites on tunas; Myxosporea, Monogenea, Digenea, Cestoda, Nematoda, Acanthocephala and Copepoda all recorded for bluefin tunas. Among them, Digenea (on gill), Nematoda (in stomach), and Copepoda (on skin surface and/or inside the opercle) are often noticed. Bell (1963) listed some worms and crustacean parasites for Pacific bluefin tuna off California as 'Smith, unpublished data', i.e., Trematoda and Nematoda from stomachs, gonads, and gills, and Copepoda from gills and skin at pectoral fin base.

Takebe et al. (2013) examined the occurrence of Trematoda (*Didymocystis wedli*) from juvenile Pacific bluefin tuna (26.8–43.9 cm TL). They found this species in specimens caught in the Sea of Japan, but not from specimens caught in the Pacific coast of Japan. Squid is thought to be an intermediate host (Jones 1991), and frequent predation on the squid (*E. chunii*) by Pacific bluefin in the Sea of Japan

(Shimose et al. 2013) is thought to cause this parasitism (Takebe et al. 2013) which changes seasonally (Kobayashi 2004).

Summary for Pacific bluefin tuna

Primary diets of 2–15 mm TL, 20–60 cm FL, 53–136 cm FL, and 176–248 cm FL sized class Pacific bluefin tuna are reported (Fig. 4.1). Larvae (2–15 mm TL) prey on small planktonic crustaceans (Uotani et al. 1990). No information is available for Pacific bluefin tuna ranging in size from 15 mm TL to 20 cm FL. Prey type changes around 25–35 cm FL from crustacean larvae or small squid (juvenile *E. chunii*) to clupeoid fishes or mesopelagic fish (*M. japonicus*) (Shimose et al. 2013). Larger juvenile Pacific bluefin, ranging in size from 53–136 cm FL, migrate to the eastern

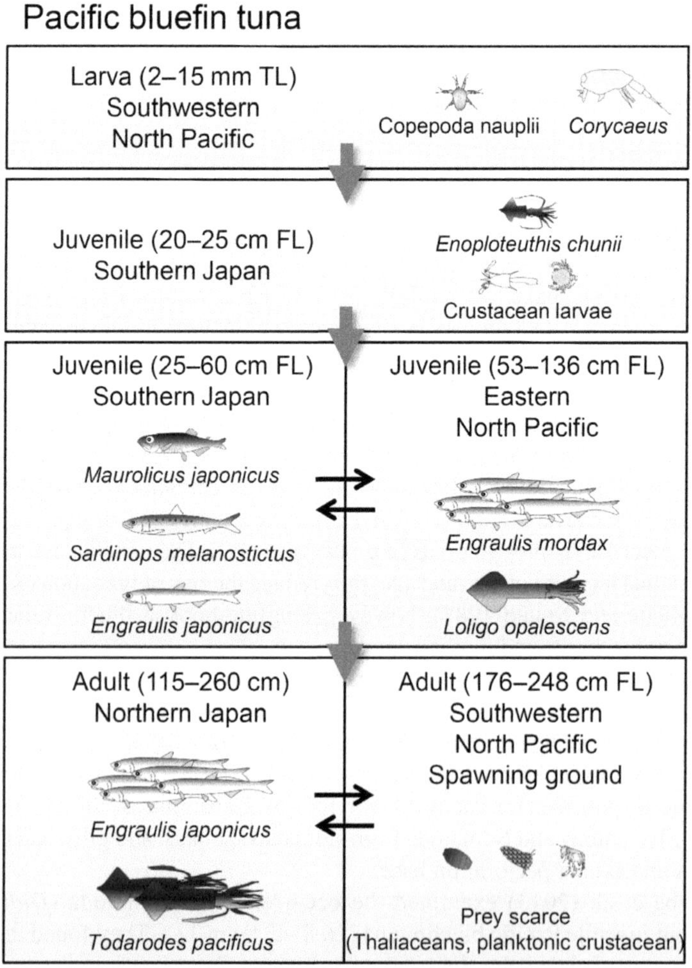

Pacific bluefin tuna

Figure 4.1. Summary of ontogenetic changes in diets of Pacific bluefin tuna. Data sources are described in the text.

North Pacific and primarily consume anchovy (*E. mordax*) (Pinkas et al. 1971). Major prey items of adult Pacific bluefin include anchovy (*E. japonicus*) and common squid (*T. pacificus*) in the Sea of Japan (S. Tanaka, personal communication). Prey of adult Pacific bluefin, with a size range of 176–248 cm FL, is reported in the southwestern North Pacific spawning ground where prey is thought to be scarce and was dominated by Thaliaceans and planktonic crustaceans (Yamauchi 2011). Seasonal changes of prey species are not well studied, and prey species are thought to differ in unstudied areas.

Atlantic Bluefin Tuna

Early works on diet

Atlantic bluefin tuna (*Thunnus thynnus*) have two separated spawning grounds (Gulf of Mexico and Mediterranean Sea), with extensive feeding grounds throughout the central North Atlantic Ocean (Fromentin and Powers 2005; Rooker et al. 2007). Consequently, the feeding patterns of Atlantic bluefin tuna have been fairly well documented over time.

Stomach contents of immature Atlantic bluefin tuna (reported 30 inches ~76 cm TL) were first reported by Nichols (1922) using a single specimen caught by trolling in the western North Atlantic off Block Island, Rhode Island. The stomach contained 50 small herring (*Clupea harengus*) approximately 10 cm in length, a single halfbeak (Hemiramphidae), and one small squid. Crane (1936) examined stomach contents of 34 Atlantic bluefin tuna (41–97 inches ~104–246 cm SL) caught by harpoon at the mouth of Casco Bay, Maine in July of 1936. The most abundant prey was silver hake (*Merluccius bilinearis*, 8–13 inches ~20–33 cm SL) found in 26 of the 34 stomachs. Squids and Northern krill (*Meganyctiphanes norvegica*) were also found in several stomachs. Fish recorded included Clupeidae (75 and 215 mm), rockfish (*Sebastes* sp., 53–117 mm) and needlefish (Belonidae, 135 mm).

De Sylva (1956) reported large squids, squid beaks, and the radulae of bottom-dwelling snails from the stomachs of Atlantic bluefin tuna off Cat Cay, Bahamas. Krumholz (1959) examined stomach contents of seven adult sized Atlantic bluefin tuna (224–254 cm TL) caught near Bimini, Bahamas in May of 1956. The most numerous prey was juvenile porcupine fish (*Diodon hystrix*) consisting of 560 in six stomachs. The salp (*Pyrosoma atlantica*, 10–13 cm) was the second most abundant prey with 87 found in four total stomachs. Vertebrae of eel-like fish and portunid crabs were also found in small numbers.

Matthews et al. (1977) examined stomach contents of 38 adult sized Atlantic bluefin tuna (158–232 cm FL) caught by longline in the western North Atlantic, along the eastern coast of the United States and east to the Azores Islands. In total, 20 fish species and seven invertebrate species were recorded with Bramidae (*Brama orcini* and *Taractes* spp.) representing the most frequently occurring family of fish. *Sargassum* associated fishes (Balistidae, *Hippocampus erectus*) and pelagic fishes (*Alepisaurus* sp., *Auxis* sp.) were also frequently found.

Eggleston and Bochenek (1990) examined stomach contents of 97 juvenile Atlantic bluefin tuna (70–132 cm FL) caught by recreational fishing off Virginia during June and July 1986. Teleost fishes were the most important prey with sand lance (*Ammodytes* spp.) as the most abundant (N = 84%). Atlantic brief squid (*Lolliguncula brevis*) was also considered important and occurred in 10% of stomachs.

Regional differences

Dragovich (1970) quantitatively examined stomach contents of 169 small (52–102 cm FL), 74 large (160–267 cm) and seven Atlantic bluefin tuna of unknown size caught by commercial longline, purse seine and tournament trolling in the western North Atlantic from March–July of 1965 and 1966. Among the 250 specimens, only 87 contained prey (excluding parasites and *Sargassum* seaweed). All prey species were pelagic including 22 fishes, 10 mollusks, eight crustaceans, and one salp taxa. Fishes were the most important in volume (76.9%) and both fishes and mollusks were found from both small and large bluefin tuna; however, crustaceans and salp were found only in large tuna. Some fishes (e.g., *Diodon* sp., juvenile *Ahlia egmontis, Scomber* sp.) were found in large numbers from a single stomach. From the Carolinas to New York along the eastern US coast (34–39°N, 67–74°W), *B. orcini* (as *Collybus drachme*) was the most frequently occurring (18.3%), and Scombridae was the most abundant in volume (34.6%). In the region north of Andros Island, Bahamas (25°N, 79°W), where 31 large tuna were caught by trolling, frequency of occurrence for Salpidae (43.8%) and *A. egmontis* (31.2%) was high, and *Diodon holocanthus* and *Dactylopterus volitans* had large volumes of 37.5 and 36.7%, respectively.

   Chase (2002) examined stomach contents of Atlantic bluefin tuna (120–299 cm curved fork length, CFL) caught by commercial and sport fishing boats at five different feeding grounds along the New England continental shelf between 1988 and 1992. From the analysis of 819 stomachs, sand lance (*Ammodytes* spp.), Atlantic herring (*C. harengus*), Atlantic mackerel (*Scomber scombrus*), squid (Cephalopoda), and bluefish (*Pomatomus saltatrix*) were important prey items among all areas of collection, but dominant items differed among areas. Several species were unique to specific regions including Atlantic menhaden (*Brevoortia tyrannus*), pollock (*Pollachius virens*), and Atlantic cod (*Gadus morhua*). Benthic fig sponge (*Suberites ficus*) and seahorse (*Hippocampus erectus*) were frequently observed. Two elasmobranchs (*Squalus acanthias, Raja* spp.) were also found.

Mediterranean sea

The Mediterranean Sea is one of the major spawning grounds of Atlantic bluefin tuna, and some fisheries target bluefin tuna throughout the year. In this area, all life history stages (larvae to spawning adults) occur and the feeding biology of juveniles to adults are well studied. Sinopoli et al. (2004) examined stomach contents of 107 juvenile Atlantic bluefin tuna (63–495 mm TL) caught by purse seine and trolling in the southern Tyrrhenian Sea during July–November between 1998–2000. Fishes and cephalopods (e.g., *Heteroteuthis dispar, Onychoteuthis banksi*) were important prey in weight, while crustaceans were also frequently observed. Fishes were important prey species in weight for small (60–200 mm TL) and medium (201–350 mm TL) sized bluefin, and the importance of cephalopods increased in percent stomach weight for larger juveniles (351–500 mm TL).

Karakulak et al. (2009) examined stomach contents of 218 adult Atlantic bluefin tuna (98.5–294 cm FL) caught by purse seine in the eastern Mediterranean during the spawning season of May–June between 2003–2006. Lanternfishes (e.g., *Hygophum benoiti*, Myctophidae, *Diaphus* sp., *Ceratoscopelus maderensis*, *Electrona risso*), carangid fish (*Trachurus* sp.) and the pelagic octopuses (*Argonauta argo*, *Tremoctopus violaceus*) were important in number and frequency of occurrence. Frequency of occurrence of meso- and bathy-pelagic fishes (Myctophidae, Chauliodontidae, Paralepididae) provided support that deep diving behavior was likely used for feeding. Battaglia et al. (2013) examined 123 adult Atlantic bluefin tuna (115–222 cm FL) caught by handline during the spring season of 2010 and 2011 in the Strait of Messina, central Mediterranean Sea. In total, 91 taxa were identified which included 54 teleosts, 20 cephalopods, and 13 crustaceans. The myctophids (*H. benoiti* and *C. maderensis*), stomiid (*Chauliodus sloani*), and the oegopsid squid (*Illex coindetii*) were important prey by index of relative importance IRI. Battaglia et al. (2013) also estimated feeding time and depth of feeding for several major prey species and noted both meso- and bathy-pelagic prey may support the diet during this reproductive season.

Stable isotope approach

Carbon ($\delta^{13}C$) and nitrogen ($\delta^{15}N$) stable isotopes are often used to examine food webs. Stomach content analysis identifies specific prey species; however, digestion processes often limit the analysis to several days. On the other hand, stable isotope values reflect the diet of the predator over a longer time scale from weeks to months. Carbon isotopes are used to estimate the source(s) of primary production (Fry and Sherr 1984), and nitrogen isotopes are often used to estimate trophic position (Peterson and Fry 1987).

Estrada et al. (2005) analyzed stable isotopes of $\delta^{13}C$ and $\delta^{15}N$ in scales, muscle tissue, and bones of adult (ca. 7 years old, n = 65) and juvenile (1 year old, n = 7) Atlantic bluefin tuna caught in the northwest Atlantic Ocean between June and October of 2001. Among some potential prey species, Atlantic herring (*C. harengus*), sand lance (*Ammodytes americanus*) and silver hake (*M. bilinearis*) were thought to be preyed upon with evidence of 3–4‰ depleted $\delta^{15}N$ relative to bluefin tuna which corresponds to one trophic level. Trophic position of juvenile bluefin tuna was similar to those of yellowfin tuna (*Thunnus albacares*) and albacore (*Thunnus alalunga*), but lower than adult bluefin tuna. Carbon isotope values suggested silver hake was a common prey item in early summer (June–July) with a shift to herring and sand lance by late summer (August–September).

Sarà and Sarà (2007) also investigated carbon and nitrogen stable isotopes of muscle tissue in Atlantic bluefin tuna from the southern Tyrrhenian Sea (central Mediterranean). Small to large specimens (n = 35, 15–225 kg) were caught by trap fishery in May, and small juveniles (n = 14, 0.7–2.2 kg) were caught with a sardine purse seine in September and October 2004. $\delta^{13}C$ was negatively correlated and $\delta^{15}N$ was positively correlated with the predator's weight. Three trophic groups were categorized with stable isotopes; juveniles (0.7–2.2 kg) relied on zooplankton and small fishes; sub-adults (15–50 kg) relied on mid-sized pelagic fishes, shrimps and cephalopods; adults (70–225 kg) relied on cephalopods and larger fishes.

Logan et al. (2011) examined stomach contents of juvenile Atlantic bluefin tuna (mainly 60–150 cm CFL) caught in the western (Mid-Atlantic Bight, n = 42) and eastern (Bay of Biscay, n = 171) Atlantic Ocean, and also analyzed stable isotopes. Horse mackerel (*Trachurus trachurus*), blue whiting (*Micromesistius poutassou*), anchovy (*Engraulis encrasicolus*) and krill (*M. norvegica*) were primary items in stomach contents from the Bay of Biscay, and relative consumption reflected annual changes of prey abundance in the area. Sand lance (*Ammodytes* spp.) was the primary item found in stomach contents in the Mid-Atlantic Bight. Mixing models derived from stable isotopes indicated that krill (Bay of Biscay) and crustaceans (Mid-Atlantic Bight) contributed more to the long term diet than what stomach content results suggested.

Varela et al. (2013) examined stomach contents of adult Atlantic bluefin tuna (n = 189, 143–262 cm FL) caught by the trap fishery in the Strait of Gibraltar in May and June from 2009–2011 and stable isotopes of muscle and liver from 116 individuals were also analyzed. Most stomachs were empty or contained only hard structures (e.g., otoliths of *Trachurus* sp., beaks of *Illex coindetii*, claws of *Polybius henslowii*), and it was suggested that bluefin tuna do not feed significantly during the pre-spawning migration. *Polybius henslowii*, *Sardina pilchardus*, *Sardinella* sp., and *Scomber colias* were found as soft (undigested) prey. Mixing model analysis showed the difference in the estimated dominant prey types among years. Based on the stable isotopes of muscle tissue, fishes were the most abundant prey group in all three years followed by squid. Liver stable isotope data supported fishes as the most dominant group in 2010 and 2011, while crustaceans were most abundant in 2009.

Impact on commercial fishes

Overholtz (2006) estimated the impact on Atlantic herring (*C. harengus*) by predation of bluefin tuna in the Gulf of Maine off Georges Bank from 1970–2002. Herring is an abundant fish preyed on by many piscivorous animals such as marine mammals, seabirds, and demersal fishes as well as bluefin tuna. Estimated mean consumption by bluefin tuna was highest at 58.0 kt (kilotons) in 1970, declined to 2.2 kt in 1982, and increased to 24.4 kt in 2002. The study estimated that bluefin tuna predation on herring was at about 7% of the total herring consumption in the region.

Butler et al. (2010) estimated daily ration of Atlantic bluefin tuna and consumption of Atlantic menhaden (*B. tyrannus*) by bluefin tuna in the continental shelf waters off North Carolina during the late fall to winter months of 2003–2006. Atlantic menhaden is an important commercial fish and is primarily preyed upon by bluefin tuna during the winter season. In weight, Atlantic menhaden accounted for 98% of medium to large (185.4–205.7 cm CFL) and 95% for giant (>205.7 cm) bluefin stomach contents. Gut fullness peaked at 12:00–13:00 for medium to large sized bluefin and 13:00–14:00 for giant bluefin, and the time required to empty was approximately 20 hours from the peak (assuming no feeding at night). Daily ration was estimated to be ca. 2% of bluefin body weight per day. Mean consumption of the menhaden by >6 year old bluefin tuna was estimated to be 3021 t using bluefin abundance data in 2005, and 10,020 t using abundances derived from 1975.

Competitors, predators and parasites

Clua and Grosvalet (2001) reported a total of 22 observations on feeding aggregations of dolphins, tunas, and seabirds in the Azores Islands in the eastern North Atlantic. During some observations, Atlantic bluefin tuna (~200 kg) and yellowfin tuna (~100 kg) preyed upon fish bait schools that were organized by dolphins. These observations indicated that dolphins and seabirds are potential competitors of Atlantic bluefin tuna by targeting similar prey, but also suggested that dolphins sometimes partially benefit tunas by organizing prey.

Predators of Atlantic buefin tuna are not well reported. Di Natale (2012) reported that several adult bluefin tuna caught in the Mediterranean Sea had natural marks of cookiecutter shark (*Isistius brasiliensis*) bites; however bites are often not lethal for adult bluefin tuna.

Parasites have been reported from a number of studies on bluefin tuna. Crane (1936) found many dark red *Caligus* sp. on the bases of the anal and caudal fin and dorsal and anal finlets of adult Atlantic bluefin tuna (ca. 246 cm SL). For example, dissection of 34 stomachs of adult Atlantic bluefin tuna (104–246 cm SL) revealed that six stomachs contained worms of the Trematoda (as *Distoma*). Eggleston and Bochenek (1990) examined stomachs of 97 juvenile Atlantic bluefin tuna (70–132 cm FL) caught off Virginia, and found one to two trematodes (*Hirudinella ventricosa*) from each of the eight stomachs (mean number = 1.14) that contained parasites. Infestation rates of trematodes are thought to be different among areas and body size of hosts, but previous studies support their presence in both juvenile and adult bluefin.

Mladineo and Bočina (2009) examined infection rates for 38 juvenile Atlantic bluefin tuna (15 ± 3.7 kg) by *Didymocystis wedli* and *Koellikerioides intestinalis* at the tuna farming facility. Bluefin tuna were caught in May of 2006 and farmed until the summer of 2006 in the Adriatic Sea. *Didymocystis wedli* were found in the gills and were common in 62% of bluefin tuna examined. In addition, *K. intestinalis* were found in the intestine and 55% were infected.

Summary for Atlantic bluefin tuna

Atlantic bluefin tuna consists of two genetically discrete populations, i.e., western population spawning in the Gulf of Mexico and eastern population spawning in the Mediterranean Sea (e.g., Block et al. 2005; Rooker et al. 2008). Both populations migrate to the central Atlantic Ocean for feeding where the feeding grounds of both populations overlap (Block et al. 2005). Feeding biology of Atlantic bluefin tuna is well studied off the U.S. coast where both western and eastern populations mix for feeding (Fig. 4.2). These studies reported that Atlantic bluefin tuna of 70 cm FL to 299 cm CFL prey on abundant schooling fishes such as silver hake (*M. bilinearis*), Atlantic herring (*C. harengus*), Atlantic menhaden (*B. tyrannus*), sand lance (*Ammodytes* spp.), Atlantic mackerel (*S. scombrus*) and also large fish (e.g., *P. saltatrix*) (Nichols 1922; Crane 1936; Chase 2002; Eggleston and Bochenek 1990; Overholtz 2006; Butler et al. 2010). Major prey items are thought to change depending on areas and the season. Oceanic pelagic fishes (Bramidae, *Alepisaurus*, *Auxis*) are abundant in bluefin stomachs in offshore U.S. waters (Matthews et al. 1977). Juvenile porcupine fish (*D. hystrix*)

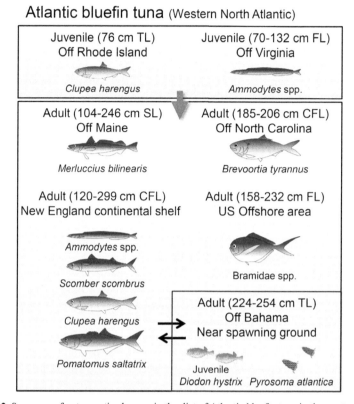

Atlantic bluefin tuna (Western North Atlantic)

Figure 4.2. Summary of ontogenetic changes in the diet of Atlantic bluefin tuna in the western Atlantic. Data sources are described in the text.

are abundant in the stomachs of Atlantic bluefin tuna off the Bahamas (near the Gulf of Mexico spawning ground) in the spawning season (May) (Krumholz 1959). In the Mediterranean Sea, where the eastern population spawns, diets of juveniles (6–50 cm TL) and adults (98–294 cm FL) are reported (Fig. 4.3). Diets of juvenile bluefin shift from fishes to squids at around 35 cm TL (Sinopoli et al. 2004), while adults prey on schooling fish (*Trachurus* sp.) and deep dwelling fishes and squids which are available in deeper layers (Karakulak et al. 2009; Battaglia et al. 2013).

Southern Bluefin Tuna

Early works on diet

Serventy (1956) is the first description on stomach contents of juvenile southern bluefin tuna (*Thunnus maccoyii*) (mainly <30 lb) in New South Wales and Tasmanian waters. Jack mackerel (*Trachurus novaezelandiae*, mainly 200–300 mm TL), mackerel (*Scomber australasicus*, 90–250 mm) and pilchards (*Sardinops neopilchardus*, 200 mm) were relatively large fish prey. The largest specimen was a 350 mm jack mackerel found from a ca. 16 kg (35 lb) bluefin tuna stomach. Anchovy

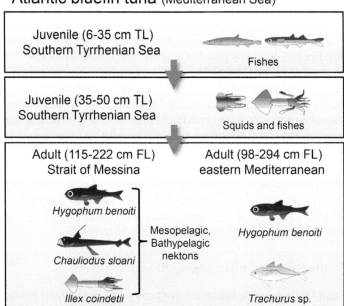

Figure 4.3. Summary of ontogenetic changes in the diet of Atlantic bluefin tuna in the Mediterranean Sea. Data sources are described in the text.

(*Engraulis australis*, 50–150 mm) was also found in large numbers. Other fishes included scad (*Trachurus declivis*, 230 mm), young barracouta (*Thyrsites atan*, 200 mm), pearl fish (*Emmelichthys nitidus*, 210 mm), morwong (*Nemadactylus* sp., 90 mm), pike (*Sphyraena novaehollandiae*, 210 mm), salmon (*Arripis trutta*), small king barracouta (*Rexea solandri*), blue sprats (*Spratelloides robustus*), small marine eels, myctophids and many larval fishes. *Nototodarus gouldi* (230 mm) was the most common cephalopod in the area, and other cephalopods found were *Enoploteuthis galaxias*, *Calliteuthis miranda*, *Argonauta nodosa* and *Octopus australis*. Larval crustacean (e.g., stomatopod) and pelagic amphipod (e.g., *Phrosina semilunata*) were also found and considered to be important food items to bluefin tuna. Siphonophores (*Diphyes*), salps and chaetognaths consisted of other invertebrate prey.

Talbot and Penrith (1963) described prey of southern bluefin tuna along South Africa based on 263 specimens. Fishes (*Merluccius capensis*, *Lepidopus caudatus*) and prawns (*Funchalia woodwardi*) were major prey items of bluefin tuna. Others included squids, tunicate (*Pyrosoma* sp.), Amphipoda, megalopa, Palinuridae. Overall, fishes represented the most important prey item with respect to percentage by weight.

Diet of larvae and juveniles

Uotani et al. (1981) examined the diet of larval southern bluefin tuna at the north-eastern Indian Ocean spawning ground from 1977–1979 along with other tuna species larvae. Larvae of southern bluefin tuna (2.11–9.55 mm TL) fed on *Corycaeus* and *Evadne*, and this trend was similar to other tuna species (albacore, yellowfin tuna and bigeye tuna *T. obesus*). Young and Davis (1990) also examined the diet of larval southern bluefin tuna at the Indian Ocean spawning ground. The authors used a selectivity index that revealed southern bluefin selectively feed on copepod nauplii and corycaeids relative to calanoids. Feeding was only found in daytime, and cannibalism was found in post-flexion larvae of southern bluefin.

Itoh et al. (2011) examined the diet of 720 juvenile southern bluefin tuna (33–86 cm FL) caught by trolling on the continental shelf off southwestern Australia during the austral summer period (December–March) from 1997–2010. Most specimens were around 50 cm (one year old) with a small number over 70 cm (two years old). Stomach contents consisted overwhelmingly of teleost fishes by number and volume with some crustaceans and squids. In fishes, pilchard (*Sardinops sagax*), anchovy (*E. australis*), chub mackerel (*S. australasicus*), and jack mackerel (*T. declivis*) were abundant. On the continental shelf, chub mackerel and jack mackerel were preyed upon near the shelf-edge and pilchard was preyed more often along coastal areas. Prey length varied in size from 5–240 mm TL with dominance of 30–50 mm TL.

Regional difference and niche segregation

Young et al. (1997) examined stomach contents of 1219 southern bluefin tuna off eastern Tasmania during the primary fishing season (May–July) from 1992–1994, and found large differences among areas and years. The body sizes of predators were smaller (40–130 cm FL) at inshore (shelf) locations and larger (74–192 cm FL) within offshore (oceanic) waters. Feeding activity was highest during morning hours (8:00–8:59) based on mean weight of prey by collection time and estimated daily ration revealed that prey of southern bluefin was more abundant at inshore locations. Major inshore prey items included carangid (*T. declivis*) and bramid (*E. nitidus*) fishes and juvenile squid (*N. gouldi*). In offshore waters, squids were more common in sub-Antarctic waters and fishes and crustaceans were more abundant in the East Australia Current. Macro-zooplankton prey (*Phronima sedentaria*) was common in stomachs of large tunas (>150 cm) at offshore areas. The primary conclusion of this study revealed that the inshore waters of eastern Tasmania are an important feeding area for immature southern bluefin tuna during the season.

Young et al. (2010) examined stomach contents of 3562 individuals consisting of 10 pelagic predatory fish species including southern bluefin tuna caught by longlines off eastern Australia from 1992–2006. Main diets of southern bluefin tuna included Carangidae, Bramidae, Emmelichthyidae (fishes, important in weight), Ommastrephidae (squid) and Hyperiidea (amphipod, important in frequency of occurrence). Southern bluefin tuna had similar dietary overlap with blue sharks (*Prionace glauca*). In fish <100 cm FL, both southern bluefin tuna and yellowfin tuna showed high diet overlap.

Summary for Southern bluefin tuna

Primary diets of 2–10 mm TL, 33–86 cm FL, and 40–192 cm FL size southern bluefin tuna are reported (Fig. 4.4). Larvae (2–10 mm TL) prey on small planktonic crustaceans (Uotani et al. 1981; Young and Davis 1990), but prey in bluefin ranging in size from 10 mm TL to 33 cm FL are not reported. Major prey items in juvenile bluefin (33–86 cm FL) are schooling fishes such as pilchards (*Sardinops* spp.), anchovy (*E. australis*), jack mackerels (*Trachurus* spp.), and chub mackerel (*S. australasicus*) (Itoh et al. 2011). Larger bluefin tuna begin preying on larger fishes (e.g., *E. nitidus*) and squids (e.g., *N. gouldi*) (Young et al. 1997). Stomach content analysis revealed that prey is less abundant in offshore waters relative to inshore waters off eastern Tasmania where larger southern bluefin tuna occur (Young et al. 1997).

Southern bluefin tuna

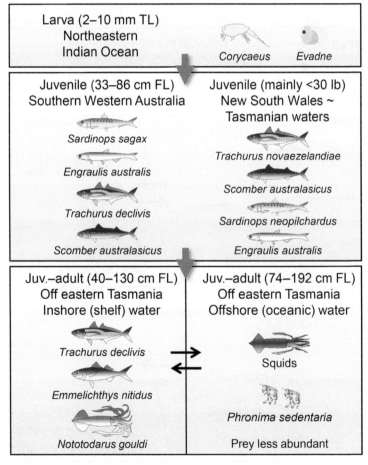

Figure 4.4. Summary of ontogenetic changes in the diet of southern bluefin tuna. Data sources are described in the text.

Common Pattern of Feeding Ecology

Ontogenetic diet change

Diets of three bluefin tuna species are similar to one another with respect to their life stages (Fig. 4.5). Larvae of Pacific bluefin tuna and southern bluefin tuna prey on small planktonic animals, such as Copepoda nauplii, *Corycaeus*, and *Evadne* (Uotani et al. 1981; 1990; Young and Davis 1990). Although prey of larval Atlantic bluefin tuna has not yet been reported for both western and eastern populations, similar planktonic organisms are expected to be preyed upon.

   Juveniles of all three bluefin tuna prey on small fishes and squids, with prey type varying among regions. Because the prey is thought to be abundant suitable sized nekton among all areas, bluefin tunas are thought to be opportunistic feeders. Even during the juvenile stage, an ontogenetic diet shift is observed (Sinopoli et al. 2004;

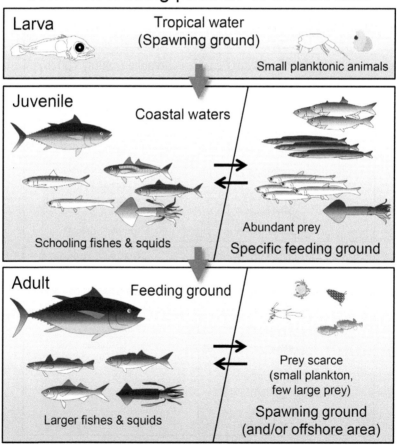

Figure 4.5. Common pattern of the ontogenetic diet change of bluefin tunas.

Shimose et al. 2013). This diet shift is thought to occur with development of swimming and diving abilities of bluefin tunas, and/or seasonal change of abundant suitable sized prey in the area. Growth of this juvenile stage is rapid and ontogenetic change and seasonal change in diet is thought to coincide. Juvenile to immature bluefin tunas migrate for feeding where schooling prey is abundant (Pinkas et al. 1971; Eggleston and Bochenek 1990; Itoh et al. 2011).

Adult bluefin tuna undergo seasonal migrations between spawning and feeding grounds. Tunas are thought to have tropical origin and spawning grounds of bluefin tunas exist in the tropical waters (Boyce et al. 2008). Tropical spawning areas are preferable for larval bluefin tunas, but are not found to be suitable for feeding for adult bluefin tunas. Stomachs carry few prey around tropical spawning grounds of Pacific bluefin (southwestern North Pacific, Yamauchi 2011). On the other hand, prey is abundant at the feeding ground in the temperate waters, i.e., the Sea of Japan for Pacific bluefin (S. Tanaka, personal communication) and the U.S. coast for Atlantic bluefin (Chase 2002; Butler et al. 2010). More quantitative information of bluefin tuna prey at unstudied areas is needed to discuss the feeding migration of bluefin tunas.

Food web around bluefin tunas

Feeding activity of larvae, juvenile and adult bluefin tunas is thought to increase during daytime hours (Uotani et al. 1990; Young and Davis 1990; Young et al. 1997; Butler et al. 2010). Food webs around bluefin tunas are summarized here based on the daytime activity in the wild (Fig. 4.6). Larval bluefin tunas are thought to be preyed on by many oceanic animals which can recognize and swallow larval bluefin tunas. Cannibalism is also thought to occur in wild populations as observed in the rearing tank (Sawada et al. 2005), although the evidence of predation on juvenile bluefin tunas in the wild is limited (e.g., Shimose et al. 2012). Cookiecutter sharks are also a type of predator on bluefin tunas, biting parts of their body and the injury can be non-lethal (T.S., unpublished data). Competitors are reported by underwater observations (Clua and Grosvalet 2001) and stomach content examinations with other species caught at the same time (e.g., Young et al. 2010; Yamauchi 2011). Major prey species of bluefin tunas are also preyed on by other tunas, billfishes, sharks, dolphins, seabirds, etc., and these animals are significant competitors of bluefin tunas. However, dolphins may organize prey and possibly benefit tunas to increase feeding success (Clua and Grosvalet 2001).

Both juvenile and adult bluefin tunas mainly prey on suitable sized schooling fishes, supplementing with other fishes and squids. Abundant schooling nektons, such as *Clupea, Sardinops, Engraulis, Trachurus, Ammodytes, Scomber* (fishes), and *Loligo* (squid) often occur in stomachs of bluefin tunas in high numbers (e.g., Pinkas et al. 1971; Eggleston and Bochenek 1990; Young et al. 1997; Chase 2002; Itoh et al. 2011). Clupeidae forms large schools near coastal areas (Whitehead 1985) and is a major prey species for all three bluefin tuna species. These abundant fishes and squids play an important role as bluefin tuna prey and vary according to location, year, and season. Abundance and residence time of bluefin tunas are thought to correlate with prey abundance thus commercial catches of tuna are theoretically influenced by prey

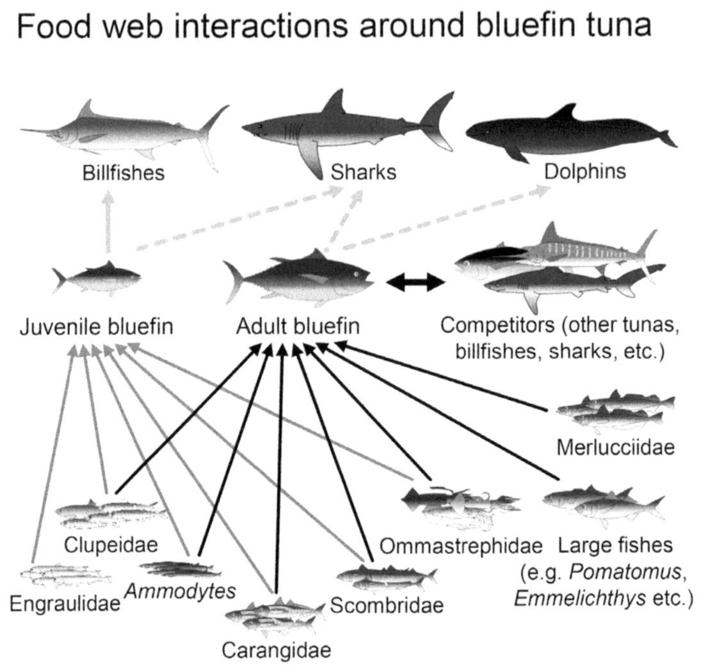

Food web interactions around bluefin tuna

Figure 4.6. Food web interactions around bluefin tunas.

abundance (Itoh 1961; Overholtz 2006). When prey abundance is low, bluefin tunas are thought to migrate to more abundant foraging grounds. In the spawning ground where natural prey is scarce, bluefin tunas prey on small planktonic species and/or make deep dives to feed on meso- or bathy-pelagic fishes and squids (e.g., Yamauchi 2011).

Summary and Conclusion

Feeding ecology of bluefin tunas has been studied for a century with most studies using stomach content analyses to identify prey items consumed at specific times/areas. These studies revealed prey compositions of specific size classes of bluefin tunas, and the differences among regions was discussed. However, time of day and fishing gears/methods are different among studies, and the relative volume of prey is not described for some studies. Such quantitative information is needed to discuss the feeding migration and/or diving behavior of bluefin tunas. Interactions of abundances between tunas and their prey are also not well studied due to the complexity involving food webs and highly migratory habits of bluefin tunas. Relative abundance of bluefin tuna prey in unstudied areas and oceanographic information are important to provide a more comprehensive understanding of the feeding ecology of bluefin tunas.

Acknowledgments

The authors thank M. Yamauchi (Okinawa Prefectural Fisheries Research and Extension Center) and S. Tanaka (Tokai University) for providing useful information of Pacific bluefin tuna diet.

References

Battaglia, P., F. Andaloro, P. Consoli, V. Esposito, D. Malara, S. Musolino, C. Pedà and T. Romeo. 2013. Feeding habits of the Atlantic bluefin tuna, *Thunnus thynnus* (L. 1758), in the central Mediterranean Sea (Strait of Messina). Helgol. Mar. Res. 67: 97–107.

Bayliff, W.H. 1994. A review of the biology and fisheries for northern bluefin tuna, *Thunnus thynnus*, in the Pacific Ocean. FAO Fish. Tech. Pap. 336: 244–295.

Bell, R.R. 1963. Synopsis of biological data on California bluefin tuna *Thunnus saliens* Jordan and Evermann 1926. FAO Fish. Rep. 6: 380–421.

Block, B.A., S.L.H. Teo, A. Walli, A. Boustany, M.J.W. Stokesbury, C.J. Farwell, K.C. Weng, H. Dewar and T.D. Williams. 2005. Electronic tagging and population structure of Atlantic bluefin tuna. Nature 434: 1121–1127.

Butler, C.M., P.J. Rudershausen and J.A. Buckel. 2010. Feeding ecology of Atlantic bluefin tuna (*Thunnus thynnus*) in North Carolina: diet, daily ration, and consumption of Atlantic menhaden (*Brevoortia tyrannus*). Fish. Bull. 108: 56–69.

Chase, B.C. 2002. Differences in diet of Atlantic bluefin tuna (*Thunnus thynnus*) at five seasonal feeding grounds on the New England continental shelf. Fish. Bull. 100: 168–180.

Chen, K.S., P. Crone and C.C. Hsu. 2006. Reproductive biology of female Pacific bluefin tuna *Thunnus orientalis* from south-western North Pacific Ocean. Fish. Sci. 72: 985–994.

Collette, B.B. and C.E. Nauen. 1983. FAO species catalogue. Scombrids of the world. An annotated and illustrated catalogue of tunas, mackerels, bonitos and related species known to date. FAO Fish. Synop. (125) 2: 137.

Clua, É and F. Grosvalet. 2001. Mixed-species feeding aggregation of dolphins, large tunas and seabirds in the Azores. Aquat. Living Resour. 14: 11–18.

Consoli, P., T. Romeo, P. Battaglia, L. Castriota, V. Esposito and F. Andaloro. 2008. Feeding habits of the albacore tuna *Thunnus alalunga* (Perciformes, Scombridae) from central Mediterranean Sea. Mar. Biol. 155: 113–120.

Crane, J. 1936. Notes on the biology and ecology of giant tuna, *Thunnus thynnus* Linnaeus, observed at Portland, Maine. Zoologica. 21: 207–213.

De Sylva, D. 1956. The food of tunas. Bull. Internat. Oceanogr. Found. 2: 37–48.

Di Natale, A. 2012. New data on the historical distribution of bluefin tuna (*Thunnus thynnus* L.) in the Arctic Ocean. Collect. Vol. Sci. Pap. ICCAT 68: 102–114.

Doi, T. 1960. On the predatory relationships among bluefin tuna and coastal fishes in the southern waters of Japan. Bull. Jpn. Soc. Sci. Fish. 26: 99–102 (in Japanese, English abstract).

Dragovich, A. 1970. The food of bluefin tuna (*Thunnus thynnus*) in the western North Atlantic Ocean. Trans. Am. Fish. Soc. 99: 726–731.

Eggleston, D.B. and E.A. Bochenek. 1990. Stomach contents and parasite infestation of school bluefin tuna, *Thunnus thynnus*, collected from the middle Atlantic bight, Virginia. Fish. Bull. 88: 389–395.

Estrada, J.A., M. Lutcavage and S.R. Thorrold. 2005. Diet and trophic position of Atlantic bluefin tuna (*Thunnus thynnus*) inferred from stable carbon and nitrogen isotope analysis. Mar. Biol. 147: 37–45.

Farley, J.H., T.L.O. Davis, J.S. Gunn, N.P. Clear and A.L. Preece. 2007. Demographic patterns of southern bluefin tuna, *Thunnus maccoyii*, as inferred from direct age data. Fish. Res. 83: 151–161.

Fournier, D.A. and J.R. Sibert. 1990. MULTIFAN a likelihood-based method for estimating growth parameters and age composition from multiple length frequency data sets illustrated using data for southern bluefin tuna (*Thunnus maccoyii*). Can. J. Fish. Aquat. Sci. 47: 301–317.

Fromentin, J.M. and J.E. Powers. 2005. Atlantic bluefin tuna: population dynamics, ecology, fisheries and management. Fish Fish. 6: 281–306.

Fry, B. and E.B. Sherr. 1984. $\delta^{13}C$ measurements as indicators of carbon flow in marine and freshwater ecosystems. Contrib. Mar. Sci. 27: 13–47.

Graham, B.S., D. Grubbs, K. Holland and B.N. Popp. 2007. A rapid ontogenetic shift in the diet of juvenile yellowfin tuna from Hawaii. Mar. Biol. 150: 647–658.

Ito, S. 1961. Fishery biology of the sardine, *Sardinops melanosticta* (T. & S.), in the waters around Japan. Bull. Japan Sea Reg. Fish. Res. Lab. 9: 1–227 (in Japanese).

Itoh, T., H. Kemps and J. Totterdell. 2011. Diet of young southern bluefin tuna *Thunnus maccoyii* in the southwestern coastal waters of Australia in summer. Fish. Sci. 77: 337–344.

Jones, J.B. 1991. Movements of albacore tuna (*Thunnus alalunga*) in the South Pacific: evidence from parasites. Mar. Biol. 111: 1–9.

Karakulak, F.S., A. Salman and I.K. Oray. 2009. Diet composition of bluefin tuna (*Thunnus thynnus* L. 1758) in the Eastern Mediterranean Sea, Turkey. J. Appl. Ichthyol. 25: 757–761.

Kishinouye, K. 1923. Contributions to the comparative study of the so-called scombroid fishes. J. Coll. Agric. Imper. Univ. Tokyo 8: 293–475.

Kobayashi, T. 2004. Parasitism of *Didymocystiswedli* to the blue fin tuna, *Thunnus thynnus* from the Sea of Japan, Yamaguchi prefecture, Japan. Bull. Yamaguchi Pref. Fish. Res. Cent. 2: 19–21 (in Japanese).

Krumholz, L.A. 1959. Stomach contents and organ weights of some bluefin tuna, *Thunnus thynnus* (Linnaeus), near Bimini, Bahamas. Zoologica. 44: 127–131.

Kubo, T., W. Sakamoto, O. Murata and H. Kumai. 2008. Whole-body heat transfer coefficient and body temperature change of juvenile Pacific bluefin tuna *Thunnus orientalis* according to growth. Fish. Sci. 74: 995–1004.

Logan, J.M., E. Rodríguez-Marín, N. Goñi, S. Barreiro, H. Arrizabalaga, W. Golet and M. Lutcavage. 2011. Diet of young Atlantic bluefin tuna (*Thunnus thynnus*) in eastern and western Atlantic foraging grounds. Mar. Biol. 158: 73–85.

Nichols, J.T. 1922. Color of the tuna. Copeia 1922: 73–74.

Matthews, F.D., D.M. Damkaer, L.W. Knapp and B.B. Collette. 1977. Food of western North Atlantic tunas (*Thunnus*) and lancet fishes (*Alepisaurus*). U.S. Dep. Comm. NOAA Tech. Rep. NMFS-SSRF 706: 1–19.

Mladineo, I. and I. Bočina. 2009. Type and ultrastructure of *Didymocystiswedli* and *Koellikerioidesintestinalis* (Digenea, Didymozoidae) cysts in captive Atlantic bluefin tuna (*Thunnus thynnus* Linnaeus, 1758). J. Appl. Ichthyol. 25: 762–765.

Munday, B.L., Y. Sawada, T. Cribb and C.J. Hayward. 2003. Diseases of tunas, *Thunnus* spp. J. Fish Dis. 26: 187–206.

Overholtz, W.J. 2006. Estimates of consumption of Atlantic herring (*Clupea harengus*) by bluefin tuna (*Thunnus thynnus*) during 1970–2002: an approach incorporating uncertainty. J. Northw. Atl. Fish. Sci. 36: 55–63.

Peterson, B.J. and B. Fry. 1987. Stable isotopes in ecosystem studies. Ann. Rev. Ecol. Syst. 18: 293–320.

Pinkas, L., M.S. Oliphant and I.L.K. Iverson. 1971. Food habits of albacore, bluefin tuna, and bonito in California waters. Calif. Dept. Fish Game Fish. Bull. 152: 1–105.

Restrepo, V.R., G.A. Diaz, J.F. Walter, J.D. Neilson, S.E. Campana, D. Secor and R.L. Wingate. 2010. Updated estimate of the growth curve of Western Atlantic bluefin tuna. Aquat. Living Resour. 23: 335–342.

Robins, J.P. 1963. Synopsis of biological data on bluefin tuna *Thunnus thynnus maccoyii* (Castelnau) 1872. FAO Fish. Rep. 6: 562–587.

Rooker, J.R., J.R. Alvarado Bremer, B.A. Block, H. Dewar, G. de Metrio, A. Corriero, R.T. Kraus, E.D. Prince, E. Rodríguez-Marín and D.H. Secor. 2007. Life history and stock structure of Atlantic bluefin tuna (*Thunnus thynnus*). Rev. Fish. Sci. 15: 265–310.

Rooker, J.R., D.H. Secor, G. De Metrio, R. Schloesser, B.A. Block and J.D. Neilson. 2008. Natal homing and connectivity in Atlantic bluefin tuna populations. Science 322: 742–744.

Sarà, G. and R. Sarà. 2007. Feeding habits and trophic levels of bluefin tuna *Thunnus thynnus* of different size classes in the Mediterranean Sea. J. Appl. Ichthyol. 23: 122–127.

Sawada, Y., T. Okada, S. Miyashita, O. Murata and H. Kumai. 2005. Completion of the Pacific bluefin tuna *Thunnus orientalis* (Temmincket Schlegel) life cycle. Aquacul. Res. 36: 413–421.

Serventy, D.L. 1956. The southern bluefin tuna *Thunnus thynnus maccoyii* (Castelnau) in Australian waters. Aust. J. Mar. Freshw. Res. 7: 1–43.

Shimose, T., T. Tanabe, K.-S. Chen and C.-C. Hsu. 2009. Age determination and growth of Pacific bluefin tuna, *Thunnus orientalis*, off Japan and Taiwan. Fish. Res. 100: 134–139.

Shimose, T., K. Yokawa, H. Saito and K. Tachihara. 2012. Sexual difference in the migration pattern of blue marlin, *Makaira nigricans*, related to spawning and feeding activities in the western and central North Pacific Ocean. Bull. Mar. Sci. 88: 231–250.

Shimose, T., H. Watanabe, T. Tanabe and T. Kubodera. 2013. Ontogenetic diet shift of age-0 year Pacific bluefin tuna *Thunnus orientalis*. J. Fish Biol. 82: 263–276.

Sinopoli, M., C. Pipitone, S. Campagnuolo, D. Campo, L. Castriota, E. Mostarda and F. Andaloro. 2004. Diet of young-of-the-year bluefin tuna, *Thunnus thynnus* (Linnaeus, 1758), in the southern Tyrrhenian (Mediterranean) Sea. J. Appl. Ichthyol. 20: 310–313.

Takebe, T., Y. Saeki, S. Masuma, H. Nikaido, K. Ide, S. Shiozawa and N. Mano. 2013. Prevalence and transmission capability of *Didymocystis wedli* (Digenea; Didymozoidae) in cage-reared young Pacific bluefin tuna *Thunnus orientalis* in the Amami area of Japan. Nippon Suisan Gakkaishi 79: 214–218 (in Japanese, English abstract).

Talbot, F.H. and M.J. Penrith. 1963. Synopsis of biological data on species of the genus *Thunnus* (sensulato) (South Africa). FAO Fish. Rep. 6: 608–646.

Tanaka, Y., M. Mohri and H. Yamada. 2007. Distribution, growth and hatch date of juvenile Pacific bluefin tuna *Thunnus orientalis* in the coastal area of the Sea of Japan. Fish. Sci. 73: 534–542.

Tiews, K. 1963. Synopsis of biological data on bluefin tuna, *Thunnus thynnus* (Linnaeus) 1758 (Atlantic and Mediterranean). FAO Fish. Rep. 6: 422–481.

Uotani, I., K. Matsuzaki, Y. Makino, K. Noda, O. Inamura and M. Horikawa. 1981. Food habits of larvae of tunas and their related species in the area northwest of Australia. Bull. Jpn. Soc. Sci. Fish. 47: 1165–1172 (in Japanese, English abstract).

Uotani, I., T. Saito, K. Hiranuma and Y. Nishikawa. 1990. Feeding habit of bluefin tuna *Thunnus thynnus* larvae in the western North Pacific Ocean. Nippon Suisan Gakkaishi 56: 713–717 (in Japanese, English abstract).

Varela, J.L., E. Rodríguez-Marín and A. Medina. 2013. Estimating diets of pre-spawning Atlantic bluefin tuna from stomach content and stable isotope analyses. J. Sea Res. 76: 187–192.

Watanabe, H. 1960. Regional differences in food composition of the tunas and marlins from several oceanic areas. Rep. Nankai Reg. Fish. Res. Lab. 12: 75–84 (in Japanese, English abstract).

Whitehead, P.J.P. 1985. FAO species catalogue. Clupeoid fishes of the world. An annotated and illustrated catalogue of the herrings, sardines, pilchards, sprats, anchovies and wolf herrings. Part 1 - Chirocentridae, Clupeidae and Pristigasteridae. FAO Fish. Synop. (125) 7: 303.

Yabe, H., N. Anraku and T. Mori. 1953. Scombroidyoungs found in the coastal seas of Aburatsu, Kyushu, in summer. Contr. Nankai Reg. Fish. Res. Lab. 11: 1–10 (in Japanese).

Yamanaka, H. and staff. 1963. Synopsis of biological data on Kuromaguro *Thunnus orientalis* (Temminck and Schlegel) 1842 (Pacific Ocean). FAO Fish. Rep. 6: 180–217.

Yamauchi, M. 2011. Fishing report of the Pacific bluefin tuna, *Thunnus orientalis*, caught by a small-scale tuna longline fishing boat and food habits of pelagic fishes around the Ryukyu Archipelago, Japan. M.S. Thesis, University of the Ryukyus, Okinawa, Japan (in Japanese, English abstract).

Yokota, T., M. Toriyama, F. Kanai and S. Nomura. 1961. Studies on the feeding habit of fishes. Rep. Nankai Reg. Fish. Res. Lab. 14: 1–234 (in Japanese, English abstract).

Young, J.W. and T.L.O. Davis. 1990. Feeding ecology of larvae of southern bluefin, albacore and skipjack tunas (Pisces: Scombridae) in the eastern Indian Ocean. Mar. Ecol. Prog. Ser. 61: 17–29.

Young, J.W., T.D. Lamb, D. Le, R.W. Bradford and A.W. Whitelaw. 1997. Feeding ecology and interannual variations in diet of southern bluefin tuna, *Thunnus maccoyii*, in relation to coastal and oceanic waters off eastern Tasmania, Australia. Environ. Biol. Fish. 50: 275–291.

Young, J.W., M.J. Lansdell, R.A. Campbell, S.P. Cooper, F. Juanes and M.A. Guest. 2010. Feeding ecology and niche segregation in oceanic top predators off eastern Australia. Mar. Biol. 157: 2347–2368.

III. Distribution and Migration

CHAPTER 5

Horizontal Movements of Pacific Bluefin Tuna

Ko Fujioka,[1,*] *Masachika Masujima,*[2] *Andre M. Boustany*[3]
and *Takashi Kitagawa*[4,5]

Introduction

Knowledge of the horizontal movements of highly migratory species such as tunas in the genus *Thunnus* is essential for understanding their life history. This information is important for population stock assessments, a prerequisite for maintaining a sustainable population. Tunas travel large distances within tropical to subarctic waters throughout their life. One of the largest, the Pacific bluefin tuna (PBF) *T. orientalis* has a wide and continuous distribution across the Pacific Ocean from the northwestern to eastern Pacific. In the Southern Pacific Ocean, PBF are also found in New Zealand waters, although less is known regarding the migration pathways to this region. Their wide seasonal migrations and size variations due to rapid growth make it difficult to analyze the entire migration pattern within one tagging study.

Since the early 1990s, electronic data storage tags (archival tags) have been deployed for PBF, the southern bluefin tuna *T. maccoyii*, and the Atlantic bluefin tuna *T. thynnus* (Gunn et al. 1994; Block et al. 1998; 2001; Kitagawa et al. 2000; 2004; Boustany et al. 2010). These recoverable tags record the free-ranging swimming depth

[1] National Research Institute of Far Seas Fisheries, Fisheries Research Agency, 5-7-1 Orido, Shimizu, Shizuoka 424-8633, Japan.
[2] National Research Institute of Fisheries Science, Fisheries Research Agency, 2-12-4 Fukuura, Kanazawa, Yokohama, Kanagawa 236-8648, Japan.
[3] Nicholas School of the Environment, Duke University, Box 90328 Durham NC 27708, USA.
[4] Atmosphere & Ocean Research Institute, The University of Tokyo, 5-1-5 Kashiwanoha, Kashiwa, Chiba 277-8564.
[5] CREST, Japan Science and Technology Agency, 4-1-8 Honcho, Kawaguchi, Saitama, 332-0012, Japan.
* Email: fuji88@affrc.go.jp

of the fish, internal and external temperatures, and ambient light levels. These data can be used to calculate the movements and long-distance migration patterns associated with oceanic conditions. The light data are processed to reconstruct daily fish location estimates (Hill 1994; Welch and Eveson 1999). The accuracy and precision of the estimated movements are improved using the recorded light levels as well as sea surface temperature and depth (Teo et al. 2004; Lam et al. 2008; Galuardi et al. 2010). Lam et al. (2008) provided a detailed explanation and a state-space model analysis using the sea surface temperature matching algorithm. The internal and external temperature data have been also used to analyze the physiological thermoregulation of tunas (Kitagawa et al. 2000; 2001; 2006b). These findings have provided important information on the possible distribution and adaptive mechanisms of PBFs.

While some knowledge of PBF migrations during the early years of their life has been gathered by previous studies, the application of archival tags on smaller juveniles is now under experimental review (Fujioka et al. 2013a,b). Prior to this study, the smallest fish tracked with archival tags was 45 cm in Fork Length (FL; Kitagawa et al. 2004). Conventional tagging experiments have demonstrated the migration patterns and population structure in the early life stages of the fish (Bayliff et al. 1991; Bayliff 1994). Although the data obtained from these experiments are limited to the date and location of release and recapture, large conventional tagging datasets can provide information on migratory dynamics of the general population. Some recent studies of larval-stage PBFs determined larval transport processes by numerical simulations (particle tracking) using sea surface temperatures and Kuroshio Current data analysis (Kitagawa et al. 2010; Masujima et al. 2013).

Here, we review the advances in the understanding of comprehensive migration dynamics in early life stages of PBFs (larval, juvenile, and immature) in the western Pacific, movement of immature fish in the eastern Pacific, and the trans-Pacific migration between the western and the eastern Pacific. In particular, we examine horizontal movements and pathways, habitat use, and residency in both coastal and open ocean waters. To illustrate the horizontal movements for these life stages, we defined six groups. These are as follows: (i) larval stage, arrival at nursery grounds from spawning areas; (ii) juvenile/immature stage, movement from the south coast of Japan (off Kochi) and (iii) movement from the East China Sea (off Nagasaki) to the mid-Pacific; (iv) immature stage, trans-Pacific migration from the western to the eastern Pacific; (v) immature stage, movement in the eastern Pacific; and (vi) immature/prespawner stage, trans-Pacific migration from the eastern to the western Pacific. PBF movement dynamics were determined by integrating the results of numerical simulations, fishery catch records, and conventional and archival tag data from the past two decades.

Arriving at Nursery Grounds

Adult PBFs spawn in the area between the Philippines and the Ryukyu Islands (around Nansei Islands) in the northwestern Pacific Ocean from April to June and in the Sea of Japan from July to August (Yabe et al. 1966; Ueyanagi 1969; Okiyama 1974; Kitagawa et al. 1995) (Fig. 5.1). In the spawning area, the ocean current system is one of the most important factors for the survival of larval PBFs. It can affect successful transportation

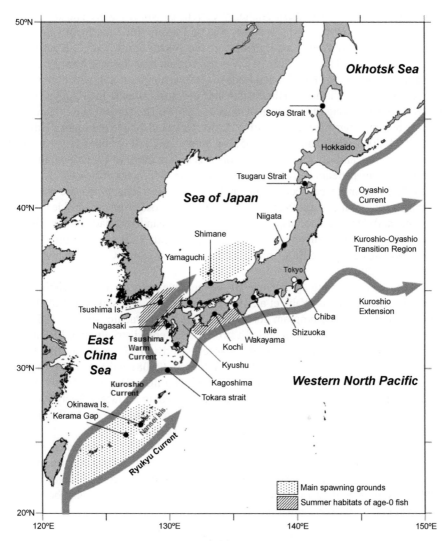

Figure 5.1. Two main spawning grounds of Pacific bluefin tuna in the area between the Philippines and the Ryukyu Islands in the northwestern Pacific Ocean and in the Sea of Japan. The larvae hatched in the northwestern Pacific Ocean south of Japan are transported northward by the Kuroshio Current toward Japan, and the juveniles appear in coastal areas off Kochi and Nagasaki prefectures during the summer of their first year. Major currents around Japan (Kuroshio Current, Ryukyu Current, Tsushima Current, and Oyashio Current) are also shown. Locations of the prefectures (fishing ports), islands, and straits of Japan are marked for reference.

to the larval nursery ground or passage through a favorable environment and availability of resources that directly affect their growth and mortality levels. Larval PBFs cannot swim effectively (Miyashita 2002; Tamura and Takagi 2009); they are passively transported from the spawning ground by advection of the ocean current (close to the sea surface). The current system in the northwestern Pacific Ocean is described here.

The Kuroshio is a western boundary current of North Pacific and one of the strongest currents in the world. It passes east of Taiwan, flows northeastward along the continental shelf of the East China Sea to the Tokara Strait south of Kyushu and then to the south of Honshu (Fig. 5.1). A warm current branches northward from the Kuroshio to the west of Kyushu and leads to the Tsushima Current, which flows into the Sea of Japan. In the area on the eastern side of Nansei Islands, the Ryukyu current flows northward along the continental slope (Fig. 5.1). The vertical current structure is more bottom intensified to the north of Okinawa Island than to the south (Ichikawa et al. 2000; 2004; Zhu et al. 2003; 2005; Nakamura 2005). It has been suggested that there is a net mean flow through the Kerama Gap into the East China Sea; this can increase the downstream Kuroshio volume transport (Morinaga et al. 1998; Zheng et al. 2008). Many mesoscale eddies, hundreds of kilometers long, propagate westward from the center of the North Pacific. Some of these mesoscale eddies collide with the Kuroshio and change their transport volume, path, or meandering patterns (Ebuchi and Hanawa 2000; 2001; Yang et al. 1999). Some mesoscale eddies in the northwestern Pacific propagate into the East China Sea through the Kerama Gap (Ichikawa 2001; Andres et al. 2008).

Some field sampling studies of PBF larvae have been performed within these oceanic current systems (Yabe et al. 1966; Okiyama 1974; Nishikawa et al. 1985). After the field research program conducted in the North Pacific between 1956 and 1981, Nishikawa et al. (1985) reported a strongly clustered distribution of PBF larvae with many zero-catch sites around the Nansei Islands. Tracking of the PBF larval patch movement has been employed by Satoh et al. (2008) and Satoh (2010); they refer to the leading research case in southern bluefin tuna (Davis et al. 1991). The larval patches were tracked using surface drifters similar to Surface Velocity Program drifter (Niiler 2001) for periods of up to one week, and the patches on the Pacific side of Nansei Islands were entrained into the mesoscale eddies.

Recently, in ocean General Circulation Models (GCMs), a numerical particle tracking method has been used in fishery oceanographic research focusing on transport, survival, and growth in the early life stages of tuna (Teo et al. 2007; Kitagawa et al. 2010; Masujima et al. 2013). The basic formula for particle tracking can be written as shown in the following equation:

$$\frac{d\vec{X}}{dt} = \vec{u}.$$

This means that at each numerical time step, traveling distance of particles (i.e., virtual eggs and larvae) can be calculated as the magnitude of the ocean current multiplied by the time step, with the traveling direction being the same as the direction of the ocean current. One of the most important features of a GCM is its spatiotemporal uniformity of data, while field research data are often distributed in restricted time and/or space. Despite the simplification of biological processes of larvae, particle tracking using GCMs provides general arguments and allows expanding scientifically based speculations.

Applying the particle tracking method to analyze the larval PBF transport, Masujima et al. (2013) have obtained numerically estimated PBF spawning sites by backtracking where the larvae were traced back to their spawning grounds. First, the larval age was estimated using a linear regression of the standard length to the age in days (Satoh 2010) plus one day for hatching. Then, at each sampling location, 10,000 particles with no swimming ability were advected horizontally backward in time. The authors used the geographical points where the larvae were sampled in field observations. The results suggested that possible PBF spawning sites were located near the sampled positions, i.e., around the Kuroshio axis and mesoscale eddies for the larvae sampled on the western and eastern side of Nansei Islands, respectively (Fig. 5.2). Kitagawa et al. (2010) have also performed a numerical tracking experiment with virtual PBF larvae initially distributed uniformly around the Nansei Islands. They have reported that the Kuroshio transport path is consistently successful in PBF recruitment. It takes 60–90 days to reach the nurseries south of Honshu from the spawning ground. On the eastern side of Nansei Islands, the virtual larvae in the mesoscale eddies also reach the nursery grounds, but less successfully than those using the Kuroshio Current (Kitagawa et al. 2010).

The combined results of field tracking and numerical particle tracking experiments show that PBF larvae (and early juveniles), even though still poor swimmers, successfully complete the migration to the nursery area off the southern coast of Honshu, utilizing the combination of mesoscale eddies and the Kuroshio Current. These results suggest that annual PBF recruitment could be affected not only by variations in the Kuroshio path or volume transport but also by the positions of cyclonic and anticyclonic mesoscale eddies (Satoh 2010). The Kuroshio Current in the East China Sea can be affected within two week periods by remotely generated eddies (Takahashi et al. 2009) or by local effects such as a monsoon (Nakamura et al. 2010) and a typhoon

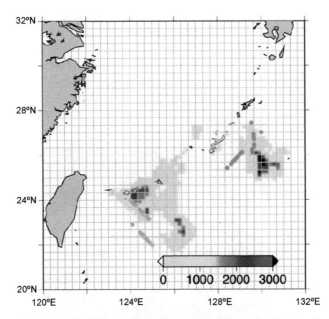

Figure 5.2. Numerically estimated PBF spawning sites obtained employing backtracking for the number of days equal to the larval age (plus one day for hatching time) from the geographical points where the larvae had been sampled in field observations (modified from Masujima et al. 2013).

crossing the Kuroshio axis (Morimoto et al. 2009). Classifying mesoscale eddies using sea surface height measurements (TOPEX/POSEIDON venture), Kakuma and Morinaga (2000) and Morinaga et al. (1998) have found that the westward velocity of mesoscale eddies is 5.4–7.2 km/day; they have also reported that not all eddies reach Nansei Islands or the East China Sea. While patterns of appearance and movement have been well investigated (Kakuma and Morinaga 2000; Yanagi et al. 2002), the westward propagation speed of eddies at 5–8 km/day indicates that the mesoscale eddies are the most stable oceanic structures during the PBF larval period (i.e., a few weeks). Thus, the relative positions of the spawning sites and oceanographic mesoscale eddies could be important factors in the successful recruitment to the PBF resources.

Tuna larvae are generally found at low densities and have a patchy distribution (Wade 1951; Strasburg 1960; Klawe 1963; Richards and Simmons 1971). From a practical point of view, PBF larvae (or juveniles) are difficult to sample in the field (Davis et al. 1991). Thus, the effect of variations in the transport of the larvae on the PBF resource in the North Pacific remains unclear. Difficulty in tracking larval PBF patches in the field is caused not only by low frequency of the patches but also by bad weather conditions such as approaching typhoons (Satoh et al. 2008). To compensate for the relative limitations of these methods, a combination of numerical simulations with field observations should be employed.

After hatching in the spawning ground of the northwestern Pacific Ocean, PBF larvae move north to the nursery areas near the coast of Japan (Bayliff 1994; Kitagawa et al. 2010). Juvenile PBFs have been found at age-0 (15–32 cm FL) (Fig. 5.3) in coastal areas of Nagasaki and Kochi prefectures (Fig. 5.1) during summer (July–August), and two–three months old tuna (Tanaka et al. 2006) are caught for farming operations by trolling around the coastal areas of western Japan. The sea surface temperatures in the Kochi summer fishery (July–September) range from 24°C to 29°C (Bayliff 1994) and in the Sea of Japan fishery off Nagasaki from 22°C to 29°C (July–October) (Itoh 2004). These coastal areas around Kochi and Nagasaki are well known as the summer habitat (nursery areas) of juvenile PBFs. Historical fishery data suggest that age-0 fish move into the nursery grounds at least at a water temperature of 24°C. Oceanic conditions such as fluctuations of the mesoscale eddies and changes in the Kuroshio Current affect the arrival time of juveniles at the nursery grounds.

Figure 5.3. Juvenile Pacific bluefin tuna (*Thunnus orientalis*) (16 cm fork length) caught by trolling off Kochi prefecture in July 2012.

Movements of Juveniles from South Coastal Areas of Japan to Offshore Areas

According to the results of integration of conventional tagging data and catch information (Bayliff et al. 1991; Bayliff 1994; Itoh 2004), juvenile PBFs near the south coast of Japan (e.g., areas off Kochi and Shizuoka) move eastward in summer and slightly westward in winter along the southern Japanese coast. In tagging experiments, large numbers of tagged age-0 PBFs (15–31 cm FL) have been released from Kochi in July–August and recovered in the following January in the same areas, and some of the tagged fish moved eastward to the longitude of 142°E between September and the next April (Bayliff et al. 1991; Itoh 2004). As the catch frequencies of age-0 fish in the south coastal areas of Japan peaks in October and then declines through the spring of the following year (Itoh 2004), some juvenile PBF remain in the south coastal areas of Japan, and others must have spread to the eastern coastal areas of Shizuoka and Chiba prefectures during summer and autumn. Just one tag has been returned from the Sea of Japan in these studies (Itoh 2004); it is unlikely the juvenile fish move to the west and north along the coast of Kyushu (Bayliff et al. 1991; Bayliff 1994; Itoh 2004).

However, when the fish were tagged (25–35 cm FL) off Shizuoka prefecture, the fish remaining in the area moved slightly to the west and were recaptured off the coast of Mie and Wakayama prefecture in winter (December–January) while some of the tagged fish moved away from the south coast to the east of Japan (Itoh 2004).

During the autumn or winter of their first or second year of life, juvenile PBFs move from the waters off Chiba prefecture to the east of Japan and towards the center of the North Pacific Ocean (35–45°N, 141°E–130°W) (Bayliff et al. 1991; Bayliff 1994). However, there is not much information about the movements from the south coast of Japan to offshore areas. Some progress in the research on the migratory ecology of juveniles arriving off Kochi has been achieved using archival tagging (Fig. 5.4) conducted by Fujioka et al. (2013a,b). The results of the daily fish position estimation off Kochi show that juvenile PBF (24.5–29.0 cm FL at release) moved eastward to the area off Wakayama Prefecture between summer and winter (Fig. 5.5). The juveniles

Figure 5.4. Juvenile Pacific bluefin tuna (23 cm fork length) with an archival tag surgically implanted in the peritoneal cavity and a conventional plastic dart tag placed on the second dorsal fin.

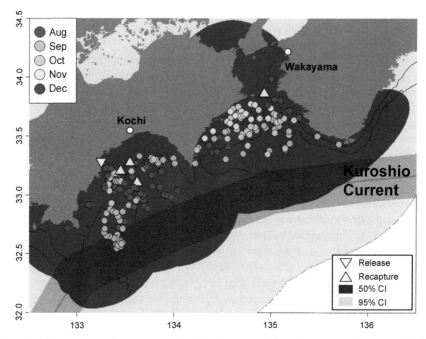

Figure 5.5. Movements of four juvenile Pacific bluefin tunas (24.5–29.0 cm FL at release) tracked with archival tags from summer to winter of 2012. The daily fish positions and the confidence intervals (CI) are calculated using the state-space extended Kalman filter statistical model. Release and recapture positions are shown as inverted triangle and triangles, respectively.

are mostly distributed between coastal areas and the warm Kuroshio Current; the offshore movement across the Kuroshio Current has not been observed. This suggests that there are limited areas within the Kuroshio Current serving as juvenile habitats at the south coast of Japan. Further analysis, focusing on the mechanism of expansion to offshore areas or spreading to the eastern Pacific, should be considered.

Movements of Juveniles from Nagasaki area in the East China Sea to the Mid-Pacific Area

Juvenile PBFs hatched in northwestern Pacific Ocean, some of which are transported to the waters off the Nagasaki coast during the summer (August–September), extend their distribution northeastward to Niigata prefecture, and are caught in the coastal areas of the Sea of Japan from October to November. Juvenile PBFs are believed to move southwest along the coastal areas through Shimane and Yamaguchi prefectures in November and December. However, some larvae/juveniles (11–28 cm FL) hatch in the Sea of Japan, which is another spawning ground (Fig. 5.1) with the sea surface temperatures ranging between 24°C and 26°C, from late August to late September (Tanaka et al. 2007). They mix with the fish hatched in the northwestern Pacific Ocean in the waters off the Nagasaki coast (winter fishing grounds in the East China Sea, October–December) (Oshima et al. 2010).

Overwintering habitats of juvenile PBFs in the East China Sea can vary, e.g., according to Kitagawa et al. (2006a), they were at the furthest south locations when the sea surface temperatures were low in 1996 and shifted north when the temperatures increased in 1998. These differences in the distribution are probably a result of La Niña and El Niño events that affect the sea surface temperatures in the East China Sea.

Movements of immature PBFs (late age-0 and early age-one fish, according to Yukinawa and Yabuta 1967 and Shimose et al. 2009) after overwintering in the East China Sea follow three distinctive movement patterns (Itoh et al. 2003; Kitagawa et al. 2004). In the first pattern, PBFs remain within the East China Sea and move along the edge of the Tsushima Warm Current in a southwest to northeasterly direction from winter to early summer (December–June) (Fig. 5.6a). In the second pattern, PBFs move into the western North Pacific in March, after staying within the East China Sea for a few months (December–February). They then move eastward along the coastal side of the Kuroshio front and into the Kuroshio–Oyashio transition region in April. This region is characterized by irregularly distributed eddies and thermohaline fronts between the Kuroshio Extension and the subarctic Oyashio front (Kawai 1972). The

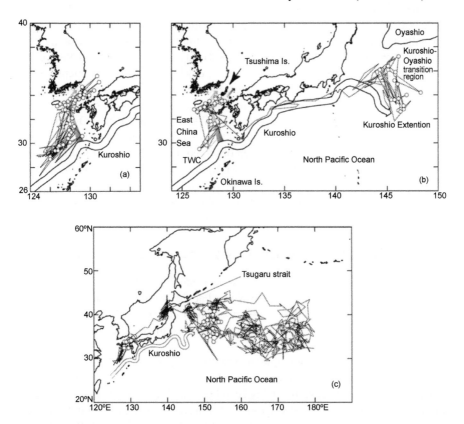

Figure 5.6. Estimated tracks of an archival tag for a Pacific bluefin tuna in East China Sea (a) and for bluefin tunas migrating into the Kuroshio–Oyashio transition region ((b) and (c). Open circles indicate positions where time-series data have also been recorded, and solid circles indicate the release site of the archival-tagged tunas (modified from Kitagawa et al. 2004).

fish remain in the southeast and northwest sections of this area from April to June (Fig. 5.6b). In the third pattern, the fish move into the Kuroshio–Oyashio transition region through the Sea of Japan and the Tsugaru Strait in November, after staying in the East China Sea for six months (December–May) and in the Sea of Japan for five months (June–October) (Fig. 5.6c). Some fish leave the Sea of Japan and migrate in August to the western North Pacific through the Soya Strait and Okhotsk Sea (Fig. 5.1), which is the north passage of the Tsugaru Strait (the north end of Hokkaido) (Inagake et al. 2001). Offshore movements to the Kuroshio–Oyashio transition region have been reported for immature PBFs between late age-0 and -one between April and November, although some individuals remain around the coastal areas of Japan during this period (Inagake et al. 2001; Itoh et al. 2003; Kitagawa et al. 2004; 2006a).

In the Kuroshio–Oyashio transition region, PBFs are mostly located on the Kuroshio front and in the eddies generated by the Kuroshio Extension (Kitagawa et al. 2004). Block et al. (1998) and Tameishi (1997) have suggested that warm-core rings or warm streamers function as migratory pathways for PBFs and anchovies. Thus, eddies and warm-core rings in the northern Kuroshio Extension might be important habitats for age-one fish in April–June. During this period, PBFs would move horizontally following the fluctuations of the Kuroshio front, where their prey is usually abundant. The PBFs spend most of the time on the warmer side of the front, but travel horizontally to the colder side to feed (Kitagawa et al. 2004). In the western North Pacific Ocean, the majority of fish shows clockwise annual movement patterns (Inagake et al. 2001). From May to August, the age-one fish move northward around the 145°E and 150°E longitude lines where the warm water usually spreads from the first or second crest of the Kuroshio Extension. When the fish reach the Oyashio front, movements eastward along the front and/or the subarctic boundary are recorded from August to October. These fish have been observed moving southward to the waters near the Shatsky Rise (30–39°N, 157–163°E) and the Emperor seamount chain in November and further southward into the Kuroshio Extension in December. Their movement paths may be dependent on availability and abundance of suitable prey associated with oceanic and bathymetric conditions.

Trans-Pacific Migration from the Western to the Eastern Pacific

Conventional tagging experiments have shown that many PBFs migrate across the ocean from the western to the eastern North Pacific in their first or second year, a distance of approximately 8,000 km (Orange and Fink 1963; Clemens and Flittner 1969; Bayliff et al. 1991), in what is referred to as the trans-Pacific migration (Bayliff et al. 1991).

The trans-Pacific migration of immature PBFs released in the East China Sea has been observed in their second year between November and January. These PBFs are caught in the eastern Pacific mostly as 70–90 cm FL (Bayliff et al. 1991). However, approximately 66 cm FL (mostly age-one) PBF have also been caught in the eastern Pacific. These fish might have migrated from the south coast of Japan (off Kochi), not the East China Sea, during their juvenile period. In fact, conventional tag data have shown differences between the timing of trans-Pacific migration from these two areas (Bayliff et al. 1991). PBF released during 1979–1988 off the south coast of Japan, in the areas off Kochi (N = 4910) and Shizuoka Prefecture (N = 2502), have been recaptured at around age-one in the eastern Pacific (N = 25 and 8, respectively). In contrast, tagged fish released in the East China Sea [i.e., off Nagasaki (N = 1884) and Kagoshima (N = 1658)] have been recaptured in the eastern Pacific, mostly at age-two (N = 39 and 38, respectively). Therefore, the timing of the trans-Pacific migration (first year or second year) could be affected by the transport to those nursery areas in their first summer (south coast of Japan or East China Sea). Because juvenile PBFs arriving in the waters off Nagasaki are passing the first winter around the East China Sea, their departure to the eastern Pacific will be delayed in comparison with the individuals arriving off Kochi.

Inagake et al. (2001) and Itoh et al. (2003) first reported recaptures of archival-tagged PBFs of 55.0–87.6 cm FL after trans-Pacific migration. Inagake et al. (2001) reported that the trans-Pacific migration path of the fish corresponds to the Subarctic Frontal Zone (Suga et al. 2003) around 42–43°N. The region to the north of 42°N has cool water masses of subarctic origin with winter temperatures below 8°C. The fish were released in the East China Sea on November 29, 1996 and migrated to the Kuroshio–Oyashio transition region in the western Pacific in May 1997 (Itoh et al. 2003). On November 11, 1997, the fish initiated their trans-Pacific migration, moving eastward (mean direction: 3.8° south of east) along the Subarctic Frontal Zone (Fig. 5.7) though they changed their direction to the southeast on December 8, 1997. The fish arrived in the eastern Pacific on January 15, 1998 (after 66 days) and were recaptured on August 1, 1998. During the trans-Pacific migration, the fish traveled further per day in cold waters with mean ambient temperature at 14.5 ± 2.9 (°C $\pm$ SD). These temperatures were significantly lower than those during other periods (17.6 ± 2.1). Mean daily distance covered by the fish during the trans-Pacific migration was also significantly higher [163.5 ± 37.8 (km $\pm$ SD)] than during other periods (130.7 ± 39.3) (Kitagawa et al. 2009; 2013). Kitagawa et al. (2009) also suggested that the trans-Pacific migration path passes through cooler waters, 9.1°C–16.2°C, where the fish cover longer distances per day than in the western and eastern marginal regions. Some previous research had shown that archival-tagged immature PBFs in the Sea of Japan maintain the same thermal difference between ambient and peritoneal cavity temperature (peritoneal cavity temperature minus ambient water temperature) in waters at 11°C–17°C as PBFs in other areas with water temperatures of 18°C–22°C (Kitagawa et al. 2002). This indicates that they actively produce heat in the cool waters of the Sea of Japan and the Subarctic Frontal Zone during their trans-Pacific migration. Sustained swimming during the migration is likely to produce substantial amounts of body heat. When combined with the thermoconservation ability of PBF, this phenomenon helps the fish to migrate through the cold waters of the Subarctic Frontal Zone.

It is unknown as to what proportions of the total PBF population consists of migratory individuals or even whether there are separate non-migratory and migratory subpopulations. The proportion of PBF making trans-Pacific migrations appears to be variable on inter-annual scales and may be tied to prey availability in the western Pacific (Polovina 1996). These important population dynamics questions could be answered by conducting long-term studies using conventional and archival tagging in the south coast of Japan and East China Sea. Such research should include stable isotope analysis to improve the understanding of migratory patterns. This type of analysis can

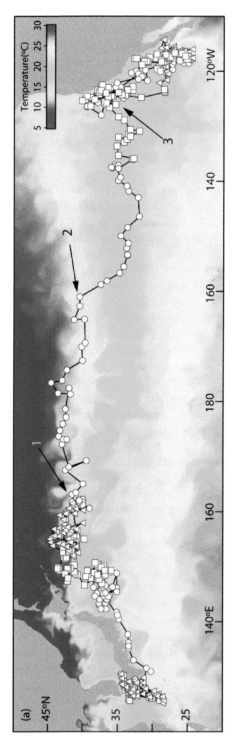

Figure 5.7. The migration pathway of an archival-tagged Pacific bluefin tuna, with sea surface temperature data for November 1997 (modified from Kitagawa et al. 2009 and 2013).

be used to examine tissue turnover rates and isotopic trophic discrimination factors to identify the overall migration patterns (Madigan et al. 2012). Such analysis, using large numbers of fish of various ages before and after the trans-Pacific migration, could help to characterize the migration dynamics.

Movements of Immature Pacific Bluefin Tuna in the Eastern North Pacific

Kitagawa (2013) has detailed the movements of immature PBFs in the eastern North Pacific, and Domeier et al. (2005), Kitagawa et al. (2007) and Boustany et al. (2010) have described the seasonal movements of immature PBF off the west coast of North America (in the eastern North Pacific). Electronically tagged, immature PBFs at 86–109 cm FL (age-two and three) were released off Baja California, Mexico in the summer of 2000–2002 (Domeier et al. 2005), 2002 (Kitagawa et al. 2007) and off the coast of California, USA, and Baja California, between August 2002 and August 2005 (Boustany et al. 2010).

Repeatable seasonal movements along the west coast of North America were demonstrated for these electronically tagged PBF. All three tagging studies showed residency off the southern coast of Baja California, Mexico during the spring (Fig. 5.8). As waters warmed, fish moved northwest into the southern California Bight during summer and of the southern and central California, USA coast in the fall (Kitagawa et al. 2007; Boustany et al. 2010). As waters cooled in the winter, most tracked PBF moved southeast, back into coastal waters off Mexico, although some fish moved offshore, to the west of central California at this time (Boustany et al. 2010; Fig. 5.8c). The seasonal movement patterns correlated with peaks in coastal upwelling-induced primary productivity and Pacific sardine (*Sardinops sagax*) availability (Domeier et al. 2005; Kitagawa et al. 2007; Boustany et al. 2010).

From late spring through autumn, PBFs were mainly located in the areas with the highest levels of primary productivity in the California Current ecosystem. Although the seasonal patterns of primary productivity and PBF movements were similar among years, there also existed inter-annual variation in both these variables. In 2002, the autumn peak in primary productivity was located further north along the California coast than in 2003–2004. Mirroring this shift, tagged PBF were found further north in the autumn of 2002 than in any of the other tracking years (Boustany et al. 2010; Block et al. 2011; Fig. 5.8a,c). The correlation between primary productivity and PBF movements in the California Current system is likely due to an interplay between food availability and temperature induced physiological limitations. In autumn, the landings of Pacific sardines (*Sardinops sagax*), an important prey item for PBF, generally increase to their highest level in central California, coinciding with rising sea surface temperature in this region (Goericke et al. 2004). This period is at the beginning of the highest sea surface temperatures in the annual cycle in the region; this might be an explanation for the displacement of PBF up the coast (Block et al. 2011). Kitagawa et al. (2007) reported that the northward movement of prey caused by high productivity and high temperatures accompanied by a weakened equatorward wind stress in the region, lead to a northward migration of PBFs.

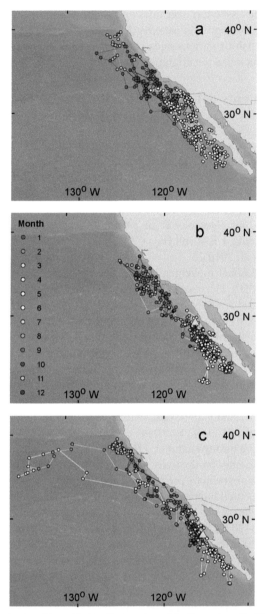

Figure 5.8. Example tracks of juvenile bluefin tuna tagged in the eastern Pacific showing seasonal movements along the Mexican and U.S. west coasts. (a) The first year after tagging for fish A0475 (89 cm CFL at release; tagged August 5, 2002); (b) A second year of movements for this same fish (A0475) showing lower maximal northern extent in 2003 vs. 2002; (c) seasonal movements of fish A0471 (112 cm CFL at release; tagged August 5, 2002) showing offshore movements in the winter before returning to the coast of North America in the spring. This fish also showed a more northerly distribution in the fall, similar to other fish tracked in 2002. Release and recapture positions are shown as color coded triangles. Modified from Boustany et al. 2010.

In mid-winter to early spring, tagged PBFs were found in areas with lower productivity than in other regions along the coast at that time of year. Boustany et al. (2010) suggested that during this period, PBF fed on aggregations of pelagic red crabs, sardines, and anchovies that preferentially spawn in areas of reduced coastal upwelling, where coastal retention of larvae is maximized.

Although tracking and catch data show that most of the PBF in the Eastern Pacific are small to medium immature fish (age one–four, less than 130 cm CFL), larger fish do occasionally reside off the west coast of North America. Bones of large bluefin tuna (over 160 cm CFL) have been collected from archeological sites of indigenous communities off the west coast of Canada dating as far back as 3000 B.C. and into the mid 19th century (Crockford 1997). In the late 19th century, there were consistent catches of large PBFs [over 100 kg (220 lbs)] reported off the coast of Southern California's Catalina Island (Boustany 2011). In more recent years, large PBFs have been rare along the west coast of North America (Foreman and Ishizuka 1990). It is unclear whether this is due to a dearth of large fish being available ocean-wide as a result of decreased stock size or if longer term climatic conditions have affected migration patterns (ISC 2014). Even catches for smaller PBF in the eastern Pacific have been highly variable on inter-annual and decadel time scales. Catches in the eastern Pacific, mainly off southern California, United States, and Baja California, Mexico, reached a peak of close to 18,000 mt in 1965 before declining in the 1980s and early 1990s and fluctuating by more than 5000 mt throughout the 20th century (Hanan 1983; Bayliff 1994).

Trans-Pacific Migration from Eastern to Western Pacific

As the only known spawning regions for PBF are in the western Pacific, fish in the eastern Pacific must make trans-Pacific migrations westward at some point if they are to reproduce. This westward migration has been observed in a number of PBF archivally tagged in the eastern Pacific. Seventeen of the archival-tagged fish released from the west coast of North America between August 2002 and August 2005 moved offshore (more than 5° from the coast; Boustany et al. 2010). The offshore movements started in December through March as coastal productivity along the central and northern California coast decreased and most tagged bluefin tuna began their southward migration along the coast (Fig. 5.8c). Seven of these offshore fish (4.5%) continued westward across the Pacific and crossed the dateline at 180° longitude, while the remainder returned to the California Current and migrated south into Mexican waters (Boustany et al. 2010; Fig. 5.8c). Fish that migrated to the western Pacific showed residency in the region of the Shatsky Rise and Emperor Seamounts before continuing to Japanese waters (Block et al. 2003; Fig. 5.9). Two of the fish that traveled to the western Pacific subsequently moved back to the eastern Pacific, and one individual then traveled back to the western Pacific where it was recaptured, representing four trans-Pacific migrations (Boustany et al. 2010).

Archival tagging studies of the trans-Pacific migration from the eastern to western Pacific revealed that PBF use a more southern migration route than do PBF moving from the western to eastern Pacific area (Block et al. 2003; Itoh et al. 2003; Kitagawa et

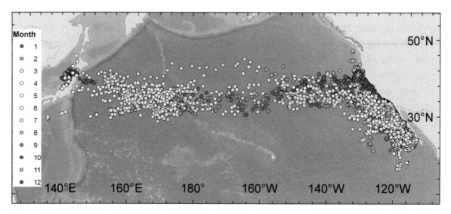

Figure 5.9. Distribution of 143 fish tracked with archival tags in the eastern Pacific (38,012 total positional estimates), color-coded by month. Movement to the western Pacific (n = 7) occurs in the winter while migrations back to the eastern Pacific (n = 2) occurred in the summer along a more northerly pathway. Modified from Boustany et al. 2010.

al. 2009; Boustany et al. 2010; Figs. 5.7, 5.9). This is likely due to migration pathways being correlated with the Transition Zone Chlorophyll Front, which is located at lower latitudes during the winter and spring when PBF migrate east to west than during the summer and fall, when PBF migrate west to east (Itoh et al. 2003; Kitagawa et al. 2009; Boustany et al. 2010). The purpose of these migrations back to the western Pacific are unclear as none of the tagged fish were tracked to a known spawning ground, and instead were found in the region around the Tsugaru Strait or in the Pacific side of Japan around 30°N and 140°E, a known feeding area for immature/prespawners (Boustany et al. 2010; Fig. 5.9). These fish might gather to their two main spawning grounds as mature adults, because many spawning adults have been caught in the Sea of Japan (over age-three) close to the Tsugaru Strait by purse seine fishery (Kanaiwa et al. 2011; Tanaka 2006) and in the waters around Nansei Islands (over age-five) by the longline fishery (Hiraoka et al. 2014; Ashida et al. 2015). However, knowledge of the movements of these large PBF (over age-three) in the western Pacific is poor, especially movements patterns of the fish into and out of their spawning grounds, and also movements between their two main spawning grounds associated with the difference of catch-age in these regions are unknown, there will be a need for future investigation.

It is possible that these east to west migrations are also mainly for feeding purposes. During the westward migration, as shown in a study by Block et al. (2003), the fish remain at seamount areas for some time in the residency phase defined by Itoh et al. (2003). However, the importance and evolutionary mechanisms of the trans-Pacific migration, which covers more than 8,000 km between the western and eastern Pacific, are still unclear.

Even less well understood are the mechanisms and routes of migration from the northern to the southern Pacific. While most PBF remain in the North Pacific, a small proportion travels to the South Pacific off the coasts of Australia and New Zealand

(Smith et al. 1994). These fish are generally large (over 160 cm) in size and are caught in fisheries mainly between April and September, with peak catches occurring between July and September (Smith et al. 2001; Murray 2005). Direct migration to New Zealand and Australian waters from the North Pacific has not been observed, but one large PBF tagged with an electronic tag was observed to have moved from the area near Japan to the Southern Hemisphere in a little over 40 days (Murray 2005). Electronic tagging studies have also been conducted in New Zealand waters and tagged fish showed movements around New Zealand and Australia northward, but no fish were observed to have crossed into the North Pacific over the tracking duration (G. Shillinger, pers. comm.). Movement patterns connecting PBFs in the Northern and Southern Hemispheres deserve further study.

Schematic model of migration pathways for PBFs at different life stages (larval, juvenile, immature, and immature/prespawner and spawner) are presented as a hypothesis in Fig. 5.10.

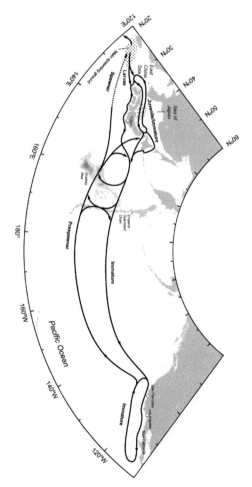

Figure 5.10. Schematic model of migration pathways of Pacific bluefin tuna.

Acknowledgments

We wish to thank Charles J. Farwell and Ethan Estess, Monterey Bay Aquarium, for their valuable advice and comments that substantially improved this manuscript. We also wish to express our thanks to Yoshinori Aoki, Atmosphere & Ocean Research Institute, The University of Tokyo, for his kind support.

References

Andres, M., J.H. Park, M. Wimbush, X.H. Zhu, K.I. Chang and H. Ichikawa. 2008. Study of the Kuroshio/ Ryukyu Current system based on satellite-altimeter and *in situ* measurements. J. Oceanogr. 64: 937–950.

Anonymous. 2002. Fisheries Agency of Japan and National Research Institute of Far Seas Fisheries. Report on 2002 Research Cruise of the R/V Shoyo-Maru (in Japanese).

Ashida, H., N. Suzuki, T. Tanabe, N. Suzuki and Y. Aonuma. 2015. Reproductive condition, batch fecundity, and spawning fraction of large Pacific bluefin tuna *Thunnus orientalis* landed at Ishigaki Island, Okinawa, Japan. Environ. Biol. Fishes 98: 1173–1183.

Bayliff, W.H. 1980. Synopsis of biological data on the northern bluefin tuna, *Thunnus thynnus* (Linnaeus, 1758), in the Pacific Ocean. IATTC Sci. Rep. 2: 261–293.

Bayliff, W.H. 1994. A review of the biology and fisheries for northern bluefin tuna, *Thunnus Thynnus*, in the Pacific Ocean. FAO Fish. Techn. Pap. 336: 244–295.

Bayliff, W.H., Y. Ishizuka and R. Deriso. 1991. Growth, movement, and attrition of northern bluefin tuna, *Thunnus Thynnus*, in the Pacific Ocean, as determined by tagging. IATTC Bull. 20: 3–94.

Block, B.A., D.P. Costa, G.W. Boehlert and R.E. Kochevar. 2003. Revealing pelagic habitat use: the tagging of Pacific pelagics program. Oceanologica Acta 25: 255–266.

Block, B.A., H. Dewar, C. Farwell and E.D. Prince. 1998. A new satellite technology for tracking the movement of Atlantic bluefin tuna. Proc. Natl. Acad. Sci. USA 95: 9384–9389.

Block, B.A., H. Dewar, S.B. Blackwell, T.D. Williams, E.D. Prince, C.J. Farwell, A. Boustany, S.L.H. Teo, A. Seitz, A. Walli and D. Fudge. 2001. Migratory movements, depth preferences, and thermal biology of Atlantic bluefin tuna. Science 293: 1310–1314.

Block, B.A., I.D. Jonsen, S.J. Jorgensen, A.J. Winship, S.A. Shaffer, S.J. Bograd, E.L. Hazen, D.G. Foley, G.A. Breed, A.L. Harrison, J.E. Ganong, A. Swithenbank, M. Castleton, H. Dewar, B.R. Mate, G.L. Shillinger, K.M. Schaefer, S.R. Benson, M.J. Weise, R.W. Henry and D.P. Costa. 2011. Tracking apex marine predator movements in a dynamic ocean. Nature 475(7354): 86–90.

Boustany, A.M. 2011. Bluefin Tuna: The State of the Science. Ocean Science. Division, Pew Environment Group, Washington, DC.

Boustany, A.M., R. Matteson, M. Castleton, C. Farwell and B.A. Block. 2010. Movements of Pacific bluefin tuna (*Thunnus orientalis*) in the Eastern North Pacific revealed with archival tags. Prog. Oceanogr. 86: 94–104.

Brill, R.W., H. Dewar and J.B. Graham. 1994. Basic concepts relevant to heat transfer in fishes, and their use in measuring the physiological thermoregulatory abilities of tunas. Environ. Biol. Fishes 40: 109–124.

Clemens, A.E. and G.A. Flittner. 1969. Bluefin tuna migrate across the Pacific Ocean. Calif. Fish Game 55: 132–135.

Crockford, S.J. 1997. Archeological evidence of large northern bluefin tuna, *Thunnus thynnus*, in coastal waters of British Columbia and northern Washington. Fish. Bull. 95: 11–24.

Davis, T.L.O., V. Lyne and G.P. Jenkins. 1991. Advection, dispersion and mortality of a patch of southern bluefin tuna larvae *Thunnus maccoyii* in the East Indian Ocean. Mar. Ecol. Prog. Ser. 73: 33–45.

Dewar, H., J.B. Graham and R.W. Brill. 1994. Studies of tropical tuna swimming performance in a large water tunnel-Thermoregulation. J. Exp. Biol. 192: 33–44.

Domeier, M.L., D. Kiefer, N. Nasby-Lucas, A. Wagschal and F. O'Brien. 2005. Tracking Pacific bluefin tuna (*Thunnus thynnus orientalis*) in the northeastern Pacific with an automated algorithm that estimates latitude by matching sea-surface-temperature data from satellites with temperature data from tags on fish. Fish. Bull. 103: 292–306.

Ebuchi, N. and K. Hanawa. 2000. Mesoscale eddies observed by TOLEX-ADCP and TOPEX/POSEIDON altimeter in the Kuroshio recirculation region south of Japan. J. Oceanogr. 56: 43–57.

Ebuchi, N. and K. Hanawa. 2001. Trajectory of mesoscale eddies in the Kuroshio recirculation region. J. Oceanogr. 57: 471–480.

Foreman, T.J. and Y. Ishizuka. 1990. Giant bluefin off Southern California, with a new California size record. California Fish and Game 76: 81–186.

Fujioka, K., H. Fukuda, S. Okamoto and Y. Takeuchi. 2013a. First record of the small (age-0) Pacific bluefin tuna (*Thunnus orientalis*) migration in the sea off Kochi revealed by archival tags. Abst. 9th Indo-Pacific Fish Conf. p. 228.

Fujioka, K., H. Fukuda, S. Okamoto and Y. Takeuchi. 2013b. Migration patterns of juvenile (age-0) Pacific bluefin tuna (*Thunnus orientalis*) in coastal nursery areas of Japan. Proc. 64th Ann. Tuna Conf. p. 40.

Galuardi, B., F. Royer, W. Golet, J. Logan, J. Nielson and M. Lutcavage. 2010. Complex migration routes of Atlantic bluefin tuna (*Thunnus thynnus*) question current population structure paradigm. Can. J. Fish. Aquat. Sci. 67: 966–976.

Goericke, R., S.J. Bograd, G. Gaxiola-Castro, J. Gomex-Valdes, R. Hooff, A. Huyer, K. D. Hyrenbach, B.E. Laveniegos, A. Mantyla, W.T. Peterson, F.B. Schwing, R.L. Smith, W.J. Sideman, E. Venrick and P.A. Wheeler. 2004. The State of the California Current, 2003–2004: a rare "Normal" year. CalCOFI. Rep. 45: 27–59.

Graham, J.B. 1983. Heat transfer. pp. 248–278. *In*: P.W. Webb and D. Weihs (eds.). Fish Biomechanics. Praeger, New York.

Graham, J.B. and K.A. Dickson. 2001. Anatomical and physiological specializations for endothermy. pp. 121–168. *In*: B.A. Block and E.D. Stevens (eds.). Tuna: Physiology, Ecology, and Evolution. Academic Press, San Diego, CA, USA.

Gunn, J.S., T. Polacheck, T.L.O. Davis, M. Sherlock and A. Betlehem. 1994. The development and use of archival tags for studying the migration, behavior and physiology of Southern bluefin tuna, with an assessment of the potential for transfer of the technology to groundfish research. Proc. ICES Symp. Fish. Migration 21: 1–23.

Hanan, D.A. 1983. Review and analysis of the bluefin tuna, *Thunnus thynnus*, fishery in the eastern North Pacific Ocean. Fish. Bull. 81: 107–119.

Hill, R.D. 1994. Theory of geolocation by light levels. pp. 227–236. *In*: B.J. LeBoeuf and R.M. Laws (eds.). Elephant Seals, Population Ecology, Behavior and Physiology. University of California Press, Berkely, CA.

Hiraoka, Y., M. Ichinokawa, K. Oshima and Y. Takeuchi. 2014. Updated standardized CPUE and size frequency for Pacific bluefin tuna caught by Japanese coastal longliners. International Scientific Committee for Tuna and Tuna-like Species in the North Pacific Ocean. Report of the Pacific Bluefin Tuna Working Group. ISC/14/PBFWG-1/02.

Ichikawa, K. 2001. Variation of the Kuroshio in the Tokara Strait induced by meso-scale eddies. J. Oceanogr. 57: 55–68.

Ichikawa, H., H. Nakamura and A. Nishina. 2000. Strong northeastward current outside of the East China Sea. pp. 284–288. *In*: E. Desa (ed.). Proceedings of the 5th Pacific Ocean Remote Sensing Conference, PORSEC Secretariat, (Indian) National Institute of Oceanography, Goa, 1.

Ichikawa, H., H. Nakamura, A. Nishina and M. Higashi. 2004. Variability of northeastward current southeast of northern Ryukyu Islands. J. Oceanogr. 60: 351–363.

Inagake, D., H. Yamada, K. Segawa, M. Okazaki, A. Nitta and T. Itoh. 2001. Migration of young bluefin tuna, *Thunnus orientalis*Temmincket Schlegel, through archival tagging experiments and its relation with oceanographic condition in the Western North Pacific. Bull. Natl. Res. Inst. Far Seas Fish. 38: 53–81.

Itoh, T. 2004. Studies on migratory ecology of Pacific bluefin tuna. Doc. Thesis, Graduate School of Agricultural and Life Sciences, University of Tokyo, Tokyo (in Japanese).

Itoh, T., S. Tsuji and A. Nitta. 2003. Migration patterns of young Pacific bluefin tuna (*Thunnus thynnus orientalis*) determined with archival tags. Fish. Bull. 101: 514–534.

International Scientific Committee for Tuna and Tuna-like Species in the North Pacific Ocean. 2014. Stock assessment of Bluefin tuna in the Pacific Ocean in 2014. Report of the Pacific Bluefin Tuna Working Group. 121 pp. (http://isc.ac.affrc.go.jp/index.html).

Kakuma, S. and K. Morinaga. 2000. Mesoscale eddies around the Okinawa islands. Report of the Experimental Station of Okinawa Prefecture pp. 52–56 (in Japanese).

Kanaiwa, M., A. Shibano and Y. Takeuchi. 2011. Estimation of length distribution for landing data of Pacific bluefin tuna in Sakai-minato port. International Scientific Committee for Tuna and Tuna-like Species in the North Pacific Ocean. Report of the Pacific Bluefin Tuna Working Group. ISC/11-1/PBFWG/04 (http://isc.ac.affrc.go.jp/index.html).

Kawai, H. 1972. Hydrography of the Kuroshio extension. pp. 235–352. *In*: H. Stommel and K. Yoshida (eds.). Kuroshio. University of Tokyo Press, Tokyo.

Kitagawa, T. 2013. Behavioral Ecology and thermal physiology of immature Pacific bluefin tuna (*Thunnus orientalis*). pp. 152–178. *In*: H. Ueda and K. Tsukamoto (eds.). Physiology and Ecology of Fish Migration. CRC Press, Florida.

Kitagawa, T., A.M. Boustany, C.J. Farwell, T.D. Williams, M.R. Castleton and B.A. Block. 2007. Horizontal and vertical movements of juvenile bluefin tuna (*Thunnus orientalis*) in relation to seasons and oceanographic conditions in the eastern Pacific Ocean. Fish. Oceanogr. 16: 409–421.

Kitagawa, T., Y. Kato, M.J. Miller, Y. Sasai, H. Sasaki and S. Kimura. 2010. The restricted spawning area and season of Pacific bluefin tuna facilitate use of nursery areas: a modeling approach to larval and juvenile dispersal processes. J. Exp. Mar. Biol. Ecol. 393: 23–31.

Kitagawa, T., S. Kimura, H. Nakata and H. Yamada. 2006b. Thermal adaptation of Pacific bluefin tuna *Thunnus orientalis* to temperate waters. Fish. Sci. 72: 149–156.

Kitagawa, T., S. Kimura, H. Nakata, H. Yamada, A. Nitta, Y. Sasai and H. Sasaki. 2009. Immature Pacific bluefin tuna, *Thunnus thynnus* orientalis, utilizes cold waters in the Subarctic Frontal Zone for trans-Pacific migration. Environ. Biol. Fish. 84: 193–196.

Kitagawa, T., H. Nakata, S. Kimura and S. Tsuji. 2001. Thermoconservation mechanism inferred from peritoneal cavity temperature recorded in free swimming Pacific bluefin tuna (*Thunnus thynnus* orientalis). Mar. Ecol. Prog. Ser. 220: 253–263.

Kitagawa, T., H. Nakata, S. Kimura and H. Yamada. 2004. Diving behavior of immature, feeding Pacific bluefin tuna (*Thunnus thynnus* orientalis) in relation to season and area: the East China Sea and the Kuroshio-Oyashio transition region. Fish. Oceanogr. 72: 149–156.

Kitagawa, T., H. Nakata, S. Kimura, T. Itoh, S. Tsuji and A. Nitta. 2000. Effect of ambient temperature on the vertical distribution and movement of Pacific bluefin tuna (*Thunnus thynnus* orientalis). Mar. Ecol. Prog. Ser. 206: 251–260.

Kitagawa, T., H. Nakata, S. Kimura, T. Sugimoto and H. Yamada. 2002. Differences in vertical distribution and movement of Pacific bluefin tuna (*Thunnus thynnus* orientalis) among areas: the East China Sea, the Sea of Japan and the western North Pacific. Mar. Freshwater Res. 53: 245–252.

Kitagawa, T., A. Sartimbul, H. Nakata, S. Kimura, H. Yamada and A. Nitta. 2006a. The effect of water temperature on habitat use of young Pacific bluefin tuna *Thunnus orientalis* in the East China Sea. Fish. Sci. 72: 1166–1176.

Kitagawa, Y., Y. Nishikawa, T. Kubota and M. Okiyama. 1995. Distribution of ichthyoplanktons in the Japan Sea during summer, 1984, with special reference to scombroid fishes. Bull. Jpn. Soc. Fish. Oceanogr. 59: 107–114 (in Japanese with English abstract).

Klawe, W.T. 1963. Observations on the spawning of four species of tuna (Neothunnus macropterus, Katsuwonus pelamis, Auxis thazard and Euthynnus lineatus) in the eastern Pacific Ocean, based on the distribution of their larvae and juveniles. Bull. Inter-Am. Trop. Tuna Comm. 6: 449–540.

Lam, C.H., A. Nielsen and J.R. Sibert. 2008. Improving light and temperature based geolocation by unscented Kalman filtering. Fish. Res. 91: 15–25.

Madigan, D.J., Litvin, S.Y., Popp, B.N., Carlisle, A.B., Farwell, C.J. and Block, B.A. 2012. Tissue turnover rates and isotopic trophic discrimination factors in the endothermic teleost, Pacific bluefin tuna (*Thunnus orientalis*). PLoS ONE 7: e49220.

Masujima, M., Y. Kato and K. Segawa. 2013 (in print). Numerical studies focusing on the early life stage of pacific bluefin tuna (*Thunnus orientalis*). Bull. Fish. Res. Agen.

Miyashita, S. 2002. Studies on the seedling production of the Pacific bluefin tuna *Thunnus thynnus orientalis*. Bull. Fish. Lab. Kinki Univ. 8: 1–171 (in Japanese with English abstract).

Morimoto, A., S. Kojima, S. Jan and D. Takahashi. 2009. Movement of the Kuroshio axis to the northeast shelf of Taiwan during typhoon events. Estuar. Coast. Shelf Sci. 82: 547–552.

Morinaga, K., N. Nakagawa, K. Osamu and B. Guo. 1998. Flow pattern of the Kuroshio west of the main Okinawa Island. Proceedings of the Japan-China Joint Symposium CSSCS, Seikai Natl. Fish. Res. Inst., Fisheries Agency of Japan, Nagasaki, Japan pp. 203–210.

Murray, T.E. 2005. The Distribution of Pacific bluefin Tuna (*Thunnus orientalis*) in the Southeast Pacific Ocean, with Emphasis on New Zealand Waters. Ministry of Fisheries. NZ Fish. Ass. Rep. 2005/42: 14 pp.

Nakamura, H. 2005. Numerical study on the Kuroshio path states in the northern Okinawa Trough of the East China Sea. J. Geophys. Res. 110: C04003.

Horizontal Movements of Pacific Bluefin Tuna 121

Nakamura, H., M. Nonaka and H. Sasaki. 2010. Seasonality of the Kuroshio path destabilization phenomenon in the Okinawa Trough: A numerical study of its mechanism. J. Phys. Oceanogr. 40: 530–550.

Neill, W.H. and E.D. Stevens. 1974. Thermal inertia versus thermoregulation in "warm" turtles and tuna. Science 184: 1008–1010.

Neill, W.H., R.K.C. Chang and A.E. Dizon. 1976. Magnitude and ecological implications of thermal inertia in skipjack tuna, Katsuwonuspelamis (Linnaeus). Environ. Biol. Fish. 1: 61–80.

Niiler, P.P. 2001. The world ccean surface circulation. pp. 193–204. *In*: J. Church, G. Siedler and J. Gould (eds.). Ocean Circulation and Climate-Observing and Modeling the Global Ocean. Academic Press, London.

Nishikawa, Y., M. Honma, S. Ueyanagi and S. Kikawa. 1985. Average distribution of larvae of oceanic species of Scombroid fishes, 1956–1981. Far Seas Fish. Res. Lab. 12: 1–99.

Okiyama, M. 1974. Occurrence of the postlarvae of bulefin tuna, *Thunnus thynnus*, in the Japan Sea. Bull. Jpn. Sea Reg. Fish. Res. Lab. 11: 9–21.

Orange, C.J. and B.D. Fink. 1963. Migration of a tagged bluefin tuna across the Pacific Ocean. Calif. Fish Game 49: 307–309.

Oshima, K., M. Ichinokawa and Y. Takeuchi. 2010. Discrimination of age-0 Pacific bluefin tuna from two different spawning grounds in length-frequency distributions. Proc. 61st Ann. Tuna Conf. p. 55.

Polovina, J.J. 1996. Decadal variation in the trans Pacific migration of northern bluefin tuna (*Thunnus thynnus*) coherent with climate induced change in prey abundance. Fish. Oceanogr. 5(2): 114–119.

Richards, W.J. and D.C. Simrnons. 1971. Distribution of tuna larvae (Pisces, Scombridae) in the northwestern Gulf of Guinea and off Sierra Leone. Fish. Bull. NOAA-NMFS. 69: 555–568.

Satoh, K. 2010. Horizontal and vertical distribution of larvae of Pacific bluefin tuna *Thunnus orientalis* in patches entrained in mesoscale eddies. Mar. Ecol. Prog. Ser. 404: 227–240.

Satoh, K., Y. Tanaka and M. Iwahashi. 2008. Variations in the instantaneous mortality rate between larval patches of Pacific bluefin tuna *Thunnus orientalis* in the northwestern Pacific Ocean. Fish. Res. 89: 248–256.

Shimose, T., T. Tanabe, K.S. Chen and C.C. Hsu. 2009. Age determination and growth of Pacific bluefin tuna, *Thunnus orientalis*, off Japan and Taiwan. Fish. Res. 100: 134–139.

Smith, P.J., A.M. Conroy and P.R. Taylor. 1994. Biochemical-genetic identification of northern Bluefin tuna *Thunnus thynnus* in the New Zealand fishery. NZ J. Mar. Freshw. Res. 28: 113–118.

Smith, P.J., L. Griggs and S. Chow. 2001. DNA identification of Pacific bluefin tuna (*Thunnus orientalis*) in the New Zealand fishery. NZ J. Mar. Freshw. Res. 35(4): 843–850.

Strasburg, D.W. 1960. Estimates of larval tuna abundance in the central Pacific. Fish. Bull. U.S. Fish Wildl. Serv. 60: 231–255.

Suga, T., K. Motoki and K. Hanawa. 2003. Subsurface water masses in the Central North Pacific Transition Region: the repeat section along the 180 degrees meridian. J. Phys. Oceanogr. 59: 435–444.

Takahashi, D., X. Guo, A. Morimoto and S. Kojima. 2009. Biweekly periodic variation of the Kuroshio axis northeast of Taiwan as revealed by ocean high-frequency radar. Cont. Shelf. Res. 29: 1896–1907.

Tameishi, H. 1997. Fisheries oceanography of warm-core ring and fish way. Bull. Jpn. Soc. Fish. Oceanogr. 61: 157–161 (in Japanese).

Tamura, Y. and T. Takagi. 2009. Morphological features and functions of bluefin tuna change with growth. Fish. Sci. 75: 567–575.

Tanaka, S. 2006. Maturation of bluefin tunain the Sea of Japan. International Scientific Committee for Tuna and Tuna-like Species in the North Pacific Ocean. Report of the Pacific Bluefin Tuna Working Group. ISC/06/PBF-WG/09.

Tanaka, Y., M. Mohri and H. Yamada. 2007. Distribution, growth and hatch date of juvenile Pacific bluefin tuna *Thunnus orientalis* in the coastal area of the Sea of Japan. Fish. Sci. 73: 534–542.

Tanaka, Y., K. Satoh, M. Iwahashi and H. Yamada. 2006. Growth-dependent recruitment of Pacific bluefin tuna *Thunnus orientalis* in the northwestern Pacific Ocean. Mar. Ecol. Prog. Ser. 319: 225–235.

Teo, S.L.H., A.M. Boustany and B.A. Block. 2007. Oceanographic preferences of Atlantic bluefin tuna, *Thunnus thynnus*, on their Gulf of Mexico breeding grounds. Mar. Biol. 152: 1105–1119.

Teo, S.L.H., A.M. Boustany, S. Blackwell, A. Walli, K.C. Weng and B.A. Block. 2004. Validation of geolocation estimates based on light level and sea surface temperature from electronic tags. Mar. Ecol. Prog. Ser. 283: 81–98.

Uda, M. 1957. A consideration on the long years trend of the fisheries fluctuation in relation to sea condition. Bull. Jpn. Soc. Sci. Fish. 23: 368–372.

Ueyanagi, S. 1969. Observations on the distribution of tuna larvae in the Indo-Pacific Ocean with emphasis on the delineation of the spawning areas of albacore, *Thunnus alalunga*. Bull. Far Seas Fish. Res. Lab. 2: 177–256.

Wade, C.B. 1951. Larvae of tuna and tuna-like fishes from the Philippine waters. Fish. Bull. 51: 445–485.

Welch, D.W. and J.P. Eveson. 1999. An assessment of light-based geoposition estimates from archival tags. Can. J. Fish. Aquat. Sci. 56: 1317–1327.

Yabe, H., S. Ueyanagi and H. Watanabe. 1966. Studies on the early life history of bluefin tuna *Thunnus thynnus* and on the larvae of the Southern bluefin tuna T. maccoyii. Rep. Nankai Reg. Fish. Res. Lab. 23: 95–129 (in Japanese).

Yanagi, T., T. Tokeshi and S. Kakuma. 2002. Eddy activities around the NanseiShoto (Okinawa Islands) revealed by TRMM. J. Oceanogr. 58: 617–624.

Yang, Y., C.T. Liu, J.H. Hu and M. Koga. 1999. Taiwan Current (Kuroshio) and impinging eddies. J. Oceanogr. 55: 609–617.

Yukinawa, M. and Y. Yabuta. 1967. Age and growth of bluefin tuna, *Thunnus thynnus* (Linnaeus), in the North Pacific Ocean. Rep. Nankai Reg. Fish. Res. Lab. 25: 1–18 (in Japanese with English abstract).

Zheng, X.T., Q.Y. Liu, H.B. Hu, Y. Miyazawa and Y.L. Jia. 2008. The study of temporal and spatial characteristics of western boundary current East of Ryukyu submarine ridge and the transport of Kuroshio in East China Sea. Acta Oceanol. Sin. 30: 1–9.

Zhu, X.H., I.S. Han, J.H. Park, H. Ichikawa, K. Murakami, A. Kaneko and A. Ostrovskii. 2003. The northeastward current southeast of Okinawa Island observed during November 2000 to August 2001. Geophys. Res. Lett. 30: 1071.

Zhu, X.H., J.H. Park and I. Kaneko. 2005. The northeastward current southeast of the Ryukyu Islands in late fall of 2000 estimated by an inverse technique. Geophys. Res. Lett. 32: L05608.

CHAPTER 6

Formation of a Pacific Bluefin Tuna Fishing Ground on Their Spawning Grounds around the Ryukyu Islands

Implication of a Relationship with Mesoscale Eddies

Itaru Ohta[1,]* and *Harumi Yamada*[2]

Introduction

The Pacific bluefin tuna (PBT), *Thunnus orientalis* (Scombridae), is widely distributed in the Pacific Ocean from the subarctic to tropical areas and over the equator to temperate areas of the Southern Hemisphere (Collette and Nauen 1983). It is known to be a highly migratory fish species (Inagake et al. 2001; Itoh et al. 2003a; Itoh 2004). Despite this wide distribution range, reproductive activity is believed to occur only in the western North Pacific Ocean, particularly in the waters around the Ryukyu Islands and in the Sea of Japan (Itoh 2004). These two areas are recognized as the spawning grounds of PBT as these mature fish migrate here during the spawning season. The spawning seasons are from April to June around the Ryukyu Islands (Chen et al. 2006) and from June to August in the Sea of Japan (Tanaka et al. 2007). The waters around the Ryukyu Islands are the major spawning grounds, and they contribute to more than 70% of the PBT recruitment around Japan (Itoh 2004). These aggregating spawning

[1] Okinawa Prefectural Fisheries Research and Extension Center, Kyan 1528, Itoman, Okinawa 901-0354, Japan.

[2] Tohoku National Fisheries Research Institute, Fisheries Research Agency, Shinhama-cho 3-27-5, Shiogama, Miyagi 985-0001, Japan.

* Email: ootaitar@pref.okinawa.lg.jp

groups have been targeted by long-line fisheries. This spawning aggregation is not only economically valuable as a fishery resource but also ecologically important for reproduction as a support to the majority of the PBT population. However, ecological information related to the spawning aggregation of PBT and fishery ground formations in this area is poorly understood.

The spawning area around the Ryukyu Islands is strongly influenced by the Kuroshio Current and mesoscale eddies (Satoh 2010). This current flows northward along the east side of Taiwan and the west side of the Ryukyu Islands (Fig. 6.1). Mesoscale eddies induce the formation of unusual oceanographic structures and significant accumulation of some marine organisms (Bakun 2006; Sabarros et al. 2009). These eddies can also be detected in a wide area using satellite images of the Sea surface Height (SSH) anomaly (Ebuchi and Hanawa 2000). Mesoscale eddies in the Kuroshio recirculation region of southern Japan are composed of cyclonic and anticyclonic eddies with diameters of 100–500 km and a temporal scale of 80 days. Typical maximum surface velocity and the SSH anomaly associated with these eddies are 15–20 cm s^{-1} and 15 cm, respectively (Ebuchi and Hanawa 2000). These eddies propagate westward at approximately 7 cm s^{-1}, and some of them coalesce with the Kuroshio Current (Ebuchi and Hanawa 2001). PBT and the Atlantic bluefin tuna *T. thynnus* (ABT) show active vertical movement related to feeding and thermoregulation that are associated with the vertical structures within water masses (Kitagawa et al. 2000; Itoh et al. 2003b; Kitagawa et al. 2004; Kitagawa et al. 2006; Teo et al. 2007b). Previous ABT research using electronic tags has demonstrated the changes in movement patterns, diving behavior, and oceanographic preferences on their breeding grounds in the Gulf of Mexico (Teo et al. 2007a,b). Thus, oceanographic

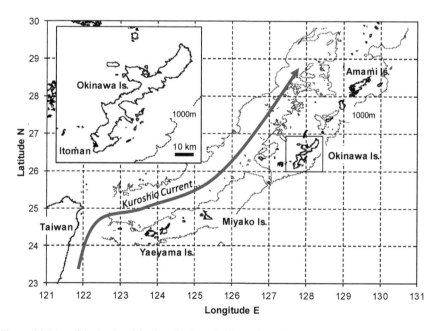

Figure 6.1. Map of the Ryukyu Islands and schematic illustration of the Kuroshio Current.

conditions may also influence the behavior of the PBT spawning groups. The SSH anomaly is an indicator that reflects vertical structures within water masses and mesoscale eddies (Bakun 2006; Ebuchi and Hanawa 2000). It provides some of the most important oceanographic information and is available via the internet. Therefore, SSH helps explain PBT distribution patterns and the formation of fishing grounds related to their behavioral responses to the ambient environment.

In this chapter, we describe the general characteristics of the fishing grounds of the PBT spawning groups around the Ryukyu Islands based on fishery information, and discuss the relationship between formation of fishery grounds and mesoscale eddies inferred from SSH anomaly images.

Analysis Procedure

Market research was conducted to monitor long-line vessel fishing activity and the catch at the port of Itoman, Okinawa Island. The Itoman port is a major port where the PBT spawning groups around the Ryukyu Islands are frequently landed (Figs. 6.1 and 6.2). Market research was conducted almost every day during the fishing period from April to June during 2000–2003. Fork Length (FL) to the nearest 1 cm and gutted Body Weight (BW: excluding gill and internal organs except gonads) to the nearest 1 kg was measured for the 1,229 PBT caught by 8–19 ton class long-line vessels. In addition, fishery information such as the date, location, number of the PBT caught during a single fishing trip (N_{PBT}), frequency of operations during a single fishing trip (F_{op}), and number of hooks per operation (N_{hook}) was collected by interviewing the captains of each long-line vessel. In total, information for 284 fishing trips within the range from N20° to N30° and E122° to 135° was collected. The Catch Per Unit Effort (CPUE) of a single fishing trip was calculated as follows:

$$\text{CPUE } (N_{PBT}/1{,}000 \text{ hooks}) = N_{PBT}/(F_{op} \cdot N_{hook}) \times 10^3$$

Figure 6.2. Pacific bluefin tuna landed in the market at Itoman port (photo by K. Maeda).

A single long-line operation targeting the PBT in this area was conducted once a day during daytime. The interview surveys revealed that F_{op} ranged from 1 to 15 (mode: 6–8) during a fishing trip. N_{hook} among vessels also varied. Many used 500–2,500 hooks per operation, but 86% of the 2,008 operations conducted used 1,200–2,000 hooks. As the locations of each operation were not far apart in a single trip, single location data (latitude and longitude with accuracy to the minute) were used to represent each fishing trip.

To examine the relationship between the PBT CPUE and mesoscale eddies, 54 SSH anomaly satellite images observed by the TOPEX/POSEIDON and ERS2 altimeter were downloaded from the Colorado University website (http://eddy.colorado.edu/ccar/data_viewer/index). The locations of each fishing trip were plotted on these satellite images on the date of halfway through a fishing trip. Each location was then classified into three SSH categories (SSHC) (low: < -5 cm, middle: between -5 and 5 cm, and high: > 5 cm) based on the color of the satellite images. The 284 CPUE and SSHC datasets were examined by two-factor (year and SSHC) analysis of variance (ANOVA). Differences in the CPUE among these factors were examined by Tukey's HSD test as a post-hoc analysis using the STATISTICA 06J package (StatSoft Inc., Tulsa, OK, USA).

Fishing Grounds and Duration of PBT in the Waters Around the Ryukyu Islands

The size of the 1,229 PBT landed at Itoman port between 2000 and 2003 ranged from 157 to 252 cm (mean ± standard deviation: 204 ± 15 cm) in FL and from 67 to 342 kg (155 ± 35 kg) in BW (Fig. 6.3). Most of the fish had mature gonads, indicating that they were large adults and had migrated to the area to spawn. The maximum size of the PBT was more than 300 cm in FL and 500 kg in BW (Chapter 3), but the largest individuals were rarely caught.

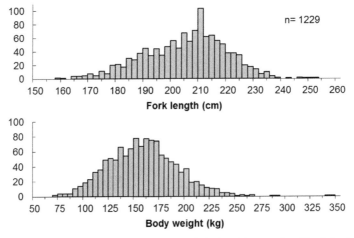

Figure 6.3. Size distribution of Pacific bluefin tuna caught by long-line vessels and landed at Itoman port from 2000 to 2003. (a) Fork length (cm) (b) Body weight (kg).

The PBT fishing season around the Ryukyu Islands is usually from April to June. Although there were annual variations in the catch and fishing activity, the peak was from early May to mid-June (Fig. 6.4). Long-line fishing operated widely around the Ryukyu Islands but more frequently on its southeast side, and fishing activity was relatively low on its northwest side where the Kuroshio Current flows (Fig. 6.5).

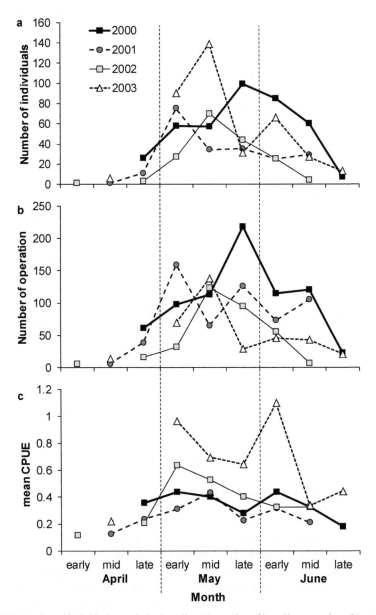

Figure 6.4. Number of individuals caught by long-line (a), number of long-line operations (b), and mean catch per unit effort (CPUE) (number of fish per 1,000 hooks) of Pacific bluefin tuna landed at Itoman port during the fishing season from 2000 to 2003.

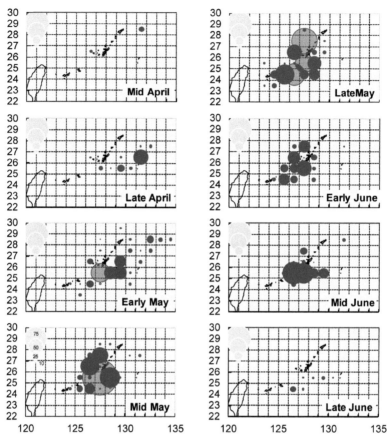

Figure 6.5. Seasonal distribution of fishing locations and Pacific bluefin tuna catch (number of individuals) in each 1 × 1 degree grid, based on data from 2000 to 2003.

These locations moved seasonally from the northeast to the southwest sides of the Ryukyu Islands and converged at the southern part of the Okinawa Island at the end of the fishing season. Similarly, CPUE shifted seasonally but was relatively higher in the southern part of the Okinawa Island throughout the fishing season (Fig. 6.6).

Relationship between CPUE of PBT and Mesoscale Eddies

To examine the relationship between the CPUE of PBT and mesoscale eddies, the CPUE of each fishing trip was plotted on the SSH anomaly images, as shown in the examples (Fig. 6.7). These images indicated that in the spawning area of the PBT, numerous cyclonic and anticyclonic eddies generally exist and vary with time. They also indicated that the locations of high CPUE were observed in cyclonic eddies. Significant differences were observed in CPUE among three SSHC (ANOVA $P < 0.01$, Table 6.1; Fig. 6.8). The mean CPUE in the area of low SSHC (mean ±

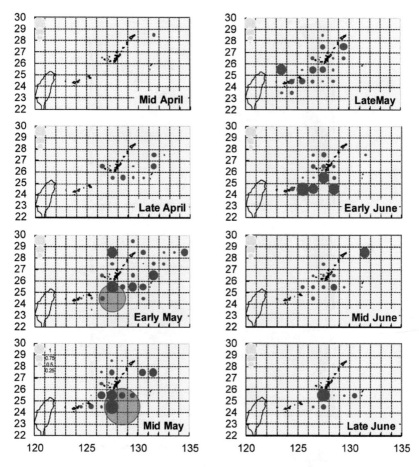

Figure 6.6. Seasonal distribution of fishing locations and mean catch per unit effort (CPUE) (number of fish per 1,000 hooks) of Pacific bluefin tuna in each 1 × 1 degree grid, based on data from 2000 to 2003.

standard error: 0.53 ± 0.04) was higher than that of middle (0.36 ± 0.03) and high SSHC (0.35 ± 0.07) (Tukey's HSD test, $P < 0.05$). This was an annual trend although the annual mean CPUE significantly differed among years (ANOVA $P < 0.001$, Table 6.1; Fig. 6.8) and was higher in 2003 than that in the other years (Tukey's HSD test, $P < 0.01$).

The results of a previous tracking study in the PBT spawning area using ultrasonic telemetry provided some profound insight into this result (Fisheries Agency of Japan 2002). This research was conducted in the spawning area in the southern end of the Ryukyu Islands in May 2001. One adult PBT fish (estimated BW: 230 kg) was tracked for 10 days (256 hours) over 1,356 km and moved southward over 1,000 km. During the time after the pathway was plotted on the SSH anomaly image, it was discovered that this fish alternately passed through a high SSH area and a low SSH area (Fig. 6.9). Vertical water temperature profiles monitored by an expendable bathythermograph

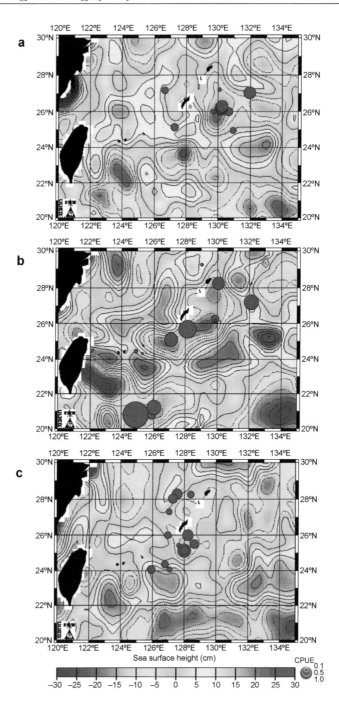

Figure 6.7. Examples of catch per unit effort (CPUE) data plots on the sea surface height (SSH) anomaly images: (a) May 6, 2000, (b) May 10, 2001, and (c) May 11, 2002.

Table 6.1. Results of two-factor analysis of variance with (year; SSHC: sea surface height category) for the catch per unit effort of the Pacific bluefin tuna. Levels of significance were $**P < 0.01$, $***P < 0.001$, and ns: not significant.

Sources of variation	df	MS	F	
Year	3	1.54779	9.4405	***
SSHC	2	0.86558	5.2795	**
Year × SSHC	6	0.18009	1.0984	ns
Residuals	272	0.16395		

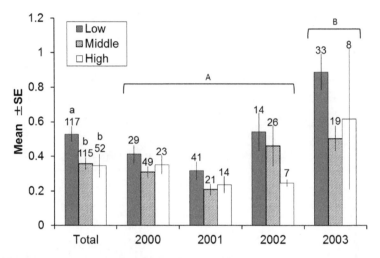

Figure 6.8. Mean catch per unit effort (CPUE) (number of fish per 1,000 hooks) (± standard error) of Pacific bluefin tuna in three sea surface height categories (SSHC: low, middle, and high) and each year (2000–2003). Statistical significance indicated by different letters (SSHC: a, b; year: A, B). Numerals indicate the number of data.

(XBT) during tracking showed remarkable differences between each SSH area. Although Sea Surface Temperature (SST) gradually increased with a decrease in latitude, water temperatures at the 200 and 300 m depth layers in the high SSH area were 2–3°C warmer than those in the low SSH area (Fig. 6.10). This observation indicated that SSH reflected a different water mass structure, and that the high and low SSH areas indicated the presence of anticyclonic and cyclonic eddies, respectively. During the 10 days of tracking, this fish demonstrated vertical movement ranging from the surface to the depths of 400 m (water temperature: 32–11°C), with repeated ascents and descents. The swimming depths of the PBT were relatively shallower in cyclonic eddies (May 23 and 26) than in anticyclonic eddies (May 21–22 and May 24–25) (Fig. 6.11). This fish spent 80–84% of the daytime in <100 m depth in cyclonic eddies, whereas 32–74% of the daytime was spent in anticyclonic eddies. Long-line targeting for the PBT is usually set at a depth of <100 m in their spawning area and is shallower than the other tuna fish (yellowfin tuna, *T. albacares* and bigeye tuna,

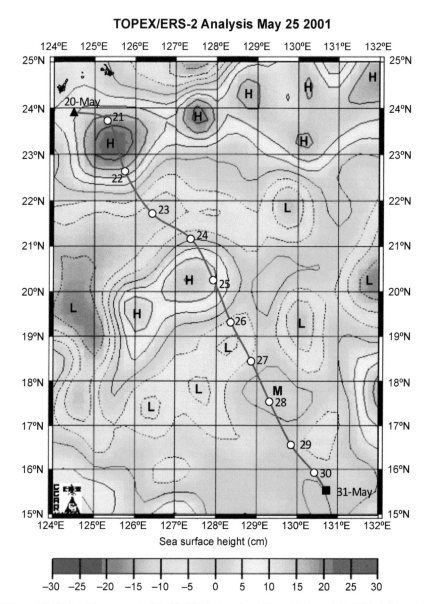

Figure 6.9. Horizontal movement of Pacific bluefin tuna #1 (estimated body weight, 230 kg) determined by acoustic telemetry from May 20 to 31, 2001 (data from Fisheries Agency of Japan 2002); drawing of the sea surface height (SSH) anomaly on the satellite image (May 25, 2001). Locations of start (black triangles), of end (black squares), and of noon of each day (white circles) during acoustic tracking are shown.

T. obesus). Therefore, an increase in encounter probability between fish and fishing gear at a depth of <100 m (i.e., gear efficiency or catchability) may be one of the reasons why the CPUE was high in cyclonic eddies.

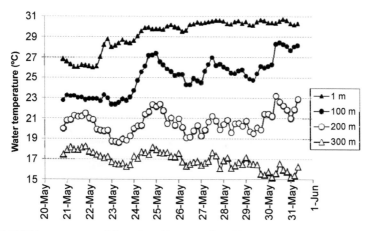

Figure 6.10. Water temperature of four selected layers monitored by expendable bathythermograph (XBT) at 4 hours intervals during acoustic tracking (data from Fisheries Agency of Japan 2002).

These behavioral changes of the PBT in cyclonic eddies may be due to feeding and spawning behavior, orientation, and/or thermal adaptation related to changes in ambient oceanographic conditions. The results of XBT monitoring indicated signs of upwelling in cyclonic eddies (Fig. 6.10). Consequently, these eddies may engender a concentration of prey organisms and increase of productivity in the shallower layer. In contrast, the PBT have high thermoconservation ability, enabling them to adapt to cooler temperate waters, suggesting that they face an overheating problem in warm waters (Kitagawa et al. 2006). Their repeated vertical movements observed during acoustic tracking (Fig. 6.11) could be behavioral thermoregulation to avoid overheating when in a warmer spawning area (Kitagawa et al. 2006). However, the ABT on the other hand show distinctive changes in diving behavior and thermal biology during the putative breeding phase (Teo et al. 2007a); they exhibit shallow oscillatory dives and increase their heat transfer coefficients around high ambient temperatures on their spawning grounds for behavioral and physiological thermoregulation (Teo et al. 2007a). These observations also suggest that shallow oscillatory dives of the PBT (Fig. 6.11), which are accompanied by frequent visits to the surface at night, may be courtship and/or spawning behavior similar to that of the ABT (Teo et al. 2007a). Furthermore, some oceanographic and mesoscale eddy preferences of the ABT have been reported; cyclonic eddies with cooler temperature waters may be important for reducing the metabolic stress in higher temperatures for the adult ABT (Teo et al. 2007b).

Larval biology may also be important for formation of spawning grounds. A previous tracking study on high density patches of the PBT larvae suggested that the positional relationship between spawning events and mesoscale eddies is important for the PBT recruitment (Satoh 2010). Patches of larvae are entrained by mesoscale eddies (Satoh 2010) that propagate westward, and some of these eddies coalesce with the Kuroshio Current (Ebuchi and Hanawa 2001). This could link the spawning area to the fishery grounds of young PBT in southern Japan through the Kuroshio Current (Satoh 2010). A previous modeling study has also indicated that the PBT spawn in restricted regions during a limited period (April–July) to maximize the chances that

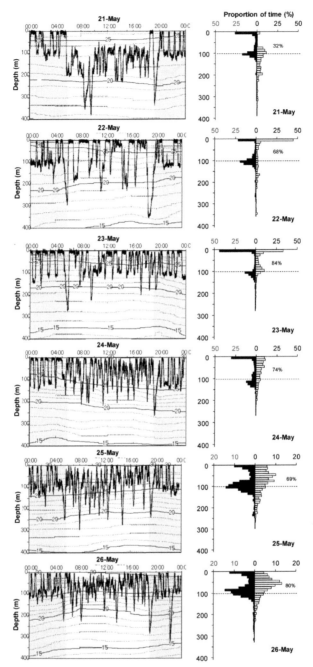

Figure 6.11. Vertical movement and water temperature profile (left) and swimming depth distribution (right) of Pacific bluefin tuna #1 determined by acoustic telemetry and expendable bathythermograph (XBT) monitoring from May 21 to 26, 2001 (data from Fisheries Agency of Japan 2002). White and black bars indicate day and night, respectively. Percentages on the figure indicate proportion of time at <100 m depth.

their larvae would enter the nursery regions of temperate waters with a lower energy cost, ensuring optimal thermal growth and survival conditions although this spawning area may be unfavorable for adults due to the warm water temperatures (Kitagawa et al. 2010). Furthermore, the growth and survival of these larvae are positively influenced by SST and food density (Satoh et al. 2013). The cohort spatial structure of the PBT larvae entrained in a mesoscale eddy would be relatively stable, and the environmental conditions of their spawning location are critically important for their growth and survival (Satoh 2006; Satoh et al. 2013). As shown by Kato et al. (2008), moderate (optimum) oceanic turbulence that exists in eddies appears to increase the encounter rate between the PBT larvae and their prey. The relationship between the ABT larval abundance and the oceanographic conditions of their spawning grounds in the Mediterranean Sea also indicate that frontal structures and anticyclonic eddies seem to have an influence on their spawning activity (Garcia et al. 2005).

Thus, both adult and larval bluefin tuna biology are related to mesoscale eddies, which may influence the formation of their spawning grounds. If cyclonic eddies produce a critical environment for the PBT larvae, these eddies may act as a cue for the adult spawning aggregation. Therefore, the high CPUE in cyclonic eddies demonstrated in this study may not only indicate an increased catchability related to vertical distribution of the PBT but also reflect an abundance of these fish associated with the spawning aggregation.

Conclusion

In this chapter, we discussed the seasonal shift of fishing grounds of the PBT spawning groups around the Ryukyu Islands, indicating that the seasonal shift was associated with their spawning migration. Furthermore, CPUE of the PBT was significantly higher in the area of lower SSH. This observation may be related to some behavioral changes of the PBT that are associated with a change in the oceanographic structure of cyclonic eddies, which was consistent with the results of a previous tracking study. Further research to verify the mechanisms of fishery and spawning ground formation related to oceanographic features will make an important contribution for sustainable fisheries for PBT.

Acknowledgements

We appreciate the late K. Yano for greatly contributing to the adult PBT tuna tracking research referred to in this study. We thank H. Uehara and M. Fukuda for their assistance with the market study. The manuscript was improved with helpful comments from K. Hirate and T. Kitagawa. The authors would like to thank Enago (www.enago.jp) for the English language review. Funding for this study was provided by the Fisheries Agency of Japan.

References

Bakun, A. 2006. Fronts and eddies as key structures in the habitat of marine fish larvae: opportunity, adaptive response and competitive advantage. Sci. Mar. 70: 105–122.

Chen, K.S., P. Crone and C.C. Hsu. 2006. Reproductive biology of female Pacific Bluefin tuna *Thunnus orientalis* from south-western North Pacific Ocean. Fish. Sci. 72: 985–994.

Collette, B.B. and C.E. Nauen. 1983. FAO species catalogue. Scombrids of the world. An annotated and illustrated catalogue of tunas, mackerels, bonitos and related species known to date. FAO Fish. Synop. 2: 125.

Ebuchi, N. and K. Hanawa. 2000. Mesoscale eddies observed by TOLEX-ADCP and TOPEX/POSEIDON altimeter in the Kuroshio recirculation region south of Japan. J. Oceanogr. 56: 43–57.

Ebuchi, N. and K. Hanawa. 2001. Trajectory of mesoscale eddies in the Kuroshio recirculation region. J. Oceanogr. 57: 471–480.

Fisheries Agency of Japan and National Research Institute of Far Seas Fisheries. 2002. Report on 2002 Research Cruise of the R/V Shoyo-Maru.

Garcia, A., F. Alemany, P. Velez-Belchi, J.L. LópezJurado, D. Cortés, J.M. de la Serna, C. González Pola, J.M. Rodriguez, J. Jansá and T. Ramirez. 2005. Characterization of the bluefin tuna spawning habitat off the Balearic archipelago in relation to key hydrographic features and associated environmental conditions. Col. Vol. Sci. Pap. ICCAT 58: 535–549.

Inagake, D., H. Yamada, K. Segawa, M. Okazaki, A. Nitta and T. Itoh. 2001. Migration of young bluefin tuna, *Thunnus* orientalis Temmincket Schlegel, through archival tagging experiments and its relation with oceanographic condition in the western North Pacific. Bull. Natl. Res. Inst. Far. Seas. Fish. 38: 53–81.

Itoh, T. 2004. Construction of new schematic illustration of migration for Pacific Bluefin tuna. pp. 254–261. *In*: T. Sugimoto (ed.). Ocean Current and Biological Resources. Seizando, Tokyo.

Itoh, T., S. Tsuji and A. Nitta. 2003a. Migration patterns of young Pacific bluefin tuna (*Thunnus orientalis*) determined with archival tags. Fish. Bull. 101: 514–534.

Itoh, T., S. Tsuji and A. Nitta. 2003b. Swimming depth, ambient water temperature preference, and feeding frequency of young Pacific bluefin tuna (*Thunnus orientalis*) determined with archival tags. Fish. Bull. 101: 535–544.

Kato, Y., T. Takebe, S. Masuma, T. Kitagawa and S. Kimura. 2008. Turbulence effect on survival and feeding of Pacific bluefin tuna *Thunnus orientalis* larvae, on the basis of a rearing experiment. Fish. Sci. 74: 48–53.

Kitagawa, T., H. Nakata, S. Kimura, T. Itoh, S. Tsuji and A. Nitta. 2000. Effect of ambient temperature on the vertical distribution and movement of Pacific bluefin tuna *Thunnus thynnus orientalis*. Mar. Ecol. Prog. Ser. 206: 251–260.

Kitagawa, T., S. Kimura, H. Nakata and H. Yamada. 2004. Diving behavior of immature, feeding Pacific bluefin tuna (*Thunnus thynnus orientalis*) in relation to season and area: the East China Sea and the Kuroshio–Oyashio transition region. Fish. Oceanogr. 13: 161–180.

Kitagawa, T., S. Kimura, H. Nakata and H. Yamada. 2006. Thermal adaptation of Pacific bluefin tuna *Thunnus orientalis* to temperate waters. Fish. Sci. 72: 149–156.

Kitagawa, T., Y. Kato, M.J. Miller, Y. Sasai, H. Sasaki and S. Kimura. 2010. The restricted spawning area and season of Pacific bluefin tuna facilitate use of nursery areas: A modeling approach to larval and juvenile dispersal processes. J. Exp. Mar. Biol. Ecol. 393: 23–31.

Sabarros, P.S., F. Ménard, J. Lévénez, E. Tew-Kai and J. Ternon. 2009. Mesoscale eddies influence distribution and aggregation patterns of micronekton in the Mozambique Channel. Mar. Ecol. Prog. Ser. 395: 101–107.

Satoh, K. 2010. Horizontal and vertical distribution of larvae of Pacific Bluefin tuna *Thunnus orientalis* in patches entrained in mesoscale eddies. Mar. Ecol. Prog. Ser. 404: 117–240.

Satoh, K., Y. Tanaka, M. Masujima, M. Okazaki, Y. Kato, H. Shono and K. Suzuki. 2013. Relationship between the growth and survival of larval Pacific bluefin tuna, *Thunnus orientalis*. Mar. Biol. 160: 691–702.

Tanaka, Y., M. Mohri and H. Yamada. 2007. Distribution, growth and hatch date of juvenile Pacific bluefin tuna *Thunnus orientalis* in the coastal area of the Sea of Japan. Fish. Sci. 73: 534–542.

Teo, S.L.H., A. Boustany, H. Dewar, M.J.W. Stokesbury, K.C. Weng, S. Beemer, A.C. Seitz, C.J. Farwell, E.D. Prince and B.A. Block. 2007a. Annual migrations, diving behavior, and thermal biology of Atlantic bluefin tuna, *Thunnus thynnus*, on their Gulf of Mexico breeding grounds. Mar. Biol. 151: 1–18.

Teo, S.L.H., A.M. Boustany and B.A. Block. 2007b. Oceanographic preference of Atlantic bluefin tuna, *Thunnus thynnus*, on their Gulf of Mexico breeding grounds. Mar. Biol. 152: 1105–1119.

CHAPTER 7

Movements and Habitat Use of Atlantic Bluefin Tuna

Steven L.H. Teo[1],* and Andre M. Boustany[2]

Introduction

Atlantic bluefin tuna share many biological characteristics with southern and Pacific bluefin tunas that have made them the focus of numerous scientific studies over the years: including large body size, hydrodynamic efficiency, endothermy, ocean basin scale migrations, and the ability to exploit a wide range of temperatures and habitats (e.g., Carey and Teal 1966; Mather et al. 1995; Block and Stevens 2001; Fromentin and Powers 2005). Their high economic value and poor stock status in recent years have added substantial impetus to bluefin tuna research. In this chapter, we review the biology of Atlantic bluefin tuna with particular emphasis on their movement and habitat use patterns. The terms 'bluefin' or 'bluefin tuna' are used for Atlantic bluefin tuna unless otherwise specified.

Both southern and Pacific bluefin tunas are thought to consist of a single stock but Atlantic bluefin tuna consists of at least two distinct stocks. The Atlantic bluefin tuna is the only bluefin with major spawning regions on opposite sides of an ocean basin (Gulf of Mexico/Caribbean and Mediterranean Sea) (Mather et al. 1995), and its stock structure is especially complex within the Mediterranean Sea. Based on recent genetic analyses, Atlantic bluefin tuna spawning in the Gulf of Mexico are comprised of a single stock but there are likely to be at least two stocks in the Mediterranean Sea (Carlsson et al. 2007; Boustany et al. 2008; Riccioni et al. 2010). Their relatively complex stock structure may be related to the movement and habitat use patterns described in this chapter. The movement patterns of Atlantic bluefin tuna became better known after electronic tagging experiments began in the 1990s (e.g., Block et al. 2005). In general, the western and eastern Atlantic stocks were mixed while on

---
[1] Southwest Fisheries Science Center, National Oceanic and Atmospheric, Administration, 8901 La Jolla Shores Drive, La Jolla, California 92037-1508, USA.
[2] Nicholas School of the Environment, Duke University, Box 90328 Durham NC 27708, USA.
* Email: steve.teo@noaa.gov

their foraging grounds in the North Atlantic but segregate when they return to their natal spawning grounds for reproduction (Fig. 7.1).

Management of Atlantic bluefin tuna is based on a two stock hypothesis. In the 2012 stock assessment by the International Commission for the Conservation of Atlantic Tunas (ICCAT) (ICCAT 2012), Atlantic bluefin tuna is assumed to consist of two stocks: a western Atlantic stock spawning in the Gulf of Mexico and an eastern Atlantic stock spawning in the Mediterranean Sea, with the stocks separated by the 45°W meridian for management purposes. Although there are likely to be at least two stocks in the Mediterranean Sea, it is assumed for assessment and management purposes that there is only a single bluefin stock in the eastern Atlantic and Mediterranean Sea. One of the major issues in the assessment of the eastern Atlantic bluefin tuna stock was the substantial under reporting of catches in the Mediterranean Sea between 1998 and 2007 (ICCAT 2012). Nevertheless, assessment results indicate an improvement in the stock status compared with previous assessments. As catches and fishing mortality on this stock have declined substantially in recent years, spawning stock biomass estimates showed a concomitant increase, even though the speed and magnitude of this increase remained uncertain. Overfishing on the eastern Atlantic stock appeared to have ended ($F_{2011} < F_{0.1}$) but the spawning stock biomass remained in an overfished condition ($SSB_{2011} < SSB_{0.1}$).

The western Atlantic bluefin tuna stock also showed some evidence of an improving status but was highly dependent on model assumptions of the spawner-recruit relationship (ICCAT 2012). Interpretation of assessment results were also complicated by the migration of eastern Atlantic bluefin to western Atlantic foraging grounds. After the rebuilding plan was adopted in 1998, the spawning stock biomass of the western stock increased by about 19% and fishing mortality on adult fish (age-9+) declined after 2003. The 2003 year class of the western stock appeared to be the largest year class since 1974, and this year class was expected to start contributing to the spawning stock biomass in 2012. However, there was still some uncertainty about whether the strength of this year class in the western Atlantic was driven primarily by increased recruitment from the Gulf of Mexico or increased migration across the Atlantic by eastern Atlantic bluefin (ICCAT 2013). There was also uncertainty about the recruitment dynamics of western Atlantic bluefin. If a low potential recruitment scenario was assumed, the western stock was found not to be experiencing overfishing ($F_{2008-2010} < F_{MSY}$) and was not in an overfished condition ($SSB_{2011} > SSB_{MSY}$). However, if a high potential recruitment scenario was assumed, the western stock would have been experiencing overfishing ($F_{2008-2010} > F_{MSY}$) and in an overfished condition ($SSB_{2011} < SSB_{MSY}$).

There have been several reviews in the past, detailing various aspects of Atlantic bluefin tuna biology. Mather et al. (1995) provided a highly detailed review of its life history, stock structure, historical fisheries, distribution, and migration patterns, based primarily on information prior to the 1980s, and is especially valuable for historical information that may now be hard to find. The U.S. National Research Council (Magnuson et al. 1994) reviewed the biology and stock status of Atlantic bluefin tuna and highlighted important avenues of research for this species, such as increasing the use of electronic tags to better understand stock structure and trans-Atlantic migrations.

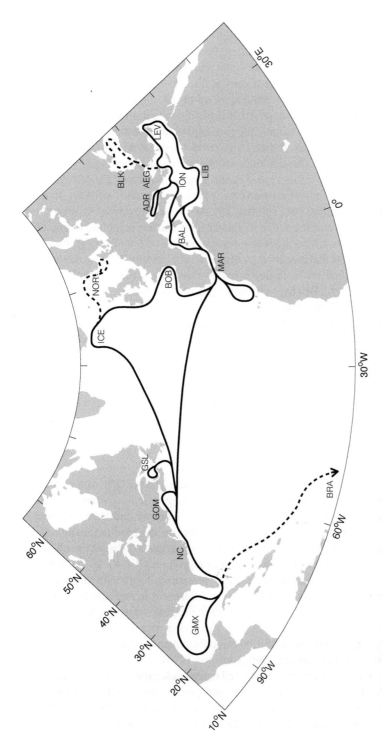

Figure 7.1. Stylized migratory pathways for bluefin tuna in the Atlantic Ocean. Areas mentioned in the text are indicated: GMX: Gulf of Mexico; NC: North Carolina; GOM: Gulf of Maine; GSL: Gulf of St. Lawrence; BRA: Brazil; ICE: Iceland; NOR: North Sea/Norwegian Sea; BOB: Bay of Biscay; MAR: Morocco; BAL: Balearic Sea; ION: Ionian Sea; LIB: Libya; ADR: Adriatic Sea; AEG: Aegean Sea; BLK: Black Sea; LEV: Levantine Sea.

Since the late 1990s, the increased use of electronic tags has vastly improved our understanding of the movements and habitat use of this species. In the 2000s, two reviews were published, with Fromentin and Powers (2005) stressing the connection of bluefin biology to current stock assessment and management issues, while Rooker et al. (2007) concentrated on the life history and stock structure.

Our aim for this chapter is to provide an up-to-date understanding of the main movement and habitat use patterns of Atlantic bluefin tuna during various life history stages, focusing largely on electronic tagging experiments, and the influence of the environment on these patterns. We will also review some aspects of the life history when it is pertinent to the discussion (e.g., stock structure, median age of maturity). It should be noted that we are only reporting the general movement patterns described by various studies and individual movement patterns can and will vary from the reported patterns, sometimes even substantially. An in-depth review of Atlantic bluefin tuna life history is beyond the scope of this chapter and we recommend one of the abovementioned reviews for a more thorough treatment of life history if readers are so inclined. We assume that readers are familiar with electronic tagging methods and refer inexperienced readers to in-depth reviews elsewhere (e.g., Gunn and Block 2001; Sippel et al. 2014). Briefly, two main types of electronic tags are often used to elucidate large-scale bluefin movements: (1) implantable archival tags (or archival tags); and (2) popup satellite archival tags (or popup tags). Archival tags are typically implanted into the bluefin's peritoneal cavity, and programmed to record and store data (e.g., ambient light, water temperature, pressure) until the tag is recovered upon recapture of the fish (i.e., fishery-dependent). Popup tags are similar to archival tags but are instead attached externally, so that the tag is able to detach from the fish at a pre-programmed date, float or 'popup' to the surface, and transmit a summary of recorded data to satellites without recapturing the fish (i.e., fishery-independent).

Stock Structure

Early evidence for multiple stocks of bluefin tuna in the Atlantic Ocean came from the large geographic discontinuity between known spawning regions, seasonal distribution of catches, and apparent distinctness in the age of maturity and spawning season. Atlantic bluefin tuna stock assessments use age-four in the Mediterranean Sea and age-nine in the Gulf of Mexico as age of maturity, suggesting distinct evolutionary strategies of bluefin tuna from the two spawning grounds (ICCAT 2012). Although recent studies have suggested revising the age of maturity for bluefin tuna from both spawning grounds (see below), it is likely that differences in reproductive strategies exist in both regions. There are also differences in the peak spawning season between the Gulf of Mexico and Mediterranean Sea. Spawning occurs in the Mediterranean from May to August and in the Gulf of Mexico from April to early July (Dicenta and Piccinetti 1980; Cort and Loirzou 1990; Richards 1990).

A number of researchers have attempted to use molecular tools to examine stock structure. Early research using molecular and protein markers failed to find any differentiation between bluefin from the eastern and western Atlantic (Edmonds and Sammons 1973; Thompson and Contin 1980; Alvarado-Bremer et al. 1999; Pujolar and

Pla 2000; Ely et al. 2002). However, these studies were limited in that they sampled relatively few fish, used slowly evolving genetic markers and/or sampled fish from feeding grounds as opposed to spawning grounds, where populations would be more likely to segregate. Subsequent genetic studies using more variable genetic markers such as microsatellites (Broughton and Gold 1997; Carlsson et al. 2004; 2007; Riccioni et al. 2010) and mitochondrial D-loop region (Carlsson et al. 2007; Boustany et al. 2008) have discerned population subdivision within Atlantic bluefin. Examination of genetic diversity suggested a relatively recent founding of bluefin in the Atlantic Ocean (<200,000 years before present), supporting the evidence of strong natal homing as genetic differentiation would not be possible over such a short time frame if even a minimal amount of gene flow was occurring among populations (Carlsson et al. 2007; Boustany et al. 2008). Both these studies and Carlsson et al. (2004) were also able to observe population subdivision within the Mediterranean Sea, with samples from the western basin of the Mediterranean (Balearic Islands, Corsica, Tyrrhenian Sea) being genetically distinct from those in the central basin when looking at young of the year fish (Adriatic Sea, Ionian Sea). All of these studies used relatively small sample sizes, so samples within each of the basins were pooled among years and individual sample sites, preventing the examination of finer scale population structure. Using larger sample sizes, Riccioni et al. (2010) were able to discern up to five genetically distinct stocks of bluefin in the western and central Mediterranean. Historical samples supported the hypothesis that these stock boundaries have been stable over time, and the notion that there is strong natal homing to specific spawning grounds in the Mediterranean Sea (Riccioni et al. 2010). The only study that examined the genetics of fish in the eastern Mediterranean basin was not able to discern eastern (Turkey) from western basin (Balearics) samples, despite using similar genetic markers as previous studies that observed differentiation between western and central basin populations (Viñas et al. 2011). This raises the possibility of a mosaic population structure, a pattern observed in other fish species within the Mediterranean Sea (De Innocentiis et al. 2004; Magoulas et al. 2006).

Recent studies using more advanced genetic markers such as Single Nucleotide Polymorphisms (SNPs) have been better able to discern population subdivision between western Atlantic and eastern Atlantic/Mediterranean Sea samples and hold great promise in their applicability for delineating population structure within the Mediterranean Sea (Albaina et al. 2013). Preliminary studies carried out under the ICCAT Atlantic-wide research program for bluefin tuna (Grande Bluefin Tuna Year Programme; GBYP) using SNPs have tentatively delineated at least three spawning populations (Gulf of Mexico, western Mediterranean, eastern Mediterranean), although the borders among these regions remain unresolved (ICCAT 2013). Understanding the population structure of bluefin within the Mediterranean Sea is of primary importance as they are currently assessed as one population (ICCAT 2012). Cryptic population sub-structuring raises the possibility of serial depletion of populations while maintaining consistently high catch rates or fishing out of smaller populations whose declines remain unobserved due to strong catches in the larger populations.

In general, results from genetic analyses have been consistent with those employing other techniques examining population structure. Using microconstituent analysis of otoliths, it has been possible to observe a high degree of natal homing (>90%) in bluefin tuna, something that would be necessary before genetic differentiation would arise (Rooker et al. 2008b; Secor et al. 2013). Electronic tagging studies have shown segregation to separate spawning sites for fish tagged in the same area, also indicative of natal homing (Block et al. 2001; 2005; Boustany et al. 2008). Importantly, no tagged fish has been observed to travel to both the Gulf of Mexico and Mediterranean Sea, providing contemporary evidence of movement patterns that would allow historical genetic differentiation to persist (Block et al. 2005; Boustany et al. 2008; Galuardi et al. 2010). All three technologies also suggest high levels of mixing among populations in the North Atlantic, with fish sampled or tagged off the East Coast of the US and Canada representing a mixture of Gulf of Mexico and Mediterranean Sea fish (Block et al. 2005; Rooker et al. 2008b; Boustany et al. 2008; Secor et al. 2013). Similarly, regions of the central Atlantic and south of Iceland appear to contain fish from multiple populations based on genetic heterogeneity of samples collected and in tracks of tagged bluefin that visited these regions and subsequently traveled to one of the spawning regions (Block et al. 2005; Carlsson et al. 2006). There is therefore no clear separation of the bluefin stocks across the 45°W meridian, as assumed for assessment and management purposes. All of these results argue for new stock assessment methodologies that can better incorporate data on complex stock structure and mixing patterns among populations on feeding grounds (Taylor et al. 2011).

Juvenile Movements

Juvenile Atlantic bluefin tuna can be broadly divided into four main groups based on distinct stock-origin and/or major movement patterns: (1) Mediterranean-resident; (2) non-residential eastern Atlantic; (3) trans-Atlantic; and (4) western Atlantic. Movements of age-0 bluefin are not well documented but they are thought to stay in their nursery grounds for several months after hatching (Mather et al. 1995). Subsequently, a portion of each age-0 cohort leaves the Mediterranean Sea during autumn, so we named these fish 'non-resident eastern Atlantic bluefin'. However, other fish remain in the Mediterranean for much, if not all, of their life history and are therefore named 'Mediterranean-resident bluefin'. It is currently unclear if Mediterranean-resident fish are a separate stock from non-resident eastern Atlantic fish because there are as yet no published studies on the genetics of bluefin exhibiting different movement patterns. In contrast, there is thought to be only a single stock of bluefin tuna spawning in the western Atlantic. However, age-1+ juveniles in the western Atlantic are a mixture of western and eastern Atlantic stocks because juveniles from the eastern Atlantic perform trans-Atlantic movements (i.e., trans-Atlantic bluefin) and mix with the juveniles from the western Atlantic on their feeding grounds (Rooker et al. 2007). Some western Atlantic bluefin are also expected to migrate across the Atlantic but most of the research on trans-Atlantic fish have focused on eastern Atlantic juvenile bluefin on western Atlantic foraging grounds due to the much larger eastern Atlantic population size.

Mediterranean-resident bluefin tuna

Early work in the 20th century suggested that Mediterranean-resident bluefin juveniles undertook relatively limited horizontal movements. These juveniles were thought to stay within or close to their nursery grounds for up to several years, until they approach maturity (Sara 1973; Mather et al. 1995). After hatching, rapidly growing age-0 bluefin were found in large schools along most of the Mediterranean coast from August through November. Observations of these age-0 fish were limited in winter due largely to bad weather limiting fishery operations, but some small-scale tagging experiments indicate that these fish tended to be in the same or nearby areas when fisheries began operations again in late winter and early spring (Sara 1973). Their movements appeared to be relatively localized but were thought to increase in scale as they grew and approached maturity.

More recently, electronic tagging experiments in the western and central Mediterranean have shown that juvenile bluefin exhibit slightly larger-scale horizontal movements than previously known but these experiments have tended to be on age-2+ fish. For example, Yamashita and Miyabe (2001) tagged 60 juvenile bluefin between 70–90 cm fork length with archival tags in the Adriatic Sea. Seven archival tags were recovered between one to seven months after release. These fish spent the majority of time (33–55%) at or near the surface, with the main peak at 0–10 m and a secondary peak at approximately 50 m (Yamashita and Miyabe 2001). However, deep dives (>510 m) were also recorded on occasion, which indicated that the fish were exiting the Adriatic Sea because depths >500 m are found only at the mouth of the Adriatic. Although geolocation algorithms from that period had large errors and movement tracks were not reported for these fish, Yamashita and Miyabe (2001) interpreted their movements to be relatively limited. However, Japanese scientists subsequently reported the recovery of an archival tag that showed bluefin moving from the Adriatic to the Ionian and Aegean Sea before being recaptured off the Libyan coast (FAO 2005). Juvenile bluefin tuna (13.5–18 kg) tagged in the northwestern Mediterranean (near Roses, Spain) similarly exhibited movements on the scale of hundreds of kilometers and remained within the western Mediterranean (Cermeño et al. 2012). A juvenile fish from the same experiment was recaptured 963 days after being released with an implanted archival tag, and stayed along the North African coast from Morocco to Tunisia for more than two years (see Fig. 3 from Cermeno et al. 2012).

A well-designed electronic tagging experiment requires a large spatial and temporal scale in order to quantify the movement patterns and rates of bluefin tuna between different areas (Sippel et al. 2014). The ICCAT-GBYP program was developed in part to address this issue for many tagging experiments, especially in the eastern Atlantic (http://www.iccat.int/GBYP/en/index.htm). A coordinated tagging experiment for juveniles in the eastern Atlantic (Bay of Biscay) and the western and central Mediterranean Sea (Strait of Gibraltar, Balearic Sea, and Tyrrhenian Sea) was conducted in 2012 as part of GBYP. Although preliminary results have been reported for juveniles in the Bay of Biscay, no results for Mediterranean-resident bluefin juveniles are as yet available (Goni et al. 2012) (http://www.iccat.int/GBYP/en/tagging.htm). More well-designed tagging experiments in coordination with fisheries observations and aerial survey activities (http://www.iccat.int/GBYP/en/asurvey.htm), will be essential for quantifying movement rates between areas. In particular, very little information is available on juveniles in the eastern Mediterranean Sea. Bluefin tuna are known to spawn in the eastern Mediterranean, with gravid bluefin adults (Karakulak et al. 2004) and bluefin larvae (Oray and Karakulak 2005) being collected from the Levantine Sea. However, there have been no published studies on juvenile movements from the area to date.

Non-resident eastern Atlantic bluefin tuna

Movements are better known for non-resident eastern Atlantic juveniles because there are several important fisheries targeting these fish in the eastern Atlantic. The majority of these fisheries occur on the Atlantic coast of Spain, Portugal, and Morocco. An unknown proportion of age-0 bluefin was hypothesized by Mather et al. (1995) to leave the Mediterranean Sea during the autumn of their first year (about 1 kg) and move to a nursery area off the Atlantic Moroccan coast. This was supported by several lines of evidence: (1) age-0 fish about 1 kg tended to disappear from the Mediterranean coast of Spain around October; (2) age-0 fish around 1.5 kg regularly appeared off the Atlantic coast of Morocco around the second half of November; and (3) large catches of age-0 fish by traps and purse seine fisheries along the western end of the Mediterranean Sea in autumn (Mather et al. 1995). These age-0 bluefin are thought to spend the remainder of their first year off the Atlantic coast of Morocco, where they were caught by hook and line, live bait, and seine fisheries (Mather et al. 1995).

At about age-one, an unknown proportion of these fish in the nursery area off Morocco migrate from the Ibero-Moroccan Bay in spring, moving along the Portuguese coast, to the Bay of Biscay in summer. Conventional tagging data from the 1970s showed that small bluefin tagged off the Atlantic Moroccan coast were recaptured off Portugal and the Bay of Biscay (Mather et al. 1995). After foraging in the Bay of Biscay during summer, these age-one fish likely moved to their wintering grounds, which are largely unknown but have been speculated to range between the Canary Islands and the Moroccan Atlantic coast, or even further south (Mather et al. 1995; Rodriguez-Marin et al. 2005). The lack of information about the wintering grounds was due to a lack of tag recaptures during winter, which may in turn be due to limited fisheries or juveniles being more spread out during this season. Most of these fish appear to return to the Bay of Biscay during subsequent summer for several years. The Bay of Biscay fishery predominantly catches age-one and two fish but smaller numbers of fish up to age-six are caught in the Bay of Biscay (Rodriguez-Marin et al. 2009). As these fish mature, a large portion of the bluefin tuna re-enter the Mediterranean, likely for spawning. In the past, when bluefin were caught in large numbers by Nordic fisheries, several fish tagged in the Atlantic side of the Spanish coast were also recaptured off the Norwegian coast (see Fig. 72 in Mather et al. 1995).

The Bay of Biscay is the most important recapture area for bluefin tagged in other eastern Atlantic and Mediterranean areas (Table 7.1) (Rodriguez-Marin et al. 2008). Bluefin at liberty for >1 year that were tagged in the Alboran Sea and Straits of Gibraltar, Balearic Islands, and the North Atlantic coast of Morocco were predominantly recaptured (40–83% of recaptures) in the Bay of Biscay. This area may

Table 7.1. Relative recapture percentage ($\frac{recovered}{tagged} \times 100$) by tagging and recapture area for **(A)** <1 year and **(B)** >1 year at liberty. Numbers in parentheses represent percentage with respect to total recaptured bluefin by tagging area. Reproduced with permission from Rodriguez-Marin et al. 2008.

A.			Recapture Area		
Tagging Area	Number of bluefin tagged	Bay of Biscay	Mediterranean Sea	South Atlantic Iberian Peninsula	West Atlantic
Bay of Biscay	7648	3.9 (99)		0.01 (0.3)	0.03 (0.7)
Alboran Sea & Gibraltar Strait	4699	0.3 (14)	1.6 (83)	0.06 (3)	
Balearics Islands Area	2418	0.2 (4)	4.2 (96)		
North Atlantic Coast of Morocco	195			7.7 (100)	
B.			Recapture Area		
Tagging Area	Number of bluefin tagged	Bay of Biscay	Mediterranean Sea	South Atlantic Iberian Peninsula	West Atlantic
Bay of Biscay	7648	1.5 (78)	0.1 (6)	0.1 (6)	0.2 (10)
Alboran Sea & Gibraltar Strait	4699	0.4 (81)	0.09 (19)		
Balearics Islands Area	2418	0.08 (40)	0.04 (20)	0.08 (40)	
North Atlantic Coast of Morocco	195	2.6 (83)		0.5 (17)	

be critical foraging habitat or that reporting and fishing mortality rates are higher. However, the proportion of each cohort that moves out of the Mediterranean Sea and feed in the Bay of Biscay is largely unknown. In recent years, electronic tags have been deployed on age-one to three bluefin in the Bay of Biscay during 2004–2009 (Table 7.2), and more recently as part of the GBYP (Goni et al. 2012). Since age-one fish are not able to carry available popup tags due to their small body size, age-one fish were tagged with archival tags, which are fishery-dependent and will therefore require some time before enough fish are recaptured to answer the research questions. To the best of our knowledge, results from recovered archival tags from age-one juveniles have not yet been reported. In contrast, seven popup tags deployed on age-three fish have successfully reported their data (including one physically recovered) (Goni et al. 2012). Due to the short time between deployment and popoff (11 to 63 days), all but one fish remained within the Bay of Biscay during that period while the remaining fish moved to the southwest tip of the U.K. but returned to the Bay of Biscay by the

Table 7.2. Deployments and recoveries from electronic tag experiments performed by AZTI-Tecnalia from 2004 to 2009 (Pers. comm., Igor Arregi, AZTI-Tecnalia).

Years	Tag type	Number of tags	Mean and range of fork length (cm)	Fishing gear	Cruise type	Number recovered
2004	Dummy	125	60.6 (55–67)	Bait boat	Opportunistic	4
2007	Archival	87	64.9 (62–70)	Bait boat	Directed	1
2005–2009	Archival	49	73.3 (60–107)	Bait boat & rod-and-reel	Opportunistic	2

end of the track. The GBYP tagging program has also deployed archival and popup tags on age-two and three bluefin in the Strait of Gibraltar but these results have not yet been made public.

There is currently little information on the vertical movements of juvenile bluefin in the eastern Atlantic because results from electronic tagging experiments on these fish have not yet been reported. Diving behavior of juveniles in the eastern Atlantic are expected to be similar to other juveniles. Juvenile bluefin generally spend the majority of time at or near the surface with diving behavior thought to be associated with diurnal changes, forage, and water temperature (e.g., Kitagawa et al. 2007; Galuardi and Lutcavage 2012). In the Bay of Biscay, juvenile bluefin are likely to exhibit similar behavior because they have been shown to feed primarily on epipelagic prey, with anchovies and krill being major prey groups when abundant, and are caught by surface fishing gear like bait boats (Logan et al. 2010; Goni et al. 2012).

Trans-Atlantic bluefin tuna

One important advance in recent years has been the recognition that a large proportion of juveniles in the western Atlantic and caught by North American fisheries (e.g., sports fishery, purse seine) were migrants from the eastern Atlantic with natal origins in the Mediterranean Sea. There has long been evidence suggesting a close link between age-one fish from the Atlantic coast of Spain and Morocco, with juveniles off the North American coast. For example, conventional tagging data since the 1970s have shown young bluefin tagged in the Bay of Biscay being recaptured by western Atlantic fisheries (Aloncle 1973; Mather et al. 1995; Rooker et al. 2007; Rodriguez-Marin et al. 2008). A cluster analysis of catch trends and size of fish caught by bluefin fisheries from both sides of the Atlantic also show a correspondence between the Bay of Biscay and western Atlantic fisheries (Fromentin 2009). However, in recent years, evidence from stable isotopes, genetics, pollutant tracers, and electronic tags have provided conclusive evidence that a large portion (in some years >50%) of juvenile bluefin in western Atlantic waters are of eastern Atlantic origin.

The earliest such evidence for the close link between eastern and western Atlantic bluefin fisheries possibly came from Aloncle (1973), who reported that out of 34 fish tagged in the Bay of Biscay (likely age-one bluefin), two fish were recaptured in U.S. coastal waters (Cape Cod and New York Bight) while three were recaptured in the eastern Atlantic. Bluefin tagging experiments in the Bay of Biscay from 1977 to 2008 showed that approximately 10% of the tag recoveries with time-at-liberty >1 year from juvenile bluefin (mostly age-one and two) were recovered from the western Atlantic (Table 7.1) (Rodriguez-Marin et al. 2008). Interestingly, fish tagged in other areas of the eastern Atlantic (Gibraltar Strait and Alboran Sea; Balearic Islands; and the North Atlantic coast of Morocco) were not recaptured in the western Atlantic (Rooker et al. 2007; Rodriguez-Marin et al. 2008).

Links between juvenile bluefin fisheries in the northwest and northeast Atlantic Ocean are also clear from the closely related catch trends and size of fish caught in these two areas. The catch and catch-at-size data from 18 fisheries in the North Atlantic and Mediterranean Sea were collated and grouped into 10 regional time series by Fromentin (2009). Where data were sufficient, the year-class contributions to catches were calculated for each fishery and compared using a cluster analysis. The analysis clearly showed that the US purse seine fishery in the Northwest Atlantic was most closely related to the bait boat fishery in the Bay of Biscay (Fig. 7.2). Although this

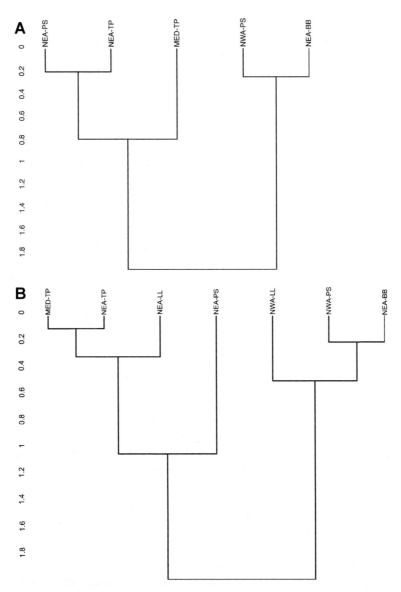

Figure 7.2. (A) Tree diagram of the cluster (dendrogram) analysis performed on the distance matrix between the total year-classes contribution to the catches of five fisheries for which time series are available from 1966 to 1979. (B) Same for the seven fisheries for which time series are available from 1971 to 1981 (MED-PS was not included because of missing catch-at-size in 1980). Acronyms: MED-TP: Mediterranean Trap; NEA-BB: Northeast Atlantic Bait boat; NEA-TP: Northeast Atlantic Trap; NEA-LL: Northeast Atlantic Longline; NEA-PS: Northeast Atlantic Purse Seine; NWA-LL: Northwest Atlantic Longline; NWA-PS: Northwest Atlantic Purse Seine. Reproduced with permission from Fromentin 2009.

result is suggestive of the important contribution of juveniles from the Bay of Biscay to U.S. juvenile bluefin fisheries, conclusive evidence of this came from recent analysis of otolith microchemistry, organo-pollutants, and electronic tagging.

In our opinion, the strongest evidence comes from stable isotope analysis of bluefin otoliths sampled from multiple age-classes from both sides of the Atlantic (Rooker et al. 2008a,b). Differences in the chemical composition of bluefin vertebrae from the western and eastern Atlantic were first demonstrated by Calaprice (1986), and were also confirmed in otolith samples by Secor and Zdanowicz (1998), and Rooker et al. (2003). Differences in the $\delta^{18}O$ and $\delta^{13}C$ values of otoliths from yearling bluefin were successfully used by Rooker et al. (2008a) to segregate the fish into their respective nursery areas in the eastern Atlantic/Mediterranean Sea and the mid-Atlantic Bight on the East Coast of the USA. Although the $\delta^{13}C$ values from both eastern and western Atlantic nursery areas overlapped substantially and lacked discriminatory power, the $\delta^{18}O$ values in the eastern samples were enriched by approximately 0.77‰ relative to the western samples (Rooker et al. 2008a).

Although Atlantic bluefin tuna were thought to migrate across the Atlantic at age-1+, yearling otolith samples from the mid-Atlantic Bight in 2001 surprisingly showed an eastern nursery signal (Rooker et al. 2008a). The physico-chemical conditions of the nursery areas in the western Atlantic in 2001 were similar to other years in the study, and the $\delta^{18}O$ values for 2001 were unlikely to be linked to temperature and salinity variability in these western nursery areas. Pacific bluefin tuna commonly migrate across the Pacific as age-0 and one fish from the western Pacific to the US and Mexican Pacific coast (Bayliff 1994), which therefore suggests that age-0 and one fish are able to complete the trans-Atlantic migration. It is therefore likely that the yearlings sampled in the western Atlantic in 2001 were spawned in the Mediterranean Sea and migrated to the western Atlantic as yearlings.

Rooker et al. (2008b) further determined the $\delta^{18}O$ and $\delta^{13}C$ concentrations from the first year of life of school-sized (<60 kg, <age 5), medium (60–140 kg, age five– nine), and giant (>140 kg, age 10+) bluefin tuna from spawning (Gulf of Mexico and Mediterranean Sea) and foraging areas (Gulf of St. Lawrence, Gulf of Maine, and Mid-Atlantic Bight), and compared them to baseline values from yearlings in known nursery areas. From 1997 to 2000, less than half of the school-sized fish in the mid-Atlantic Bight were from nursery areas in the western Atlantic (school-sized: 42.6 ± 7.2%), which means that trans-Atlantic juvenile migrants, likely non-resident eastern Atlantic bluefin from the Bay of Biscay, play an important role in supporting the juvenile bluefin fisheries in the western Atlantic (Fig. 7.3).

An analysis of organochlorine tracers in Atlantic bluefin tuna have also demonstrated the high proportion of eastern-origin juvenile fish in the northwestern Atlantic (Dickhut et al. 2009). Marine samples from the Mediterranean Sea had relatively low levels of chlordane compounds (cis-chlordane, trans-chlordane, cis-nonachlor, and trans-nonachlor) relative to polychlorinated biphenyls (PCBs) as compared to samples from the western North Atlantic (e.g., McKenzie et al. 1999; Loganathan et al. 2001; Deshpande et al. 2002; Stefanelli et al. 2002). Dickhut et al. (2009) therefore used chlordane/PCB ratios as chemical tags for bluefin tuna feeding in the Mediterranean Sea versus the northwest Atlantic. Muscle tissue samples were collected from bluefin of various age classes in the U.S. mid-Atlantic Bight

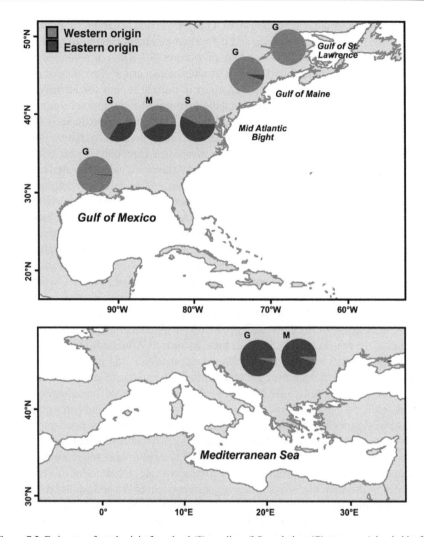

Figure 7.3. Estimates of natal origin for school (S), medium (M), and giant (G) category Atlantic bluefin tuna from spawning areas (Mediterranean Sea, Gulf of Mexico) and foraging areas (Gulf of St. Lawrence, Gulf of Maine, Mid Atlantic Bight). Contribution rates (%) were determined by comparing milled otolith cores (corresponds to yearling period) to a baseline sample from eastern and western nurseries (east + west = 100%). Assignment to either eastern or western nursery was based on maximum likelihood estimations. Standard deviations (SD) were expressed as percentages of estimated proportions. Size classes were approximated based on weight or age (actual or derived from length): giant (>140 kg, ≥ age 10 years), medium (60 to 140 kg, age five to nine years), and school (<60 kg, <age 5 years) category bluefin tuna. Percentage contribution of 'western population' and standard deviation (SD) around estimated proportion per region and size category: Gulf of Mexico [giant: 99.3% (SD 1.7%), n = 42]; Mediterranean Sea [giant: 4.2% (SD 3.1%), n = 94; medium: 4.2% (SD 4.4%), n = 38]; Gulf of St. Lawrence [giant:100% (SD 0.0%), n = 38]; Gulf of Maine [giant: 94.8% (SD 5.3%), n = 72]; Mid Atlantic Bight [giant: 64.9% (SD 21.9%), n = 12; medium: 55.7% (SD 10.4%), n = 56; school: 42.6% (SD 7.2%), n = 86]. Reproduced with permission from Rooker et al. 2008.

(young-of-year to large schooling fish), Gulf of Mexico (medium to giant), and the Tyrrhenian Sea between Sardinia and the Italian mainland (young-of-year to giant). Similar to Rooker et al. (2008b), a high proportion (33–83%) of juveniles in the US mid-Atlantic Bight were found to be of Mediterranean origin. One note of caution in using chemical or stable isotopic signatures in muscle tissues for identifying stock origin is that the signal is subject to metabolic turnover and dilution through growth. The turnover time for the trans-nonachlor/PCB153 ratio was approximately 10 months to 1.6 years depending on the age of fish arriving in the northwest Atlantic (Dickhut et al. 2009). Interestingly, 13% (5 out of 38) of samples from fish >1 year of age in the Mediterranean Sea clearly showed a northwest Atlantic signal of elevated nonachlor/PCB ratios. These were therefore likely to be trans-Atlantic bluefin that migrated to the Northwest Atlantic foraging grounds for a period of time before returning to natal spawning grounds in the Mediterranean Sea. Although nonachlor/PCB ratios alone cannot identify the origin of these fish, this interpretation is supported by the evidence from stable isotopes, genetics, and electronic tags that non-resident eastern Atlantic and western Atlantic bluefin mix on their foraging grounds in the North Atlantic but sort to their respective spawning grounds in the Gulf of Mexico/Caribbean and Mediterranean Sea.

Rooker et al. (2008b) demonstrated that as these trans-Atlantic fish feed and mature in the western North Atlantic, the bluefin begin to return to their natal spawning grounds in the Mediterranean Sea over time. In the mid-Atlantic Bight, where all three size classes of fish are found (schooling-sized, medium, and giant), the proportion of eastern-origin fish decreases with increasing size and age (Fig. 7.3). Using otolith stable isotope analysis, $42.6 \pm 7.2\%$ of schooling-sized fish were found to be of western origin, while $55.7 \pm 10.4\%$ and $64.9 \pm 21.9\%$ of medium and giant fish, respectively, were of western-origin. While other plausible hypotheses cannot be ruled out, the most likely reason for this pattern is that once these trans-Atlantic bluefin of eastern origin become mature and migrate back to their natal spawning area for reproduction, the fish tend to remain in the eastern Atlantic after spawning. Evidence of this behavioral tendency is also seen in electronic tagging studies of juveniles and adults on both sides of the Atlantic (Block et al. 2005; Aranda et al. 2013). Not surprisingly, the central North Atlantic region straddling the 45°W is an area with a mixture of both eastern and western-origin fish but the amount of mixing can vary substantially (Carlsson et al. 2006; Rooker et al. 2014). Although results from otolith microchemistry and organochlorine pollutants have clearly identified the mixing of eastern and western Atlantic stocks on their feeding grounds, these methods can only elucidate movements over large spatial scales.

Detailed movements of Atlantic bluefin tuna were elucidated through electronic tagging experiments beginning in the 1990s. Much of the initial work was performed by research programs based in the western Atlantic but a number of recent electronic tagging experiments were also conducted in the eastern Atlantic and Mediterranean Sea. Early electronic tagging studies focused on tagging bluefin in the western Atlantic (Lutcavage et al. 2000; Block et al. 2001) and were therefore expected to primarily involve western Atlantic bluefin tuna. However, it quickly became clear that a large proportion of the fish tagged in the western Atlantic were of eastern Atlantic origin (Fig. 7.4) (Block et al. 2005). Approximately 42% (26 out of 62) of the fish that

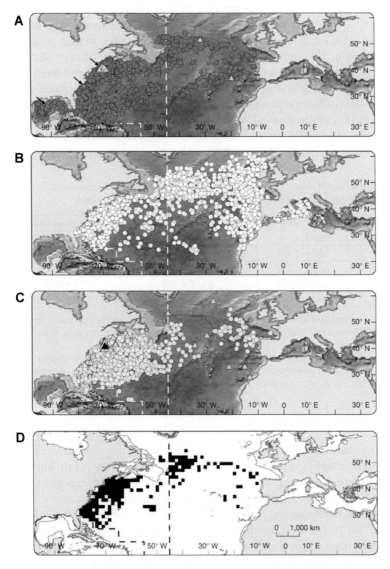

Figure 7.4. Positions of Atlantic bluefin tuna electronically tagged at three western Atlantic locations (arrows) during 1996–2004. Circles represent locations based on deployment positions, light-based and SST-based geolocation estimates, and PAT tag satellite endpoint positions. (A) Fish classified as western breeders (10 archival tags, 26 PAT tags, 219 ± 27 cm curved fork length (CFL) at release, median time at large 579 days). (B) Fish classified as potential eastern breeders (23 archival tags, 3 PAT tags, 207 ± 17 cm CFL at release, median time at large 926 days). (C) Fish that did not visit a known ICCAT breeding ground (53 archival tags, 215 PAT tags, 202 ± 16 cm CFL at release, median time at large 141 days). (D) Spatial overlap of western and eastern breeders identified in A and B. The dashed line in all panels indicates the current ICCAT management boundary (458W meridian) and western Atlantic breeding zone. Triangles represent recapture locations of electronically tagged fish; the black triangle denotes (n = 35) recaptures. Reproduced with permission from Block et al. 2005.

eventually ventured into a known spawning ground, migrated into the Mediterranean Sea (Table 7.3). If we assume that bluefin exhibit natal spawning migrations, this would mean that a large proportion (~40%) of the fish in the western Atlantic were of eastern origin. Although this hypothesis has now been supported by studies using otolith microchemistry (Rooker et al. 2008b) and organo-chlorine pollutants (Dickhut et al. 2009), these results were surprising at that time because Atlantic bluefin tuna stock assessments had assumed that the trans-Atlantic migration rate was relatively small and did not substantially affect the population dynamics of either stock. Instead, trans-Atlantic migration rates were found to be high enough that a large proportion of fish caught in the western Atlantic fisheries were likely of eastern origin.

Table 7.3. Probability of Atlantic bluefin tuna being located in the western management unit after being tagged in the western Atlantic.[a]Reproduced with permission from Block et al. 2005.

Days at large	Probability of fish being located in the western management unit[b]		
	Western (36)	Eastern (26)	Neutral (268)
1–180	0.994–0.982	0.933–0.900	0.996–0.984
181–360	0.962–0.934	0.606–0.542	0.951–0.927
361–720	0.887–0.844	0.347–0.288	0.954–0.927
>720	1.000–1.000	0.082–0.050	0.898–0.857

[a] Each fish was identified as a western spawner, eastern spawner or neutral fish, based on the criteria described in Block et al. 2005. Fish were released at three locations in the western Atlantic (Fig. 7.4, arrows).

[b] The probabilities shown are the 95% confidence intervals. The numbers of fish used to make these estimates are shown in parentheses. All geoposition data, inclusive of geolocation estimates based on light and SSTs, release and recapture points were used.

Western Atlantic bluefin tuna

Before their natal migrations back to their respective spawning grounds, the horizontal movements of juveniles in the western Atlantic are basically indistinguishable between those of western and eastern-origin (e.g., Block et al. 2005; Galuardi and Lutcavage 2012). The Tag-A-Tiny experiment is an electronic tagging experiment focusing on young juvenile bluefin (mostly ages two–five) in the western Atlantic (Galuardi and Lutcavage 2012). The DNA samples from these fish are currently being analyzed to identify their natal origin (Galuardi and Lutcavage 2012; Reeb et al. unpubl. data), which will allow direct comparisons of the movements of western and eastern origin fish. However, these analyses have not yet been completed and the movements of eastern versus western-origin juveniles cannot be directly compared unless the fish are tracked long enough for them to mature and return to their natal spawning grounds (e.g., Block et al. 2005) or are recaptured and their otoliths sampled for microconstituent analysis (a rare occurrence). Nevertheless, the currently available evidence can be examined for consistency with the hypothesis that both stocks show similar movements and habitat use in the western Atlantic. If the alternative hypothesis that trans-Atlantic fish behave very differently from western Atlantic fish, one would expect to observe two or more distinct patterns of juvenile movements in the western Atlantic. However, electronic tagging studies like Block et al. (2005), and Galuardi and Lutcavage (2012) have generally not shown this for juveniles in the western Atlantic.

Western Atlantic bluefin tuna are primarily spawned in the Gulf of Mexico between April and the end of June, and thought to leave the Gulf of Mexico after spending several months in nursery areas in the Gulf of Mexico, with the peak of migration through the Straits of Florida around July and August (Mather et al. 1995). Movements of these age-0 fish are largely unknown after passing through the Straits of Florida but age-0 fish have been observed by fishermen along the U.S. coast from Florida to Cape Hatteras (Mather et al. 1974). By summer and autumn, age-0 and one bluefin are caught by recreational fisheries off the coast of New Jersey and Massachusetts in summer and fall (Mather et al. 1995; Rooker et al. 2008a).

The spatial distribution of juveniles in the western Atlantic varied by season and was generally more coastal than adult bluefin (Block et al. 2005; Galuardi and Lutcavage 2012). In summer (July–September), juveniles were primarily distributed in the mid-Atlantic Bight and the Gulf of Maine, staying close to the shelf break and Gulf stream frontal regions (Fig. 7.5). As summer transitioned into autumn (October–December), the juveniles began to move southward and were primarily in the mid-Atlantic Bight, with the Cape Hatteras area being an area of higher bluefin concentration. In winter (January–March), juvenile fish continued their general movements southward along the North Carolina coast. Juvenile bluefin began their movements northward in spring (April–June), and were back in the mid-Atlantic Bight and Gulf of Maine by the end of June. These general movement patterns describe the core spatial distributions of juvenile fish but some individual fish do exhibit movements that differ from the general case. The spatial extent of juvenile distribution also varied greatly by season, with the largest 95% utilization distribution in February–April (>10,000 km²) and the smallest in July–October (<3000 km²) (Galuardi and Lutcavage 2012). Juvenile bluefin are substantially more coastally distributed than large adults (Fig. 7.6). Bluefin tuna <200 cm predominantly moved along the North American Atlantic coast but fish ≥200 cm were distributed over large regions of the North Atlantic including feeding aggregations in the central Atlantic (Walli et al. 2009) and the spawning grounds of the Gulf of Mexico and Mediterranean Sea (Teo et al. 2007b).

Like other juvenile bluefin tuna, western Atlantic juveniles spent the majority of the time at or near the surface but occasionally dove to several hundred meters (Galuardi and Lutcavage 2012). Interestingly, there were no detectable diel differences in mean depth throughout all four seasons but substantial vertical habitat compression towards the surface was apparent in the summer months and expansion during the winter months. The vertical habitat compression during summer appears to be associated with strong, shallow thermoclines in the Gulf of Maine and mid-Atlantic continental shelf waters during this period (Lawson et al. 2010; Galuardi and Lutcavage 2012). Changes in vertical and horizontal movements in relation to water masses occupied by the fish can perhaps be most easily seen by examining the detailed depth, water temperature, and geolocation data from individual archival tags (Fig. 7.7).

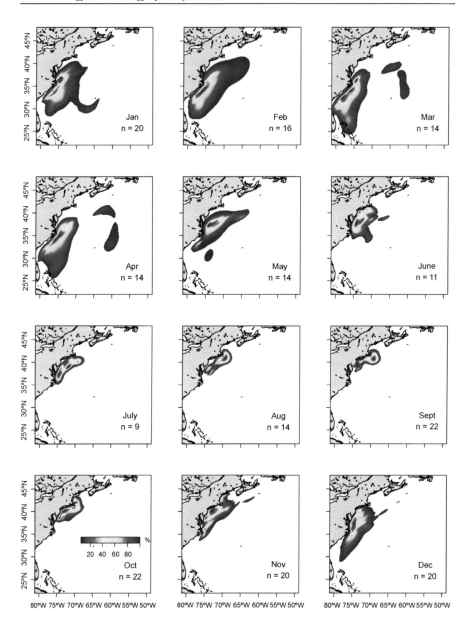

Figure 7.5. Utilization distributions aggregated for all juvenile Atlantic bluefin tuna tagged with popup tags for each month. Core-use areas are spatially constrained in summer months (July–Sept.) and are more dispersed in winter months (Jan.–March). Fall months show a southern migration along the shelf break and increase in spatial dispersal while spring months show the reverse trend. Reproduced with permission from Galuardi and Lutcavage 2012.

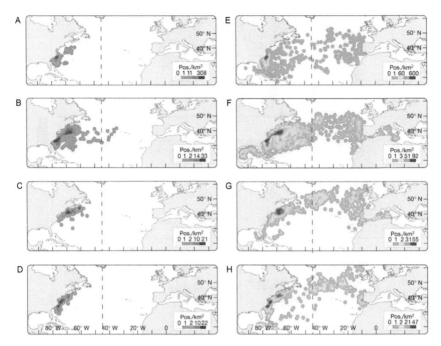

Figure 7.6. Seasonal distribution by size of Atlantic bluefin tuna that were tagged in the western Atlantic and measured before release. (A-D) Less than 200 cm CFL. (A) Winter; (B) spring; (C) summer; (D) autumn. (E-H) Greater than or equal to 200 cm CFL. (E) Winter; (F) spring; (G) summer; (H) autumn. The dashed line in each panel indicates the current ICCAT management boundary (45°W meridian). High kernel densities indicate seasonal hot spots where western-tagged Atlantic bluefin tuna spent the majority of time from 1996 to 2004. Only fish that were measured were used in this analysis. A western or eastern growth model was applied to obtain daily length after tagging. (A) n = 101, mean size at release 192 ± 9 cm CFL. (B) n = 56, 192 ± 6 cm CFL. (C) n = 22, 192 ± 7 cm CFL. (D) n = 13, 187 ± 8 cm CFL. (E) n = 162, 219 ± 14 cm CFL. (F) n = 167, 220 ± 13 cm CFL. (G) n = 97, 225 ± 15 cm CFL. (H) n = 49, 227 ± 15 cm CFL. Pos., positions. Reproduced with permission from Block et al. 2005.

Adult Movements

One of the key changes that occur when bluefin tuna mature into adults is the initiation of migrations between their foraging and spawning grounds. As detailed above, western Atlantic and non-resident eastern Atlantic bluefin overlap substantially on North Atlantic foraging grounds but subsequently sort to their respective spawning areas (Block et al. 2005; Rooker et al. 2008b; Boustany et al. 2008). After mature fish complete their spawning migrations, they tend to remain on the same half of the Atlantic Ocean, albeit with some exceptions. For example, after a non-residential eastern Atlantic bluefin has spawned in the Mediterranean and returned to the North Atlantic, it is less likely to forage along the North American coast (Table 7.3) (Block et al. 2005). The foraging grounds of larger and older fish also appear to expand into cooler waters in comparison with juveniles, after completing their spawning migration. For example, some giant adult bluefin of western-origin spawn in the Gulf of Mexico and then migrate up the North American coast and forage in the Gulf of St. Lawrence (Rooker et al. 2008b; Wilson et al. 2011).

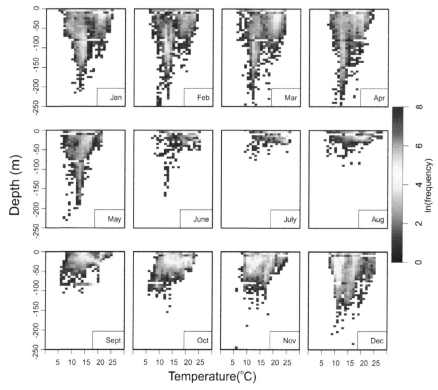

Figure 7.7. Recovered implanted archival tag showing external temperature (A, C, and E) and corresponding reconstructed migration (B, D and F) referenced by season. Colored bars in panels A, C and E correspond to the season color on the corresponding maps (panels B, D and F). Clear seasonal differences can be seen through the 30 months data were collected. When the fish inhabited well mixed water the depth patterns were more variable, with deep excursions and deeper mean depths. In summer months, in well stratified water, this fish had a more shallow depth distribution. Reproduced with permission from Galuardi and Lutcavage 2012.

Spawning migrations

Western Atlantic bluefin tuna primarily spawn in the Gulf of Mexico in April–June (Mather et al. 1995; Schaefer 2001) but larval bluefin have also been found in areas just outside the Gulf of Mexico, including the Florida Straits and Gulf Stream waters (McGowan and Richards 1989), and the western Caribbean (Muhling et al. 2011a). Western Atlantic bluefin tuna spawning in the Gulf of Mexico were likely spawned in the Gulf of Mexico (Carlsson et al. 2007; Rooker et al. 2008b; Boustany et al. 2008) and associated with feeding aggregations in North Carolina, mid-Atlantic Bight, and New England waters during winter and spring (Stokesbury et al. 2004; Teo et al. 2007b). These fish can migrate to their spawning grounds in the Gulf of Mexico from approximately December to June and exit the Gulf of Mexico by the end of June (Block et al. 2005; Teo et al. 2007b; Galuardi et al. 2010). However, migrations into the Gulf of Mexico and spawning are thought to peak in April–June (Mather et al. 1995; Schaefer 2001).

As adult bluefin migrate through the Florida Straits to their spawning grounds, they shift from a migratory entry phase with an emphasis on movement towards a spawning location, to a breeding phase with behaviors associated with courtship and spawning, and finally a migratory exit phase out of the Gulf of Mexico similar in many respects to the entry phase (Teo et al. 2007b). This is similar to behavior of other species of fish, which changes dramatically during courtship and spawning [e.g., Pacific bonito (Magnuson and Prescott 1966), striped parrotfish (Colin 1978), eyed flounder (Konstantinou and Shen 1995), and Pacific halibut (Seitz et al. 2005)]. The entry and exit phases of bluefin migrating into and out of the Gulf of Mexico are characterized by deep diving and rapid, directed movement paths (Figs. 7.8 and 7.9) (Teo et al. 2007b). Deep diving during the entry and exit phases most likely represents an attempt to thermoregulate or reduce the energetic cost of swimming against the Loop Current, which is very warm and rapidly advects out of the Gulf of Mexico. However, the reduced cost of locomotion from deep diving would only apply to the entry phase because it would likely be energetically cheaper to swim within the Loop

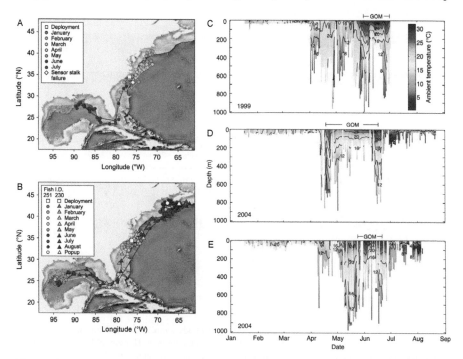

Figure 7.8. Geolocation tracks (A, B) and daily water temperature profiles (C, D, and E) of three Atlantic bluefin that were tagged off North Carolina and migrated to the Gulf of Mexico in the spawning season. (A and C) Bluefin 98–512 was tagged with an archival tag on January 17, 1999, and light and water temperature sensors failed on July 2, 1999. (B, D and E) Bluefins 03–251 (circles and D) and 03–230 (triangles and E) were both tagged with popup tags off North Carolina on January 16, 2004. Horizontal black lines over temperature profiles indicate periods in the Gulf of Mexico. Modified with permission from Teo et al. 2007b.

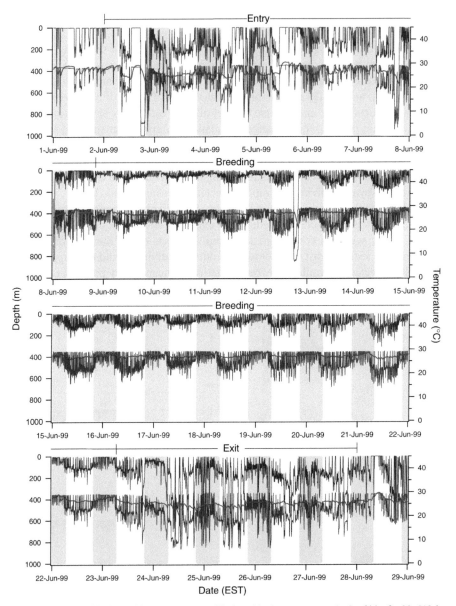

Figure 7.9. Depth (black), ambient temperature (blue) and body temperature (red) of bluefin 98–512 in June 1999 (inside Gulf of Mexico). Vertical gray bars indicate nighttime. Horizontal black lines indicate entry, breeding, and exit phases. Reproduced with permission from Teo et al. 2007b.

Current during the exit phase. Therefore, the deep diving in the exit phase suggests that this behavior is, at least in part, an attempt to avoid the warm surface waters and thermoregulate. Nevertheless, other explanations like foraging for prey or avoidance of predators near the surface are also plausible (Guinet et al. 2007).

The breeding phase is sandwiched between the entry and exit phases and can be identified from distinctive changes in diving behavior, horizontal movements, and thermal biology. During the breeding phase, the bluefin remain in relatively shallow depths and exhibit shallow, oscillatory dives at night (Fig. 7.9). The mean daily maximum depths during the breeding period (203 ± 76 m) are significantly shallower (P < 0.001, Tukey-Kramer) than the entry (568 ± 50 m) and exit phases (580 ± 144 m), which are not significantly different from each other (P > 0.05, Tukey-Kramer) (Teo et al. 2007b). At the same time, the fish are more residential during the breeding phase with significantly less linear tracks and smaller distances traveled per day (Teo et al. 2007b). These changes in horizontal movements are consistent with the fish exhibiting periods of rapid migration to and from the spawning areas but becoming more resident once they reach the spawning area (Prince et al. 2005).

Vertical movements recorded by the tags during the breeding phase are likely indicative of courtship and/or spawning behavior. The nighttime shallow diving behavior, accompanied by frequent visits to the surface, are in concordance with estimates of spawning times of other tuna species in the wild (McPherson 1991; Nikaido et al. 1991; Schaefer 1996; 1998) and captive populations of spawning Pacific bluefin tuna (S. Masuma, Japan Sea Farming Association, pers. comm.) and yellowfin tuna (Schaefer 2001). In addition, acoustic tracking experiments have also shown that Pacific bluefin tuna exhibit similar diving behavior on their breeding grounds in the East China Sea (NRIFSF 2002; Kitagawa et al. 2006). Changes in diving behavior during the breeding phase are accompanied by changes in their thermal biology (Teo et al. 2007b). Teo et al. (2007b) hypothesized that, at the high temperatures in the Gulf of Mexico, bluefin tuna may no longer need low heat transfer for heat conservation, and instead may be depending more on changes in diving behavior for thermoregulation.

Bluefin are thought to spawn extensively in various regions of the Mediterranean Sea but there are several documented major spawning grounds. Best known are spawning grounds around the Balearic Islands in the western Mediterranean, and the Tyrrhenian Sea and waters around Malta in the central Mediterranean Sea (e.g., Mather et al. 1995; Nishida et al. 1998; Heinisch et al. 2008). Bluefin eggs and larvae have also been found in the Levantine Sea in the eastern Mediterranean but substantially less work has been done on this spawning ground (Karakulak et al. 2004; Oray and Karakulak 2005). The spawning season in the Mediterranean Sea is slightly later than the Gulf of Mexico, ranging from May to July, and progresses from the eastern to western Mediterranean. Largely thought to be due to the seasonal east-west trend in warming sea surface temperatures, spawning begins in the eastern Mediterranean (Levantine Sea; May–June) before the central Mediterranean (Tyrrhenian Sea and Malta area; June), and the western Mediterranean (Balearic Islands; June–July) (Heinisch et al. 2008).

Spawning behavior of non-resident eastern Atlantic bluefin in the western Mediterranean Sea (Aranda et al. 2013) appear to be similar to fish spawning in the Gulf of Mexico (Teo et al. 2007b). During the spawning season in the Balearic Islands, bluefin exhibit non-directed movements in the spawning area, suggesting residency associated with spawning (Fig. 7.10). The fish also have shallow depth distributions and a distinctive diving behavior between midnight and sunrise during most nights when they are around the Balearic Islands. This distinctive diving behavior is characterized

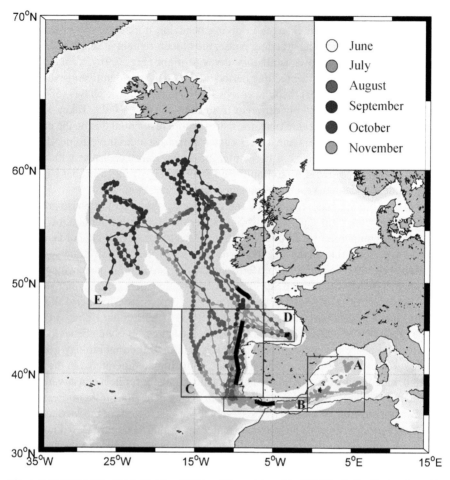

Figure 7.10. Estimated paths (with 50 and 95% confidence intervals) of 13 Atlantic bluefin tuna tagged in early June, 2009–2011 (≥45 days at liberty). Five successive regions throughout the migratory pathways between the western Mediterranean and the North Atlantic Ocean are distinguished (A-E, black boxes): Balearic area (A) Strait of Gibraltar (B) western Iberian coast (C) Bay of Biscay (D) and North Atlantic area (E) Bold black lines represent five-day coverage of tag #39 track in each of these regions. Reproduced with permission from Aranda et al. 2013.

by high-frequency shallow diving with the majority of the time spent at depths <40 m but punctuated with brief oscillatory dives below the mixed layer (Fig. 7.11), which appear similar to that exhibited by spawning bluefin in the Gulf of Mexico (Fig. 7.9). As the fish exit the Mediterranean Sea through the Strait of Gibraltar after spawning, they exhibit directed movements and deep diving behavior similar to the bluefin leaving the Gulf of Mexico (Aranda et al. 2013). A similar deep diving behavior was also observed for fish entering the Mediterranean Sea from the Atlantic Ocean (Wilson and Block 2009). Due to the similarities of diving behaviors between the bluefin spawning

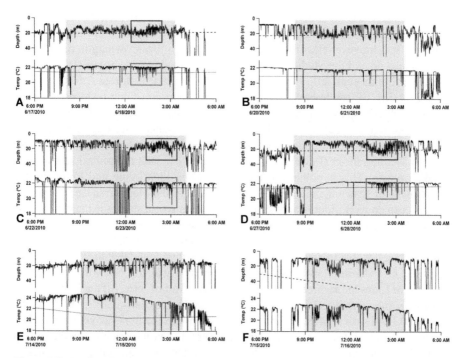

Figure 7.11. Depth and water temperature time series from tag #25 in the Balearic area. High-frequency shallow dives (HFSD profiles) were detected between midnight and sunrise (A, C, D); this pattern consisted of frequently repeated shallow dives (blue squares) below the bottom limit of the mixed layer (dashed line), which resulted in thermal oscillations (red squares) about the thermocline (dotted lines). Vertical profiles displaying deeper scattered dives and no clear oscillatory pattern (non-HFSD profiles) alternated with daily series of consecutive HFSD profiles (B); this pattern occurred several successive days following the last HFSD profile (E, F). Reproduced with permission from Aranda et al. 2013.

around the Balearic Islands and in the Gulf of Mexico, similar physiological and/or ecological drivers, like thermoregulation and predator avoidance (e.g., orcas) (Guinet et al. 2007), are likely to be influencing these diving behaviors.

Aranda et al. (2013) managed to capture video of the courtship and spawning behaviors during their tagging work (see video S3 in the supplementary information of Aranda et al. 2013). Based on the eyewitness accounts by Aranda et al. (2013), "courtship started with grouping of individuals near the sea surface and a few males closely pursuing a female. The fish then took on a darker background colour that enhanced their striped pattern. Eventually, as the spawners released several series of gametes, they shook the caudal fin strenuously to spread and mix them in the water to facilitate fertilisation".

Electronic tagging studies have been conducted in the central and eastern Mediterranean basins but spawning behavior and migrations of these fish, which are more likely to be Mediterranean-resident bluefin, have not yet been described (e.g., De Metrio et al. 2002; 2004; 2005; Cermeño et al. 2012). As stated above, recent genetic studies have found significant stock structure between the western (Balearic Islands, and Tyrrhenian Sea) and central Mediterranean (Ionian and Adriatic Seas, and Tunisia) (Carlsson et al. 2004; 2006; Boustany et al. 2008; Riccioni et al. 2010) but it is currently unknown if stock structure is related to post-spawning migratory patterns. It is however clear that a large proportion of fish spawning in the Balearic Islands area do exit the Mediterranean Sea after spawning is complete (De Metrio et al. 2002; Aranda et al. 2013), and are therefore non-resident eastern Atlantic bluefin. Based on this, we can hypothesize that a high proportion of non-residential eastern Atlantic bluefin spawn in the western Mediterranean (Balearic Islands) and that this proportion declines towards the eastern Mediterranean Sea (e.g., Aranda et al. 2013). Conversely, the proportion of Mediterranean-resident bluefin spawning in the eastern Mediterranean (Levantine Sea) may be higher than the central and western Mediterranean. This hypothesis could be tested by identifying and following the movements of spawning adults in all three regions of the Mediterranean but this has not yet been done. Besides Aranda et al. (2013), other recent studies in the western and central Mediterranean have also deployed electronic tags on fish over the spawning season (Cermeño et al. 2012; Fromentin and Lopuszanski 2013). Although these studies were not able to detect spawning behavior, they found that the tagged bluefin remained in the Mediterranean during their entire time at liberty. Some fish tagged near the Gulf of Lions made directed movements to the Gulf of Sirte, off the Libyan coast, which is a known spawning ground (Fromentin and Lopuszanski 2013; Quílez-Badia et al. 2013). However, detailed information from these fish are as yet unavailable. Only one electronic tagging study has been reported in the eastern Mediterranean basin (Levantine Sea) (De Metrio et al. 2004) but the tags experienced high rates of premature tag shedding and spawning behavior was not reported.

Post-spawning migrations

Although Atlantic bluefin tuna are able to migrate large distances, the core foraging area for western-origin adult fish appear to be the North American coast. After leaving the spawning grounds, western Atlantic adult bluefin predominantly move northwards to their foraging grounds along the North American coast. Their post-spawning movements tend to be along the continental shelf, slope, and Gulf Stream margin, and relatively high densities of bluefin appear in summer off the northeastern coast of the U.S. and the Canadian Maritime provinces, ranging from the Massachusetts coast to the Gulf of St. Lawrence (Figs. 7.4 and 7.6) (e.g., Stokesbury et al. 2004; Block et al. 2005; Teo et al. 2007b; Galuardi et al. 2010; Vanderlaan et al. 2011; Wilson et al. 2011). Furthermore, a portion of adult western Atlantic bluefin appear to forage and aggregate in the central North Atlantic in the area around 40°W, east of the Flemish Cap (Block et al. 2005; Walli et al. 2009). Within these foraging areas, the fish display diving behavior with significantly shallower diving depths and higher dive frequencies as compared to times in transit (Walli et al. 2009). Visceral temperatures of fish within these areas show a significantly higher variance that occurs independently of ambient temperature variability, which suggest an increase in visceral warming events likely caused by higher feeding activity. These areas are likely critical foraging habitats with sufficient available prey to satisfy the energetic needs of bluefin while still remaining within their preferred ambient temperatures.

Interestingly, there appears to be a firm link between the very large bluefin found in both the Gulf of Mexico and the Gulf of St. Lawrence (Diaz 2011; Vanderlaan et al. 2011; Wilson et al. 2011). An analysis of otolith stable isotopes found that >99% of giant bluefin in the waters within or adjacent to the Gulf of St. Lawrence were of western-origin throughout the past few decades (1970s; 1980s; and 2000s) (Schloesser et al. 2010). The δ^{13}C and δ^{18}O values from the otolith cores also showed little variation among the decades and regions, which suggests that the stock composition of adults migrating to the Gulf of St. Lawrence has been relatively stable and that these fish are closely related to fish spawning in the Gulf of Mexico. However, neither the proportion of spawners in the Gulf of Mexico that migrate to the Gulf of St. Lawrence nor the variability of that migration rate is known. Although the vast majority of giant bluefin tuna in the Gulf of St. Lawrence appears to be linked to spawning grounds in the western Atlantic (Gulf of Mexico or Caribbean/Bahamas), it should be noted this does not mean all giant bluefin from the Gulf of St. Lawrence spawn in the Gulf of Mexico. For example, one out of 14 giant bluefin tuna tagged in the Gulf of St. Lawrence migrated to the Mediterranean Sea (Wilson et al. 2011).

The link between adult bluefin in the Gulf of Mexico and the Gulf of St. Lawrence may be indicative of ecological niche expansion as the fish grow in size. Increased body size, coupled with highly efficient counter-current heat exchangers and high cardiac performance at low temperatures leads to an increased tolerance of cooler temperatures in the waters in and around the Gulf of St. Lawrence (Carey and Teal 1966; Mather et al. 1995; Block and Stevens 2001; Blank et al. 2004). Some of the coldest water temperatures (0.04–0.08°C) recorded by fully functioning electronic tags were from adult bluefin at the entrance to the Gulf of St. Lawrence (Walli et al. 2009). Although these <1°C records are rare, these fish commonly encountered water temperatures of around 2–4°C during deep dives off Nova Scotia and west of the Flemish Cap, due to the influence of the Labrador Current.

In late autumn and winter, after adults move southwards along the North American coast, bluefin aggregations are found in the waters off North Carolina and the South Atlantic Bight (Block et al. 2005; Teo et al. 2007b; Walli et al. 2009; Galuardi et al. 2010). In these waters, adults are mixed with juveniles from both the western and eastern Atlantic (Rooker et al. 2008b). Some adults subsequently leave the area around the end of winter and move northwards to mid-Atlantic Bight and Scotian Shelf waters for several months before moving to their spawning grounds towards the end of spring (Teo et al. 2007b). However, other adults move directly to the Gulf of Mexico in winter (as early as December), before the main spawning season, while others venture into the central Atlantic before returning to spawning grounds (Block et al. 2005; Walli et al. 2009; Galuardi et al. 2010). These western-origin adult bluefin mix with eastern-origin fish on these central Atlantic foraging grounds (Rooker et al. 2014) and can cross the 45°W meridian several times over the course of one or more years (Block et al. 2005; Galuardi et al. 2010). It is currently unclear if these differences in movement patterns are due to extant metapopulations or individual variability (Galuardi et al. 2010).

A portion of large (>200 cm; 8+ years in age) bluefin exhibited a trans-Atlantic shift in residence from the western to the eastern Atlantic as adults (Walli et al. 2009). These fish were most likely to be non-resident eastern Atlantic bluefin that migrated to the western Atlantic as juveniles. Although these were large bluefin, they did not exhibit spawning migrations prior to moving to the eastern Atlantic. In the eastern Atlantic, the most important aggregation for these fish appeared to be off the coast of the Iberian Peninsula, where trans-Atlantic fish spent considerable amounts of time (126 ± 75 days) (Walli et al. 2009). These fish aggregating off the Iberian Peninsula were subsequently tracked to known spawning grounds in the western and central Mediterranean basins.

The Atlantic coast of the Iberian Peninsula is an important foraging ground for juveniles and adults. Several studies (e.g., Block et al. 2005; Walli et al. 2009; Quílez-Badia et al. 2013; Aranda et al. 2013) have shown that after bluefin tuna visit the spawning grounds in the western and central Mediterranean Sea during June–July, the fish leave the Mediterranean and spend considerable amounts of time along the Iberian Peninsula during fall, winter, and the subsequent spring (Fig. 7.10; Fig. 3 in Block et al. 2005; Fig. 5 in Walli et al. 2009). In addition, these fish aggregated off the Azores, Ireland, Morocco, and remote offshore locations over the mid-Atlantic Ridge during different seasons (Walli et al. 2009). These aggregations also overlap substantially with the foraging grounds of juveniles (see juvenile movements), which suggests that similar ocean conditions and prey availability may be aggregating bluefin in these areas.

The movement patterns of Mediterranean-resident adult bluefin are slowly becoming better known with several recently completed or ongoing tagging experiments (e.g., De Metrio et al. 2004; 2005; Cermeño et al. 2012; Goni et al. 2012; Fromentin and Lopuszanski 2013) (http://www.iccat.int/GBYP/en/tagging.htm). There are numerous fishing grounds within the Mediterranean Sea for extant and historical fisheries that indicate bluefin aggregations but the lack of large-scale, historical tagging data to link these aggregations have made it very difficult to determine their migratory patterns (Mather et al. 1995). Electronic tagging experiments are beginning to show that these Mediterranean-resident bluefin have highly complex movement patterns but their general movement patterns have been difficult to discern. Based on electronic tag data from the western Mediterranean, both juvenile and adults appear to aggregate in an area offshore of the Gulf of Lions for several months in summer, fall, and possibly winter. The adults in this aggregation appear to be linked to the known spawning grounds in the Gulf of Sirte, off the Libyan coast (Fromentin and Lopuszanski 2013; Quílez-Badia et al. 2013) but detailed data from the Gulf of Sirte is still lacking. In addition, the channel between Corsica and Sardinia is a well-known fishing ground for bluefin tuna that appear to be large enough to be adult. Electronic tagging experiments on these fish indicate that they predominantly remain around the area but it is unclear if these fish are large juveniles or adults that spawn in the general vicinity (De Metrio et al. 2002). However, other tagging experiments indicate that some of these fish may be transients moving between the Tyrrhenian Sea and the northwest Mediterranean Sea (Quílez-Badia et al. 2013).

In the central Mediterranean, electronic tagging experiments in the Adriatic Sea (Yamashita and Miyabe 2001; Tudela et al. 2011) have suggested that adults spend substantial amounts of time within the Adriatic Sea, especially during the summer and fall. The deeper areas within the relatively Adriatic Sea, like the Jabuka Pit/Fossa di Pomo and South Adriatic Pit, appear to be areas of higher residency (Quílez-Badia et al. 2013). Upon leaving the Adriatic, the bluefin tend to stay within nearby areas, like the Ionian, Aegean, and Tyrrhenian Seas, as well as the Libyan coast. However, longer term tracks have indicated more complex movement patterns, with some fish moving to the western Mediterranean, around the Gulf of Lion and the Balearic Islands (see Quilez-Badia et al. 2013).

Substantially less tagging work have been done in eastern Mediterranean Sea but an electronic tagging experiment in the northern Levantine Sea show that adult bluefin in the eastern Mediterranean tend to remain within that basin (De Metrio et al. 2004). However, the tags did not remain attached on the fish for very long, with the longest duration being 67 days. Nevertheless, several of the tagged bluefin reached the Aegean Sea after about one to two months at liberty and more may have migrated to other areas of the Mediterranean Sea if long enough tracks were recorded.

Current uncertainties

Results from electronic tagging studies have also generated substantial debate on the reproductive biology of western Atlantic bluefin. Stock assessments of the western Atlantic bluefin stock (e.g., ICCAT 2012) have assumed a knife-edge 100% maturity at eight years of age based on the work of Baglin (1982) [currently adjusted to nine years of age based on a new growth curve by Restrepo (2011)]. However, the majority of age-8+ fish tagged in the western Atlantic do not migrate into known western Atlantic spawning grounds during spawning season (e.g., Lutcavage et al. 1999; Block et al. 2005; Galuardi et al. 2010). Three hypotheses have been proposed to explain this: (1) age of maturity for western Atlantic bluefin is later than eight years of age (Block et al. 2005); (2) age-8+ bluefin are mature but are not obligate annual spawners (Secor 2007); and (3) age-8+ bluefin are spawning but spawning is occurring outside of documented spawning grounds and seasons (Lutcavage et al. 1999; Galuardi et al. 2010).

In an early electronic tagging study, Lutcavage et al. (1999) hypothesized that western Atlantic bluefin spawned in the middle of the Atlantic Ocean but this hypothesis is not well supported. Lutcavage et al. (1999) reported that the vast majority (17 out of 20) of popup tags deployed on giant bluefin (190–263 cm) reported endpoint locations in the mid-Atlantic between Bermuda and the Azores during the spawning season (April–June). This was surprising because this postulated spawning area is far from the traditional spawning area in the Gulf of Mexico and Caribbean Sea. However, subsequent studies on similar-sized fish produced different results, with the majority of adults not residing in the mid-Atlantic during the spawning season (e.g., Stokesbury et al. 2004; Block et al. 2005; Sibert et al. 2006; Teo et al. 2007b; Galuardi et al. 2010). This difference may be due to the use of early generation popup tags that lacked pressure sensors and premature release software in the Lutcavage et al. (1999) study. Popup tags often detach from the fish before the programmed popup date (premature release), and pressure sensors and premature release software allow researchers to more easily and accurately detect premature releases. Stokesbury et al. (2004) found that popup tags that remained attached to fish produced geolocations that were distributed along the North American coast and the Gulf of Mexico but tags that had detached prematurely and drifted on the ocean surface for substantial amounts of time (123–319 days) showed similar endpoint locations to Lutcavage et al. (1999). Therefore, it was likely that the endpoint locations in the mid-Atlantic were due to premature detachment of the tags followed by a drift towards the mid-Atlantic Gyre over time. In addition, the lack of a pressure sensor on those early generation popup tags made the detection of premature release difficult and unreliable. However, this does not mean that bluefin tuna do not move into the mid-Atlantic. Even with current generations of popup tags where premature detachment is well detected, some giant bluefin (presumed to be mature) do not appear to visit documented spawning grounds during the spawning season (Galuardi et al. 2010). Out of nine giant bluefin (>200 kg) with tracks that encompassed some portion of the spawning season, two bluefin did not enter the Gulf of Mexico for the entire spawning season and one popped off during the middle of the spawning season without previously entering the Gulf of Mexico. Assuming that these bluefin were mature adults and obligate annual spawners, Galuardi et al. (2010) concluded that bluefin were likely spawning outside of documented spawning grounds.

Histological and endocrinological studies in the Northwest Atlantic have also raised the possibility that western Atlantic bluefin mature earlier than age-nine. Goldstein et al. (2007) reported 'mature' fish as young as age seven in the Gulf of Maine based on histological examinations. However, none of the females sampled had ovaries in the final stages of development (stage four–five in this study) indicative of an actively mature female (Goldstein et al. 2007). Given this and the relatively low gonadal-somatic index ratios (0.58–0.86) it seems unlikely that these fish were either actively spawning or would be doing so anytime soon, highlighting the non-standardized use of the word 'mature'. Subsequent histological and endocrinological analyses of bluefin tuna in the Northwest Atlantic found that all fish >134 cm CFL had maturing testes and oocytes, but there was no indication of near term spawning activity (ICCAT 2013). In our opinion, it is clear that a portion of giant bluefin tuna do move into the mid-Atlantic, even during spawning season, but the core distribution is primarily along the North American coast and the primary spawning ground for the western Atlantic remains the Gulf of Mexico and the adjacent Caribbean Sea (e.g., Block et al. 2005; Galuardi et al. 2010). The possibility of spawning outside the Gulf of Mexico and Caribbean Sea cannot be excluded but current evidence does not support this hypothesis. It is also possible that, like Pacific bluefin tuna, younger Atlantic bluefin tuna may spawn in a different area than older bluefin (Galuardi et al. 2010). However, stronger evidence like egg or larval samples is necessary to establish a novel spawning region (e.g., Muhling et al. 2011a) but is unfortunately difficult and expensive to obtain. Larval surveys in the mid-Atlantic region have not found bluefin tuna larvae (Fromentin and Powers 2005).

Secor (2007) raised the possibility that Atlantic bluefin are non-obligate annual spawners and skip spawning in some years. However, the available evidence from electronic tagging experiments, albeit highly limited, suggests that once bluefin start making spawning migrations, they continue to do so annually. Only very limited data are available because the fish must first be tracked to their spawning grounds, thereby establishing that the fish are mature, and then continue to be tracked for another year or more so that the time series extends over at least two spawning seasons. Nevertheless, Teo et al. (2007b) reported that bluefin 98–512 and A0532 migrated into the Gulf of Mexico over consecutive spawning seasons for 1999–2000 and 2003–2005 respectively. In addition, Block et al. (2005) reported that bluefin 603 exhibiting consecutive migrations to the spawning grounds near the Balearic Islands during the spawning season for three consecutive years (2001–2003) (Fig. 3 in Block et al. 2005). To the best of our knowledge, no study has yet reported a bluefin that after spawning in one year, did not do so during the next spawning season. It is possible that some bluefin tuna skip spawning in the wild (Secor 2007; Rideout and Tomkiewicz 2011), especially during years of high environmental stress, but there is no strong evidence that this is common. Overall, the balance of available evidence suggests that wild Atlantic bluefin tuna spawn annually and do not regularly skip spawning.

In contrast, there is more support for the hypothesis that the age of 50% maturity for western Atlantic bluefin is substantially greater than age-eight. In order to use electronic tagging data to estimate a maturity ogive, the proportion of fish of each age/size class that visited a known western Atlantic spawning ground must first be calculated from tags that have remained on bluefin past the end of spawning season (e.g., past July 1 in western Atlantic bluefin), and fit to a logistic curve. Using a large database of electronic tag data, preliminary work using this methodology suggested that the age of 50% maturity for western Atlantic bluefin is approximately age-15 (ICCAT 2013). This result is also supported by an analysis of size composition data from the Japanese longline fleet that operated in the Gulf of Mexico during the late 1970s and early 1980s. Diaz (2011) converted the size composition data into age compositions using the growth equation of Restrepo et al. (2011) and used a catch curve approach to estimate the maturity ogive. A maturity ogive with the age of 50% maturity at 15.8 years best fit the size composition data, given the assumptions in the model. The results of this analysis are sensitive to assumptions on mortality and selectivity but the similarity of the estimated ages of 50% maturity from two independent studies using very different methods suggest that age of 50% maturity for western-origin bluefin is likely to be closer to age-15 than age-eight. Moreover, the study by Baglin (1982) used an inadequate experimental design for establishing an age of maturity and it is not appropriate to conclude that age-8+ bluefin are 100% mature from this work (Schaefer 2001).

It is also important to note that the term 'age of maturity' is often used in an inconsistent and inappropriate manner (Schaefer 2001). Some researchers assume knife-edge maturity and report the age of the youngest sampled fish exhibiting maturity as the age of maturity (minimum age at maturity); while other researchers assume a maturity ogive and use the term for the age at 50% maturity. Reporting or using the minimum age at maturity as the age of maturity is inappropriate and misleading, and it is instead more important to report the age at 50% maturity (Schaefer 2001). Corriero et al. (2005) examined histological samples of bluefin gonads from the Mediterranean Sea and reported a length at 50% maturity of 103.6 cm FL (age-three) and that all fish >135 cm FL were mature. However, it should be noted that all samples from Corriero et al. (2005) were collected from within the Mediterranean Sea during the spawning period and may therefore be relevant for only for Mediterranean-resident bluefin. The sampling design may also be biased towards mature fish by sampling on the spawning grounds and not including non-resident eastern Atlantic bluefin. In the western Atlantic, studies reporting the age or size at 50% maturity are still lacking for western Atlantic bluefin. Instead, Baglin (1982) examined the gross morphology of gonads from western-origin bluefin that were three to seven years of age and found negligible gonadal development for all the fish, with a maximum gonadosomatic index of 1.75% for a single seven year old fish. In contrast, giant bluefin tuna in the Gulf of Mexico and Straits of Florida (190–264 cm FL) had well-developed gonads during April–May (Baglin 1982). There is clearly an urgent need for a well-designed study to re-examine the age of maturity of western Atlantic bluefin tuna. Ideally, a maturity ogive should be developed from the histological examination of gonadal samples from a representative spatial and temporal distribution of the stock (Schaefer 2001). The gonadal samples should not be sampled primarily from the spawning grounds because that would bias the sample towards mature fish. For example, even if all age-eight fish in the spawning grounds are mature, not all age-eight fish are on the spawning grounds.

Spawning and Foraging Habitat

Being large pelagic predators, the habitat of tunas are largely determined by biotic and abiotic environmental conditions. There has therefore been substantial scientific interest in how environmental conditions affect the distribution of Atlantic bluefin tuna, especially on their spawning and foraging grounds (e.g., Sund et al. 1981). Several data sources and analytical approaches have been taken to answer these questions. Bluefin distributions are determined from either traditional fishery data like longline CPUE (e.g., Teo and Block 2010) and larval tows (e.g., Muhling et al. 2010; García et al. 2013), or from electronic tags (e.g., Teo et al. 2007a), and related to environmental conditions using either qualitative or quantitative models (e.g., Royer et al. 2005; Walli et al. 2009). Some of these models can be used to develop predictive maps of likely habitat given current or future environmental conditions (e.g., Teo and Block 2010; Muhling et al. 2010). For example, near real-time habitat maps may be used to direct fishing vessels away from high bycatch areas to reduce the fishing impact on non-target species (Hobday and Hartmann 2006), help determine the impacts of marine pollutants on important spawning habitat (Muhling et al. 2012b), or examine the potential impacts of future climate change on spawning habitat (Muhling et al. 2011b).

Spawning habitat

Studies on how environmental conditions affect the spawning habitat of Atlantic bluefin tuna have concentrated on the spawning grounds in the Gulf of Mexico and the Balearic Islands in the western Mediterranean. Spawning season for all three species of bluefin tunas tends to be relatively short compared to tropical tunas (Schaefer 2001). The brevity of the spawning period may increase the importance of adult bluefin tuna using environmental cues that correspond to enhanced larval growth and survival, like warm temperatures, but is balanced by the physiological limits of the adults (Teo et al. 2007b). Due to the bluefin tuna's ability to conserve heat, Sharp and Vlymen (1978) hypothesized that the fish may be vulnerable to overheating in warm waters during bouts of intense activity, like spawning in the Gulf of Mexico.

In the Gulf of Mexico, spring plankton surveys targeting larval bluefin tuna have been conducted by NOAA Fisheries since 1977. These surveys are used to develop an index of larval abundance for use with stock assessments and are probably the longest, and most consistent time-series on larval bluefin distribution in the world. As part of this survey, Richards et al. (1989) observed that bluefin larvae were associated with the Loop Current frontal zone in the eastern Gulf of Mexico. This led Bakun (1996) to hypothesize that Atlantic bluefin tuna may be preferentially using the Loop Current frontal zone for spawning because meanders in the Loop Current as it approaches the Florida Straits create alternating zones of positive and negative vorticity, which in turn generate alternating zones of enhanced production and retention, respectively. Although the western and central Gulf of Mexico have been now shown to be more important than the Loop Current for spawning habitat (Nishida et al. 1998; Teo et al. 2007a; Muhling et al. 2010), the processes of enrichment, concentration, and retention are no doubt still critical in determining the success of reproduction.

Recently, Muhling et al. (2010) used the same larval time series to define favorable habitat for bluefin larvae. The presence or absence of bluefin larvae at a particular tow station was related to a suite of biotic and abiotic environmental variables (e.g., water temperature, salinity, plankton volume, wind speed, moon phase) using a non-parametric classification tree model. The model was tuned more to avoid false positives (expected presence of bluefin larvae in unfavorable habitat) rather than false negatives (expected absence of bluefin larvae in favorable habitat) because the aim of the study was to define potentially favorable habitat. Therefore, the model likely predicted more favorable habitat than is actually available *in situ*. In general, Muhling et al. (2010) found that favorable habitat for bluefin larvae were found in moderately warm waters across the northern Gulf of Mexico, with the exception of Loop Current waters and the warm core eddies spun off by the Loop Current, and high chlorophyll, continental shelf waters (Fig. 7.12). In addition, bluefin larvae were more likely to be found during darker moon phases but that could be due to net avoidance by the larvae during tows with more lunar illumination. Although this study did not examine the influence of important oceanographic variables like sea surface height anomalies, subsequent studies using remotely-sensed environmental variables (e.g., Muhling et al. 2012a) found similar results because Muhling et al. (2010) had already managed to elucidate the presence of warm core eddies through variables like temperature and salinity.

These results were consistent with electronic tag (Teo et al. 2007a) and longline CPUE data (Teo and Block 2010) from the Gulf of Mexico. Adult bluefin tuna had significant preference for areas over the continental slope, with moderate SSTs, moderate eddy kinetic energy, low surface chlorophyll concentrations and moderate wind speeds (Teo et al. 2007a). Teo et al. (2007a) used negative binomial models to fit bluefin longline CPUE in the Gulf of Mexico to a suite of environmental variables, and found that bluefin CPUE was significantly higher in cyclonic (cool core) mesoscale eddies. These results suggest that bluefin tuna appear to prefer spawning outside of the Loop Current (high eddy kinetic energy, warm SSTs) and continental shelf waters (high surface chlorophyll concentration). Within the western and central Gulf of Mexico, they appear to prefer areas with mesoscale eddies, especially cyclonic eddies (Fig. 7.13). Cyclonic (cool core) and anti-cyclonic (warm core) mesoscale eddies are generated by or pinched off from the Loop Current, which travel from east to west (Dietrich and Lin 1994) and produces an area with moderate eddy kinetic energy. These cyclonic eddies are associated with enhanced primary and secondary production (e.g.,

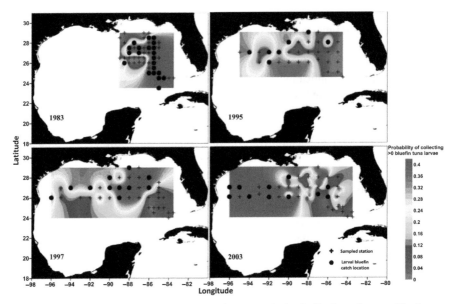

Figure 7.12. Predicted probabilities from classification tree analysis of collecting at least one bluefin tuna larvae across all sampled stations in cruise leg 2 of 1983, 1990 and 1997, and cruise leg 1 of 2003. All sampled stations, and larval bluefin tuna catch locations, are shown. Probabilities were kriged between stations to aid interpretation, but this should not be taken as a means to predict habitat between stations. Reproduced with permission from Muhling et al. 2010.

Gasca 1999; Wormuth et al. 2000; Gasca et al. 2001). In addition, these areas with mesoscale eddies generally have alternating cyclonic and anticyclonic eddies, which results in alternating zones of enhanced production and retention likely to enhance larval bluefin survival.

In the western Mediterranean spawning area around the Balearic Islands, bluefin larvae were collected during annual larval surveys in June and July from 2001 to 2005, as part of the TUNIBAL project (e.g., Alemany et al. 2010; Reglero et al. 2012; García et al. 2013). As expected, the distribution of bluefin larvae was highly variable and heavily influenced by environmental variability (Reglero et al. 2012). Spawning around the Balearic islands appear to occur in offshore mixed waters, near the frontal zone between inflowing surface Atlantic water masses and higher salinity Mediterranean surface waters (Fig. 7.14) (Alemany et al. 2010). Although these fish prefer to spawn in SSTs between 21.5 to 26.5, adult bluefin tuna appear to start spawning once SSTs reach 20.5°C (Alemany et al. 2010), which is substantially cooler than the 24°C that was previously thought to be the lower limit for bluefin tuna spawning (e.g., Schaefer 2001). Nevertheless, larval growth was found to be highest in 2003, when SSTs were relatively high but surface chlorophyll and planktonic prey concentrations were relatively low (García et al. 2013). Therefore, warm ambient temperatures enhance larval growth and survival, but high primary productivity and prey concentrations do not appear to do so. This may indicate that bluefin tunas have evolved to spawn in

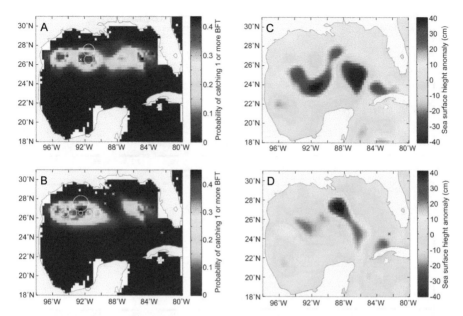

Figure 7.13. (A, B) Expected probability of catching bluefin tuna. Colors indicate the expected probability of catching one or more bluefin tuna in the Gulf of Mexico on May 15, (A) 2002 and (B) 2005. Circles indicate actual relative bluefin tuna CPUE for May 2002 and 2005. Crosses indicate locations where at least one longline set was deployed but no fish were caught. (C, D) Map of sea surface height anomalies. Colors indicate sea surface height anomalies on 15 May (C) 2002 and (D) 2005. Modified with permission from Teo and Block 2010.

these warm, low productivity waters, which may also reduce predator concentrations (Bakun 2006). A comparative study of bluefin larvae habitat on the spawning grounds in the Gulf of Mexico and around the Balearic Islands found that larvae from both spawning grounds were more likely to be found in warm, low chlorophyll waters with moderate currents and retentive characteristics (Muhling et al. 2013).

Foraging habitat

Once on their foraging grounds, tuna and other pelagic predators have long been known to be closely associated with frontal features (e.g., Sund et al. 1981; Laurs et al. 1984). However, the fine-scale relationships between these predators and frontal features, and the biophysical mechanisms are not completely understood. Fine-scale analysis of the association of bluefin tuna schools in the Mediterranean Sea and the Northwest Atlantic to frontal features were conducted by Royer et al. (2004) and Schick et al. (2004) by using aerial survey data and spatial statistics. The spatial structure of observed schools in the Gulf of Lions and the Gulf of Maine were examined and related to chlorophyll (Royer et al. 2004) and sea surface temperature frontal features (Royer et al. 2004; Schick et al. 2004). Both studies found that the association of bluefin schools with the frontal features were on a mesoscale (10s of km) rather than fine-scale (<10 km),

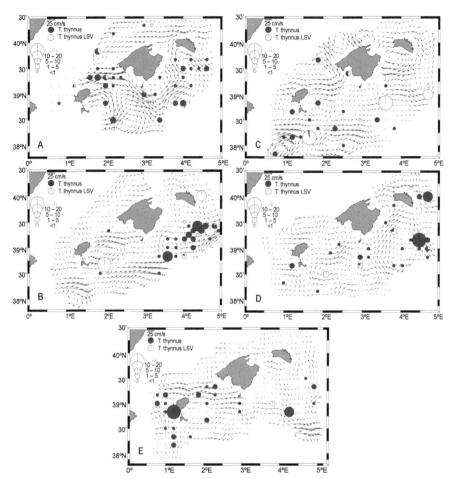

Figure 7.14. Spatial distribution and density of yolk-sac and post-lecitothrophic Atlantic bluefin (*Thunnus thynnus*) larvae in each TUNIBAL survey (A-E; 2001–2005). Densities correspond to number of larvae 100 m⁻³. Arrows represent the direction of geostrophic currents. Reproduced with permission from Alemany et al. 2010.

which suggest that the bluefin were not directly using the biophysical characteristics of the fronts as cues but were more likely attracted by prey items in the area (Royer et al. 2004; Schick et al. 2004).

On an ocean-basin scale, several regions in the North Atlantic were highlighted by Walli et al. (2009) as critical foraging habitat for bluefin tuna, where they seasonally exhibit substantial residence times and increased feeding events (Fig. 7.15). These regions and residency periods are: (1) the North Carolina coast from mid-October to as late as mid-May, depending on the year, with peak residency from December through March; (2) the Northwest Atlantic (Gulf of Maine, Georges Banks, Nova Scotia, and presumably Gulf of St. Lawrence) from early March to late December, with peak residency from June through October; (3) the Northwest Corner (an area east of the Flemish Cap in the central North Atlantic) from April through December, with peak residency in June; and (4) the Iberian Peninsula from September to December as well

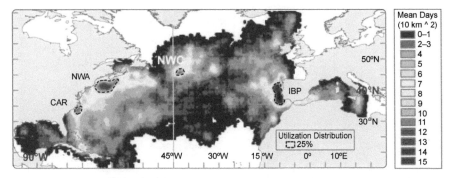

Figure 7.15. Normalized kernel density grid of number of daily geolocations weighted by number of fish tracked per unit area. Black, dotted line outlines 25% utilization distributions, showing four regions of high residency throughout the North Atlantic. Acronyms: CAR, Carolina; NWA, Northwest Atlantic; NWC, Northwest Corner; IBP, Iberian Peninsula. Modified with permission from Walli et al. 2009.

as May. The presence of bluefin in the Northwest Atlantic, Northwest Corner, and the Carolina coast were significantly correlated with increased primary productivity in these areas, as indicated by surface chlorophyll concentration, but not the Iberian Peninsula (Walli et al. 2009). However, the effect of sea surface temperature on the presence of bluefin in these areas was more mixed, with sea surface temperature being positively correlated with the bluefin presence in the Northwest Atlantic but negatively correlated in the North Carolina waters (Fig. 7.16).

The location and timing of residency in these areas appears to coincide with favorable environmental conditions and high prey availability (Walli et al. 2009). For example, Boustany et al. (2001) showed that bluefin presence in Carolina waters coincides with a large number of prey species, especially Atlantic menhaden (*Brevoortia tyrannus*), spot (*Leiostomus xanthurus*) and Atlantic croaker (*Micropogonias undulates*), which spawn in this region between November and March in waters of 18–24°C. These spawning prey species represents a relatively predictable food supply during winter, and stomach content analyses have shown a strong preference for Atlantic menhaden (89–99% by weight) by bluefin in the area during winter (Butler 2009). Interestingly, bluefin tends to leave the Carolina coast when SSTs warm to >21°C (Boustany 2006), which helps to explain the negative correlation between bluefin presence and SSTs in this area. As water temperatures warm, the thermal gradient across the Carolina shelf breaks down, which is thought to reduce the concentration of prey (Checkley et al. 1999).

Bluefin tuna start arriving in the Northwest Atlantic area around the end of March and appear to stay in warmer, weakly stratified, off-shelf areas in spring before moving into the stratified waters of the continental shelf area, as those waters warmed up during early summer and autumn (Walli et al. 2009; Lawson et al. 2010). The fish also displayed different diving behavior when they were on versus off the shelf, likely due to the different water column structure (Lawson et al. 2010). During summer and fall, bluefin dove more frequently (up to 180 dives d^{-1}) and faster (descent rates up to 4.1 m s^{-1}) in the stratified waters over the continental shelf than in spring, when the fish

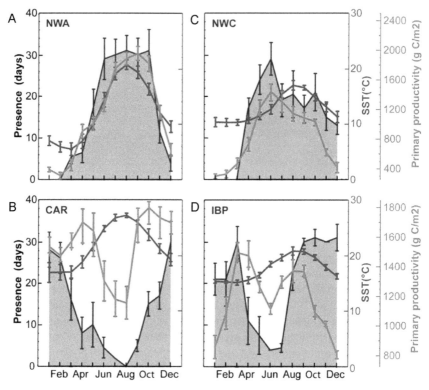

Figure 7.16. Mean (±SD) monthly number of days that bluefin tuna were present (1996–2005) within high use areas (gray shaded) in relation to mean (±SD) monthly level of primary productivity (green line) and sea surface temperature (blue line). (A) Northwest Atlantic (n = 32) (B) Northwestern Corner (n = 5) (C) Carolina (n = 52) (D) Iberian Peninsula (n = 4). Reproduced with permission from Walli et al. 2009.

were in weakly stratified off-shelf waters. In addition, dive duration and depth were also reduced substantially in off-shelf waters during spring (median: 0.45 hours and 77.0 m) versus in on shelf waters during summer (median: 0.16 hours and 24.9 m). The onshore movements as spring turns to summer appear to follow the migrations of potential prey (e.g., herring, mackerel, and squid) from offshore waters across Georges Bank into the Gulf of Maine (Garrison et al. 2002). Analyses of bluefin diet have demonstrated a strong link to aggregations of mackerel (*Scomber scombrus*), herring (*Clupea harengus*), various squids and sandlances (*Ammodytes americanus*) (Eggleston and Bochenek 1990; Lutcavage et al. 2000; Chase 2002; Estrada et al. 2005).

Starting in spring, some bluefin tuna move to the Northwest Corner region in the central North Atlantic, where they can reside for up to eight months (Walli et al. 2009). This region has very productive spring phytoplankton blooms associated with mesoscale eddies and meanders that enrich and concentrate the primary productivity (Behrenfeld and Falkowski 997; Li 2002). Therefore, bluefin presence in the Northwest Corner appears to be closely related to the primary productivity of the region (Fig. 7.16). The North Atlantic Current (extension of the Gulf Stream in the North Atlantic) goes past the southern part of this region, and bluefin tuna appears to feed on the cold side and rest on the warm side of this strong frontal zone (Walli et al. 2009). The fish moved back and forth across the fronts on a weekly basis and exhibited deeper dives on the cold side of the fronts during the day, potentially to access mesopelagic fish species, coupled with sharp changes in body temperature likely indicating ingestion of cold prey (Kitagawa et al. 2000; 2004). However, while in the warm waters of the North Atlantic Drift, the fish would exhibit reduced diving behavior with no diel difference in diving behavior.

The Iberian Peninsula appears to be critical foraging habitat for trans-Atlantic adult bluefin, as well as bluefin exiting the Mediterranean Sea after spawning (Walli et al. 2009; Aranda et al. 2013). Adult bluefin presence is more common in this area during, spring, autumn, and winter. Upwelling and primary productivity peaks in this area during spring and autumn (Smyth et al. 2001; Joint et al. 2002; Tilstone et al. 2003), which attracts mackerel, blue whiting and large spawning aggregations of sardines (Carrera et al. 2001; Bode et al. 2004). This area is also a critical foraging habitat for age-0 and one juvenile bluefin after leaving their nursery grounds in the Mediterranean Sea (Rodriguez-Marin et al. 2008) (see above).

Druon et al. (2011) and Damalas and Megalofonou (2012) have developed habitat models of bluefin foraging habitat in the Mediterranean Sea. Druon et al. (2011) developed an ad hoc procedural model of foraging and spawning habitat in the Mediterranean Sea, using rules developed from previous studies. Observations of bluefin from aerial surveys (Bonhommeau et al. 2010), electronic tagging (Fromentin 2010), and logbooks (from a French purse seiner) were assembled (most observations came from the western and, to a lesser extent, central Mediterranean). It was assumed that bluefin habitat was defined by the vicinity of bluefin observations to specific oceanic features believed to be relevant for either foraging (frontal features and range of chlorophyll concentration) or spawning (increases in sea surface temperatures over various time frames, minimum sea surface temperature, and range of chlorophyll concentration). The minimum gradients of sea surface temperature and chlorophyll concentration that characterize fronts relevant for the potential foraging habitat were found to be $0.11°C$ km^{-1} and 0.0053 mg m^{-3} km^{-1} respectively. The optimal range of surface chlorophyll concentrations for bluefin foraging habitat is $0.11–0.34$ mg m^{-3} (Druon et al. 2011). Due to the ad hoc and non-statistical nature of the modeling framework, the uncertainties of the estimated habitat parameters and possible model process errors are unknown. Therefore, these results should be interpreted with caution.

Generalized additive models were used to examine the relationships between a suite of environmental variables and bluefin CPUE from commercial longliners in the eastern Mediterranean (Ionian, Aegean, and Levantine Seas) (Damalas and Megalofonou 2012). The probability of catching bluefin was higher at higher sea surface temperatures ($>22°C$) and around the full moon. If bluefin were caught, the CPUE was more likely to be higher as sea surface temperatures approach $26°C$ but decline thereafter, and also as the lunar phase approached the full moon.

Surprisingly little research has been done to examine the influences of environmental conditions on Atlantic bluefin foraging habitat using electronic tag data and appropriate statistical models. State-space movement models are often used to estimate the location of bluefin tuna from electronic tag data and estimate movement parameters (e.g., Sibert et al. 2003; Royer et al. 2005; Nielsen and Sibert 2007; Lam et al. 2008). One way of analyzing the movements of fish in relation to environmental conditions is to incorporate environmental variables into the state-space movement model to explain observed changes in behavior (i.e., changes in estimated movement parameters) (e.g., Royer et al. 2005; Patterson et al. 2008; 2009). However, such studies for Atlantic bluefin tuna are lacking.

Enigmatic Aggregations

One of the most intriguing aspects of bluefin movement and habitat use patterns is the relatively sudden (and enigmatic) appearance and/or disappearance of bluefin tuna fisheries in various geographical regions. Three well known cases involve Atlantic bluefin tuna fisheries: (1) the Japanese longline fishery off Brazil in the 1950s and 1960s; (2) Nordic bluefin fisheries in the Norwegian and North Seas in the 20th century between the 1920s and 1970s; and (3) Turkish bluefin fisheries in the Black Sea prior to the 1980s. Available data for all three cases are generally limited to catch and effort due to the historical nature of these fisheries, which have perhaps added to the mystery. Nevertheless, the enigmatic nature of these aggregations has inspired research to understand them better.

Among these three examples, the bluefin tuna caught by the Japanese longline fleet in the Atlantic Ocean off Brazil is arguably the most puzzling. This fishery started fishing operations in the waters off Brazil in 1956, primarily targeting yellowfin and albacore tuna for the canning industry (Takeuchi et al. 2009). These longline vessels initially caught only small numbers of bluefin tuna but the bluefin catch rapidly increased from 1959, reaching a peak of about 13000 t in 1964 (Takeuchi et al. 1999) and possibly even more (Fonteneau 2009). Just as quickly, the catches rapidly decreased and reached negligible levels by 1970. Bluefin tuna was probably not an important target species for the Japanese distant water longline fleet during this period due to the lack of 'super freezers' at that time (Takeuchi et al. 1999; Fonteneau 2009). Although bluefin tuna is synonymous today with the sashimi market, 'super-freezers' (freezers that froze fish at $-55°C$) that made these fish acceptable for the sashimi market, were only available on longliners around the end of the 1960s (Miyake 2005). These fish were therefore likely sold to canneries for relatively low prices. Currently, the longline effort in the tropical waters off Brazil remains high but bluefin tuna catch is negligible (Takeuchi et al. 2009). The main bluefin hotspot occurred around 10°N–10°S, and 30–40°W, just off the Brazilian equatorial coast (although bluefin catch occurred off the African coast in some years), and was highly seasonal (CPUE peaks in March–April and September–October) (Takeuchi et al. 2009).

Several aspects of this fishery are puzzling: (1) were these fish off Brazil a separate stock or were they part of extant eastern and/or western Atlantic stocks; (2) where do these fish go after seasonally leaving Brazilian waters; (3) what were the fish doing off Brazil; and (4) what was the cause of CPUE decline (Mather et al. 1995; Takeuchi et al. 1999; 2009; Fromentin and Powers 2005; Fonteneau 2009; Fromentin et al. 2014). It is not possible to conclusively determine if these bluefin are a separate genetic stock or related to the extant western and eastern Atlantic stocks because no archives of genetic samples of these fish have been found so far. However, Takeuchi et al. (2009) used Japanese longline CPUE data from that period to postulate that the movements of these fish may be more similar to the western stock. The bluefin were hypothesized to be off Brazil in March–April before moving north to the Caribbean and Bahamas in May–June for possible spawning, and then moving further north to New England waters in July–August. At the end of the summer, in September–October, the fish then came back south to Brazil in September–October and appeared to move further south along the Brazilian coast towards temperate waters off Argentina in October–February. This hypothesized movement pattern is supported by limited tagging data, with two tags from bluefin tagged in the Bahamas being recaptured off Brazil and Argentina (Mather et al. 1995). On the other hand, several fish tagged in the Bahamas were recaptured in the eastern Atlantic, including Norwegian waters. However, Fonteneau (2009) pointed out that the size of bluefin in Brazilian waters appeared to be smaller (average weight of 180 kg based on limited size samples) than the fish caught in Norwegian waters, which were well in excess of 200 kg. Therefore, Takeuchi et al. (2009) speculated that the Brazilian bluefin were likely a metapopulation that either merged with others, or have become extinct.

Bluefin tuna are generally thought to primarily occupy temperate waters, only traveling into subtropical waters like the Gulf of Mexico to spawn so it is surprising that large numbers were caught in the tropical waters off Brazil. Brazilian waters did not appear to be a spawning ground because limited sampling (N = 32) of gonads suggested that the fish in the area were not reproductively active (Takeuchi et al. 1999). It is possible that the fish were simply passing through Brazilian waters on their way towards cooler waters nearer Argentina. If the movement patterns proposed by Takeuchi et al. (2009) are correct, these fish may have been maximizing food resources by moving into the Northern Hemisphere in May–August and the Southern Hemisphere in October–February, which is similar to the 'endless summer' migratory pattern of sooty shearwaters (Shaffer et al. 2006). The loss of this metapopulation or substock of bluefin tuna makes it difficult to understand their biology. Nevertheless, it is important to understand the processes and events that led to the loss of the Brazilian bluefin so as to help reduce the chances of such events happening in the future, especially if the primary causes are related to fishing pressure or other anthropogenic effects. Due to the lack of detailed information, it will be difficult to make firm conclusions about this. Takeuchi et al. (1999) suggested six hypotheses that might explain the sudden increase and subsequent collapse of bluefin catch off Brazil but concluded that no single hypothesis is well supported by the evidence. The least objectionable hypotheses appear to be that either the Brazilian fish were a metapopulation or substock with an unusual migratory pattern that was eliminated by high fishing pressure or that the Brazilian fish were a subset of western Atlantic bluefin that moved into the Southern Hemisphere for a limited period of time due to environmental conditions during that period.

Recently, Fromentin et al. (2014) developed an environmental niche model of Atlantic bluefin habitat and found that the South Atlantic contains suitable habitat and, more interestingly, the western equatorial Atlantic had relatively favorable environmental conditions during the 1960s, but not after (Fig. 7.17). The suitable environmental conditions for Brazilian waters during the 1960s may have allowed

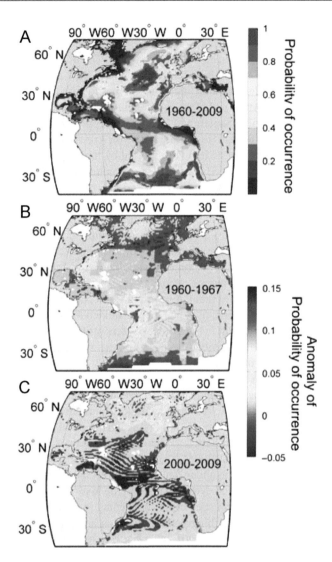

Figure 7.17. Maps of the probabilities of Atlantic Bluefin tuna (ABFT) occurrence deduced from the NPPEN niche model (see Material and Methods): (A) for the entire period (1960– 2009); (B) map of the anomalies of the probabilities of ABFT occurrence during the 'Brazilian episode' (computed as the map of ABFT occurrence during 1960–1967 minus the median probabilities calculated in each pixel from 1960 to 2009); and (C) same as (B) for the period 2000–2009. Reproduced with permission from Fromentin et al. 2014.

the bluefin to migrate from the northern spawning grounds into the South Atlantic, acting as an 'ecological bridge', which has since broken down.

More is known about the collapse of Nordic bluefin fisheries in the 1960s than the Brazilian bluefin episode. There were important Nordic fisheries (predominantly Norwegian) in the Norwegian and North Seas from the 1920s till the 1970s (Tangen

2009), with catches peaking at approximately 18000 tons per year in 1952. However, the catch consistently dropped to <1000 tons by 1968, and remained low afterwards. This collapse is particularly interesting in recent years because bluefin tuna have not come back into the Norwegian and North Seas even though they are relatively abundant off the Faroe Islands and herring stocks are now abundant in the North Sea (Fonteneau 2009). Several hypotheses for this collapse have been put forward: (1) change in migratory routes; (2) recruitment failure due to the environment or overfishing; and (3) substock extinction (Fromentin 2009). Fromentin (2009) examined the catch and size composition of the most important bluefin fisheries in the 20th century that operated in the North Atlantic and Mediterranean Sea to elucidate the likely causes of the collapse of the Nordic bluefin fisheries. Most importantly, Fromentin (2009) showed that these fish appeared to be most closely related to the bluefin caught in Spanish trap fisheries in the 1950s and 60s but this relationship appeared to have broken down in the 1970s. In addition, these fish appeared to be partially related to Northwest Atlantic and Mediterranean trap fisheries (Fig. 7.2). Therefore, during the peak of the Nordic fisheries in the 1950s and 60s, a large portion of the non-resident eastern Atlantic bluefin probably migrated between spawning grounds in the Mediterranean Sea and the Norwegian and North Seas, which were important feeding grounds during that period due to the herring in the area. In addition, a portion of bluefin caught in Northwest Atlantic trap fisheries also used the Norwegian and North Seas as feeding grounds. However a simple causal relationship between the collapse of Nordic herring stocks and the bluefin fisheries is not supported because the respective regime shifts occurred in different periods. In addition, the bluefin collapse occurred suddenly in 1963 while the herring stocks declined over two decades. Fromentin (2009) concluded that the collapse in the Nordic bluefin fisheries is probably due to a complex interaction of environmental, trophic, and fishing processes, which affected bluefin migration patterns and possibly recruitment. Based on these results, recent genetic evidence (Carlsson et al. 2004; 2007; Boustany et al. 2008; Riccioni et al. 2010), and the discovery of a spawning ground in the eastern Mediterranean Sea (Karakulak et al. 2004; Oray and Karakulak 2005), Fromentin (2009) supported the hypothesis that the Atlantic bluefin tuna is likely made up of at least three populations: (1) a stock that migrates between feeding grounds in the North Atlantic and spawning grounds in the western and central Mediterranean Sea (i.e., non-resident eastern Atlantic bluefin); (2) a more resident stock both feeding and spawning in the Mediterranean Sea, especially the central and eastern Mediterranean Sea (i.e., Mediterranean-resident bluefin); and (3) a western Atlantic stock that feeds primarily in the Northwestern Atlantic and spawns in the Gulf of Mexico (western Atlantic bluefin).

The Black Sea has a substantially lower salinity than oceanic waters, with salinity ranging from approximately 18 ppt in the surface layer to 22 ppt below the pycnocline (Tomczak and Godfrey 2003). It is therefore surprising that bluefin tuna were seasonally common and supported a Turkish trap fishery in the Black Sea (Mather et al. 1995; Karakulak and Oray 2009) because they are generally thought to strongly prefer waters with oceanic salinity. Bluefin tuna used to migrate into the Black Sea from the Sea of Marmara from April until September, peaking in July–August (Karakulak and Oray 2009). Bluefin tuna are thought to be highly conservative in spawning area choice and spawn only in warm, oceanic waters (Schaefer 2001) so it is very surprising that there is some evidence, albeit highly limited, that bluefin tuna may have moved into the Black Sea to spawn (MacKenzie and Mariani 2012). Several egg and larval surveys conducted by Soviet scientists between 1933 and 1957 reported the presence of bluefin eggs and/or larvae in the north-central part of the Black Sea in July–August (Mather et al. 1995; Zaitsev and Mamaev 1997; MacKenzie and Mariani 2012). Other authors have cast doubt on the identification of these early surveys because more recent surveys in 1993 failed to collect any eggs or larvae (Piccinetti-Manfrin et al. 1995). However, this lack of bluefin egg and larval samples in the 1993 survey is not surprising because it was no longer common for bluefin to enter the Black Sea by 1993. Other evidence supporting the spawning hypothesis is that gonad samples from fish entering the Black Sea were ripening or fully ripe but were spent when taken from fish leaving the Black Sea (Mather et al. 1995). Nevertheless, the Black Sea substock was likely not a large population and mostly supported a fishery in the low hundreds of tons (Karakulak and Oray 2009). If these fish were migrating into the Black Sea and spawning on a regular basis, they must have developed some interesting physiological adaptations to deal with the low salinity. For example, MacKenzie and Mariani (2012) showed that bluefin eggs would have had a very low density of ≤ 1012 kg m^{-3} in order to be in the upper 10 m of the water column, given the summer salinities found in the Black Sea, which is lower than any previously recorded density of bluefin eggs. However, they also pointed out that other scombrid fish, including swordfish do spawn in the Black Sea and other fish species produce eggs whose density are adapted to local conditions (Gordina et al. 2001). It is therefore not out of the question that some bluefin spawned in the Black Sea but it is unknown if this was a viable population or a 'sink' population, which depended on recruitment from other spawning areas to maintain the population. Unfortunately, there has been relatively little research done on this historical population and it may never be known if they spawned in the Black Sea and if so, what adaptations to the challenging environment were present. We consider the question of whether bluefin tuna successfully spawned in the Black Sea to be an open question, and will likely remain so. In addition, the causes of the collapse of the Turkish trap fishery in the Black Sea is unknown and one can speculate on some combination of changes in the environment, migratory pathways, recruitment failure, and fishing pressure.

These three enigmatic aggregations illustrate how little is actually known about bluefin tuna biology and how they continue to surprise and intrigue scientists. Bluefin tuna likely have historical migratory and habitat use patterns beyond those currently exhibited by extant stocks. The causes for the collapse of these fisheries remains unclear but the main plausible hypotheses can be aggregated into several main categories: (1) these bluefin are part of extant stocks and the collapse is due to changes in migratory patterns or contractions in home range, which are in turn due to a complex interaction of environment, trophic, population dynamic, and fishery processes; (2) these bluefin were a distinct metapopulation or substock from extant stocks with distinct migratory and habitat use patterns but is currently extinct due to overfishing and/or recruitment failure, which may be in turn due to a complex interaction of various processes, including environmental and fishing pressure; or (3) these bluefin were a distinct metapopulation or substock from extant stocks with distinct migratory and habitat use patterns but have merged into larger extant stocks.

Concluding Remarks

Although much of the biology of Atlantic bluefin tuna remains enigmatic and even basic biological parameters like age of maturity remain uncertain, there have nevertheless been substantial advances in understanding their movements and habitat use over the past decade or so. Electronic tags have been the preeminent tool in enabling those advances. However, there has been relatively little consideration of appropriate experimental design for many tagging experiments, which reduces the usefulness of the data in integrated stock assessment models (Sippel et al. 2014). In addition, most tagging experiments have tended to lack the scale necessary to tackle questions that cover the entire range of Atlantic bluefin tuna. It is therefore important to pool resources and share data between various laboratories. Such data sharing would in particular help modelers collaborate with biologists to develop appropriate models from the electronic tagging data available. The ICCAT-GBYP program is perhaps an important step towards the end. We look forward to the next decade of bluefin tuna research.

Acknowledgements

We would like to thank all authors who kindly agreed to have their figures and tables reproduced in this book chapter. There were also many useful discussions with the authors, which greatly improved this manuscript. We would also like to thank H. Dewar, T. Sippel, and W. Yamamoto for reviewing a draft of this manuscript, and T. Kitagawa and S. Kimura for providing the opportunity for us to write this book chapter and bearing with us throughout the long writing process. The opinions contained in this manuscript are solely the personal opinions of the authors and do not constitute the official position of NOAA.

References

Albaina, A., M. Iriondo, I. Velado, U. Laconcha, I. Zarraonaindia, H. Arrizabalaga, M.A. Pardo, M. Lutcavage, W.S. Grant and A. Estonba. 2013. Single nucleotide polymorphism discovery in albacore and Atlantic bluefin tuna provides insights into worldwide population structure. Anim. Genet. 44: 678–692.

Alemany, F., L. Quintanilla, P. Velez-Belchí, a. García, D. Cortés, J.M. Rodríguez, M.L. Fernández de Puelles, C. González-Pola and J.L. López-Jurado. 2010. Characterization of the spawning habitat of Atlantic bluefin tuna and related species in the Balearic Sea (western Mediterranean). Prog. Oceanogr. 86: 21–38.

Aloncle, H. 1973. Bluefin tuna tagging in the Bay of Biscay. Collect Vol. Sci. Pap. ICCAT 1: 445–458.

Alvarado-Bremer, J.R., I. Nasseri and B. Ely. 1999. A provisional study of northern bluefin tuna populations. Collect Vol. Sci. Pap. ICCAT 49: 127–129.

Aranda, G., F.J. Abascal, J.L. Varela and A. Medina. 2013. Spawning behaviour and post-spawning migration patterns of Atlantic bluefin tuna (*Thunnus thynnus*) ascertained from satellite archival tags. PLoS One 8: e76445.

Baglin, R.E. 1982. Reproductive biology of western Atlantic bluefin tuna. Fish Bull. 80: 121–134.

Bakun, A. 1996. Patterns in the Ocean. California Sea Grant College System.

Bakun, A. 2006. Fronts and eddies as key structures in the habitat of marine fish larvae: opportunity, adaptive response. Sci. Mar. 70S2: 105–122.

Bayliff, W.H. 1994. A review of the biology and fisheries for the northern bluefin tuna, *Thunnus thynnus*, in the Pacific Ocean. pp. 244–295. *In*: R.S. Shomura, J. Majkowski and S. Langi (eds.). Interactions of Pacific Tuna Fisheries. FAO Fisheries Technical Paper 336/2. Food and Agriculture Organization of the United Nations, Rome, Italy.

Behrenfeld, M.J. and P.G. Falkowski. 1997. Photosynthetic rates derived from satellite-based chlorophyll concentration. Limnology Oceanogr. 42: 1–20.

Blank, J.M., J.M. Morrissette, A.M. Landeira-Fernandez, S.B. Blackwell, T.D. Williams and B.A. Block. 2004. *In situ* cardiac performance of Pacific bluefin tuna hearts in response to acute temperature change. J. Exp. Biol. 207: 881–890.

Block, B.a., H. Dewar, S.B. Blackwell, T.D. Williams, E.D. Prince, C.J. Farwell, a. Boustany, S.L. Teo, a. Seitz, a. Walli and D. Fudge. 2001. Migratory movements, depth preferences, and thermal biology of Atlantic bluefin tuna. Science 293: 1310–4.

Block, B.A. and E.D. Stevens. 2001. Tuna: Physiology, Ecology, and Evolution. Academic Press, San Diego.

Block, B.A., S.L.H. Teo, A. Walli, A.M. Boustany, M.J.W. Stokesbury, C.J. Farwell, K.C. Weng, H. Dewar and T.D. Williams. 2005. Electronic tagging and population structure of Atlantic bluefin tuna. Nature 434: 1121–1127.

Bode, A., M.T. Alvarez-Ossorio, P. Carrera and J. Lorenzo. 2004. Reconstruction of trophic pathways between plankton and the North Iberian sardine (Sardina pilchardus) using stable isotopes. Sci. Mar. 68: 165–178.

Bonhommeau, S., H. Farrugio, F. Poisson and J. Fromentin. 2010. Aerial surveys of bluefin tuna in the western Mediterranean Sea: retrospective, prospective, perspectives. Collect Vol. Sci. Pap. ICCAT 65: 801–811.

Boustany, A.M. 2006. An Examination of Population Structure, Movement Patterns and Environmental Preferences in Northern Bluefin tuna. Stanford University.

Boustany, A., D.J. Marcinek, J.E. Keen, H. Dewar and B.A. Block. 2001. Movements and temperature preference of Atlantic bluefin tuna (*Thunnus thynnus*) off North Carolina: a comparison of acoustic, archival and pop-up satellite tagging. pp. 89–108. *In*: J.R. Sibert and J.L. Nielsen (eds.). Electronic Tagging and Tracking in Marine Fishes. Kluwer Academic Publishers, Dordrecht.

Boustany, A.M., C.A. Reeb and B.A. Block. 2008. Mitochondrial DNA and electronic tracking reveal population structure of Atlantic bluefin tuna (*Thunnus thynnus*). Mar. Biol. 156: 13–24.

Broughton, R.E. and J.R. Gold. 1997. Microsatellite development and survey of variation in northern bluefin tuna (*Thunnus thynnus*). Mol. Mar. Biol. Biotechnol. 6: 308–314.

Calaprice, J.R. 1986. Chemical variability and stock variation in northern Atlantic bluefin tuna. Collect Vol. Sci. Pap. ICCAT 254: 222–254.

Carey, F.G. and J.M. Teal. 1966. Heat conservation in tuna fish muscle. Proc. Natl. Acad. Sci. USA 56: 1464–1469.

Carlsson, J., J.R. Mcdowell, J.E.L. Carlsson and J.E. Graves. 2006. Genetic heterogeneity of Atlantic bluefin tuna caught in the eastern North Atlantic Ocean south of Iceland. 1117.

Carlsson, J., J.R. McDowell, J.E.L. Carlsson and J.E. Graves. 2007. Genetic identity of YOY bluefin tuna from the eastern and western Atlantic spawning areas. J. Hered. 98: 23–8.

Carlsson, J., J.R. McDowell, P. Díaz-Jaimes, J.E.L. Carlsson, S.B. Boles, J.R. Gold and J.E. Graves. 2004. Microsatellite and mitochondrial DNA analyses of Atlantic bluefin tuna (*Thunnus thynnus thynnus*) population structure in the Mediterranean Sea. Mol. Ecol. 13: 3345–56.

Carrera, P., M. Meixide, C. Porteiro and J. Miquel. 2001. Study of the blue whiting movements around the Bay of Biscay using acoustic methods. Fish Res. 50: 151–161.

Cermeño, P., S. Tudela, G. Quilez-Badia, S.S. Trapaga and E. Graupera. 2012. New data on bluefin tuna migratory behavior in the western and central Mediterranean Sea. Collect Vol. Sci. Pap. ICCAT 68: 151–162.

Chase, B.C. 2002. Differences in diet of Atlantic bluefin tuna (*Thunnus thynnus*) at five seasonal feeding grounds on the New England continental shelf. Fish Bull. 100: 168–180.

Checkley, D.M., Peter B. Ortner, F.E. Werner, L.R. Settle and Shailer R. Cummings. 1999. Spawning habitat of the Atlantic menhaden in Onslow Bay, North Carolina. Fish Oceanogr. 8(Suppl.) 2: 22–36.

Colin, P.L. 1978. Daily and summer-winter variation in mass spawning of the striped parrotfish, Scarus croicensis. Fish Bull. 76: 117–124.

Corriero, A., S. Karakulak, N. Santamaria, M. Deflorio, D. Spedicato, P. Addis, S. Desantis, F. Cirillo, A. Fenech-Farrugia, R. Vassallo-Agius, J.M. Serna, Y. Oray, A. Cau, P. Megalofonou and G. De Metrio. 2005. Size and age at sexual maturity of female bluefin tuna (*Thunnus thynnus* L. 1758) from the Mediterranean Sea. J. Appl. Ichthyol. 21: 483–486.

Cort, J.J. and B. Loirzou. 1990. Reproduction—eastern Atlantic and Mediterranean. pp. 99–101. *In*: D. Clay (ed.). World Bluefin Meeting, May 25–31, 1990. La Jolla, California.

Damalas, D. and P. Megalofonou. 2012. Discovering where bluefin tuna, *Thunnus thynnus*, might go: using environmental and fishery data to map potential tuna habitat in the eastern Mediterranean Sea. Sci. Mar. 76: 691–704.

Deshpande, A.D., A.F. Draxler, V.S. Zdanowicz, M.E. Schrock and A.J. Paulson. 2002. Contaminant levels in the muscle of four species of fish important to the recreational fishery of the New York Bight Apex. Mar. Pollut. Bull. 44: 164–171.

Diaz, G.A. 2011. A revision of western Atlantic bluefin tuna age of maturity derived from size samples collected by the Japanese longline fleet in the Gulf of Mexico (1975–1980). Collect Vol. Sci. Pap. ICCAT 66: 1216–1226.

Dicenta, A. and C. Piccinetti. 1980. Comparison between the estimated reproductive stocks of bluefin tuna (*T. thynnus*) of the Gulf of Mexico and western Mediterranean. Collect Vol. Sci. Pap. ICCAT 9: 442–448.

Dickhut, R.M., A.D. Deshpande, A. Cincinelli, M.a. Cochran, S. Corsolini, R.W. Brill, D.H. Secor and J.E. Graves. 2009. Atlantic bluefin tuna (*Thunnus thynnus*) population dynamics delineated by organochlorine tracers. Environ. Sci. Technol. 43: 8522–7.

Druon, J.-N., J.-M. Fromentin, F. Aulanier and J. Heikkonen. 2011. Potential feeding and spawning habitats of Atlantic bluefin tuna in the Mediterranean Sea. Mar. Ecol. Prog. Ser. 439: 223–240.

Edmonds, P.H. and J.I. Sammons. 1973. Similarity of genic polymorphism of tetrazolium oxidase in bluefin tuna (*Thunnus thynnus*) from the Atlantic coast of France and the western North Atlantic. J. Fish Res. Board Canada 30: 1031–1032.

Eggleston, D.B. and E.A. Bochenek. 1990. Stomach contents and parasite infestation of school bluefin tuna *Thunnus thynnus* collected from the Middle Atlantic Bight, Virginia. Fish Bull. 88: 389–395.

Ely, B., D.S. Stoner, J.M. Dean, Bremer J.R. Alvarado, S. Chow, S. Tsuji, T. Ito, K. Uosaki, P. Addis, A. Cau, E.J. Thelen, W.J. Jones, D.E. Black, L. Smith, K. Scott, I. Naseri and J.M. Quattro. 2002. Genetic analysis of Atlantic northern bluefin tuna captured in the northwest Atlantic Ocean and the Mediterranean Sea. Collect Vol. Sci. Pap. ICCAT 54: 372–376.

Estrada, J.a., M. Lutcavage and S.R. Thorrold. 2005. Diet and trophic position of Atlantic bluefin tuna (*Thunnus thynnus*) inferred from stable carbon and nitrogen isotope analysis. Mar. Biol. 147: 37–45.

Fonteneau, A. 2009. Atlantic bluefin tuna: 100 centuries of fluctuating fisheries. Collect Vol. Sci. Pap. ICCAT 63: 51–68.

Fromentin, J.-M. 2009. Lessons from the past: investigating historical data from bluefin tuna fisheries. Fish Fish 10: 197–216.

Fromentin, J. 2010. Tagging bluefin tuna in the Mediterranean Sea: challenge or mission: impossible? Collect Vol. Sci. Pap. ICCAT 65: 812–821.

Fromentin, J.-M. and D. Lopuszanski. 2013. Migration, residency, and homing of bluefin tuna in the western Mediterranean Sea. ICES J. Mar. Sci.: doi:10.1093/icesjms/fst157.

Fromentin, J.-M. and J.E. Powers. 2005. Atlantic bluefin tuna: population dynamics, ecology, fisheries and management. Fish Fish 6: 281–306.

Fromentin, J.-M., G. Reygondeau, S. Bonhommeau and G. Beaugrand. 2014. Oceanographic changes and exploitation drive the spatio-temporal dynamics of Atlantic bluefin tuna (*Thunnus thynnus*). Fish Oceanogr. 23: 147–156.

Galuardi, B and M. Lutcavage. 2012. Dispersal routes and habitat utilization of juvenile Atlantic bluefin tuna, *Thunnus thynnus*, tracked with mini PSAT and archival tags. PLoS One 7: e37829.

Galuardi, B., F. Royer and W. Golet. 2010. Complex migration routes of Atlantic bluefin tuna (*Thunnus thynnus*) question current population structure paradigm. Can J. Fish Aquat. Sci. 67: 966–976.

García, a., D. Cortés, J. Quintanilla, T. Rámirez, L. Quintanilla, J.M. Rodríguez and F. Alemany. 2013. Climate-induced environmental conditions influencing interannual variability of Mediterranean bluefin (*Thunnus thynnus*) larval growth. Fish Oceanogr. 22: 273–287.

Garrison, L.P., W. Michaels, J.S. Link and M.J. Fogarty. 2002. Spatial distribution and overlap between ichthyoplankton and pelagic fish and squids on the southern flank of Georges Bank. Fish Oceanogr. 11: 267–285.

Gasca, R. 1999. Siphonophores (Cnidaria) and summer mesoscale features in the Gulf of Mexico. Bull. Mar. Sci. 65: 75–89.

Gasca, R., I. Castellanos and D.C. Biggs. 2001. Euphausiids (Crustacea, Euphausiacea) and summer mesoscale features in the Gulf of Mexico. Bull. Mar. Sci. 68: 397–408.

Goldstein, J., S. Heppell, A. Cooper, S. Brault and M. Lutcavage. 2007. Reproductive status and body condition of Atlantic bluefin tuna in the Gulf of Maine, 2000–2002. Mar. Biol. 151: 2063–2075.

Goni, N., I. Arregui, M.D. Godoy, J.M. de la Serna, I. Onandia and E. Belda. 2012. Revised final report on the activities of the ICCAT/GBYP – Phase 3 Tagging Program (2012). Madrid, Spain.

Gordina, A., E. Pavlova, E. Ovsyany, J. Wilson, R. Kemp and A.S. Romanov. 2001. Long-term changes in Sevastopol Bay (the Black Sea) with particular reference to the ichthyoplankton and zooplankton. Estuar Coast Shelf Sci. 52: 1–13.

Guinet, C., P. Domenici, R. de Stephanis, L. Barrett-Lennard, J. Ford and P. Verborgh. 2007. Killer whale predation on bluefin tuna: exploring the hypothesis of the endurance-exhaustion technique. Mar. Ecol. Prog. Ser. 347: 111–119.

Gunn, J. and B.A. Block. 2001. Advances in acoustic, archival, and satellite tagging of tunas. pp. 167–224. *In*: B.A. Block and E.D. Stevens (eds.). Tuna: Physiology, Ecology, and Evolution. Academic Press, San Diego.

Heinisch, G., A. Corriero, A. Medina, F.J. Abascal, J.-M. Serna, R. Vassallo-Agius, A.B. Ríos, A. García, F. Gándara, C. Fauvel, C.R. Bridges, C.C. Mylonas, S.F. Karakulak, I. Oray, G. Metrio, H. Rosenfeld and H. Gordin. 2008. Spatial-temporal pattern of bluefin tuna (*Thunnus thynnus* L. 1758) gonad maturation across the Mediterranean Sea. Mar. Biol. 154: 623–630.

Hobday, a.J. and K. Hartmann. 2006. Near real-time spatial management based on habitat predictions for a longline bycatch species. Fish Manag. Ecol. 13: 365–380.

ICCAT. 2012. Report of the 2012 Atlantic bluefin tuna stock assessment session. Madrid, Spain - September 4 to 11, 2012.

ICCAT. 2013. Report of the 2013 bluefin meeting on biological parameters review. Tenerife, Spain - May 7 to 13, 2013.

ICCAT. 2013. Report of the 2013 Bluefin Meeting on Biological Parameters Review. Tenerife, Spain. May 7–13.

Innocentiis, S. De, A. Lesti, S. Livi, A.R. Rossi, D. Crosetti and L. Sola. 2004. Microsatellite markers reveal population structure in gilthead sea bream *Sparus auratus* from the Atlantic Ocean and Mediterranean Sea. Fish Sci. 70: 852–859.

Joint, I., S.B. Groom, R. Wollast, L. Chou, G.H. Tilstone, F.G. Figueiras, M. Loijens and T.J. Smyth. 2002. The response of phytoplankton production to periodic upwelling and relaxation events at the Iberian shelf break: estimates by the 14C method and by satellite remote sensing. J. Mar. Syst. 32: 219–238.

Karakulak, F.S. and I.K. Oray. 2009. Remarks on the fluctuations of bluefin tuna catches in Turkish waters. Collect Vol. Sci. Pap. ICCAT 63: 153–160.

Karakulak, B.S., I. Oray, A. Corriero, M. Deflorio, N. Santamaria, S. Desantis and G. De Metrio. 2004. Evidence of a spawning area for the bluefin tuna (*Thunnus thynnus* L.) in the eastern Mediterranean. J. Appl. Ichthyol. 20: 318–320.

Kitagawa, T., A.M. Boustany, C.J. Farwell, T.D. Williams, M.R. Castleton and B.A. Block. 2007. Horizontal and vertical movements of juvenile bluefin tuna (*Thunnus orientalis*) in relation to seasons and oceanographic conditions in the eastern Pacific Ocean. Fish Oceanogr. 16: 409–421.

Kitagawa, T., S. Kimura, H. Nakata and H. Yamada. 2004. Diving behavior of immature, feeding Pacific bluefin tuna (*Thunnus thynnus orientalis*) in relation to season and area: the East China Sea and the Kuroshio-Oyashio transition region. Fish Oceanogr. 13: 161–180.

Kitagawa, T., S. Kimura, H. Nakata and H. Yamada. 2006. Thermal adaptation of Pacific bluefin tuna, *Thunnus orientalis*, to temperate waters. Fish Sci. 72: 149–156.

Kitagawa, T., H. Nakata, S. Kimura, T. Itoh, S. Tsuji and A. Nitta. 2000. Effect of ambient temperature on the vertical distribution and movement of Pacific bluefin tuna *Thunnus thynnus orientalis*. Mar. Ecol. Prog. Ser. 206: 251–260.

Konstantinou, H. and D.C. Shen. 1995. The social and reproductive behavior of the eyed flounder, Bothus ocellatus, with notes on the spawning of *Bothus lunatus* and *Bothus ellipticus*. Environ. Biol. Fishes 44: 311–324.

Lam, C.H., A. Nielsen and J.R. Sibert. 2008. Improving light and temperature based geolocation by unscented Kalman filtering. Fish Res. 91: 15–25.

Laurs, R.M., P.C. Fiedler and D.R. Montgomery. 1984. Albacore tuna catch distributions relative to environmental features observed from satellites. Deep Sea Res. Part A Oceanogr. Res. Pap. 31: 1085–1099.

Lawson, G., M. Castleton and B. Block. 2010. Movements and diving behavior of Atlantic bluefin tuna *Thunnus thynnus* in relation to water column structure in the northwestern Atlantic. Mar. Ecol. Prog. Ser. 400: 245–265.

Li, W.K.W. 2002. Macroecological patterns of phytoplankton in the northwestern North Atlantic Ocean. Nature 419: 154–157.

Logan, J.M., E. Rodríguez-Marín, N. Goñi, S. Barreiro, H. Arrizabalaga, W. Golet and M. Lutcavage. 2010. Diet of young Atlantic bluefin tuna (*Thunnus thynnus*) in eastern and western Atlantic foraging grounds. Mar. Biol. 158: 73–85.

Loganathan, B.G., K.S. Sajwan, J.P. Richardson, C.S. Chetty and D.A. Owen. 2001. Persistent organochlorine concentrations in sediment and fish from Atlantic coastal and brackish waters off Savannah, Georgia, USA. Mar. Pollut. Bull. 42: 246–250.

Lutcavage, M.E., R.W. Brill, G.B. Skomal, B.C. Chase, J.L. Goldstein and J. Tutein. 2000. Tracking adult North Atlantic bluefin tuna (*Thunnus thynnus*) in the northwestern Atlantic using ultrasonic telemetry. Mar. Biol. 137: 347–358.

Lutcavage, M.E., R.W. Brill, G.B. Skomal, B.C. Chase and P.W. Howey. 1999. Results of pop-up satellite tagging of spawning size class fish in the Gulf of Maine: do North Atlantic bluefin tuna spawn in the mid-Atlantic? Can J Fish Aquat. Sci. 56: 173–177.

MacKenzie, B.R. and P. Mariani. 2012. Spawning of bluefin tuna in the Black Sea: historical evidence, environmental constraints and population plasticity. PLoS One 7: e39998.

Magnuson, J.J., B.A. Block, R. Deriso, J.R. Gold, W.S. Grant, T.J. Quinn II, S.B. Saila, L. Shapiro L and E.D. Stevens. 1994. An Assessment of Atlantic Bluefin Tuna. National Academies Press, Washington D.C.

Magnuson, J.J. and J.H. Prescott. 1966. Courtship, locomotion, feeding, and miscellaneous behaviour of pacific bonito (*Sarda chiliensis*). Anim. Behav. 14: 54–67.

Magoulas, A., R. Castilho, S. Caetano, S. Marcato and T. Patarnello. 2006. Mitochondrial DNA reveals a mosaic pattern of phylogeographical structure in Atlantic and Mediterranean populations of anchovy (*Engraulis encrasicolus*). Mol. Phylogenet. Evol. 39: 734–746.

Mather, F.J., J.M. Mason and A.C. Jones. 1995. Historical document: Life History and Fisheries of Atlantic Bluefin tuna. NOAA Tech. Memo NMFS-SEFSC: 165.

Mather, F.J., B.J. Rothschild, G.J. Paulik and W.H. Lenarz. 1974. Analysis of migrations and mortality of bluefin tuna, *Thunnus thynnus*, tagged in the northwestern Atlantic ocean. Fish Bull. 72: 900–914.

McGowan, M.F. and W.J. Richards. 1989. Bluefin tuna, *Thunnus thynnus*, larvae in the Gulf Stream off the Southeastern United States: satellite and shipboard observations of their environment. Fish Bull. 87: 615–631.

McKenzie, C., B.J. Godley, R.W. Furness and D.E. Wells. 1999. Concentrations and patterns of organochlorine contaminants in marine turtles from Mediterranean and Atlantic waters. Mar. Environ. Res. 47: 117–135.

McPherson, G.R. 1991. Reproductive biology of yellowfin tuna in the eastern Australian Fishing Zone, with special reference to the north-western Coral Sea. Aust. J. Mar. Freshw. Res. 42: 465–477.

Metrio, G. De, G.P. Arnold, B.A. Block, M. Deflorio, M. Cataldo, P. Megalofonou, S. Beemer, C. Farwell and A. Seitz. 2002. Behaviour of post-spawning Atlantic bluefin tuna tagged with pop-up satellite tags in the Mediterranean and eastern Atlantic. Collect Vol. Sci. Pap. ICCAT 54: 415–424.

Metrio, G. De, G.P. Arnold, B.A. Block, P. Megalofonou, M. Lutcavage, I. Oray and M. Deflorio. 2005. Movements of bluefin tuna (*Thunnus thynnus* L.) tagged in the Mediterranean Sea with pop-up satellite tags. Collect Vol. Sci. Pap. ICCAT 58: 1337–1340.

Metrio, G. De, I. Oray, G.P. Arnold, M. Lutcavage, M. Deflorio, J.L. Cort, S. Karakulak, N. Anbar and M. Ultanur. 2004. Joint Turkish-Italian research in the eastern Mediterranean: bluefin tuna tagging with pop-up satellite tags. Collect Vol. Sci. Pap. ICCAT 56: 1163–1167.

Miyake, P.M. 2005. A brief history of the tuna fisheries of the world. pp. 23–50. *In*: W.H. Bayliff, J.I. de Leiva Moreno and J. Majokowski (eds.). Management of Tuna Fishing Capacity: Conservation and Socio-economics. Food and Agriculture Organization of the United Nations, Rome, Italy.

Muhling, B.A., J.T. Lamkin, J.M. Quattro, H. Ryan, M.A. Roberts, M.A. Roffer and K. Ramírez. 2011a. Collection of larval bluefin tuna (*Thunnus thynnus*) outside documented western Atlantic spawning grounds. Bull. Mar. Sci. 87: 687–694.

Muhling, B.A., J.T. Lamkin and W.J. Richards. 2012a. Decadal-scale responses of larval fish assemblages to multiple ecosystem processes in the northern Gulf of Mexico. Mar. Ecol. Prog. Ser. 450: 37–53.

Muhling, B.A., J.T. Lamkin and M.a. Roffer. 2010. Predicting the occurrence of Atlantic bluefin tuna (*Thunnus thynnus*) larvae in the northern Gulf of Mexico: building a classification model from archival data. Fish Oceanogr. 19: 526–539.

Muhling, B.A., S.K. Lee, J.T. Lamkin and Y. Liu. 2011b. Predicting the effects of climate change on bluefin tuna (*Thunnus thynnus*) spawning habitat in the Gulf of Mexico. ICES J. Mar. Sci. 68: 1051–1062.

Muhling, B.A., P. Reglero, L. Ciannelli, D. Alvarez-Berastegui, F. Alemany, J.T. Lamkin and M.A. Roffer. 2013. Comparison between environmental characteristics of larval bluefin tuna *Thunnus thynnus* habitat in the Gulf of Mexico and western Mediterranean Sea. Mar. Ecol. Prog. Ser. 486: 257–276.

Muhling, B.a., M.a. Roffer, J.T. Lamkin, G.W. Ingram, M.a. Upton, G. Gawlikowski, F. Muller-Karger, S. Habtes and W.J. Richards. 2012b. Overlap between Atlantic bluefin tuna spawning grounds and observed Deepwater Horizon surface oil in the northern Gulf of Mexico. Mar. Pollut. Bull. 64: 679–687.

Nielsen, A. and J.R. Sibert. 2007. State-space model for light-based tracking of marine animals. Can J Fish Aquat. Sci. 64: 1055–1068.

Nikaido, H., N. Miyabe and S. Ueyanagi. 1991. Spawning time and frequency of bigeye tuna, Thunnus obesus. Bull. Natl. Res. Inst. Far. Seas Fish 28: 47–74.

Nishida, T., S. Tsuji and K. Segawa. 1998. Spatial data analyses of Atlantic bluefin tuna larval surveys in the 1994 ICCAT BYP. Collect Vol. Sci. Pap. ICCAT 48: 107–110.

NRIFSF. 2002. Report on the 2002 research cruise of the R/V Shoyo Maru. National Research Institue of Far Seas Fisheries, Shimizu, Japan.

Oray, I.K. and F.S. Karakulak. 2005. Further evidence of spawning of bluefin tuna (*Thunnus thynnus* L., 1758) and the tuna species (*Auxis rochei* Ris., 1810, *Euthynnus alletteratus* Raf., 1810) in the eastern Mediterranean Sea: preliminary results of TUNALEV larval survey i. J Appl. Ichthyol. 21: 236–240.

Patterson, T.a., M. Basson, M.V. Bravington and J.S. Gunn. 2009. Classifying movement behaviour in relation to environmental conditions using hidden Markov models. J Anim. Ecol. 78: 1113–23.

Patterson, T.a., L. Thomas, C. Wilcox, O. Ovaskainen and J. Matthiopoulos. 2008. State-space models of individual animal movement. Trends Ecol. Evol. 23: 87–94.

Piccinetti-Manfrin, G., G. Marano, G. De Metrio and C. Piccinetti. 1995. An attempt to find eggs and larvae of bluefin tuna (*Thunnus thynnus*) in the Black Sea. Collect Vol. Sci. Pap. ICCAT 44: 316–317.

Prince, E.D., R.K. Cowen, E.S. Orbesen, S.A. Luthy, J.K. Llopiz, D.E. Richardson and J.E. Serafy. 2005. Movements and spawning of white marlin (*Tetrapturus albidus*) and blue marlin (*Makaira nigricans*) off Punta Cana, Dominican Republic. Fish Bull. 103: 659–669.

Pujolar, J.M. and C. Pla. 2000. Genetic differentiation between north-west Atlantic and Mediterranean samples of bluefin tuna (*Thunnus thynnus*) using isozyme analysis. Collect Vol. Sci. Pap. ICCAT 51: 882–890.

Quílez-Badia, G., P. Cermeño, S. Tudela, S.S. Trapaga and E. Graupera. 2013. Spatial movements of bluefin tuna revealed by electronic tagging in the Mediterranean Sea and in Atlantic waters of Morocco in 2011. Collect Vol. Sci. Pap. ICCAT 69: 435–453.

Reglero, P., L. Ciannelli, D. Alvarez-Berastegui, R. Balbín, J. López-Jurado and F. Alemany. 2012. Geographically and environmentally driven spawning distributions of tuna species in the western Mediterranean Sea. Mar. Ecol. Prog. Ser. 463: 273–284.

Restrepo, V.R., G.A. Diaz, J.F. Walter, J.D. Neilson, S.E. Campana, D. Secor and R.L. Wingate. 2011. Updated estimate of the growth curve of Western Atlantic bluefin tuna 342: 335–342.

Riccioni, G., M. Landi, G. Ferrara, I. Milano, A. Cariani, L. Zane, M. Sella, G. Barbujani and F. Tinti. 2010. Spatio-temporal population structuring and genetic diversity retention in depleted Atlantic bluefin tuna of the Mediterranean Sea. Proc. Natl. Acad. Sci. U.S.A. 107: 2102–7.

Richards, W.J. 1990. Results of a review of the US bluefin tuna larval assessment with a brief response. Collect Vol. Sci. Pap. ICCAT 32: 240–247.

Richards, W.J., T. Leming, M.F. McGowan, J.T. Lamkin and S. Kelley-Fraga. 1989. Distribution of fish larvae in relation to hydrographic features of the Loop Current boundary in the Gulf of Mexico. *In*: J.H.S. Blaxter, J.C. Gamble and H. von Westernhagen (eds.). The Early Life History of Fish: the 3rd ICES Symposium, Bergen, 3–5 October 1988. ICES, Copenhagen, Denmark.

Rideout, R.M. and J. Tomkiewicz. 2011. Skipped spawning in fishes: more common than you might think. Mar. Coast Fish. Dyn. Manag. Ecosyst. 3: 176–189.

Rodriguez-Marin, E., J.M. Ortiz De Urbina, E. Alot, J.L. Cort, J.M. de la Serna, D. Macias, C. Rodriguez-Cabello, M. Ruiz and X. Valeiras. 2009. Tracking bluefin tuna cohorts from east Atlantic Spanish fisheries since the 1980s. Collect Vol. Sci. Pap. ICCAT 63: 121–132.

Rodriguez-Marin, E., C. Rodriguez-Cabello, J.M. de La Serna, E. Alot, J.L. Cort and M. Quintans. 2008. Bluefin tuna (*Thunnus thynnus*) conventional tagging carried out by the Spanish Institute of Oceanography (IEO) in 2005 and 2006. Results and analysis including previous tagging activities. Collect Vol. Sci. Pap. ICCAT 62: 1182–1197.

Rodriguez-Marin, E., C. Rodriguez-Cabello, J.M. de La Serna, J.L. Cort, E. Alot, J.C. Rey, V. Ortiz De Zarate, J.L. Gutierrez and E. Abad. 2005. A review of bluefin tuna juveniles tagging information and mortality estimation in waters around the Iberian Peninsula. Collect Vol. Sci. Pap. ICCAT 58: 1388–1402.

Rooker, J.R., J.R. Alvarado-Bremer, B.A. Block, H. Dewar, G. de Metrio, A. Corriero, R.T. Kraus, E.D. Prince, E. Rodríguez-Marín and D.H. Secor. 2007. Life history and stock structure of Atlantic bluefin tuna (*Thunnus thynnus*). Rev. in Fish. Sci. 15: 265–310.

Rooker, J.R., H. Arrizabalaga, I. Fraile, D.H. Secor, D.L. Dettman, N. Abid, P. Addis, S. Deguara, F.S. Karakulak, A. Kimoto, M.N. Santos, H. Kaia and D. Macias. 2014. Crossing the line: migratory and homing behaviors of Atlantic bluefin tuna. Mar. Ecol. Prog. Ser. 504: 265–276.

Rooker, J., D. Secor, G. De Metrio, A. Kaufman, Ríos A. Belmonte and V. Ticina. 2008a. Evidence of trans-Atlantic movement and natal homing of bluefin tuna from stable isotopes in otoliths. Mar. Ecol. Prog. Ser. 368: 231–239.

Rooker, J.R., D.H. Secor, G. De Metrio, R. Schloesser, B.A. Block and J.D. Neilson. 2008b. Natal homing and connectivity in Atlantic bluefin tuna populations. Nature 322: 742–744.

Rooker, J.R., D.H. Secor, V.S. Zdanowicz, G. De Metrio and L.O. Relini. 2003. Identification of Atlantic bluefin tuna (*Thunnus thynnus*) stocks from putative nurseries using otolith chemistry. Fish Oceanogr. 12: 75–84.

Royer, F., J. Fromentin and P. Gaspar. 2004. Association between bluefin tuna schools and oceanic features in the western Mediterranean. Mar. Ecol. Prog. Ser. 269: 249–263.

Royer, F., J.M. Fromentin and P. Gaspar. 2005. A state-space model to derive bluefin tuna movement and habitat from archival tags. Oikos 109: 473–484.

Sara, R. 1973. Sulla biologia dei tonni (*Thunnus thynnus* L.) modelli di migrazione ed osservazioni sui meccanismi di migrazione e di comportamento. Boll di Pesca, Piscic e Idrobiol. 28: 217–243.

Schaefer, K.M. 1996. Spawning time, frequency, and batch fecundity of yellowfin tuna, *Thunnus albacares*, near Clipperton Atoll in the eastern Pacific Ocean. Fish Bull. 94: 98–112.

Schaefer, K.M. 1998. Reproductive biology of yellowfin tuna (*Thunnus albacares*) in the Eastern Pacific Ocean. Inter. Am. Trop. Tuna Comm. Bull. 21: 205–272.

Schaefer, K.M. 2001. Reproductive biology of tunas. pp. 225–270. *In*: B.A. Block and E.D. Stevens (eds.). Tuna: Physiology, Ecology, and Evolution, 1st edn. Academic Press, San Diego.

Schick, R.S., J. Goldstein and M.E. Lutcavage. 2004. Bluefin tuna (*Thunnus thynnus*) distribution in relation to sea surface temperature fronts in the Gulf of Maine (1994–96). Fish Oceanogr. 13: 225–238.

Schloesser, R.W., J.D. Neilson, D.H. Secor and J.R. Rooker. 2010. Natal origin of Atlantic bluefin tuna (*Thunnus thynnus*) from Canadian waters based on otolith $\delta^{13}C$ and $\delta^{18}O$. Can J Fish Aquat. Sci. 67: 563–569.

Secor, D.H. 2007. Do some Atlantic bluefin tuna skip spawning? Collect Vol. Sci. Pap. ICCAT 60: 1141–1153.

Secor, D.H., J.R. Rooker, J.D. Neilson and D. Busawon. 2013. Historical Atlantic bluefin tuna stock mixing within fisheries off the United States, 1976–2012. Collect Vol. Sci. Pap. ICCAT 69: 938–946.

Secor, D.H. and V.S. Zdanowicz. 1998. Otolith microconstituent analysis of juvenile bluefin tuna (*Thunnus thynnus*) from the Mediterranean Sea and Pacific Ocean. Fish Res. 36: 251–256.

Seitz, A.C., B.L. Norcross, D. Wilson and J.L. Nielsen. 2005. Identifying spawning behavior in Pacific halibut, *Hippoglossus stenolepis*, using electronic tags. Environ. Biol. Fishes 73: 445–451.

Shaffer, S.A., Y. Tremblay, H. Weimerskirch, D. Scott, D.R. Thompson, P.M. Sagar, H. Moller, G.A. Taylor, D.G. Foley, B.A. Block and D.P. Costa. 2006. Migratory shearwaters integrate oceanic resources across the Pacific Ocean in an endless summer. Proc. Natl. Acad. Sci. USA 103: 12799–802.

Sharp, G.D. and W.J. Vlymen. 1978. The relation between heat generation, conservation, and the swimming energetics of tunas. pp. 213–232. *In*: G.D. Sharp and A.E. Dizon (eds.). The Physiological Ecology of Tunas. Academic Press, San Diego.

Sibert, J.R., M.E. Lutcavage, A. Nielsen, R.W. Brill and S.G. Wilson. 2006. Interannual variation in large-scale movement of Atlantic bluefin tuna (*Thunnus thynnus*) determined from pop-up satellite archival tags. Can J Fish Aquat. Sci. 63: 2154–2166.

Sibert, J.R., M.K. Musyl and R.W. Brill. 2003. Horizontal movements of bigeye tuna (Thunnus obesus) near Hawaii determined by Kalman filter analysis of archival tagging data. Fish Oceanogr. 12: 141–151.

Sippel, T., J.P. Eveson, B. Galuardi, C. Lam, S. Hoyle, M. Maunder, P. Kleiber, F. Carvalho, V. Tsontos, S.L.H. Teo, A. Aires-da-Silva and S. Nicol. 2014. Using movement data from electronic tags in fisheries stock assessment: A review of models, technology and experimental design. Fish Res. 163: 152–160.

Smyth, T.J., P.I. Miller, S. Groom and S. Lavorel. 2001. Remote sensing of sea surface temperature and chlorophyll during Lagrangian experiments at the Iberian margin. Prog. Oceanogr. 51: 269–281.

Stefanelli, P., A. Ausili, G. Ciuffa, A. Colasanti, S. Di Muccio and R. Morlino. 2002. Investigation of Polychlorobiphenyls and Organochlorine Pesticides in Tissues of Tuna (*Thunnus Thunnus Thynnus*) from the Mediterranean Sea in 1999. Bull. Environ. Contam. Toxicol. 69: 800–807.

Stokesbury, M.J.W., S.L.H. Teo, A. Seitz, R.K.O. Dor and B.A. Block. 2004. Movement of Atlantic bluefin tuna (*Thunnus thynnus*) as determined by satellite tagging experiments initiated off New England. 1987: 1976–1987.

Sund, P.N., M. Blackburn and F. Williams. 1981. Tunas and their environment in the Pacific Ocean: a review. Oceanogr. Mar. Biol. An. Annu. Rev. 19: 443–512.

Takeuchi, Y., K. Oshima and Z. Suzuki. 2009. Inference on nature of Atlantic bluefin tuna off Brazil caught by the Japanese longline fishery around the early 1960s. Collect Vol. Sci. Pap. ICCAT 63: 186–194.

Takeuchi. Y., A. Sudal and Z. Suzukil. 1999. Review of information on large bluefin tuna caught by Japanese longline fishery off Brasil from the late 1950s to the early 1960s. Collect Vol. Sci. Pap. ICCAT 49: 416–428.

Tangen, M. 2009. The Norwegian Fishery for Atlantic Bluefin Tuna *. Collect Vol. Sci. Pap. ICCAT 93: 79–93.

Taylor, N.G., M.K. McAllister, G.L. Lawson, T. Carruthers and B.A. Block. 2011. Atlantic bluefin tuna: a novel multistock spatial model for assessing population biomass. PLoS One 6: e27693.

Teo, S.L.H. and B.A. Block. 2010. Comparative influence of ocean conditions on yellowfin and Atlantic bluefin tuna catch from longlines in the Gulf of Mexico. PLoS One 5: e10756.

Teo, S.L.H., A.M. Boustany and B.a. Block. 2007a. Oceanographic preferences of Atlantic bluefin tuna, *Thunnus thynnus*, on their Gulf of Mexico breeding grounds. Mar. Biol. 152: 1105–1119.

Teo, S.L.H., A. Boustany, H. Dewar, M.J.W. Stokesbury, K.C. Weng, S. Beemer, A.C. Seitz, C.J. Farwell, E.D. Prince and B.a. Block. 2007b. Annual migrations, diving behavior, and thermal biology of Atlantic bluefin tuna, *Thunnus thynnus*, on their Gulf of Mexico breeding grounds. Mar. Biol. 151: 1–18.

Thompson, H.C. and R.F. Contin. 1980. Electrophoretic study of Atlantic bluefin tuna (*Thunnus thynnus*) from the eastern and western North Atlantic Ocean. Bull. Mar. Sci. 30: 727–731.

Tilstone, G.H., F.G. Figueiras, L.M. Lorenzo and B. Arbones. 2003. Phytoplankton composition, photosynthesis and primary productivity during different hydrographic conditions at the Northwest Iberian upwelling system. Mar. Ecol. Prog. Ser. 252: 89–104.

Tomczak, M. and J.S. Godfrey. 2003. Regional Oceanography: an Introduction, PDF edition.

Tudela, S., S.S. Trápaga, P. Cermeño, E. Hidas, E. Graupera and G. Quilez-Badia. 2011. Bluefin tuna migratory behavior in the western and central Mediterranean Sea revealed by electronic tags. Collect Vol. Sci. Pap. ICCAT 66: 1157–1169.

Vanderlaan, A.S.M., B.A. Block, J. Chasse, A. Hanke, M.E. Lutcavage, S.G. Wilson and J.D. Neilson. 2011. Initial investigations of environmental influences on Atlantic bluefin tuna catch rates in the southern Gulf of St. Lawrence. Collect Vol. Sci. Pap. ICCAT 66: 1204–1215.

Viñas, J., A. Gordoa, R. Fernández-Cebrián, C. Pla, Ü Vahdet and R.M. Araguas. 2011. Facts and uncertainties about the genetic population structure of Atlantic bluefin tuna (*Thunnus thynnus*) in the Mediterranean. Implications for fishery management. Rev. Fish Biol. Fish. 21: 527–541.

Walli, A., S.L.H. Teo, A. Boustany, C.J. Farwell, T. Williams, H. Dewar, E. Prince and B.a. Block. 2009. Seasonal movements, aggregations and diving behavior of Atlantic bluefin tuna (*Thunnus thynnus*) revealed with archival tags. PLoS One 4: e6151.

Wilson, S. and B. Block. 2009. Habitat use in Atlantic bluefin tuna *Thunnus thynnus* inferred from diving behavior. Endanger Species Res. 10: 355–367.

Wilson, S.G., G.L. Lawson, M.J.W. Stokesbury, A. Spares, A.M. Boustany, J.D. Neilson and B.A. Block. 2011. Movements of Atlantic bluefin tuna from the Gulf of St. Lawrence to their spawning grounds. Collect Vol. Sci. Pap. ICCAT 66: 1247–1256.

Wormuth, J.H., P.H. Ressler, R.B. Cady and E.J. Harris. 2000. Zooplankton and micronekton in cyclones and anticyclones in the northeast Gulf of Mexico. Gulf Mex. Sci. 18: 23–34.

Yamashita, H. and N. Miyabe. 2001. Report of bluefin tuna archival tagging conducted by Japan in 1999 in the Adriatic Sea. Collect Vol. Sci. Pap. ICCAT 52: 809–823.

Zaitsev, Y. and V. Mamaev. 1997. Biological Diversity in the Black Sea: A Study of Change and Decline. United Nations Publications, New York.

CHAPTER 8

Distribution and Migration–Southern Bluefin Tuna (*Thunnus maccoyii*)

Alistair J. Hobday, Karen Evans, J. Paige Eveson,
Jessica H. Farley, Jason R. Hartog, Marinelle Basson* and
Toby A. Patterson

Introduction

Southern bluefin tuna (*Thunnus maccoyii*, SBT) are a highly migratory tuna species that occur in waters of the eastern Atlantic, Indian and South-west Pacific Oceans between 30°S and 50°S (Caton 1991). Mature fish migrate to a single spawning ground located in the North-east Indian Ocean between Java and Australia (Farley and Davis 1998; Patterson et al. 2008). After a larval stage, juveniles leave the spawning ground, move down the west coast of Australia and reach the south coast by age one where they spend the austral summer. Up until approximately five years of age, juvenile fish undertake annual cyclical migrations in which they generally spend austral summers in the Great Australian Bight (GAB), and move east as far as New Zealand or west as far as South Africa during the winter (Bestley et al. 2008; Basson et al. 2012). Fish older than five years rarely return to the GAB and disperse widely across the southern oceans from the western Atlantic across the Indian Ocean to the Tasman Sea. SBT are estimated to reach maturity at 10–12 years and can live to more than 40 years (Shimose and Farley this volume).

Although the species is long-lived and highly fecund, characteristics such as slow growth, late onset of maturity, the presence of a single spawning ground and a highly migratory behaviour (exposing the stock to national and international commercial

CSIRO Oceans and Atmosphere Flagship, Hobart, Tasmania 7004, Australia.
* Email: Alistair.hobday@csiro.au

fishing fleets) led to overexploitation and dramatic reductions in population size (Anon 2011). Between 1960 and the present, the global stock of SBT declined to less than 10% of its unfished biomass, leading to successive reductions in commercial quotas for fishing nations (Anon 2011). Today, SBT are fished commercially by Australia and Japan, with smaller catches taken by vessels from New Zealand, Taiwan, Indonesia, Korea, South Africa, the European Union and the Philippines. Commercial fishing for SBT is managed internationally by the Commission for the Conservation of Southern Bluefin Tuna (CCSBT), which determines global total allowable catch and allocates national quotas to member nations (http://www.ccsbt.org/site/commission_role. php). Within Australia, the species is listed as threatened both federally and within a number of states and is listed as critically endangered by the International Union for Conservation of Nature (IUCN). Recent assessments suggest the population may be recovering, and under the CCBST, harvest strategies are in place to ensure more robust population sizes by 2020 (Anon 2013).

Broad scale information on the distribution of the species has traditionally been based on catch data, fishery-dependent tagging programs, and larval surveys. More detailed insights into the horizontal and vertical movements of individuals have come from recent studies using a variety of electronic tags. In this chapter we describe the global and regional distribution and migration patterns relative to the life cycle of SBT, including information on how these patterns have been determined. We discuss management of the species in the context of the spatial dynamics of the species and potential changes to population distributions and movements in relation to climate change. We conclude with a discussion of the remaining uncertainty regarding distribution and movement patterns, and research required to address these uncertainties.

Distribution and Movement of Early Life Stages

Larval distributions

Genetic studies suggest that there is a single SBT population (Grewe et al. 1997). Larval studies support this understanding, with larvae collected from a single region in the Indian Ocean between Java and North-west Australia (Nishikawa et al. 1985; Davis et al. 1989) (Fig. 8.1). The spawning area is large, but has not been precisely surveyed or defined, and a lack of data from Indonesian territorial waters results in substantial uncertainty on the northern limits of the spawning area. The surface ocean characteristics in the general region have been shown to be particularly stable compared to the rest of the Indian Ocean in terms of consistent sea surface temperatures (SST), low eddy kinetic energy and low chlorophyll *a*, with a high frequency of low intensity SST and chlorophyll fronts (Nieblas et al. 2014). These characteristics are consistent with the loophole hypothesis of Bakun and Broad (2003) in which poor larval feeding conditions are balanced by reduced predation. Recent comparative work shows that tuna larvae feeding success is low in this region (Llopiz and Hobday 2014), supporting earlier work that suggested that food is limiting for SBT larval growth in the region (Young and Davis 1990). Net sampling of larvae in the spawning region has shown that SBT larvae (4–7 mm) are largely restricted to the mixed layer, and move towards

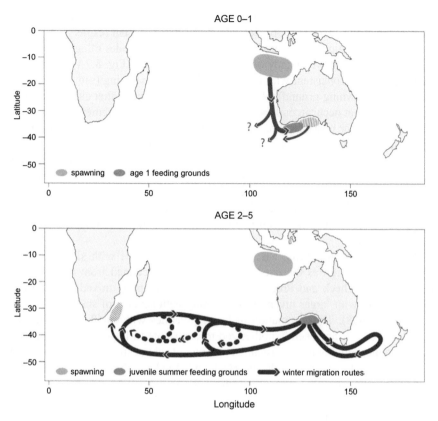

Figure 8.1. General migration patterns and distribution of juvenile southern bluefin tuna with age one movements from the spawning ground (upper panel) and age two–five movements from the Great Australian Bight (lower panel).

the surface during the day (Davis et al. 1990), probably to feed. These small larvae are passive and drift with surface water masses (Davis et al. 1991). As larvae grow (the larval stage is estimated to last 20 days—Jenkins and Davis 1990) and become capable of independent movement, an unknown fraction is entrained in the poleward flowing Leeuwin Current (Fig. 8.1).

Post-larval juvenile movement and distribution

Post-larval, juvenile SBT move down the west coast of Australia, carried in part by the Leeuwin Current. On reaching the south-west coast they are generally restricted to the continental shelf (Hobday et al. 2009a) where they predominately occur in warm waters adjacent to colder upwelling cells (Fujioka et al. 2010). Nutrient-rich and cool sub-Antarctic water periodically intrudes onto the southern Western Australian shelf, leading to elevated chlorophyll concentrations and prey densities (Ward et al. 2006). By December and January each year, age-one fish (approximately 50 cm in length) are found off southern Western Australia, with some age-two fish (mean size 79 cm) also present (Itoh and Tsuji 1996; Hobday et al. 2009a). Recent results from tagging programs in the region found that in some years two size modes of age-one fish are present while in other years only one size mode is present (Fig. 8.2A). Two modes of age-one fish is consistent with two peak periods of spawning (see: Migration of adults to the spawning ground). However, one mode suggests that either successful spawning does not occur in both periods every year, or that movement into southern Australia varies for post-larval fish from the two spawning periods, perhaps due to oceanographic influences on entrainment into the Leeuwin Current.

It has been suggested that not all juveniles move directly south from the spawning region to South-west Australian waters, but that some larvae and juveniles are carried from the spawning ground towards the west, perhaps in the south equatorial current (Harden-Jones 1984). Alternatively, juveniles may swim west from southern Western Australia into the Indian Ocean (Murphy 1977; Murphy and Majkowski 1981; Hobday et al. 2009a).

In South-west Australia, SBT generally form schools with similarly sized conspecifics, although as the overall population size declined from the 1960's to the 1990's (Polacheck and Preece 2001), schools containing mixed sizes became more prevalent with larger juveniles associating with schools of smaller fish (Dell and Hobday 2008). The association of larger juveniles with smaller juveniles that remain in the west has been hypothesized to contribute to lower numbers of these larger juveniles moving to the east and a measure of school size similarity may be of use as a potential population indicator (Dell and Hobday 2008). Small age-two fish consistently occur in the waters off southern Western Australia, likely members of the smaller age-one cohort, with the larger age-one fish having moved into the GAB by their second year (Fig. 8.2B).

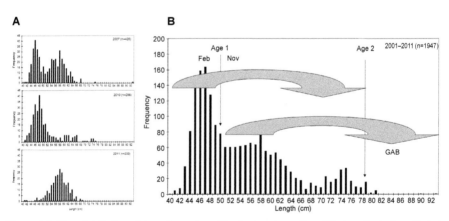

Figure 8.2. Size distribution of juvenile southern bluefin tuna in southern Western Australia collected during tagging studies in the austral summer (A) for three years showing two age-one cohorts in 2007, a small age-one cohort in 2010 and a large age-one cohort in 2011, and (B) all fish collected across the years 2001–2011. The average size for age-one and age-two fish is shown, together with an inferred spawning month (Feb or November). Note the absence of a large age-two cohort in southern Western Australia, which may have moved to the Great Australian Bight (GAB). Hobday, unpubl. data.

Whilst in waters off South-west Australia, schools of SBT can be free-swimming or associated with topographic features (termed 'lumps') distributed on the continental shelf (Hobday and Campbell 2009). Interestingly, habitat partitioning between several pelagic species seems to occur at these features, with yellowtail kingfish occurring closest to the feature, followed by bonito, and then SBT occurring in the outer radius. Skipjack tuna in the region showed avoidance of features. Such fine scale distributional information may inform spatial management, and allow fishers to avoid or target particular species over small spatial scales (Hobday and Campbell 2009).

Due to their small size, the movements of age-one SBT have generally been studied with conventional tags and via acoustic tags and subsequent monitoring with large numbers of listening stations running across the continental shelf and around topographic features in the South-west Australian region (Hobday et al. 2009a). Data collected over almost 10 years has demonstrated that both the residence time and migration routes inshore and offshore across the continental shelf of age-one and age-two SBT along the south coast varies between years (Hobday et al. 2009a; Honda et al. 2010). The strength of the warm Leeuwin Current into southern Western Australia may influence the movement timing and foraging habitat of juvenile SBT in this region (Hobday et al. 2009a; Fujioka et al. 2012). Seasonal and interannual changes in the strength of the Leeuwin Current lead to thermal differences and potential changes in food availability between tropical and temperate waters. Movements of juvenile SBT into waters further east have been observed to increase as temperatures increase, with fish leaving the region when temperatures exceeded 20°C, a temperature indicative of the leading edge of the Leeuwin Current in this region (Fujioka et al. 2012). When the Leeuwin Current was narrow and restricted to the shelf edge, the distribution of SBT in inshore waters was not affected. In contrast, long distance eastward movements frequently occurred when the Leeuwin Current intrusion was spread wide over the continental shelf. This suggests that juvenile SBT move quickly out of local foraging habitats defined by cool, sub-tropical and temperate waters ahead of the tropical Leeuwin Current intrusion, despite these waters not being physiologically limiting. Movements are likely driven instead by changes in prey availability resulting from changes in oceanographic conditions (Fujioka et al. 2012).

An overall habitat utilization model for the southern Western Australia shelf, termed the productivity-distribution hypothesis, has been proposed for juvenile SBT on the basis of movements observed in age-one and two individuals (Honda et al. 2010). As oceanographic conditions vary between years, driven by varying Leeuwin Current strength and coastal upwelling, environmental quality (i.e., food availability) in the region also varies, with the greatest variability observed at the shelf edge. When the environmental quality for SBT is good, higher prey densities occur at the shelf edge, and subsequently SBT are distributed to a greater extent at the shelf edge (Honda et al. 2010). In poor quality years, the inshore environment represents improved foraging relative to the shelf environment (Fig. 8.3). There is a suggestion that small fish may always be more common at the shelf edge as they are excluded from the inshore waters in the poor years (Honda et al. 2010), but this hypothesis remains unresolved. In order to establish if this productivity-distribution hypothesis is reflective of drivers for juvenile SBT movements and distribution in the waters of

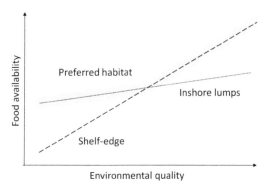

Figure 8.3. Proposed productivity-distribution hypothesis explaining distribution of juvenile SBT in southern Western Australia. Food availability varies more strongly as a function of environmental quality at the shelf edge (dashed line) compared to inshore lumps (dotted line). The preferred habitat across the range of environmental conditions is shaded.

South-western Australia, detailed distribution and behavioural data for both predator and prey species is required (e.g., Itoh et al. 2011).

Collectively, data from conventional and electronic tagging experiments have shown that at the end of the austral summer, juvenile SBT of age-one either remain in waters off South-western Australia, move east and into the GAB, or move west and into the Indian Ocean, although the proportion for each pathway has not been determined. It has been assumed that the majority of juvenile SBT move to the GAB by the following summer.

Juvenile SBT in the GAB

The waters of the GAB are an important summer feeding ground for juvenile SBT aged two–five years (Basson et al. 2012; Bestley et al. 2009; Fig. 8.1). The GAB region comprises a very wide (up to 260 km) and shallow (<160 m) continental shelf which extends some 1300 km along the coastline of southern Australia and covers an area of approximately 150,000 km².

Shelf circulation in the GAB is dominated by the Leeuwin Current and the west wind drift in the winter, and by the Flinders Current and winds from the south-east in the summer. By some descriptions, the waters of the GAB are oligotrophic with low productivity (Condie and Dunn 2006), which seems paradoxical given the high density of many predators, including SBT, in the region (see a review of the GAB ecosystem in Rogers et al. 2013). During winter, the Leeuwin Current extends into the western GAB and spreads across the shelf of the GAB as a body of warm water (~20°C), with relatively low salinity. The Leeuwin Current reinforces the wind-driven geostrophic coastal current directed to the east and interaction with the offshore Flinders Current results in a series of eddies in the eastern GAB enhancing local cross-shelf water exchange (Rogers et al. 2013). During the summer, the Leeuwin Current weakens, resulting in little to no penetration of waters associated with the current into the GAB. Winds from the east drive small upwelling cells, particularly in the eastern and far western GAB, bringing cold nutrient-rich waters onto the shelf (Middleton and Bye 2007) and support pelagic food webs in the GAB (Rogers et al. 2013).

Considerable effort has been expended on understanding the abundance, distribution and migration patterns of juvenile SBT in the GAB, motivated in part by the large surface fishery that operates there, relatively easy and reliable access to part of a widely spread population, and also broader scale declines in the population. A number of programs involving the deployment of conventional tags on juvenile SBT in the GAB have occurred since the 1950's providing information on movements and distribution of juveniles both within and outside of the GAB, as well as providing estimates of mortality and growth rates of individual SBT (Fig. 8.4). In the early 1990's, to better understand the movements of juvenile tuna, CSIRO initiated extensive development of archival tags that could provide quantitative data on how fish use the environment (Gunn et al. 1994). This work, in conjunction with related research developments elsewhere in the world, has resulted in the availability of highly efficient archival tags for marine species (e.g., Gunn and Block 2001; Nielsen et al. 2009) and has provided important insights into SBT movements (Gunn 1999; Bestley et al. 2008; Bestley et al. 2009; Basson et al. 2012). A large multi-nation multi-region archival tagging study has revealed even more variation in the movements of juvenile SBT (Basson et al. 2012). Some 568 tags were released on age two–four SBT in five ocean regions (from the western Indian Ocean to New Zealand) and over 75 tags (13%) have been recaptured to date, and demonstrated the potential power of these tags to provide the data necessary to start integrating spatial patterns into the assessment of pelagic resources (e.g., Nielsen et al. 2009).

Archival tags have been critical to understanding the timing of juvenile SBT migrations, showing that juveniles start to appear in the GAB around November/ December (Bestley et al. 2009; Basson et al. 2012). Whilst in the GAB, juvenile SBT aggregate in large schools, which spend substantial time in the upper 100 m of the water column mostly during the day with deeper average depths observed in individuals

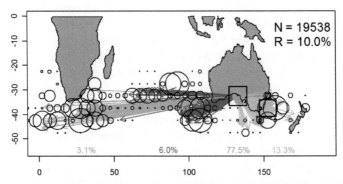

Figure 8.4. Movement of juvenile (age two) southern bluefin tuna from the Great Australia Bight as determined from recapture of 19,538 conventional tags released in the 1990's. The colours correspond to regions in which the recapture occurred, with percentages showing the recovery of tags in each area. A total of 10% of tagged fish were recaptured as age three or four year old fish. Circles show fishing effort in the southern ocean by all fleets.

during the night (Bestley et al. 2009). It is this surfacing behaviour that commercial purse seine operations take advantage of by using spotter planes to locate and target schools of juvenile SBT.

Data on the surface distribution of SBT in the GAB between January and March has also been collected since 1993 as part of a scientific aerial survey. The survey uses experienced aerial tuna spotters to locate SBT schools while searching along pre-set transect lines and to estimate the biomass in each school. The primary purpose of the survey is to provide an annual index of juvenile SBT relative abundance in the GAB (Cowling et al. 2003; Eveson et al. 2014). Since 2002, data on the surface distribution of SBT in the GAB have also been collected by commercial tuna spotters whilst engaged in purse seine operations (Basson and Farley 2014). Both the aerial survey and commercial spotting data sets show that the highest densities of SBT schools are usually found in a band inside and parallel to the continental shelf break, although the precise location varies (Fig. 8.5). In recent years, the area of highest school density has moved from the central GAB (between ~130° and 133°E) to the east (~134°E) following the shelf break (Basson and Farley 2014), although inshore areas around lumps and reefs continue to be important for small/young SBT.

Data from archival tagging experiments suggest that during the summer in the GAB, feeding success of juvenile tuna is high (Bestley et al. 2010), as they take advantage of enhanced productivity supporting spawning aggregations of small pelagic fishes (Ward et al. 2006). This enhanced productivity supports high juvenile growth rates (Bestley et al. 2010). Archival tags that are internally implanted in the peritoneal cavity close to the stomach of SBT record changes in internal body temperature that can be used as indicators of feeding events (Gunn et al. 2001; Bestley et al. 2008) and can be used to estimate relative intake size (Bestley et al. 2008). Individual foraging success has been observed to be highly variable with feeding predominantly occurring during the day and in particular, around dawn (Bestley et al. 2008), supporting assumptions that SBT are visual predators. Foraging occurs both during residence and migration phases and during migrations phases particularly so during directed movement (Bestley et al. 2008; 2010). Variation observed in feeding success may reflect differences in the availability of prey or the prey type targeted by individuals. Whilst foraging in the GAB, juvenile SBT move rapidly between inshore and shelf break habitats (Davis and Stanley 2002; Willis and Hobday 2007), avoiding cool upwelled water, likely associated with prey movements. Simulations of movements based on data from acoustic tagging experiments suggest that short movements between inshore topographic features are common—likely reflecting inshore foraging (Willis and Hobday 2007). Individuals have been observed to exhibit short-term school fidelity, suggesting that in the GAB schools break up and reform relatively often (Willis and Hobday 2007). High variability in foraging success observed in juvenile SBT whilst in the GAB suggests that movements may also be driven by social factors rather than individual-based decision making (Gunn and Block 2001) and that particular topographic features may be important for resting or for use as navigational references (Cowling et al. 2003; Willis and Hobday 2007; Bestley et al. 2008).

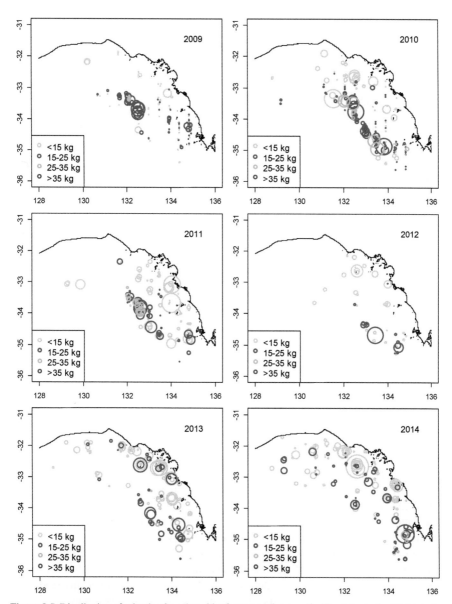

Figure 8.5. Distribution of schools of southern bluefin tuna varies over the Great Australian Bight during the austral summer (Jan–Mar) based on aerial survey data for 2009–2014.

Movements of juvenile SBT outside the GAB

Towards the end of the austral summer when most juveniles begin to leave the GAB, many undertake occasional exit forays before permanent departure for the winter grounds. Migration schedules are highly variable with individuals departing from the

GAB as early as March and as late as August (Bestley et al. 2009; Basson et al. 2012). The majority of fish move west to the Indian Ocean, with the remainder moving east to the Tasman Sea (Basson et al. 2012). Movements to the west or east are not always direct, with some fish spending time in the waters around Tasmania before migrating onwards (see example movement tracks in Fig. 8.6). During the austral winter-spring period, SBT that have moved into the Indian Ocean are widely distributed between about 30–45°S. Some individuals that migrate into the Indian Ocean continue to move further west toward South Africa, while others remain resident in the mid-Indian Ocean (Basson et al. 2012). Long-term archival tag deployments on juvenile SBT have shown that individuals do not always migrate to the same area or even the same ocean each winter, with some spending the winter in the Indian Ocean one year and then in the Tasman Sea the following year (Basson et al. 2012).

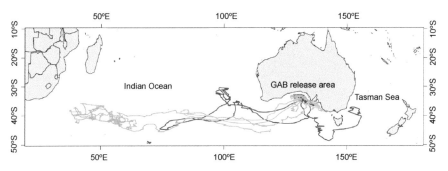

Figure 8.6. Example of three juvenile southern bluefin tuna inter-oceanic migration tracks estimated using state space models using methods described in Basson et al. (2012) and Pedersen et al. (2011).

The coldest waters inhabited by juvenile SBT lie in the most southerly latitudes where water is deeply mixed, between the southern edge of the sub-tropical front and the northern edge of the sub-Antarctic front, known as the Sub-Antarctic Mode Water (Belkin and Gordon 1996). Surface waters may be only 12°C, declining to 7°C below a sharp thermocline at ~100 m (Bestley et al. 2008). Overall, SBT do not distribute to maintain preferred depth or temperature ranges, but rather show highly plastic behaviours in response to changes in their environment (Bestley et al. 2009). In analyses of the statistical patterns of movement in relation to SST, Patterson et al. (2009) found indications that more rapid movement was associated with cooler water and that residency periods coincided with warmer phases. However this study was based on a small sample size and further investigation with data from more individuals is required to ascertain whether this pattern holds more generally.

Changes in the migration patterns of juvenile SBT have been recorded between the 1990's and 2000's, with fewer SBT moving east out of the GAB and into the Tasman Sea in the 2000's than occurred previously (Basson et al. 2012). Declining populations have been suggested as a possible driver for changes in the migration pathways used by juvenile SBT; however, changing environmental conditions may also play a role given that local environmental conditions, particularly SST and chlorophyll, have been found to influence their movement and habitat selection (Basson et al. 2012).

The long distance migrations of SBT have piqued the curiosity of many scientists, yet a mechanistic physiological explanation is lacking. Given that SBT, along with other bluefin species, undertake migrations over thousands of kilometres (Gunn and Block 2001) and can return to the same locations annually, sensory mechanisms and behavioural strategies that increase the accuracy of long-distance movements should be strongly favoured by natural selection (Willis et al. 2009). Archival tag data from the bluefin tuna species often show distinctly shaped ascents and descents at dawn and dusk known as spike dives (Gunn and Block 2001; Willis et al. 2009). Explanations that have been proposed for spike diving by tuna include (1) locating the base of the mixed layer, (2) surveying prey fields, (3) performing a geomagnetic survey for navigation, (4) undertaking a general environmental survey of the water column, and (5) foraging (Gunn and Block 2001). Willis et al. (2009) posit that spike dives may represent a behaviour related to navigation, as they are similar among fish, and are mirror images at dawn and dusk. Anatomical evidence for elaboration of the pineal organ, which is light mediated and has been implicated in navigation in other vertebrates also supports a navigational role in SBT (Willis et al. 2009). Additional experimental work is needed to determine the importance of the pineal gland and the magnetic field in tuna navigation—a difficult task given the size and physiology of these animals.

Distribution and Movement of Older Life Stages

Distribution of sub-adults and adults

Relatively little is known about movements of sub-adult SBT (ages five–10) due to their widely dispersed nature and the subsequent logistical difficulties in catching and tagging these age groups Commercial catch data suggests these fish disperse throughout southern temperate waters. In the Australian region, both sub-adult and adult SBT occur seasonally during the austral winter throughout the Tasman Sea. Sub-adults are also known to occur in the waters east of South Africa and can undertake movements around the south of Africa and into the Atlantic Ocean (CSIRO, unpubl. data). There is some evidence that sub-adults and potentially young, mature adults remain in the Tasman Sea year-round (Evans et al. 2012). In waters around Tasmania, sub-adults have been observed to forage on a high diversity of fish, cephalopods and crustaceans. Similar to juveniles, feeding has been estimated to occur predominantly during the day and in particular at dawn (Young et al. 1997).

Sub-adult and adult SBT caught in the Tasman Sea and tagged with pop-up satellite archival tags demonstrate temperature preferences for waters of 18–20°C and waters <250 m, although spend time at depths >600 m and demonstrate diel variation in diving behaviour for periods of time (Patterson et al. 2008). Within this data set, significant time amounts of time spent in waters >250 m and <18°C was limited to three localized regions: the eastern Tasman Sea, the western Tasman Sea and north-west of Tasmania at 135–145°E.

Whilst in the Tasman Sea region, sub-adults and adults encounter several water masses: the southward flowing East Australia Current, sub-Antarctic water, and the Tasman Sea which is bounded to the north by the Tasman Front. Along the east coast of Australia, the seasonal cycle of SBT habitat is strongly influenced the East Australia Current (EAC; Ridgway and Godfrey 1997). The EAC is a southward flowing western boundary current between 18°S and 42°S (Tasmania), and is a region of intense eddy activity. The current is generally stronger and closer to the coast in summer (December–March) than in winter. In warmer than average years, when the EAC moves further south, thermal habitats preferred by sub-adults and adults are compressed to the south, while in cooler years, preferred thermal habitats are found further north than usual (Hobday et al. 2011).

Migration of adults to the spawning ground

Adults tagged with pop-up satellite tags (PSAT) in the Tasman Sea region have been observed to begin to move south and into waters around Tasmania towards the end of the austral spring/beginning of summer (Fig. 8.7). They then move across the south of Australia and pass around the south-west of Australia moving north along the western coastline of Australia to the spawning ground (Patterson et al. 2008; Evans et al. 2012).

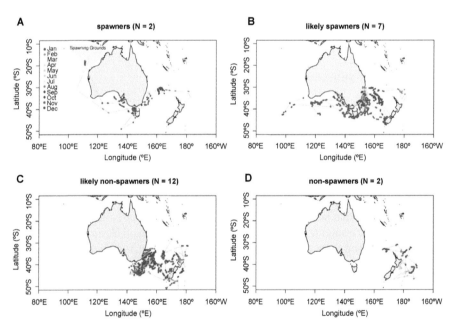

Figure 8.7. Movement paths of adult southern bluefin tuna categorized by putative spawning behaviour. (A) spawners showing movements from the tagging region to the spawning grounds (defined in blue); (B) likely spawners which made large westward migrations; (C) likely non-spawners remained in the Tasman Sea region until late in the spawning season and (D) non-spawners which remained resident in the Tasman for a full spawning cycle. Source Evans et al. 2012.

Once movement west was initiated, migration between waters around Tasmania and the spawning ground is directed and relatively quick (on the order of ~110 days).

Similar to juveniles, migration schedules of adults are highly variable, with individuals departing the Tasman Sea from September through to December (Patterson et al. 2008). Choice of winter foraging ground may, however, influence both onset of departure and subsequent arrival times on the spawning ground. Individuals tagged in the Tasman Sea have been estimated to arrive at the spawning ground in the second half of the season, potentially contributing to the second peak in abundances observed on the spawning ground (Evans et al. 2012). While SBT are captured by longline fisheries on the spawning ground in almost all months of the year, peaks in abundance have been observed to occur during October and February, also suggesting two peaks in spawning activity (Farley and Davis 1998).

While some younger fish have been captured on the spawning ground (e.g., age five), the age at which 50% of fish are mature is estimated to be around 10–12 years (Davis and Farley 2001; Gunn et al. 2008; Shimose and Farley this volume). It is unclear what proportion of sub-adult SBT migrate to the spawning ground and what proportion of young mature fish remain on winter foraging grounds rather than migrating to the spawning ground in each year. It is also unknown how long individuals remain on the spawning ground, or if adults continue to migrate annually to the spawning ground throughout their lifetime. Reproductive studies of female SBT from the spawning ground suggest that once females finish spawning, they leave the spawning ground immediately (Farley and Davis 1998). Commercial catch data suggests they move south into southern temperate waters to forage, although the pathways for this movement are unknown.

Catches of SBT in the spawning ground region suggest that individuals may vary in their vertical distributions. For example, smaller individuals are more likely to be caught deeper than larger individuals (Davis and Farley 2001). The mechanisms behind this apparent size partitioning on the spawning ground are unknown, but it is thought that a number of factors may be associated with this behaviour. It may be related to spawning activity, which is restricted to surface waters (Davis and Farley 2001). The inference is that larger fish spawn for longer than smaller fish and are likely to spend more time in shallower waters making them more susceptible to surface fishing gear than smaller fish (Davis and Farley 2001). Larger fish may also have more developed thermoregulatory capabilities than smaller fish, which might result in larger fish being able to spend more time in warmer surface waters than smaller fish (Davis and Farley 2001). Data from tagging experiments on spawning fish are limited, pop-up archival satellite tags deployed on adults have only been able to provide limited data from the regions of the spawning ground, but suggest that temperatures as high as 30°C are experienced by individuals (Patterson et al. 2008; Evans et al. 2012; Fig. 8.8). Electronic tagging on the spawning ground has also demonstrated that SBT can move rapidly back to temperate waters (Itoh et al. 2002).

PSAT data are additionally limited because they are summarized onboard the tag prior to transmission via satellite, so identification and definition of spawning has not been established. Nevertheless, individuals migrating to the region of the spawning ground were observed to switch to more surface orientated behaviour, spending greater amounts of time in waters <150 m. Time in shallower waters was punctuated by dives

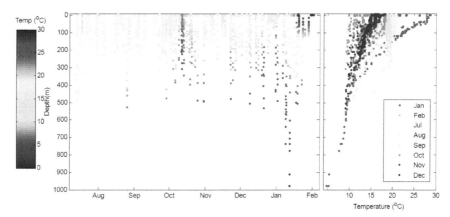

Figure 8.8. Temperature depth data from a pop-up satellite tag on an SBT which undertook a migration to the spawning grounds. The left panel shows the data through time with colours representing water temperature at depth (y-axis). The right hand panel aggregates this data by month and shows the variability in water masses encountered by this individual. Source: Patterson et al. (2008).

to >500 m, suggesting the potential for thermoregulatory behaviour (Patterson et al. 2008) in response to high surface water temperatures. Further data on the behaviour of adult SBT while on the spawning ground is needed if key descriptors of spawning, residence on the spawning ground and contributions to annual egg production are to be established.

Management Implications based on Distribution and Movements

Despite rapid advances in both electronic tagging technologies and the methods used to detail the movements of pelagic species, direct application of data from electronic tagging deployments into fishery management applications is still rare (Hobday et al. 2009b; Sippel et al. 2014). In this regard, SBT is one of the few species where information on the movement, distribution and habitat preference based on electronic tagging has supported management (Hobday et al. 2014).

In eastern Australia, information on habitat preference (surface and subsurface water temperatures) in combination with data from electronic tag deployments on sub-adults and adults is used to generate real-time habitat maps, indicating areas of high, medium and low occurrence probability (Hobday and Hartmann 2006; Hobday et al. 2009c; Fig. 8.9). Fisheries managers use this information on the likely presence of sub-adult and adult SBT to create several zones of varying access by the longline fishery. The aim of varying access to these zones is to reduce unwanted capture of SBT by fishers targeting yellowfin tuna (*Thunnus albacares*) that do not have quota to catch SBT and restricting access to areas of high SBT probability to those fishers that have quota for SBT (Hobday et al. 2009c; Hobday et al. 2010). The spatial allocation of the management zones is updated every two weeks, based on results from the model (see http://www.afma.gov.au/managing-our-fisheries/fisheries-a-to-z-index/southern-bluefin-tuna/notices-and-announcements/). Information on preferred habitat as derived

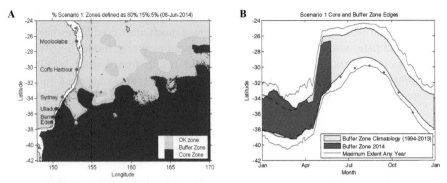

Figure 8.9. (A) Distribution of SBT habitat zones (Core = 80% probability of tuna, buffer = 15%, and OK = 5%) in eastern Australia based on the Hobday et al. (2011) habitat prediction model for June 6, 2014. (B) Annual climatology showing the mean position of the buffer zone throughout the year (yellow band) based on data from 1994 to 2013. Blue lines indicate the maximum northerly (5%) and southerly extent (5%) of buffer habitat that occurred in any year in the period considered. The position of the buffer zone in the current year is depicted by the red band. Red stars show the boundary of the core zone for the one to five month forecasts.

from electronic tag data has also been used to make seasonal forecasts of likely sub-adult and adult SBT distributions up to three months into the future (Hobday et al. 2011; Fig. 8.9). Managers use this forecast information on projected SBT distribution as a guide for future fishery interactions, rather than any direct decision making.

A similar habitat prediction approach is being implemented in the GAB, with juvenile SBT distributions forecast up to three months in advance, based on habitat preferences determined from electronic tag data (Eveson et al., in press; www.cmar. csiro.au/gab-forecasts/index.html). The forecasts aim to assist fishers to plan vessel deployments for upcoming fishing operations in an effort to improve the economic efficiency of the fishery and does not allow increased catches in this quota limited fishery.

Similar to other widely distributed species, information on SBT movements and distribution is important for improving population assessments (Eveson et al. 2012; Evans et al. 2014). The current operating model used by the CCSBT to develop the management procedure for SBT is non-spatial (Hillary et al. 2014); however, the CCSBT has recognized that a spatial model would be desirable, and the development of such a model has been highlighted as a future priority (Anon 2012). Spatial models require data for estimating fish movement, with tagging data often being the best or only source. Conventional tag data may not be sufficient for describing and estimating movement at the spatial and temporal scales required by the spatial model, particularly since movement can be confounded with mortality, tag reporting rates and the distribution of the fishing fleet; the more detailed information on movement provided by electronic tags can potentially be very useful in this regard. Eveson and Basson (unpublished data) found this to be true for SBT when they applied a spatial tag-recapture model for estimating natural mortality, fishing mortality and movement to data from juvenile SBT tagged in the 2000's. With only conventional tagging data, the movement estimates were inconsistent with our knowledge of juvenile SBT migration patterns and fishing mortality estimates in some regions were highly unrealistic. Including archival tag data in the model produced much more credible estimates, as well as estimates with greater precision. The current operating model for SBT only includes conventional tag-recapture data, but the results from Eveson and Basson (unpublished data) indicate that it will be important for a spatial version of the model to also include archival tag data.

Potential Changes in Movements and Distributions Under a Changing Climate

Projected changes in the physical oceanography of the ocean have the potential to alter the habitat suitability and distribution of SBT in many regions. This may result in changes to the timing of migration (phenology), dispersal of larvae, use of the continental shelf by individuals, and the spatial and temporal use of the spawning region. Changes in the Indian Ocean have been documented with waters off western Australia warming and the Leeuwin Current weakening slightly (Pearce and Feng 2007). Limited warming and southward movement of the frontal zones has also been documented in the Indian Ocean (e.g., Pearce and Feng 2007; Rolland et al. 2010). Changes in other variables, such as chlorophyll, have been reported (e.g., Boyce et al. 2010), but remain contentious (Mackas et al. 2011). The GAB has been reported to experience minor warming, however, the strength and persistence of this is difficult to discern from inter-annual variability over the same period of time (Lough and Hobday 2011). Along eastern Australia, the EAC has been observed to have strengthened and warmed, and is considered one of the fastest warming areas in the Southern Hemisphere (Ridgway 2007; Hobday and Pecl 2014).

Warmer waters on the spawning ground and changing productivity associated with changes to the vertical structure of the ocean may result in the spawning region becoming even more unproductive. Continued changes in oxygen content in deeper ocean layers have been reported from the Indian Ocean (Stramma et al. 2010) and if this area expands in the future it may also restrict SBT to shallower warmer waters on the spawning ground, which may have physiological consequences on the larger fish. In southern Western Australia, changes in the strength of the Leeuwin Current (Pearce and Feng 2007) and wind-driven coastal upwelling cells may impact on the productivity of seasonal feeding grounds for juveniles. The direction of change (positive/negative), however, is uncertain as most global climate models currently cannot conduct projections at the spatial resolutions required for regions such as the GAB (Hobday and Lough 2011), and downscaling is required (e.g., Hartog et al. 2011). If changes in productivity are negative, the spatial and temporal extent of utilization of waters off south-western Australia and in the GAB by SBT may also decline. While the GAB has warmed in recent years (Lough and Hobday 2011), climate projections suggest that winds will intensify, resulting in increased upwelling and local productivity (Hobday and Lough 2011), thereby potentially producing increased favourable conditions for SBT.

Projections based on statistically downscaled global climate models suggest that on the east coast of Australia, preferred thermal habitats for sub-adult and adult

SBT will be found further south (Hobday 2010; Hartog et al. 2011). In this region, the overlap of sub-adult and adult SBT with yellowfin tuna is projected to decrease slightly in the future (by 2060). This may result in a decrease in unwanted bycatch of SBT in the longline fishery (Hartog et al. 2011). In the southern Ocean, the frontal zone between the sub-Antarctic waters and temperate waters may move southwards, and if primary production is light-limited, the growing season may be shortened and the food chain less productive. Overall, impacts of climate change on SBT remain speculative, and are generally based on logical scenarios rather than any direct evidence. Ongoing monitoring is required to detect any response to climate change, and investigating SBT responses to climate variability may offer the greatest insight at this time (Hobday and Evans 2013).

Remaining Unknowns and Prospects for the Future

Currently, there is no single whole-of-life movement model for SBT. Spatial models for certain stages of the SBT life cycle have been developed, informed by fisheries and tagging data (e.g., Eveson et al. 2012). Integration of data on the movements and preferred habitats across different life stages will allow more holistic assessments of the impacts of external forcings on the population such as those associated with fishing and climate for example. In this chapter, we have identified a range of questions that still remain regarding the distribution and movements of SBT at different life stages. We suggest five areas that deserve additional attention as summarized here:

- Improved understanding of the movement pathways for larvae on and off the spawning ground, the spatial extent of the spawning ground, and seasonal and inter-annual variability in the spatial extent of the spawning ground. This information will be important for determining environmental influences on larval survival and in assessing spatial and temporal overlaps of fishing effort with adults on the spawning ground.

- Determination of influences on movement patterns on age one and two SBT in southern Western Australia, including resolving migration pathways after their first austral summer. This information will be important for assessing the proportion of the population entering southern Australian waters and identifying other important regions for the first year of growth and survival.

- Improved understanding of the movements and habitat use of sub-adult fish after they leave the GAB and before they reach maturity. Catch data from fisheries suggests the fish occupy a wide range of habitats, but the behaviour, physiological performance and habitat preference of SBT in these regions is unknown. This information will be important for establishing life history parameters for this component of the population, factors influencing onset of maturity and establishment of annual migrations to the spawning ground and resolving the proportion of the young, mature component of the population that skips migrations to the spawning ground (and therefore does not contribute to the spawning biomass).

- Experimental exploration of the physiological basis for navigation during the basin scale migrations performed by SBT using both captive and wild fish (see

Willis et al. 2009). While this may not be an easy task, large scale navigation is a remarkable behaviour, and will continue to stimulate the interest of many biologists, even if direct relevance to management is limited.

- Resolving physiological triggers for spawning migrations, and the behaviour of adults on the spawning ground and associated reproductive activity. This information will be important for establishing robust reproductive parameters for the population and determining robust measures of the spawning stock biomass.

Addressing these issues will require innovative data collection approaches, interdisciplinary studies, collaboration between various nations, and a commitment to long term studies. While the population of SBT is expected to recover to 20% of unfished biomass under most modelled scenarios of the current management scheme (Hillary et al. 2014), improved understanding of the movement behaviours of particular age groups and changes in the movement behaviours of the species across life stages in response to a changing climate will assist future management of this valuable and iconic species as new threats and challenges emerge.

Acknowledgements

Southern bluefin tuna research has been conducted by a substantive number of CSIRO scientists, in particular Russ Bradford, Thor Carter, Naomi Clear, Tim Davis, John Gunn, Tom Polacheck, Mark Rayner, Clive Stanley, Kevin Williams and Jock Young. Results described here have been supported by research funds provided by the Australian Fisheries Management Authority, Australian SBT industry, the CCSBT, CSIRO Marine and Atmospheric Research, the Department of Agriculture, Forestry and Fisheries, the New Zealand Ministry of Fisheries, the Japanese Society for the Promotion of Science, and the SBT Recruitment Monitoring Programme. We would like to thank the many skippers and crews of fishing vessels and fisheries observers who have contributed to tagging and data collection programs and collaborators from Japan, including Ryo Kawabe and Tomoyuki Itoh, and New Zealand, including Shelton Harley and Howard Reid, for their contributions to these research programs.

References

Anon. 2011. Report of the 16th Meeting of the Scientific Committee. Commission for the Conservation of Southern Bluefin Tuna. 19–28 July 2011, Bali, Indonesia. Available at: http://www.ccsbt.org/userfiles/file/docs_english/meetings/meeting_reports/ccsbt_18/report_of_SC16.pdf.

Anon. 2012. Report of the 17th Meeting of the Scientific Committee. Commission for the Conservation of Southern Bluefin Tuna. 27–31 August 2012, Tokyo, Japan. Available at: http://www.ccsbt.org/userfiles/file/docs_english/meetings/meeting_reports/ccsbt_19/report_of_SC17.pdf.

Anon. 2013. Report of the 18th Meeting of the Scientific Committee. Commission for the Conservation of Southern Bluefin Tuna. 7 September 2013, Canberra, Australia. Available at: http://www.ccsbt.org/userfiles/file/docs_english/meetings/meeting_reports/ccsbt_20/report_of_SC18.pdf.

Bakun, A. and K. Broad. 2003. Environmental 'loopholes' and fish population dynamics: comparative pattern recognition with focus on El Niño effects in the Pacific. Fish. Oceanog. 12(4-5): 458–473.

Basson, M., A.J. Hobday, J.P. Eveson and T.A. Patterson. 2012. Spatial Interactions Among Juvenile Southern Bluefin Tuna at the Global Scale: A Large Scale Archival Tag Experiment FRDC Report 2003/002.

Basson, M. and J.H. Farley. 2014. A standardised abundance index from commercial spotting data of southern bluefin tuna (*Thunnus maccoyii*): random effects to the rescue. PLoS ONE 9(12): e116245. doi:10.1371/journal.pone.0116245

Belkin, I.M. and A.L. Gordon. 1996. Southern Ocean fronts from the Greenwich meridian to Tasmania. J. Geophys. Res. 101(C2): 3675–3696.

Bestley, S., T.A. Patterson, M.A. Hindell and J.S. Gunn. 2008. Feeding ecology of wild migratory tunas revealed by archival tag records of visceral warming. J. Anim. Ecol. 77: 1223–1233.

Bestley, S., J.S. Gunn and M.A. Hindell. 2009. Plasticity in vertical behaviour of migrating juvenile southern bluefin tuna (*Thunnus maccoyii*) in relation to oceanography of the south Indian Ocean. Fish. Oceanog. 18: 237–254.

Bestley, S., T.A. Patterson, M.A. Hindell and J.S. Gunn. 2010. Predicting feeding success in a migratory predator: integrating telemetry, environment, and modeling techniques. Ecology 91: 2373–2384.

Boyce, D.G., M.R. Lewis and B. Worm. 2010. Global phytoplankton decline over the past century. Nature 466: doi:10.1038/nature09268.

Caton, A.E. 1991. Review of aspects of Southern Bluefin Tuna biology, population and fisheries. pp. 181–357. *In*: R.B. Deriso and W.H. Bayliff (eds.). World Meeting on Stock Assessment of Bluefin Tunas: Strengths and Weaknesses. Special report. La Jolla, California: Inter-American Tropical Tuna Commission, Vol. 7.

CCSBT. 2013. Review of Taiwan SBT Fishery of 2011/2012. CCSBT-ESC/1309/SBT Fisheries-Taiwan.

Condie, S.A. and J.R. Dunn. 2006. Seasonal characteristics of the surface mixed layer in the Australasian region: implications for primary production regimes and biogeography. Mar. Freshw. Res. 57: 569–590.

Cowling, A., A. Hobday and J. Gunn. 2003. Development of a fishery-independent index of abundance for juvenile southern bluefin tuna and improvement of the index through integration of environmental, archival tag and aerial survey data. CSIRO Marine Research. FRDC Report 96/118 & 99/105.

Davis, T.L.O. and C.A. Stanley. 2002. Vertical and horizontal movements of southern bluefin tuna (*Thunnus maccoyii*) in the Great Australian Bight observed with ultrasonic telemetry. Fish. Bull. 100: 448–465.

Davis, T.L.O., G.P. Jenkins and J.W. Young. 1990. Diel patterns of vertical distribution in larvae of southern bluefin *Thunnus maccoyii*, and other tuna in the East Indian Ocean. Mar. Ecol. Prog. Ser. 59(1-2): 63–74.

Davis, T.L.O., G.P. Jenkins, M. Yukinawa and Y. Nishikawa. 1989. Tuna larvae abundance: comparative estimates from concurrent Japanese and Australian sampling programs. Fish. Bull. 87: 976–981.

Davis, T.L.O., V. Lyne and G.P. Jenkins. 1991. Advection, dispersion and mortality of a patch of southern bluefin tuna larvae *Thunnus maccoyii* in the East Indian Ocean. Mar. Ecol. Prog. Ser. 73(1): 33–45.

Dell, J. and A.J. Hobday. 2008. School-based indicators of tuna population status. ICES J. Mar. Sci. 65: 612–622.

Evans, K., T.A. Patterson, H. Reid and S.J. Harley. 2012. Reproductive schedules in Southern Bluefin Tuna: are current assumptions appropriate? PLoS ONE 7: e34550.

Eveson, J.P., A.J. Hobday, C.M. Spillman, K.M. Rough and J.R. Hartog (in press). Seasonal forecasting of tuna habitat in the Great Australian Bight. Fisheries Research.

Eveson, P., J. Farley and M. Bravington. 2014. The aerial survey index of abundance: updated results for the 2013/14 fishing season. Report CCSBT-ESC/1409/18 (Rev.1) prepared for the CCSBT Extended Scientific Committee for the 19th Meeting of the Scientific Committee 1–6 September, Auckland, New Zealand.

Eveson, J.P., M. Basson and A.J. Hobday. 2012. Using electronic tag data to improve parameter estimates in a tag-based spatial fisheries assessment model. Can. J. Fish. Aquat. Sci. 69(5): 869–883, 10.1139/f2012-026.

Farley, J.H. and T.L.O. Davis. 1998. Reproductive dynamics of southern bluefin tuna, *Thunnus maccoyii*. Fish. Bull. 96(2): 223–236.

Farley, J., T.L.O. Davis, J.S. Gunn, N.P. Clear and A.L. Preece. 2007. Demographic patterns of southern bluefin tuna, *Thunnus maccoyii*, as inferred from direct age data. Fish. Res. 83: 151–161.

Fujioka, K., A.J. Hobday, R. Kawabe, K. Miyashita, T. Itoh and Y. Takao. 2010. Interannual variation in summer habitat use by juvenile southern bluefin tuna (*Thunnus maccoyii*) in southern Western Australia. Fish. Oceanog. 19(3): 183–195.

Fujioka, K., A.J. Hobday, R. Kawabe, K. Miyashita, Y. Takao, O. Sakai and T. Itoh. 2012. Departure behaviour of juvenile southern bluefin tuna (*Thunnus maccoyii*) from southern Western Australia temperate waters in relation to the Leeuwin Current. Fish. Oceanog. 21(4): 269–280.

Grewe, P.M., N.G. Elliott, B.H. Innes and R.D. Ward. 1997. Genetic population structure of southern bluefin tuna (*Thunnus maccoyii*). Mar. Biol. 127(4): 555–561.

Gunn, J. and B. Block. 2001. Advances in acoustic, archival and satellite tracking of tunas. pp. 167–223. *In*: B.A. Block and E.D. Stevens (eds.). Tuna: Physiology, Ecology and Evolution. Academic Press, New York.

Gunn, J., J. Hartog and K. Rough. 2001. The relationship between food intake and visceral warming in southern bluefin tuna (*Thunnus maccoyii*). Can we predict from archival tag data how much a tuna has eaten? pp. 109–130. *In*: J.R. Sibert and J.L. Nielsen (eds.). Electronic Tagging and Tracking in Marine Fisheries. Kluwer Academic Publishers, Netherlands.

Gunn, J., T. Polacheck, T. Davis, M. Sherlock and A. Betlehem. 1994. The development and use of archival tags for studying the migration, behaviour and physiology of southern bluefin tuna, with an assessment of the potential for transfer of the technology to groundfish research. Proceedings of the ICES Mini Symposium on Fish Migration 21: 23.

Gunn, J. 1999. From plastic darts to pop-up satellite tags. pp. 55–60. *In*: D.A. Hancock, D.C. Smith and J.D. Koehn (eds.). Fish Movement and Migration. Australian Society for Fish Biology.

Harden Jones, F.R. 1984. A view from the Ocean. pp. 1–26. *In*: J.D. McCleave, G.P. Arnold, J.J. Dodson and W.H. Neill (eds.). Mechanisms of Migration in Fishes. Plenum Press, New York.

Hartog, J., A.J. Hobday, R. Matear and M. Feng. 2011. Habitat overlap of southern bluefin tuna and yellowfin tuna in the east coast longline fishery—implications for present and future spatial management. Deep Sea Res. Part II 58: 746–752.

Heupel, M.R., J.M. Semmens and A.J. Hobday. 2006. Automated acoustic tracking of aquatic animals: scales, design and deployment of listening station arrays. Mar. Freshw. Res. 57(1): 1–13.

Hillary, R., A. Preece and C. Davies. 2014. Assessment of stock status of southern bluefin tuna in 2014 with reconditioned operating model. Prepared for the 19th CCSBT Extended Scientific Committee held in Auckland, New Zealand 1st-6th of September 2014. CCSBT-ESC/1409/21.

Hobday, A.J., R. Kawabe, Y. Takao, K. Miyashita and T. Itoh. 2009a. Correction factors derived from acoustic tag data for a juvenile southern bluefin tuna abundance index in southern Western Australia. pp. 405–422. *In*: J. Neilsen, H. Arrizabalaga, N. Fragoso, A. Hobday, M. Lutcavage and J. Sibert (eds.). Proceedings of the Second Symposium on Tagging and Tracking Marine Fish with Electronic Devices. Springer Publishing, Netherlands.

Hobday, A.J. and K. Hartmann. 2006. Near real-time spatial management based on habitat predictions for a longline bycatch species. Fish Manag. Ecol. 13: 365–380.

Hobday, A.J. 2010. Ensemble analysis of the future distribution of large pelagic fishes in Australia. Prog. Oceanog. 86(1-2): 291–301.

Hobday, A.J. and G. Campbell. 2009. Topographic preferences and habitat partitioning by pelagic fishes in southern Western Australia. Fish. Res. 95: 332–340.

Hobday, A.J. and J. Lough. 2011. Projected climate change in Australian marine and freshwater environments. Mar. Freshw. Res. 62: 1000–1014.

Hobday, A.J. and K. Evans. 2013. Detecting climate impacts with oceanic fish and fisheries data. Clim. Change 119: 49–62.

Hobday, A.J., C.M. Spillman, J.R. Hartog and J.P. Eveson. in press. Seasonal forecasting for decision support in marine fisheries and aquaculture. Fish. Oceanog.

Hobday, A.J., N. Flint, T. Stone and J.S. Gunn. 2009c. Electronic tagging data supporting flexible spatial management in an Australian longline fishery. pp. 381–403. *In*: J. Nielsen, J.R. Sibert, A.J. Hobday, M.E. Lutcavage, H. Arrizabalaga and N. Fragosa (eds.). Tagging and Tracking of Marine Animals with Electronic Devices II. Reviews: Methods and Technologies in Fish Biology and Fisheries, Vol. 9, Springer, Netherlands.

Hobday, A.J. and G.T. Pecl. 2014. Identification of global marine hotspots: sentinels for change and vanguards for adaptation action. Rev. Fish Biol. Fisher. 24: 415–425.

Hobday, A.J., J. Hartog, C. Spillman and O. Alves. 2011. Seasonal forecasting of tuna habitat for dynamic spatial management. Can. J. Fish. Aquat. Sci. 68: 898–911.

Hobday, A.J., J.R. Hartog, T. Timmis and J. Fielding. 2010. Dynamic spatial zoning to manage southern bluefin tuna capture in a multi-species longline fishery. Fish. Oceanog. 19(3): 243–253.

Hobday, A.J., S.M. Maxwell, J. Forgie, J. McDonald, M. Darby, K. Seto, H. Bailey, S.J. Bograd, D.K. Briscoe, D.P. Costa, L.B. Crowder, D.C. Dunn, S. Fossette, P.N. Halpin, J.R. Hartog, E.L. Hazen, B.G. Lascelles, R.L. Lewison, G. Poulos and A. Powers. 2014. Dynamic ocean management: Integrating scientific and technological capacity with law, policy and management. Stanford Environmental Law Journal 33(2): 125–165.

Hobday, A., H. Arrisabalaga, N. Fragoso, J. Sibert, M. Lutcavage and J. Nielsen. 2009b. Applications of electronic tagging to understanding marine animals—Preface. *In*: J. Nielsen, J.R. Sibert, A.J. Hobday et al. (eds.). Tagging and Tracking of Marine Animals with Electronic Devices II. Reviews: Methods and Technologies in Fish Biology and Fisheries. Springer, Netherlands 9: v–xvi.

Honda, K., A.J. Hobday, R. Kawabe, N. Tojo, K. Fujioka, Y. Takao and K. Miyashita. 2010. Age-dependent distribution of juvenile southern bluefin tuna (*Thunnus maccoyii*) on the continental shelf off southwest Australia determined by acoustic monitoring. Fish. Oceanog. 19: 151–158.

Itoh, T. and S. Tsuji. 1996. Age and growth of juvenile southern bluefin tuna *Thunnus maccoyii* based on otolith microstructure. Fisheries Science 62(6): 892–896.

Itoh, T., H. Kemps and J. Totterdell. 2011. Diet of young southern bluefin tuna *Thunnus maccoyii* in the southwestern coastal waters of Australia in summer. Fish. Sci. 77: 337–344.

Itoh, T., H. Kurota, N. Takahashi and S. Tsuji. 2002. Report of 2001/2002 spawning ground surveys. Report CCSBT-SC/0209/20 prepared for the CCSBT Scientific Committee for the 7th Meeting of the Scientific Committee 9–11 September, Canberra, Australia.

Jenkins, G.P. and T.L.O. Davis. 1990. Age, growth rate, and growth trajectory determined from otolith microstructure of southern bluefin tuna *Thunnus maccoyii* larvae. Mar. Ecol. Prog. Ser. 63: 93–104.

Llopiz, J.K. and A.J. Hobday. 2014. A global comparative analysis of the feeding dynamics and environmental conditions of larval tunas, mackerels, and billfishes. Deep Sea Res. II: http://dx.doi.org/10.1016/j.dsr2.2014.05.014.

Lough, J.M. and A.J. Hobday. 2011. Observed climate change in Australian marine and freshwater environments. Mar. Freshw. Res. 62: 984–999.

Middleton, J.F. and J.A.T. Bye. 2007. A review of the shelf-slope circulation along Australia's southern shelves: Cape Leeuwin to Portland. Prog. Oceanog. 75: 1–41.

Murphy, G.I. 1977. A new understanding of southern bluefin tuna. Austr. Fish. 36: 2–6.

Murphy, G.I. and J. Majkowski. 1981. State of the southern bluefin tuna population: fully exploited. Austr. Fish. 40: 20–23.

Nieblas, A.E., H. Demarcq, K. Drushka, B.M. Sloyan and S. Bonhommeau. 2014. Front variability and surface ocean features of the presumed southern bluefin tuna spawning grounds in the tropical southeast Indian Ocean. Deep Sea Res. II: http://dx.doi.org/10.1016/j.dsr2.2013.11.007.

Nielsen, J., J.R. Sibert, A.J. Hobday, M.E. Lutcavage, H. Arrizabalaga and N. Fragosa (eds.). 2009. Tagging and Tracking of Marine Animals with Electronic Devices. Reviews: Methods and Technologies in Fish Biology and Fisheries. Springer, Netherlands.

Nishikawa, Y., M. Honma, S. Ueyanagi and S. Kikawa. 1985. Average distribution of larvae of oceanic species of scombrid fishes, 1956–1981. Far Seas Fish. Res. Lab. S Ser. 12: 1–99.

Patterson, T.A., K. Evans, T.I. Carter and J.S. Gunn. 2008. Movement and behaviour of large southern bluefin tuna (*Thunnus maccoyii*) in the Australian region determined using pop-up satellite archival tags. Fish. Oceanog. 17: 352–367.

Patterson, T.A., M. Basson, M. Bravington and J.S. Gunn. 2009. Classifying movement behaviour in relation to environmental conditions using hidden Markov models. J. Anim. Ecol. 78: 1113–1123.

Pedersen, M.W., T.A. Patterson, U.H. Thygesen and H. Madsen. 2011. Estimating animal behavior and residency from movement data. Oikos 120(9): 1281–1290.

Pearce, A. and M. Feng. 2007. Observations of warming on the Western Australian continental shelf. Mar. Freshw. Res. 58: 914–920.

Polacheck, T. and A. Preece. 2001. An integrated statistical time series assessment of the southern bluefin tuna stock based on the catch at age data. Commission for the Conservation of Southern Bluefin Tuna, CCSBT-SC/0108/13.

Ridgway, K.R. 2007. Long-term trend and decadal variability of the southward penetration of the East Australian Current. Geophy. Res. Lett. 34: L13613, doi:10.1029/2007GL030393.

Ridgway, K.R. and J.S. Godfrey. 1997. Seasonal cycle of the East Australian Current. J. Geophys. Res. 102(C10): 22921–22936.

Rolland, V., H. Weimerskirch and C. Barbraud. 2010. Relative influence of fisheries and climate on the demography of four albatross species. Glob. Change Biol. 16: 1910–1922.

Rogers, P., T. Ward, P. van Ruth and A. Williams. 2013. Physical processes, biodiversity and ecology of the Great Australian Bight region: a literature review. CSIRO Report, Australia. 199 pp.

Shimose, T. and J.H. Farley. this volume. Age, Growth and Reproductive Biology of Bluefin Tunas.

Sippel, T., J.P. Eveson, B. Galuardi, C. Lam, S. Hoyle, M. Maunder, P. Kleiber, F. Carvalho, V. Tsontos, S.L.H. Teo, A. Aires-da-Silva and S. Nicol. 2014. Using movement data from electronic tags in fisheries stock assessment: A review of models, technology and experimental design. Fish. Res.: http://dx.doi.org/10.1016/j.fishres.2014.04.006.

Stramma, L., S. Schmidtko, L.A. Levin and G.C. Johnson. 2010. Ocean oxygen minima expansions and their biological impacts. Deep Sea Res. I 57: 587–595.

Ward, T.M., L.J. Mcleay, W.F. Dimmlich, P.J. Rogers, S. McClatchie, R. Matthews, J. Kampf and P.D. Van Ruth. 2006. Pelagic ecology of a northern boundary current system: effects of upwelling on the production and distribution of sardine (*Sardinops sagax*), anchovy (*Engraulis australis*) and southern bluefin tuna (*Thunnus maccoyii*) in the Great Australian Bight. Fish. Oceanog. 15: 191–207.

Willis, J. and A.J. Hobday. 2007. Influence of upwelling on movement of southern bluefin tuna (*Thunnus maccoyii*) in the Great Australian Bight. Mar. Freshw. Res. 58: 699–708.

Willis, J., J. Phillips, R. Muheim, F.J. Diego-Rasilla and A.J. Hobday. 2009. Spike dives of juvenile southern bluefin tuna (*Thunnus maccoyii*): a navigational role? Behav. Ecol. Sociobiol. 64: 57–68.

Young, J.W. and T.L.O. Davis. 1990. Feeding ecology of larvae of southern bluefin, albacore and skipjack tunas (Pisces: Scombridae) in the eastern Indian Ocean. Mar. Ecol. Prog. Ser. 61(1-2): 17–29.

Young, J.W., T.D. Lamb, D. Le, R.W. Bradford and A.W. Whitelaw. 1997. Feeding ecology and interannual variations in diet of southern bluefin tuna, *Thunnus maccoyii*, in relation to coastal and oceanic waters off eastern Tasmania, Australia. Env. Biol. Fish. 50: 275–291.

CHAPTER 9

Understanding Bluefin Migration Using Intrinsic Tracers in Tissues

Daniel J. Madigan

Introduction

Bluefin tunas are well known for their long-distance migrations. Early tagging studies using simple 'floy' or 'spaghetti' tags first indicated that bluefin tuna could cross entire ocean basins (Clemens and Flittner 1969; Bayliff et al. 1991; Bayliff 1994). While these studies provided valuable life history information, the dates and locations of tag deployment and recovery were the only migratory data provided by these simple tags. The recent advent of electronic tagging technology has transformed our understanding of bluefin tuna migrations (Carey 1983; Gunn et al. 1994; Lutcavage et al. 1999; Lutcavage et al. 2000; Stevens et al. 2000; Block et al. 2001; Block 2005; Wilson et al. 2005; Kitagawa et al. 2007; Teo et al. 2007; Willis and Hobday 2007; Patterson et al. 2008; Kitagawa et al. 2009; Boustany et al. 2010; Fujioka et al. 2010). The data provided by electronic tags have allowed researchers to better understand migration dynamics such as trans-Pacific migrations in Pacific bluefin tuna and the extent of mixing of multiple stocks of Atlantic bluefin tuna. They have allowed for analysis of diving behavior, residency times, and even feeding success using surgically-implanted archival tags.

Electronic tagging studies do, however, have certain limitations. Tagging studies can be daunting logistically and financially, requiring the purchase of expensive technology and the necessary shiptime and manpower to capture live tunas, safely implant tags internally and/or externally, and subsequently process large quantities of data. Tagging of small tunas (young-of-the-year) has proven to be difficult due to tag size. Finally, tagging studies provide movement information that is only prospective from the time and location of tagging. In other words, for a given tagging region,

School of Marine and Atmospheric Sciences, Stony Brook University, Stony Brook, New York, USA 11790.

electronic tags will, upon recovery, tell researchers where those tagged tuna went to, but provide no information on where those individual tunas had come from. Such retrospective information can be extremely useful in ascertaining the migratory origin of individual tunas, which can improve understanding of stock mixing and the dynamics of, for example, trans-oceanic migrations.

Various chemical tracer techniques have emerged that can provide such retrospective information. Organic compounds (e.g., organochlorines and PCBs) (Dickhut et al. 2009), radionuclides (Madigan et al. 2012a; Madigan et al. 2013), stable isotope composition of hard parts such as otoliths (Rooker et al. 2001; Rooker et al. 2008; Schloesser et al. 2010; Secor 2010), and stable isotope analysis of soft tissues such as muscle and liver (Logan 2009; Madigan et al. 2014) have all been used to infer retrospective movements of bluefin tunas. Unlike radioactive isotopes and organic compounds, which in the above cases are anthropogenic contaminants and thus subject to change in their regional and temporal concentrations, stable isotopes are naturally-occurring and the isotopic values of marine taxa are largely based on local oceanographic conditions, which form the local isotopic 'baseline'. Thus large stable isotope differences between oceanographically distinct regions, such as oligotrophic versus upwelling waters, are unlikely to dramatically change over time. Stable isotope analysis (SIA) thus provides a tool that can potentially be used *ad infinitum* to assess migrations between oceanographic regions. This chapter uses the soft tissue SIA approach in Pacific bluefin tuna as a case study to demonstrate how SIA can be applied to tuna migration studies, and shows how other tracer techniques can ground-truth the SIA approach.

Bulk Stable Isotopes in the North Pacific Ocean

SIA uses the ratio of a rare, heavier isotope to a lighter, more common isotope of a specific element and is expressed as parts per mille (‰). In ecological studies using soft tissues (e.g., muscle or liver), the elements most commonly used for SIA are carbon ($\delta^{13}C$) and nitrogen ($\delta^{15}N$). In early SIA studies, $\delta^{13}C$ was generally used for ascertaining food sources and $\delta^{15}N$ for estimating trophic level, as $\delta^{13}C$ was thought to fractionate (change in value from prey to consumer) minimally, while $\delta^{15}N$ was demonstrated to fractionate appreciably between trophic steps (Post 2002; Fry 2006). Recent studies have demonstrated that fractionation dynamics are not so simple, nor are they the same across all taxa (Robbins et al. 2005; Caut et al. 2009; Hussey et al. 2010; Hussey et al. 2014). Furthermore, different regions (within terrestrial or aquatic systems) can have different isotopic 'baselines', or values at the base of food webs (e.g., primary producers; phytoplankton in the open ocean) (Post 2002). These differences propagate up food webs, so phytoplankton in one marine region may differ markedly from phytoplankton in another due to 'baseline' differences; as such, SI values of zooplankton, and zooplanktivores (e.g., sardine or anchovy) would differ between regions, as would sardine consumers (e.g., tunas), and so on. For migratory predators this regional effect is particularly important, as predators take time to reflect their local diet SI values. As such, the SI ratios of a predator in a given system may more reflect where it has recently been than where it presently is.

This of course makes possible the use of SIA to infer the migratory patterns of highly migratory marine animals such as tunas. The primary factor dictating the isotopic baseline for $\delta^{15}N$ in oceanic food webs is the source of nitrogen for primary producers (Montoya et al. 2002). In general, N_2-fixation dominates in oligotrophic regions, while upwelled, deep-water nitrate is more available in high upwelling regions. In oligotrophic waters where N_2-fixation dominates, the $\delta^{15}N$ values of primary producers are demonstrably lower by several parts per mille than the $\delta^{15}N$ values of primary producers in upwelling areas, where upwelled nitrate is more highly utilized by primary producers (Montoya et al. 2002; Navarro et al. 2013).

Such strong regional differences in baseline $\delta^{15}N$ values occur in the North Pacific Ocean. In general, the Eastern Pacific Ocean (EPO), and particularly the California Current Large Marine Ecosystem (CCLME), experiences high upwelling and consequently high $\delta^{15}N$ values. The Central Pacific Ocean (CPO) is largely oligotrophic, with accordingly low baseline $\delta^{15}N$ values. The western Pacific Ocean experiences moderate seasonal upwelling, and accordingly has intermediate baseline $\delta^{15}N$ values relative to the CPO and EPO. These differences have been demonstrated in ocean circulation-biogeochemistry isotope models (Navarro et al. 2013), prey and predator taxa from various regions of the NPO (Carlisle et al. 2012; Madigan et al. 2014), and an NPO isoscape that used the $\delta^{15}N$ values of tuna muscle tissue to demonstrate regional differences in $\delta^{15}N$ values, which corresponded highly with regional oceanography (Graham et al. 2010). The propagation up food webs of baseline $\delta^{15}N$ differences between the WPO and EPO is demonstrated by $\delta^{15}N$ values of pelagic prey from both regions (Minami et al. 1995; Takai et al. 2000; Mitani et al. 2006; Takai et al. 2007; Miller et al. 2010; Madigan et al. 2012b). Thus a tuna feeding on pelagic prey would be expected to have significantly different tissue $\delta^{15}N$ values in the WPO versus the EPO (Fig. 9.1). A Pacific bluefin tuna migrating between these regions may reflect

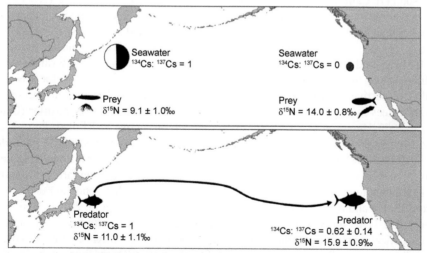

Figure 9.1. Differences in pelagic prey $\delta^{15}N$ values between the western and eastern Pacific Ocean (top panel) and consequent expected $\delta^{15}N$ in a pelagic predator, Pacific bluefin tuna *Thunnus orientalis*. Differences in radiocesium concentrations in the two ocean basins, and in Pacific bluefin tuna, are shown; see section on complementary chemical tracers. From Madigan et al. 2014.

the $\delta^{15}N$ values of its previous location. However, discerning long-term residents and recent migrants, and quantifying the timing of migrations, requires specific parameters of isotopic fractionation and turnover in tunas, which can often only be ascertained with controlled laboratory experiments.

Laboratory Experiments: Calibrating the Tool

The interpretation of SI values of wild tunas requires some knowledge of the dynamics of stable isotopes in the study organism. The primary parameters that are relevant to ecological studies are trophic discrimination (or trophic enrichment) factor (TDF or TEF) and turnover time: the rate at which tissues change in isotopic value(s) following a change in diet. As stated previously, SI values increase from prey to predator due to fractionation. As such the $\delta^{13}C$ and $\delta^{15}N$ values of tuna tissues will be some predictably higher value than those of their prey. Tunas (and all predators) also take time to reflect their new diet after a diet switch (which in wild tunas is often due to migration to a new oceanic region). The turnover time of $\delta^{13}C$ and $\delta^{15}N$ in various tissues can allow application of 'isotopic clock' techniques (Phillips and Eldridge 2006; Klaassen et al. 2010) to tuna SI values, giving estimates not only of migration origin but also of migration timing.

It can of course be a difficult task to perform diet switch experiments on large pelagic predators, as they are difficult to keep in captivity and experimental periods may need to be on the order of years for predators to reach isotopic 'steady-state' with diet. Captive Pacific bluefin tuna (PBFT) at the Tuna Research and Conservation Center (TRCC) of Hopkins Marine Station and the Monterey Bay Aquarium (Farwell 2001) allowed for a long-term study of these parameters in growing PBFT (Madigan et al. 2012c). Small (60–70 cm CFL) were captured by hook-and-line off the coast of southern California and transported to the TRCC. PBFT arrived at the TRCC with relatively low (~12‰) muscle $\delta^{15}N$ values. They were fed a controlled diet that was higher than their initial muscle $\delta^{15}N$ value (~14‰) and held in captivity from 1–2914 days. These experimental conditions provided estimates of TDF for muscle tissue in PBFT (1.9 and 1.8‰ for $\delta^{15}N$ and $\delta^{13}C$, respectively). Importantly, these values were different than the values that are still often applied generally (3.4 and 0.4‰ for $\delta^{15}N$ and $\delta^{13}C$, respectively; Post et al. 2002) to all taxa in ecological studies, though inaccurate TDF values can highly effect SIA-based inferences of diet and trophic position (Caut et al. 2009; Hussey et al. 2014). These experiments on captive PBFT also allowed for half-life estimates for turnover of $\delta^{13}C$ (255 days) and $\delta^{15}N$ (167 days) in PBFT muscle (Fig. 9.2). This showed that several years are required for PBFT muscle tissue SI values to reach steady-state with local feeding conditions. Other studies on farmed Atlantic bluefin tuna found muscle $\delta^{13}C$ and $\delta^{15}N$ TDF values of –0.16‰ and 1.64‰ (Varela et al. 2011) and a muscle $\delta^{15}N$ TDF value of 1.7‰ (Vizzini et al. 2010). Turnover rates and TDFs for SI values in liver tissue have also been examined, as liver is a metabolically-active tissue that is often used to provide more recent estimates of diet. While turnover in liver was indeed faster, SI values overall were found to be highly variable in this tissue, suggesting caution in making strong quantitative inferences based on SI values of tuna liver tissue (Madigan et al. 2012c). These contributions have all

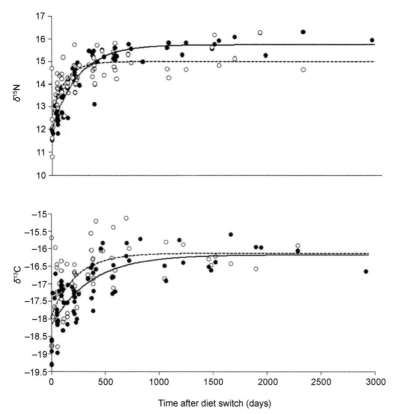

Figure 9.2. Changes in δ^{15}N and δ^{13}C values of white muscle (filled circles) and liver (open circles) over time in Pacific bluefin tuna held in captivity. Lines represent exponential fit to data for muscle (solid line) and liver (dashed line). Note that asymptote (steady-state) for all isotope-tissue groups occurred after 500 days. From Madigan et al. 2012c.

been vital to interpretation of wild tuna SI data, but the demonstrated inconsistencies between studies and across bluefin species indicate that more laboratory studies are needed as SIA studies on wild tunas expand.

Complementary Chemical Tracers

Ecological or migration studies using bulk SIA (the analysis of whole tissues) is inferential by nature. As described above, the SI values of a predator's tissue will be affected by current diet composition (and potential changes in current diet composition), TDF, and migration effects (previous diet). In migratory predators, the SI values of tissue may be simultaneously influenced by all of these factors. Combining SIA with electronic tagging data (Carlisle et al. 2012; Seminoff et al. 2012) or comparing SI values of a highly migratory species to a trophically similar, resident species (Madigan et al. 2014) can help strengthen SIA-based inferences of migration patterns. However, when using SIA to infer migration patterns, other chemical tracers are likely the most powerful complement to ground-truth inferences made from SIA alone. Once validated using independent chemical tracers (which may be ephemeral or cost-prohibitive, making long-term studies difficult), bulk SIA can then be used to infer past movements from large and/or long-term sample sets of tuna tissue.

In the case of PBFT, both anthropogenic and naturally-occurring chemical tracers have been utilized as complementary analyses to ground-truth SIA inferences. The release of large amounts of radionuclides from the Fukushima Dai-ichi nuclear power plant in Japan, which peaked in April 2011, created a point-source of radionuclides into the WPO. Concentrations near Japan were affected by local currents, particularly the Kuroshio Current (Buesseler et al. 2011; Buesseler et al. 2012). The region of high radionuclide concentration off Japan, including radiocesium, corresponded to an area that seems to be highly used by PBFT before their trans-Pacific migrations (Kitagawa et al. 2009; Buesseler et al. 2011). Preliminary testing of only 15 PBFT captured in the EPO (CCLME) in 2011 showed Fukushima-derived radionuclides in all 15 fish sampled (Madigan et al. 2012a). Subsequent validation of the technique in 2012 (Madigan et al. 2013) allowed the Fukushima 'signal' to be used to unequivocally distinguish recent migrants from the WPO from long-term (>1 year) residents of the CCLME (Madigan et al. 2014) (Figs. 9.1 and 9.3). This ephemeral tracer (Fukushima-derived radiocesium), used in conjunction with SIA in the same fish, allowed more rigorous statistical treatment of SIA data in PBFT (Madigan et al. 2014).

A newer, powerful approach to SIA has recently emerged in the form of Amino Acid Compound-Specific Isotopic Analysis (AA-CSIA) (McClelland and Montoya 2002; Chikaraishi et al. 2009; Lorrain et al. 2009; Chikaraishi et al. 2010; Olson et al. 2010). AA-CSIA utilizes the premise that $\delta^{15}N$ values of some amino acids

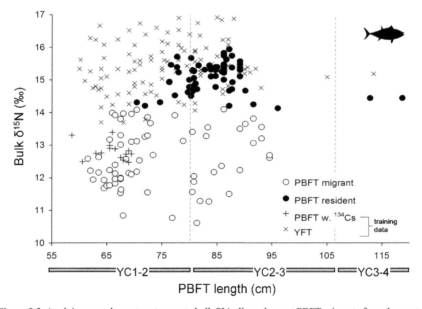

Figure 9.3. Applying complementary tracers to bulk SIA allowed recent PBFT migrants from the western Pacific Ocean to be distinguished from >1 year residents of the EPO. PBFT carrying Fukushima-derived radiocesium was used to identify SI values of recent migrants, while EPO resident yellowfin tuna were used as proxies for a tuna at steady-state with EPO prey. From Madigan et al. 2014.

fractionate (increase) highly between trophic levels ('trophic' amino acids; alanine, valine, leucine, isoleucine, proline, and glutamic acid), while other amino acids show little or no trophic fractionation between trophic steps ('source' amino acids; glycine, serine, and phenylalanine) (McClelland and Montoya 2002). Bulk $\delta^{15}N$ values thus represent an average of the $\delta^{15}N$ values of highly fractionating trophic AAs and the relatively stable source AAs. In the case of PBFT, WPO migrants show low bulk $\delta^{15}N$ values and EPO resident $\delta^{15}N$ values are significantly higher. Using AA-CSIA allowed confirmation that this is in fact an effect of migration and not of trophic differences by demonstrating that inferred 'migrant' PBFT had much lower source AA $\delta^{15}N$ values than inferred 'resident' PBFT (Madigan et al. 2014) (Fig. 9.4). AA-CSIA is significantly more time-intensive and costly than bulk SIA. However, by applying AA-CSIA to a subset of samples, the relative contribution of trophic versus migration effects to observed bulk SI values can be quantified. In conjunction with geographically constrained tracers such as the radionuclides described above, AA-CSIA provides validation that observed differences in bulk $\delta^{15}N$ values are a result of trophic effects, migration effects, or both.

AA-CSIA is a growing field, and it can be applied to almost any bulk SIA study to ascertain the drivers behind differences in $\delta^{15}N$ values in the study species. Other chemical tracers, such as organic compounds that have been used to infer migration patterns of Atlantic bluefin tuna (Dickhut et al. 2009) serve as examples of clever applications of geographically distinct concentrations of measurable compounds that can help elucidate bluefin movement dynamics. If used to ground-truth SIA patterns and inferences, these approaches could provide valuable validation of the SIA approach, which is a relatively convenient and cost-effective technique. Researchers should continue to seek independent methods for inferring migration patterns that, when combined with SIA, facilitate and validate ongoing and future studies of tuna migratory dynamics using a bulk SIA approach.

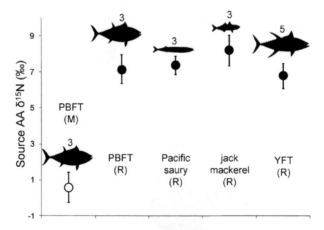

Figure 9.4. Amino acid-compound specific isotopic analysis (AA-CSIA) can be a powerful component to SIA studies, and can distinguish migration from trophic effects on observed $\delta^{15}N$ values. $\delta^{15}N$ values of 'source' AAs glycine, serine, and phenylalanine were much lower in Pacific bluefin that recent migrated from the western Pacific than were source AA $\delta^{15}N$ values of eastern Pacific resident Pacific bluefin, yellowfin, and juvenile Pacific saury and jack mackerel. Number above each icon indicates sample size.

SIA-Based Insights into Pacific Bluefin Tuna Movements

SIA has provided novel insights into the migratory dynamics of PBFT. Due to tagging data, it is known that some PBFT migrate from the WPO to the EPO, spend one to several years feeding in the EPO, and then return west to spawn (Bayliff et al. 1991; Bayliff 1994; Boustany et al. 2010). However, the ages at which PBFT migrate to the EPO, the contribution of older (>2 years old) WPO PBFT to the EPO group, and the timing of migrations in young PBFT (0–2 years old) from the WPO to the EPO remained largely unquantified, due to a current lack of large-scale tagging efforts of young PBFT in the WPO and the inability of electronic tags to provide retrospective information.

The application of SIA to CCLME-captured PBFT revealed that most PBFT migrate to the EPO at year-class (YC) 1–2 (Madigan et al. 2014) (Fig. 9.5). This corroborates inferences made from previous studies (Bayliff et al. 1991). However, an appreciable proportion (33%) of PBFT were shown to migrate to the EPO at YC 2–3 (Fig. 9.5). The older year class (3–4) was residential to the EPO for >1 year, though sample size was small ($n = 2$). Preliminary results on a much larger sample size of YC 3–5 PBFT suggest that very few of these fish migrated from the WPO in the past year (Madigan et al., in prep.). These results, based on bulk SIA, were supported both by the Fukushima radionuclide tracer (Fig. 9.3) and AA-CSIA (Fig. 9.4). This suggests that the numbers of larger, older (>3 years old) PBFT in the EPO depend entirely on the number of smaller fish that migrate to the EPO and the survivorship of that group. As such, recreational and commercial fisheries that target larger PBFT in the EPO may expect this size class to burgeon if fishing pressure on smaller PBFT is reduced (Madigan et al. 2014).

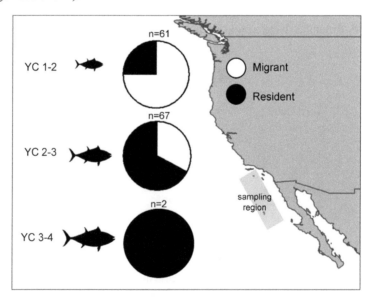

Figure 9.5. In Pacific bluefin tuna, Bulk SIA revealed differences in proportions of recent migrants from the western Pacific Ocean to >1 year residents of the eastern Pacific Ocean. Proportion of migrants to residents decreased with increasing year class of Pacific bluefin tuna. From Madigan et al. 2014.

Laboratory-based experiments allowed for estimates of the timing that WPO migrants entered the CCLME. Isotopic clock estimates of migration revealed three pulses of PBFT migration into the CCLME: one at ages 0.5–1 year; one at 1–1.5 years; and one at ~2 years (Madigan et al. 2014). These results are for a study period of three years (2008–2010) and for a relatively limited sample size of PBFT (n = 130). However this isotopic clock approach, now validated using independent tracers and AA-CSIA, can be applied to future samples of PBFT in the CCLME. Since samples from harvested PBFT are highly available, particularly due to the strong recreational fishery in southern California, the potential exists to ascertain the migratory dynamics of hundreds to thousands of PBFT in the CCLME. Changes in migratory dynamics may be expected due to changes in the PBFT commercial purse-seine fishery in the EPO, physical (SST) and biological (prey availability) variables, and atypical pulse events. For example, in 1988, ~1000 PBFT (all males) ranging from 50 to 458 kg were captured off southern California (Foreman and Ishizuka 1990). This was a rare event in the EPO, and at the time neither electronic tagging technology nor retrospective chemical tracer methods were applied to ascertain whether these PBFT were transient migrants to the EPO or if they were long-term residents. The analyses outlined here would allow the origin of such PBFT to be ascertained in the case of a similar event. It has also been argued that the proportion of PBFT that migrate from the WPO to the EPO is based upon WPO prey availability (Polovina 1996). The potential exists to use SIA in PBFT to track trans-Pacific movements over years or decades to better understand the drivers behind their trans-Pacific migrations.

To fully understand the trans-Pacific exchange of PBFT by size class, similar studies in the WPO are critical. SIA studies of pelagic predators in the WPO have shown that many pelagic species (yellowfin and skipjack tunas, dolphinfish, mako and blue sharks) have consistently lower muscle $\delta^{15}N$ values than the same species in the EPO (Takai et al. 2007; Hisamichi et al. 2010; Madigan et al. 2014), corresponding with regional differences in $\delta^{15}N$ baselines. However, similar to the EPO, PBFT sampled in the WPO showed a wide range of $\delta^{15}N$ values (10.8–16.4‰) (Hisamichi et al. 2010), though individual fish size was not reported in this study. These results potentially indicate a mix of WPO residents (>1 year) and recent EPO migrants in sampled WPO PBFT, which is in agreement with expectations based on their migratory dynamics. These SIA results for PBFT in the WPO suggests that the approach outlined here could be used on both sides of the North Pacific Ocean. A Pacific basin-wide SIA approach could lend valuable insights into the trans-Pacific migratory dynamics of PBFT in the NPO, and a long-term study would lend insight into the variability in these dynamics.

Caveats, Conclusions and Future Work

Context is perhaps the most important consideration when undertaking studies using SIA. This is especially important in studies of migratory predators, where movement, diet switches, physiology, and feeding preferences all may influence observed SIA values. In all wild studies, context in the form of $\delta^{15}N$ and $\delta^{13}C$ values of the local (predator sampling location) baselines (be it phytoplankton, zooplankton, or local prey of the study animal) should be ascertained to assess whether the predator 'fits' the isotopic composition of its local environment. For example, preliminary results for small (60–70 cm) PBFT in the CCLME showed $\delta^{15}N$ values between 11 and 13‰, consistent with krill and copepods in the CCLME and well below other similarly-sized CCLME predators which had $\delta^{15}N$ values between 14 and 17‰ (Madigan et al. 2012b). Traditional application of $\delta^{15}N$ to assess trophic level of these small PBFT would have determined that these pelagic predators are phytoplanktivorous. Considering these PBFT SI values in the context of local CCLME prey, and with knowledge of PBFT diet, made it clear that these small PBFT were reflecting not local prey, but prey from another region—in this case, the WPO. Understanding of PBFT migratory patterns from extensive tagging data supported this conclusion, and allowed instead for the application of PBFT WM $\delta^{15}N$ values to better understand their migrations.

In addition, comparative local 'controls' (in the form of SI values of local taxa for which habitat and/or diet is well understood) should be utilized when this possibility exists, and these controls will be specific to the study species and the study question(s). As mentioned above, SIA values of PBFT in the EPO were known to be influenced by migration from the WPO. However, extensive tagging studies have demonstrated that yellowfin tuna (YFT) caught in the CCLME are largely residential to the CCLME, making fairly extensive north-south migrations but rarely migrating long distances from the CCLME system (Schaefer et al. 2007). As such, the SI values of YFT in the CCLME can be assumed to represent a residential CCLME tuna; this assumption can be used in part to distinguish residential and migrant PBFT in statistical analyses such as discriminant analysis (Madigan et al. 2014). When possible, such comparisons (comparing SIA values of a known highly migratory species to a known residential species) allow estimates of SI values based on local feeding to be constrained. Additionally, SIA performed on animals that have either been electronically tagged or have some intrinsic chemical tracer (other than SI values) allows the SIA approach to be ground-truthed by independent approaches.

Much of the power of the SIA approach is logistical. Small amounts (0.5 g) of muscle tissue are required for analyses, thus samples can be collected non-lethally via biopsy punches or from fisheries-captured individuals. In many cases, samples of pelagic predators are much more easily accessed from carcasses than from live animals. As such, fish markets, processing plants, recreational and commercial fishing vessels, and research cruises are all viable platforms for a successful SIA sampling regime. The cost of processing samples for $\delta^{15}N$ and $\delta^{13}C$ analyses (generally US$ 8 –20/sample) are within most research budgets. Though the data obtained from SIA are coarse compared to electronic tagging data, one can expect bulk SIA data for ~100 individuals per archivally tagged (~US$ 1500/tag) individual and ~300 individuals per PSAT-tagged (~US$ 4500/tag) individual, based on general current tag costs. The SIA approach thus allows for analysis on a numerical scale that is difficult to achieve with other methods.

In PBFT, the bulk SIA approach has shown that while many individual PBFT migrate from the WPO to the EPO in their first year, the EPO fishery is also subsidized by older individuals. It has also provided a means to estimate the timing of migration, and to study changes in this timing over multiple years and changing ocean conditions. This work is still in its infancy, and to reach its full potential a collaborative effort in the WPO and the EPO would greatly advance the current knowledge of migratory dynamics. This could not be more timely, as PBFT have only recently been shown to be highly overfished (ISC 2012). Collaborative Pacific basin-wide studies using SIA are the logical next step to fill in crucial knowledge gaps regarding the trans-Pacific migrations of PBFT. The approach used here can easily be applied to any other pelagic species that traverses oceanographically distinct regions, especially when complemented with other techniques that strengthen SIA-based inferences.

Acknowledgments

This work relied on sampling efforts, analysis suggestions and assistance, and insights from countless individuals. Aaron Carlisle provided valuable edits. I thank Owyn Snodgrass, Brian Popp, Zosia Baumann, Luis Rodriguez, Aaron Carlisle, Rob Dunbar, Heidi Dewar, Charles Farwell, Barbara Block, Steve Litvin, Fio Micheli, Alex Norton, Nicholas Fisher, and the F/V *Shogun* for their assistance, knowledge, and hard work to help accomplish work on a species that is so difficult to find, collect, sample, and understand. This material is partially based upon work supported by the National Science Foundation Postdoctoral Research Fellowship in Biology under Grant No. 1305791.

References

Bayliff, W.H. 1994. A review of the biology and fisheries for northern bluefin tuna, *Thunnus thynnus*, in the Pacific Ocean. FAO Fish. Tech. Pap. 336: 244–295.

Bayliff, W.H., Y. Ishizuka and R. Deriso. 1991. Growth, movement, and attrition of northern bluefin tuna, *Thunnus thynnus*, in the Pacific Ocean, as determined by tagging. Inter-Am. Trop. Tuna Comm. Bull. 20: 3–94.

Block, B.A. 2005. Physiological ecology in the 21st century: advancements in biologging science. Integr. Comp. Biol. 45: 305–320.

Block, B.A., H. Dewar, S.B. Blackwell, T.D. Williams, E.D. Prince, C.J. Farwell, A. Boustany, S.L.H. Teo, A. Seitz, A. Walli and D. Fudge. 2001. Migratory movements, depth preferences, and thermal biology of Atlantic bluefin tuna. Science 293: 1310–1314.

Boustany, A.M., R. Matteson, M. Castleton, C. Farwell and B.A. Block. 2010. Movements of pacific bluefin tuna (*Thunnus orientalis*) in the Eastern North Pacific revealed with archival tags. Prog. Oceanogr. 86: 94–104.

Buesseler, K., M. Aoyama and M. Fukasawa. 2011. Impacts of the Fukushima nuclear power plants on marine radioactivity. Environ. Sci. Technol. 45: 9931–9935.

Buesseler, K.O., S.R. Jayne, N.S. Fisher, I.I. Rypina, H. Baumann, Z. Baumann, C.F. Breier, E.M. Douglass, J. George, A.M. Macdonald, H. Miyamoto, J. Nishikawa, S.M. Pike and S. Yoshida. 2012. Fukushima-derived radionuclides in the ocean and biota off Japan. Proc. Natl. Acad. Sci. USA 109: 5984–5988.

Carey, F.G. 1983. Experiments with free-swimming fish. pp. 57–68. *In*: P.G. Brewer (ed.). Oceanography. Springer-Verlag, New York, USA.

Carlisle, A.B., S.L. Kim, B.X. Semmens, D.J. Madigan, S.J. Jorgensen, C.R. Perle, S.D. Anderson, T.K. Chapple, P.E. Kanive and B.A. Block. 2012. Using stable isotope analysis to understand migration and trophic ecology of northeastern Pacific white sharks (*Carcharodon carcharias*). PLoS ONE 7: e30492.

Caut, S., E. Angulo and F. Courchamp. 2009. Variation in discrimination factors (Δ^{15}N and Δ^{13}C): the effect of diet isotopic values and applications for diet reconstruction. J. Appl. Ecol. 46: 443–453.

Chikaraishi, Y., N.O. Ogawa, Y. Kashiyama, Y. Takano, H. Suga, A. Tomitani, H. Miyashita, H. Kitazato and N. Ohkouchi. 2009. Determination of aquatic food-web structure based on compound-specific nitrogen isotopic composition of amino acids. Limnol. Oceanogr-Meth. 7: 740–750.

Chikaraishi, Y., N.O. Ogawa and N. Ohkouchi. 2010. Further evaluation of the trophic level estimation based on nitrogen isotopic composition of amino acids. pp. 37–51. *In*: N. Ohkouchi, I. Tayasu and K. Koba (eds.). Earth, Life, and Isotopes. Kyoto University Press, Kyoto, Japan.

Clemens, H.B. and G.A. Flittner. 1969. Bluefin tuna migrate across the Pacific Ocean. Calif. Fish Game, Fish Bull. 55: 132–135.

Dickhut, R.M., A.D. Deshpande, A. Cincinelli, M.A. Cochran, S. Corsolini, R.W. Brill, D.H. Secor and J.E. Graves. 2009. Atlantic Bluefin Tuna (*Thunnus thynnus*) Population Dynamics Delineated by Organochlorine Tracers. Environ. Sci. Technol. 43: 8522–8527.

Farwell, C. 2001. Tunas in captivity. pp. 391–412. *In*: B.A. Block and E.D. Stevens (eds.). Tuna: Physiology, Ecology, and Evolution. Academic Press, San Diego, CA, USA.

Foreman, T.J. and Y. Ishizuka. 1990. Giant bluefin tuna off southern California, with a new California size record. Calif. Fish Game, Fish Bull. 76: 181–186.

Fry, B. 2006. Stable Isotope Ecology. Springer-Verlag, New York.

Fujioka, K.O., A.J. Hobday, R.Y.O. Kawabe, K. Miyashita, K. Honda, T. Itoh and Y. Takao. 2010. Interannual variation in summer habitat utilization by juvenile southern bluefin tuna (*Thunnus maccoyii*) in southern Western Australia. Fish. Oceanogr. 19: 183–195.

Graham, B.S., P.L. Koch, S.D. Newsome, K.W. McMahon and D. Aurioles. 2010. Using Isoscapes to Trace the Movements and Foraging Behavior of Top Predators in Oceanic Ecosystems. pp. 299–318. *In*: J.B. West, G.J. Bowen, T.E. Dawson and K.P. Tu (eds.). Isoscapes. Springer, Netherlands.

Gunn, J., T. Polacheck, T. Davis, M. Sherlock and A. Betlehem. 1994. The development and use of archival tags for studying the migration, behaviour and physiology of southern bluefin tuna, with an assessment of the potential for transfer of the technology to groundfish research. ICES CM.

Hisamichi, Y., K. Haraguchi and T. Endo. 2010. Levels of mercury and organochlorine compounds and stable isotope ratios in three tuna species taken from different regions of Japan. Environ. Sci. Technol. 44: 5971–5978.

Hussey, N.E., M.A. MacNeil and A.T. Fisk. 2010. The requirement for accurate diet-tissue discrimination factors for interpreting stable isotopes in sharks. Hydrobiologia 654: 1–5.

Hussey, N.E., M.A. MacNeil, B.C. McMeans, J.A. Olin, S.F.J. Dudley, G. Cliff, S.P. Wintner, S.T. Fennessy and A.T. Fisk. 2014. Rescaling the trophic structure of marine food webs. Ecol. Lett. 17: 239–250.

ISC. 2012. Pacific Bluefin Stock Assessment. International Scientific Committee for Tuna and Tuna-like Species in the North Pacific Ocean.

Kitagawa, T., A.M. Boustany, C.J. Farwell, T.D. Williams, M.R. Castleton and B.A. Block. 2007. Horizontal and vertical movements of juvenile bluefin tuna (*Thunnus orientalis*) in relation to seasons and oceanographic conditions in the eastern Pacific Ocean. Fish. Oceanogr. 16: 409–421.

Kitagawa, T., S. Kimura, H. Nakata, H. Yamada, A. Nitta, Y. Sasai and H. Sasaki. 2009. Immature Pacific bluefin tuna, *Thunnus orientalis*, utilizes cold waters in the Subarctic Frontal Zone for trans-Pacific migration. Environ. Biol. Fish. 84: 193–196.

Klaassen, M., T. Piersma, H. Korthals, A. Dekinga and M.W. Dietz. 2010. Single-point isotope measurements in blood cells and plasma to estimate the time since diet switches. Funct. Ecol. 24: 796–804.

Logan, J. 2009. Tracking Diet and Movement of Atlantic Bluefin Tuna (*Thunnus thynnus*) using Carbon and Nitrogen Stable Isotopes. University of New Hampshire.

Lorrain, A., B. Graham, F. Ménard, B. Popp, S. Bouillon, P. van Breugel and Y. Cherel. 2009. Nitrogen and carbon isotope values of individual amino acids: a tool to study foraging ecology of penguins in the Southern Ocean. Mar. Ecol. Prog. Ser. 391: 293–306.

Lutcavage, M.E., R.W. Brill, G.B. Skomal, B.C. Chase, J.L. Goldstein and J. Tutein. 2000. Tracking adult North Atlantic bluefin tuna (*Thunnus thynnus*) in the northwestern Atlantic using ultrasonic telemetry. Mar. Biol. 137: 347–358.

Lutcavage, M.E., R.W. Brill, G.B. Skomal, B.C. Chase and P.W. Howey. 1999. Results of pop-up satellite tagging of spawning size class fish in the Gulf of Maine: do North Atlantic bluefin tuna spawn in the mid-Atlantic? Can. J. Fish. Aquat. Sci. 56: 173–177.

Madigan, D.J., Z. Baumann, A.B. Carlisle, D.K. Hoen, B.N. Popp, H. Dewar, O.E. Snodgrass, B.A. Block and N.S. Fisher. 2014. Reconstructing trans-oceanic migration patterns of Pacific bluefin tuna using a chemical tracer toolbox. Ecology 95: 1674–1683.

Madigan, D.J., Z. Baumann and N.S. Fisher. 2012a. Pacific bluefin tuna transport Fukushima-derived radionuclides from Japan to California. Proc. Natl. Acad. Sci. USA 109: 9483–9486.

Madigan, D.J., Z. Baumann, O.E. Snodgrass, H.A. Ergül, H. Dewar and N.S. Fisher. 2013. Radiocesium in Pacific bluefin tuna *Thunnus orientalis* in 2012 validates new tracer technique. Environ. Sci. Technol. 47(5): 2287–2294.

Madigan, D.J., A.B. Carlisle, H. Dewar, O.E. Snodgrass, S.Y. Litvin, F. Micheli and B.A. Block. 2012b. Stable isotope analysis challenges wasp-waist food web assumptions in an upwelling pelagic food web. Sci. Rep. 2: e654.

Madigan, D.J., S.Y. Litvin, B.N. Popp, A.B. Carlisle, C.J. Farwell and B.A. Block. 2012c. Tissue turnover rates and isotopic trophic discrimination factors in the endothermic teleost, Pacific bluefin tuna (*Thunnus orientalis*). PLoS ONE 7: e49220.

McClelland, J.W. and J.P. Montoya. 2002. Trophic Relationships and the Nitrogen Isotopic Composition of Amino Acids in Plankton. Ecology 83: 2173–2180.

Miller, T.W., R.D. Brodeur, G. Rau and K. Omori. 2010. Prey dominance shapes trophic structure of the northern California Current pelagic food web: evidence from stable isotopes and diet analysis. Mar. Ecol. Prog. Ser. 420: 15–26.

Minami, H., M. Minagawa and H. Ogi. 1995. Changes in stable carbon and nitrogen isotope ratios in sooty and short-tailed shearwaters during their northward migration. Condor 97: 565–574.

Mitani, Y., T. Bando, N. Takai and W. Sakamoto. 2006. Patterns of stable carbon and nitrogen isotopes in the baleen of common minke whale *Balaenoptera acutorostrata* from the western North Pacific. Fish. Sci. 72: 69–76.

Montoya, J.P., E.J. Carpenter and D.G. Capone. 2002. Nitrogen Fixation and Nitrogen Isotope Abundances in Zooplankton of the Oligotrophic North Atlantic. Limnol. and Oceanogr. 47: 1617–1628.

Navarro, J., M. Coll, C.J. Somes and R.J. Olson. 2013. Trophic niche of squids: Insights from isotopic data in marine systems worldwide. Deep-Sea Res. Part II: 95: 93–102.

Olson, R.J., B.N. Popp, B.S. Graham, G.A. López-Ibarra, F. Galván-Magaña, C.E. Lennert-Cody, N. Bocanegra-Castillo, N.J. Wallsgrove, E. Gier, V. Alatorre-Ramírez, L.T. Ballance and B. Fry. 2010. Food-web inferences of stable isotope spatial patterns in copepods and yellowfin tuna in the pelagic eastern Pacific Ocean. Prog. Oceanogr. 86: 124–138.

Patterson, T.A., K. Evans, T.I. Carter and J.S. Gunn. 2008. Movement and behaviour of large southern bluefin tuna (*Thunnus maccoyii*) in the Australian region determined using pop-up satellite archival tags. Fish. Oceanogr. 17: 352–367.

Phillips, D. and P. Eldridge. 2006. Estimating the timing of diet shifts using stable isotopes. Oecologia 147: 195–203.

Polovina, J.J. 1996. Decadal variation in the trans-Pacific migration of northern bluefin tuna (*Thunnus thynnus*) coherent with climate-induced change in prey abundance. Fish. Oceanogr. 5: 114–119.

Post, D.M. 2002. Using stable isotopes to estimate trophic position: models, methods, and assumptions. Ecology 83: 703–718.

Robbins, C.T., L.A. Felicetti and M. Sponheimer. 2005. The effect of dietary protein quality on nitrogen isotope discrimination in mammals and birds. Oecologia 144: 534–540.

Rooker, J.R., D.H. Secor, G. De Metrio, R. Schloesser, B.A. Block and J.D. Neilson. 2008. Natal Homing and Connectivity in Atlantic Bluefin Tuna Populations. Science 322: 742–744.

Rooker, J.R., D.H. Secor, V.S. Zdanowicz and T. Itoh. 2001. Discrimination of northern bluefin tuna from nursery areas in the Pacific Ocean using otolith chemistry. Mar. Ecol. Prog. Ser. 218: 275–282.

Schaefer, K., D. Fuller and B. Block. 2007. Movements, behavior, and habitat utilization of yellowfin tuna (*Thunnus albacares*) in the northeastern Pacific Ocean, ascertained through archival tag data. Mar. Biol. 152: 503–525.

Schloesser, R.W., J.D. Neilson, D.H. Secor and J.R. Rooker. 2010. Natal origin of Atlantic bluefin tuna (*Thunnus thynnus*) from Canadian waters based on otolith $\delta^{13}C$ and $\delta^{18}O$. Can. J. Fish. Aquat. Sci. 67: 563–569.

Secor, D.H. 2010. Is otolith science transformative? New views on fish migration. Environ. Biol. Fish. 89: 209–220.

Seminoff, J.A., S.R. Benson, K.E. Arthur, T. Eguchi, P.H. Dutton, R.F. Tapilatu and B.N. Popp. 2012. Stable isotope tracking of endangered sea turtles: validation with satellite telemetry and $\delta^{15}N$ analysis of amino acids. PLoS ONE 7: e37403.

Stevens, E.D., J.W. Kanwisher and F.G. Carey. 2000. Muscle temperature in free-swimming giant Atlantic bluefin tuna (*Thunnus thynnus* L.). J. Therm. Biol. 25: 419–423.

Takai, N., N. Hirose, T. Osawa, K. Hagiwara, T. Kojima, Y. Okazaki, T. Kuwae, T. Taniuchi and K. Yoshihara. 2007. Carbon source and trophic position of pelagic fish in coastal waters of south-eastern Izu Peninsula, Japan, identified by stable isotope analysis. Fish. Sci. 73: 593–608.

Takai, N., S. Onaka, Y. Ikeda, A. Yatsu, H. Kidokoro and W. Sakamoto. 2000. Geographical variations in carbon and nitrogen stable isotope ratios in squid. J. Mar. Biol. Assoc. UK 80: 675–684.

Teo, S.H., A. Boustany, H. Dewar, M.W. Stokesbury, K. Weng, S. Beemer, A. Seitz, C. Farwell, E. Prince and B. Block. 2007. Annual migrations, diving behavior, and thermal biology of Atlantic bluefin tuna, *Thunnus thynnus*, on their Gulf of Mexico breeding grounds. Mar. Biol. 151: 1–18.

Varela, J., A. Larrañaga and A. Medina. 2011. Prey-muscle carbon and nitrogen stable-isotope discrimination factors in Atlantic bluefin tuna (*Thunnus thynnus*). J. Exp. Mar. Biol. Ecol. 406: 21–28.

Vizzini, S., C. Tramati and A. Mazzola. 2010. Comparison of stable isotope composition and inorganic and organic contaminant levels in wild and farmed bluefin tuna, *Thunnus thynnus*, in the Mediterranean Sea. Chemosphere 78: 1236–1243.

Willis, J. and A.J. Hobday. 2007. Influence of upwelling on movement of southern bluefin tuna (*Thunnus maccoyii*) in the Great Australian Bight. Mar. Freshw. Res. 58: 699–708.

Wilson, S.G., M.E. Lutcavage, R.W. Brill, M.P. Genovese, A.B. Cooper and A.W. Everly. 2005. Movements of bluefin tuna (*Thunnus thynnus*) in the northwestern Atlantic Ocean recorded by pop-up satellite archival tags. Mar. Biol. 146: 409–423.

CHAPTER 10

Otolith Geochemical Analysis for Stock Discrimination and Migratory Ecology of Tunas

Yosuke Amano,[1] *Jen-Chieh Shiao,*[2] *Toyoho Ishimura,*[3]
Kazuki Yokouchi[4] and *Kotaro Shirai*[5,*]

Background

For the appropriate stock management of highly migratory fish species, it is essential to collect ecological information, such as species distribution and migration patterns, and understand their subpopulation structure and inter-population interactions in a quantitative manner. It is particularly important to identify the marine areas in which the populations contributing to reproduction originate, as well as the migration route they take before reproduction, so that an optimal management strategy can be determined. Ecological information on many migratory fish species, including tunas (*Thunnus* spp.), have been obtained by several approaches such as the mark-recapture method, bio-logging method, and genetic population structure analysis, to determine individual migratory patterns and population structure.

---

[1] Tohoku National Fisheries Research Institute, Fisheries Research Agency, Shiogama, Miyagi 985-0001, Japan.

[2] Institute of Oceanography, National Taiwan University, Roosevelt Road, Taipei 10617, Taiwan.

[3] National Institute of Technology, Ibaraki College, Nakane, Hitachinaka, Ibaraki 312-8508, Japan.

[4] National Research Institute of Aquaculture / National Research Institute of Fisheries Science, Fisheries Research Agency, Fukuura, Kanazawa, Yokohama, Kanagawa 236-8648, Japan.

[5] International Coastal Research Center, Atmosphere and Ocean Research Institute, The University of Tokyo, Kashiwanoha, Kashiwa, Chiba 277-8564, Japan.

* Email: yosukeamano@affrc.go.jp

Because each method has advantages and disadvantages for their application to ecological studies, the scale of phenomenon that can be analyzed and the optimal spatiotemporal scale may vary among methods. For example, mark-recapture is the most common and simplest method, but the estimates obtained may be greatly affected by performance of the marking procedure itself and the recapture rate, which is dependent on fishing effort. The bio-logging method is useful for obtaining information on detailed individual migratory patterns, but marking small fish or a large number of individuals is technically demanding and costly. Genetic population structure analysis is effective only when there are detectable levels of genetic differentiation among populations being studied and is not very effective in identifying individual migration routes and interactions that do not directly contribute to reproduction.

This chapter describes a novel approach to complement the conventional techniques by introducing analysis of the elemental and stable isotopic composition in the fish otolith as a natural tag for reconstructing the migratory ecology of fish. The otolith is a calcium carbonate structure (mainly as aragonite crystals) in the fish inner ear of ectodermal origin, and is responsible for balance and hearing. A total of three pairs of otoliths—the sagitta, lapillus, and asteriscus—are present bilaterally in the inner ear. The otolith core is formed during the early eyed stage of embryonic development and grows incrementally in the otic vesicle of the inner ear, which is filled with endolymph. The otolith grows proportionally with the fish's growth. As the fish grows, the otolith forms incremental rings, generally on a daily or yearly basis according to the biological circadian rhythm or annual cycle, and has therefore long been used in fisheries science as an indicator of fish age. At least 30 different elements, including sodium (Na), potassium (K), manganese (Mn), strontium (Sr), and barium (Ba), are known to be incorporated into accreting otoliths (Campana 1999). As the otolith is a non-cellular tissue, elements once deposited undergo almost no change throughout the fish's life.

Because the elemental composition of the otolith varies with temperature, salinity, and elemental composition of environmental water at the time of otolith accretion, as well as with individual physiological condition and somatic growth rate (Townsend et al. 1992; Fowler et al. 1995; Tsukamoto and Nakai 1998; Campana 1999; Elsdon et al. 2008), the otolith contains time-series information on the environmental and physiological conditions experienced by the individual throughout its life. Thus, analyzing the otolith elemental and stable isotopic composition along its growth axis, from core to margin, will allow us to identify individual migration routes and populations. Since natural tags do not require artificial marking and are not affected by recapture rate, the use of the otolith composition as a natural tag is highly advantageous in that fish origin and life trajectory can be estimated from all samples collected. Another advantage is that this approach can detect ecological differences between populations that cannot be detected by the genetic approach (Perrier et al. 2011). These characteristics indicate that the use of otolith elemental and stable isotopic composition in a study of fish autecology would be highly effective in providing new insights into the migratory ecology of fish.

The effectiveness of otolith elemental and stable isotopic composition as a natural tag for analyzing subpopulation structures and estimating individual migration histories has been extensively verified and demonstrated in various fish species living in rivers, estuaries, and the ocean (Campana 1999; Elsdon and Gillanders 2003; Sturrock et al. 2012; Walther and Limburg 2012). A number of reviews have been published on the elemental composition of the otolith, the mechanism of compositional variations, and its application in fish ecological research. Campana (1999) reviewed the mechanism of variations in otolith elemental composition as well as its application in fish ecological research. Elsdon and Gillanders (2003) discussed environmental and biological factors that may contribute to variations in otolith elemental composition. Elsdon et al. (2008) summarized the information required for otolith research, such as methods for otolith sampling and statistical analysis of data, with a special focus on assumptions and methods required for applied research. Kerr and Campana (2013) described a population discrimination method using the otolith as well as other hard tissues, such as scales, elasmobranch vertebrae, fin rays, and spines.

Many of the previous ecological studies using otolith elemental composition concern diadromous fish species which migrate across both oceanic and riverine habitats or freshwater fish species. Gillanders (2005) reviewed the otolith elemental composition in freshwater and diadromous fish species. Walther and Limburg (2012) have reviewed geochemical factors contributing to variations in the otolith elemental composition of diadromous fish and described methods for analyzing them. More recently, Sturrock et al. (2012) proposed a unique 'the hard and soft acid-base theory', in an attempt to establish an integrated system to categorize the mechanisms of variations in the otolith elemental composition of fully marine fish species. Readers interested in the general topic of otolith markers are advised to refer to these well-written review articles.

However, only a few otolith-based ecological studies of tunas have been reported and have mainly focused on age estimation and validation of otolith elemental composition. Rooker et al. (2007) briefly described the analysis of tuna otoliths as part of the review on the stock status and ecology of Atlantic bluefin tuna (*Thunnus thynnus*). The tuna spends its entire life in seawater and is rarely exposed to freshwater or brackish water throughout its life. This means that they are less exposed to physicochemical changes of environmental water than species living in the river or in brackish water regions, which is likely to be reflected in a weaker environmental signature in otolith chemical composition. Furthermore, because tunas show growth-associated development of body temperature adjustment capability, it is necessary to take into account body temperature adjustment as a potential contributing factor to the composition of elemental and stable isotopic composition in the otolith, which may complicate the interpretation of analysis results compared to other marine fish species.

In addition, when considering the application of otolith minor elemental and stable isotopic composition to an ecological study, it is important to note that the requirements vary depending on the objectives of each study, such as subpopulation discrimination or estimation of migration routes. If discrimination of subpopulations is the only objective of the study, the parameter can be used as long as the composition varies among subpopulations, regardless of the cause of the variation. However, if estimation of migration routes is also the purpose elements and stable isotope having constant partition coefficients throughout the entire life history of the relevant species should be used as a marker. Additionally, oceanographic, physicochemical, and environmental conditions should be quantitatively correlated with otolith composition at the spatiotemporal scale of the migratory ecology being studied. Thus, many preconditions (or appropriate assumptions) must be satisfied before otolith minor elemental and stable isotopic composition can be utilized, and the feasibility of technical application should be fully verified in advance.

The purpose of this chapter is to provide guidance for the use of otolith elemental and stable isotopic composition in ecological studies of tunas. This chapter focuses on quantitatively describing the process and degree of variations of each marker, primarily by describing its geochemical aspects. The types of factors that are responsible for variations in otolith elemental and stable isotopic composition will be discussed based on the latest biological and geochemical findings. Additional detailed discussions are provided regarding things to be considered when interpreting data. Some previous attempts of applying otolith elemental and stable isotopic composition in ecological studies of tunas are also presented. Finally, some of our perspectives are made as to the selection of optimal otolith elements and analysis methods required for providing new insights into the ecological studies of tunas.

Basics of Elements and Stable Isotopes

For otolith minor elemental composition to be used for the determination of the population structure of fish, the composition must be different among populations. Furthermore, for the estimation of migration routes, each compositional variation must be related to each marine area on the migration route being studied. Factors contributing to differences in otolith minor elemental composition among populations include environmental (exogenous) and individual biological (endogenous) factors. Environmental change itself may also affect the biological activities of an individual and thereby indirectly affect the elemental composition of the otolith.

When an exogenous substance is incorporated into the otolith, preferential selection of certain elements and stable isotopes may occur via a physiological process connecting it to the external environment. This process consists of four major steps: change in seawater, change between seawater and blood/endolymph, change between endolymph and otolith, and existing impurities in the otolith. In addition, multiple processes and factors affect elemental variations due to environmental and biological factors in each of the above four steps (Fig. 10.1). An accurate understanding of the mechanism of these variations in otolith elemental and stable isotopic composition caused by environmental and biological factors is essential for its application to more detailed ecological studies. However, this mechanism of variation remains to be fully elucidated (e.g., Campana 1999; Elsdon and Gillanders 2003; Sturrock et al. 2012; Walther and Limburg 2012). The multiple factors involved in the incorporation of elements and stable isotopes and the species-specific variations in the mechanism among fish species complicate the elucidation of the mechanism.

The processes strongly influenced by environmental factors and those largely influenced by biological factors contributing to variations in otolith elemental and stable isotopic composition are suggested here, by considering the ecology of tunas and physicochemical characteristics of the oceanic regions of their habitat. Since it is not realistic to list and describe all elements, those elements and stable isotopes frequently

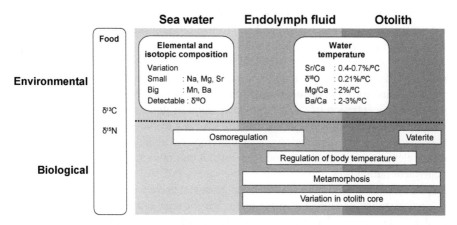

Figure 10.1. Schematic diagram of fractionation mechanisms, steps, and factors on elemental and isotopic compositions in fish otolith.

reported in the literature and contained in the otolith at relatively high concentrations are selected and discussed for the mechanism of compositional variations and the incorporation process on a step-by-step basis. Furthermore, as conventional markers are not effective in identifying migration routes in oceanic regions, a novel marker that shows promise for use in the estimation of migration routes in the oceanic regions will be proposed later.

Effect of environmental change

Change in seawater composition

The simplest factor that may provide information about the migration route of fish is the change in the composition of elements and stable isotopes in environmental water. The ideal otolith marker is one that has a unique composition in each water mass and whose otolith composition is known to be proportionally related to its composition in environmental water.

What types of elements and isotopes are suited for this purpose depends largely on the life history of each fish species. For example, the life history of diadromous fish involves a significant change in habitats, between seawater and freshwater. The distribution of freshwater fish is subject to strong geographical restrictions and limited to an inland body of water subject to substantial environmental change. The elemental and stable isotopic composition in river water may be unique to each river, depending on the geological properties of the watershed. The major and minor elements contained in river water have been summarized in detail by Meybeck (2003) and Gaillardet et al. (2003), respectively. For diadromous and freshwater fish species, a number of indices have been identified whose otolith composition is closely correlated with the level of those elements in environmental waters. Natural tags widely known to be useful for estimating the migration routes of these species include Sr/Ca ratio, Ba/Ca ratio, and the composition of stable sulfur (S) and Sr isotopes.

The magnitude and pattern of variations in major and minor elements in seawater are summarized in detail by Bruland and Lohan (2003). The elemental composition of the oceanic environment inhabited by tunas is less variable than those in riverine and coastal regions. It is often arguable whether an investigation of the mechanism of variations in otolith elemental composition should focus on the concentration itself of elements or the element/calcium ratio (Brown and Severin 2009). The mean seawater concentration of Ca, a main component of the otolith, is 10.3 mmol kg^{-1}, ranging from 10.1 mmol kg^{-1} in low-salinity seawater to 10.3 mmol kg^{-1}, and is nearly homogeneous in oceanic regions. It is therefore reasonable to think that both the minor elemental concentration and element/calcium ratio in the seawater are almost equivalent parameters.

The Sr concentration in seawater, the most extensively analyzed element in fish ecological studies, is nearly homogeneous in oceanic regions at 90 μmol kg^{-1}. de Villiers et al. (1994) performed a high-precision analysis of Sr concentration and Sr/Ca ratio in seawater and reported that the mean Sr concentration normalized for a salinity of 35‰ was 87.4 μM, with a variation of approximately 2.8% (86.5–89.2 μM), and that the mean Sr/Ca ratio was 8.539 mmol mol^{-1}, with a variation of approximately 1.4% (8.49–8.65 mmol mol^{-1}), suggesting very little variation. Brown and Severin (2009) found that the Sr/Ca ratio in environmental water is the major factor for variation in the corresponding ratio in the otoliths of freshwater and diadromous fish species, whereas the magnitude of variation in the corresponding ratio in the otolith of marine fish is greater than that in seawater, and concluded that variation in Sr/Ca ratio in the otolith of marine fish is mainly due to physiological factors. Even in coastal regions, the Sr/Ca ratio in seawater is negligibly affected by a different Sr/Ca ratio in freshwater. Brown and Severin (2009) examined whether the Sr/Ca ratio in seawater is dependent on salinity based on the corresponding ratios in various rivers and found that the variation in Sr/Ca ratio is negligible in oceanic seawater with a salinity over 33‰, which corresponds to tuna habitats. As described later and in Table 10.4, the variation range of Sr/Ca ratio in the otolith of tunas is greater than that in the elemental composition of seawater, suggesting that otolith Sr/Ca ratio does not reflect the corresponding ratio in seawater.

Major and minor elements found in seawater, such as Na, K, lithium (Li), boron (B), magnesium (Mg), and rubidium (Rb), have been analyzed in some studies due to their relatively high concentrations in the otolith. However, these elements show a so-called conservative distribution pattern, meaning that just like Sr, their concentrations in environmental water are constant, except in an extreme environment (Millero 2003). It is therefore unlikely that these elements can be informative markers for estimating the migration routes of fish. However, as described later, these elements may be affected by the indirect effect of environmental change on the organism or by biological factors, and thus may be applicable in population discrimination.

Elements having a nutrient-like distribution pattern, such as nickel (Ni), copper (Cu), zinc (Zn), germanium (Ge), cadmium (Cd), and Ba, as well as those showing a scavenged distribution, such as Mn, aluminum (Al), and cobalt (Co), have not only a vertical but also a horizontal gradient in concentration. This suggests that these elements may have unique composition patterns in each marine area. If their composition in seawater would be reflected in the otolith's composition, they could be used as markers for estimating migration routes.

Since these elements, especially Ni, Cu, Zn, Cd, Mn and Co, are also contained in biologically active substances and thus are likely to be strongly influenced by the metabolic or physiological function of fish, the composition of these elements in seawater is probably not quantitatively reflected by that in the otolith. Moreover, Ge and Al, as well as the six elements listed above, are often contained at low concentrations in the otolith and are susceptible to contamination during analysis, and therefore their composition in the otolith may not always be measured properly.

Among the elements listed here, the Ba/Ca ratio appears to be most promising for use in ecological studies of tunas. Ba belongs to group II in the periodic table, the same group as Ca, and is readily incorporated into aragonite due to its similar chemical properties and behaviors to Ca and Sr (Dietzel et al. 2004; Gaetani and Cohen 2006). Ba in seawater is present at a relatively high concentration of 110 nmol kg^{-1} on average, with a relatively wide variation range of 30–150 nmol kg^{-1} (Bruland and Lohan 2003) and has unique composition patterns in each water mass (Jacquet et al. 2005; 2007; Hoppema et al. 2010). The linearity between Ba/Ca ratio in the otolith and that in environmental water has been verified in various fish species, suggesting that Ba, like Sr, is less susceptible to biological factors than other elements (Bath et al. 2000; Milton and Chenery 2001; Elsdon and Gillanders 2003; Walther and Thorrold 2006).

In addition to the elemental and isotopic composition of seawater, chemical parameters such as salinity, dissolved oxygen and pH may also vary among marine areas. These variations are unlikely to directly lead to variations in the elemental composition of otolith, but may have indirect effects via biological factors or cause complexation and chemical speciation changes, leading to altered amounts of ions that can be incorporated into the tuna body. However, since the magnitude of variations in these chemical parameters of seawater appears to be smaller than that of variations in the corresponding parameters *in vivo*, the effect of these chemical parameters on the elemental composition of the tuna otolith is likely to be negligible.

The oxygen stable isotopic ratio ($\delta^{18}O$: the abundance ratio of ^{18}O to ^{16}O), as described below, has a variation range of about 1‰ across different marine areas (LeGrande and Schmidt 2006). By comparison, the difference in $\delta^{18}O$ between the Sea of Japan and Pacific Ocean, which are major spawning grounds for Pacific bluefin tuna (*Thunnus orientalis*) (Chen et al. 2006; Tanaka et al. 2007), is as low as about 0.5‰ (Kitagawa et al. 2013). As detailed later, the $\delta^{18}O$ appears to be more greatly influenced by water temperature than by variation in seawater composition; however, the latter variation is not negligible and deserves due attention. Seawater $\delta^{18}O$ depends on the terrestrial influx of freshwater and precipitation, which has a lower $\delta^{18}O$ value, and evaporation of water from seawater (water vapor with low $\delta^{18}O$ composition), leading to a positive linear correlation between $\delta^{18}O$ and salinity. It should, however, be noted that salinity itself does not affect the $\delta^{18}O$ value of calcium carbonate.

In view of the varying chemical composition of seawater and the purpose of estimating the migration route of marine fish, only the Ba/Ca ratio is considered to meet the requirement of a proportional relationship between their concentrations in the otolith and those in environmental water. However, elements require time to pass through physiological barriers before being incorporated into fish otoliths (Milton and Chenery 2001; Elsdon and Gillanders 2003; Yokouchi et al. 2011). As tuna is a highly migratory fish and may migrate from one marine area to another while elements are not completely yet incorporated from seawater into the otolith, the otolith composition of a given element may not match the composition of the element in seawater at the time of formation of the relevant part of the otolith.

Effect of water temperature

Water temperature is the most important physical factor to be considered when analyzing otolith elemental and isotopic composition. With a steep gradient of water temperature along the directions of depth and latitude in oceanic regions, the temperature-dependent elemental and isotopic composition of the otolith is likely to be applicable to the estimation of fish migration patterns. Fractionation by temperature is a fluid-solid interaction occurring between the endolymph and the otolith. Therefore, when changes in water temperature (or endolymph temperature) are reflected in the otolith, then there should be no time lag. Among temperature-dependent elemental and isotopic compositions, Mg/Ca, Sr/Ca, and Ba/Ca ratios, as well as the $\delta^{18}O$ have been extensively discussed in various fish species. The temperature dependency of these element/Ca and $\delta^{18}O$ appears largely due to the empirical relationship that the composition of these elements/isotopes in an inorganically produced aragonite is temperature-dependent.

Calcium carbonate can form two thermodynamically stable mineral phases: trigonal calcite and orthorhombic aragonite, whereas hexagonal vaterite is thermodynamically unstable. Both the partitioning of minor elements into calcium carbonate and its mechanism have been well studied for calcite (Tesoriero and Pankow 1996; Watson 1996; 2004; Rimstidt et al. 1998; Curti 1999). Ions smaller than the radius of the Ca ion (1.00 Å) are less likely to be incorporated into inorganic calcite and ions larger than 1.1 Å are also seldom incorporated. In contrast, divalent cations with ionic radii larger than that of Ca, such as Sr and Ba, are more readily incorporated into aragonite, while the Mg ion, whose radius is smaller than that of Ca, is hardly to be incorporated (Gaetani and Cohen 2006). Although little is known about the partitioning of other elements, these findings obtained for inorganic calcium carbonate may be helpful in roughly estimating the otolith composition of elements and their variation patterns. The temperature dependency of partitioning of minor elements into inorganic aragonite has been demonstrated for Sr/Ca (Kinsman and Holland 1969; Dietzel et al. 2004; Gaetani and Cohen 2006), Mg/Ca (Gaetani and Cohen 2006), and Ba/Ca ratios (Dietzel et al. 2004; Gaetani and Cohen 2006) (Table 10.1). Of these, the Sr/Ca ratio has been associated with consistent results in all reports, and thus may be a temperature-dependent elemental ratio quantitatively controlled by a physiochemical process. However, its temperature dependency is relatively low at 0.4–0.7%/°C at 20°C. Mg/Ca and Ba/Ca ratios are negatively correlated with temperature, with respective temperature dependencies of 2%/°C and 2–3%/°C at 20°C, and thus are more temperature-dependent than the Sr/Ca ratio. However, care should be taken for the quantitative interpretation of these data due to the scarcity of data and inconsistent results regarding temperature dependency among reports.

The $\delta^{18}O$ of otolith calcium carbonate depends on the temperature and the $\delta^{18}O$ of the surrounding water at the time of crystallization, and thus can be an environmental marker used for the reconstruction of experienced thermal history. The first systematic

Table 10.1. Expected ranges of chemical compositions of aragonite produced under pure inorganic processes caused by environmental variation which Pacific bluefin tuna will experience in its life history.

	Reported temperature dependency of distribution coefficient		Element concentration in SW[a] (μmol kg^{-1})	Me/Ca ratio in SW[a] (mmol mol^{-1})	Distribution coefficient			Expected Me/Ca (mmol mol^{-1})			Relative variation per 1°C at 20°C (%)	Relative variation caused by SW composition[b] (%)	Relative variation caused by temperature[c] (%)	Reference
					14°C	20°C	26°C	14°C	20°C	26°C				
Mg/Ca	$\ln K^{MgCa} = (1930/T) - 13.1$	Min.	53000	5146	0.00170	0.00148	0.00130	8.76	7.64	6.69	2.25	-	27	Gaetani and Cohen (2006)
		Max.	53000	5146				8.76	7.64	6.69				
Sr/Ca	$K = 1.2488 - 0.00447T$ [d]	Min.	86.5	8.40	1.19	1.16	1.13	9.97	9.75	9.53	0.38	3.9	4.5	Kinsman and Holland (1969)
		Max.	90.0	8.74				10.37	10.14	9.91				
	$D_{Sr} = 1.27 - 0.0052127T$	Min.	86.5	8.40	1.20	1.17	1.13	10.05	9.79	9.53	0.45	3.9	5.4	Dietzel et al. (2004)
		Max.	90.0	8.74				10.46	10.19	9.91				
	$\ln K^{SrCa} = (605/T) - 1.89$	Min.	86.5	8.40	1.24	1.19	1.14	10.44	10.00	9.60	0.70	3.9	8.5	Gaetani and Cohen (2006)
		Max.	90.0	8.74				10.87	10.41	9.98				
Ba/Ca	$D_{Ba} = 2.31 - 0.03455T$	Min.	0.03	0.0029	1.83	1.62	1.41	0.0053	0.0047	0.0041	2.13	80	26	Dietzel et al. (2004)
		Max.	0.15	0.0146				0.0266	0.0236	0.0206				
	$\ln K^{BaCa} = (2913/T) - 9.0$	Min.	0.03	0.0029	3.16	2.57	2.10	0.0092	0.0075	0.0061	3.39	80	41	Gaetani and Cohen (2006)
		Max.	0.15	0.0146				0.0460	0.0374	0.0306				

	Reported temperature dependency of fractionation factor		$\delta^{18}O$ of SW[a] (‰, relative to VSMOW)	Expected $\delta^{18}O$ (‰, relative to VPDB)			Variation per 1°C at 20°C (‰)	Variation caused by SW composition (‰)	Variation caused by temperature[e] (‰)	Reference
				14°C	20°C	26°C				
calcite $\delta^{18}O$	$1000\ln \alpha = 18.03 (1000/T) - 32.42$	Min.	-1.00	-1.03	-2.31	-3.54	0.21	1.00	2.52	Kim and O'Neil (1997)
		Max.	0.00	-0.03	-1.31	-2.54				
aragonite $\delta^{18}O$ [f]	$1000\ln \alpha = 17.88 (1000/T) - 31.14$	Min.	-1.00	-0.27	-1.54	-2.76	0.21	1.00	2.50	Kim et al. (2007a)
		Max.	0.00	0.73	-0.54	-1.77				

a) Range of seawater compositions were estimated based on the possible migration route of Pacific bluefin tuna. Ca concentration is assumed to be 10.3 mmol/kg. See text for more details.
b) (maximum value - minimum value)/(maximum value) x 100, calculation based on value at 20°C
c) (value at 14°C - value at 26°C)/(value at 20°C) x 100, calculation based on maximum concentration
d) Calculated by simple linear regression of reported averaged data
e) $\delta^{18}O$ value at 14°C - $\delta^{18}O$ value at 26°C
f) Isotopic values of aragonite are calculated using an acid fractionation factor at 25°C as reported in Kim et al. (2007a, 2007b)

analysis of the δ^{18}O-based temperature dependency of inorganic aragonite was reported by Grossman and Ku (1986). Kim et al. (2007a) experimentally synthesized aragonite to derive a temperature conversion equation and reported values almost in complete keeping with the measured data of the natural samples reported by Grossmann and Ku (1986). The study revealed that the δ^{18}O of carbonates formed by inorganic precipitation is about 1‰ of isotopic fractionation per 4°C. This means that if the δ^{18}O of the surrounding water is constant, the δ^{18}O of calcium carbonate can be used as an indicator of temperature. This parameter has been extensively used in environmental information analyses using coral, shellfish, and foraminifera. More details on the study of stable isotopes of biogenic carbonates can be found in Rohling and Cooke (1999).

Below is a reminder of issues to be considered when measuring isotopes in aragonite, which are not considered in some of the studies that examine the δ^{18}O of the otolith. Carbon (C) and oxygen stable isotopic ratios can be determined by analyzing the δ^{13}C (the abundance ratio of ^{13}C to ^{12}C) and δ^{18}O (the abundance ratio of ^{18}O to ^{16}O) of CO_2 gas generated by the following phosphoric acid reaction:

$$CaCO_3 + H_3PO_4 \rightarrow CaHPO_4 + H_2O + CO_2$$

Stable isotopic ratio mass spectrometry is usually used for this analysis, with values for isotopes being normalized to international standard NBS19 according to the Vienna Pee Dee Belemnite (VPDB) scale (Brand et al. 2014 and references therein). The guidelines for presenting the values of isotopes are provided in Coplen (2011). It has been shown that the δ^{18}O values of aragonite and vaterite are approximately 0.6‰ higher than that of a calcite formed in the same seawater (Tarutani et al. 1969; Grossman and Ku 1986; Kim and O'Neil 1997). Kim et al. (2007b) have demonstrated that calcite and aragonite have different isotopic fractionation factors during phosphoric acid reaction in an isotopic ratio analysis and that the factors are temperature-dependent. This suggests that the δ^{18}O of carbon dioxide generated from calcite by phosphoric acid reaction is significantly different from that generated from aragonite with the identical isotopic composition, and the magnitude of the difference is dependent on reaction temperature. This means that when determining the δ^{18}O value of otolith aragonite as compared to a calcite standard, values may substantially vary depending on reaction temperatures and isotopic fractionation factors used in each laboratory. This is particularly important when comparing δ^{18}O values with those from previous studies, since the temperature during the phosphoric acid reaction and the isotopic fractionation factors used for calculation vary among studies. In future studies on otolith δ^{18}O, the reaction temperature during analysis and the isotopic fractionation factor used for calculation should be specified.

For the purpose of verifying the water temperature dependency of inorganic carbonates described above, an example is given to see how strongly the otolith composition is influenced by water temperature. If the otolith elemental and isotopic ratios of tunas are purely dependent on physiochemical processes alone, each ratio should be within the range predicted from its temperature dependency and the possible range of water temperature experienced by tuna. Table 10.1 summarizes the elemental and isotopic composition of inorganically produced aragonite estimated from the water temperature dependency of each elemental and isotopic ratio and their composition in oceanic water within the possible range of water temperature experienced by bluefin tuna.

Given that the estimated range of water temperature experienced by Pacific bluefin tuna through the spawning, hatching, larval, and juvenile stages is about $26 \pm 2°C$ (Kimura et al. 2010) and the optimal water temperature for adult Pacific bluefin tuna is 14–19°C, the temperature range is estimated to be 12–21°C (Kitagawa et al. 2000). Although they dive to deeper depths in search of food and are exposed to lower water temperatures on a temporary basis, Pacific bluefin tuna spend the majority of their life at a depth corresponding to their optimal temperature range (Kitagawa et al. 2007). It should, however, be noted that the body temperature of Pacific bluefin tuna is not completely synchronized with the external environmental temperature, an unusual trait in fish, and therefore the temperature condition at the time of otolith formation does not necessarily reflect the seawater temperature (Carey et al. 1984; Kitagawa et al. 2001; 2006). Although the body temperature of Pacific bluefin tuna varies according to seawater temperature, the difference between seawater temperature and body temperature becomes more body weight-dependent as Pacific bluefin tuna grow, the relationship expressed as $1.668 \times W (kg)^{0.334}$ (Kitagawa et al. 2006). In view of these facts, the range of water temperature experienced by Pacific bluefin tuna otolith is estimated to be 14–26°C.

As seen from the comparisons shown in Table 10.1 and Table 10.4 the variation ranges of the Sr/Ca and Mg/Ca ratios in the tuna otolith are larger than that of seawater composition or variations caused by changing water temperature. As for the Ba/Ca ratio, the variation in seawater composition of these elements is greater than the effect of temperature change. Deviation of the values of otolith composition from those of inorganic carbonates indicates the effect of biological factors. Because water temperature also affects somatic growth rate and various other biological factors, biological processes triggered by environmental change may represent the major factor contributing to the temperature dependency of the Sr/Ca and Mg/Ca ratios in the otolith.

In contrast, since otolith $\delta^{18}O$ is strongly affected by water temperature, the relationship between otolith $\delta^{18}O$ and water temperature is not dramatically different from that between the oxygen isotopic ratio in inorganic aragonite and water temperature. In fact, $\delta^{18}O$ in several fish species, including tunas, is considered to depend on isotopic equilibrium, as in inorganic aragonite (Patterson et al. 1993; Radtke et al. 1996; 1998; Thorrold et al. 1997; Høie et al. 2004). Kitagawa et al. (2013) analyzed the otolith $\delta^{18}O$ of Pacific bluefin tuna larvae experimentally reared at six different water temperatures (23–28°C, salinity 35‰) and found that otolith $\delta^{18}O$ values precisely reflect changes in environmental water temperature and thus can be a sensitive marker of water temperature experienced by larvae of this species. This suggests that the distribution of water temperature in ocean regions, which may show a steep gradient in the horizontal or vertical directions, is reflected in the otolith $\delta^{18}O$ of tuna larvae, which can greatly contribute to the estimation of spawning grounds.

In summary, variations in otolith elemental composition occur mainly through biological processes, and thus cannot be used for reconstructing the history of experienced water temperature. Whereas, the $\delta^{18}O$ can be a highly quantitative, useful indicator of temperature that reflects both water and body temperature changes.

Effect of elemental origins, food, and water

Even though the amount of major elements incorporated into the otolith via food is generally smaller than that incorporated directly from seawater, the body element composition of the fish may be influenced by food-derived elements. Kerr and Campana (2013) have claimed, based on the findings reported by Buckel et al. (2004) and Marohn et al. (2009), that food-derived Na, Mg, K, and Ca, make only a small contribution to otolith composition. In contrast, biologically active elements, such as Ni, Cu, Zn, Cd, Mn, and Co, and biologically concentrated elements, such as Cd, lead (Pb), and mercury (Hg), are likely not only directly incorporated into the body from seawater, but also via food, suggesting their non-negligible effect on the elemental composition of the otolith.

In fact, the absolute concentration as well as relative element to calcium ratio of Cu, Mn, Zn, and selenium (Se) in marine fish blood are more than 10 times higher than those in seawater, suggesting an active uptake of these elements (Table 10.2). If the otolith composition of these elements reflects their composition in food, then the parameter is likely to be influenced by the structure and characteristics of the biological community in each marine area and thus may be used to narrow down marine areas inhabited by tuna, although otolith variation nevertheless also occurs through indirect processes. The pathway and mechanism of the uptake of these bio-elements into the otolith remain largely unknown and warrant further investigation.

Variations caused by biological factors

Change in endolymph composition

The uptake of elements from seawater into the fluid in direct contact with the otolith (i.e., the endolymph, or calcifying fluid) occurs through multiple steps involving seawater, plasma, and endolymph (Campana 1999). Elemental composition may vary from step to step and the pattern of compositional variation differs from element to element. Evidence has suggested that the transfer of elements from seawater into fish plasma is primarily regulated by osmotic regulation in the gills (Kalish 1991; Campana 1999; Hwang and Lee 2007). It appears that the uptake of some elements is also strongly influenced by nutritional intake via food (Watanabe et al. 1997). Elemental partitioning during the seawater-plasma phase is thus strongly influenced not only by physiological/ecological factors, such as age, somatic growth rate, maturation, and metamorphosis, but also by external environmental factors, such as water temperature and feeding.

Table 10.2 summarizes the relationship between selected elemental compositions (Li, Na, Mg, Cl, K, Ca, Mn, Cu, Zn, Se, Sr, and Ba) in blood plasma of several marine fishes and elemental concentrations in seawater, as reported by Strrock et al. (2013). Concentrations of conservative-type elements (i.e., Li, Na, Mg, Cl, K, Ca, and Sr) are lower in the plasma than in seawater, while the concentrations of biological elements (Mn, Cu, Zn, and Se) are more than 10 times higher. When expressed in a ratio to calcium concentration, the concentrations of biological elements in the plasma are

Table 10.2. Elemental concentrations and its Metal/Ca ratios in seawater, marine fish, male and female plaice blood plasma. Enrichment factor was calculated from the value of blood plasma composition divided by seawater composition. Reference ranges in male and female plaice blood plasma referenced from Sturrock et al. (2013) and seawater composition referenced from Bruland and Lohan (2003).

	Li	Li/Ca	Na	Na/Ca	Mg	Mg/Ca	Cl	Cl/Ca
	ppm	mmol mol⁻¹	ppm	mmol mol⁻¹	ppm	mmol mol⁻¹	ppm	mmol mol⁻¹
Seawater	0.180	2.515	10759	45437	1288	5146	19357	53010
Marine fish blood plasma composition (range)	0.05 (0.006)	2.240	4043 (109)	54679	34 (5)	438	5875 (209)	51524
Blood plasma enrichment factor	0.278	0.891	0.376	1.203	0.027	0.085	0.304	0.972
All plaice blood plasma composition (RSD, %)	0.065 (11)	2.974			26 (26)	346		
All enrichment factor	0.363	1.183			0.021	0.067		
Male plaice blood plasma composition (Reference range)	0.063 (0.05-0.07)	2.945			27 (19-41)	357		
Male enrichment factor	0.352	1.171			0.021	0.069		
Female plaice blood plasma composition (Reference range)	0.067 (0.06-0.07)	2.809			27 (21-45)	320		
Female enrichment factor	0.370	1.117			0.021	0.062		

	K	K/Ca	Ca	Mn	Mn/Ca	Cu	Cu/Ca
	ppm	mmol mol⁻¹	ppm	ppm	mmol mol⁻¹	ppm	mmol mol⁻¹
Seawater	399	990	413	0.00002	0.00003	0.00019	0.00029
Marine fish blood plasma composition (range)	287 (125)	2284	128 (15)			1.21 (1)	5.920
Blood plasma enrichment factor	0.720	2.306	0.312			6347	20327
All plaice blood plasma composition (RSD, %)	42 (62)	344	126 (16)	0.021 (50)	0.121	0.736 (25)	3.661
All enrichment factor	0.107	0.348	0.307	1274	4148	3861	12569
Male plaice blood plasma composition (Reference range)	45 (14-110)	371	124 (108-149)	0.017 (0.009-0.041)	0.098	0.766 (0.489-1.093)	3.901
Male enrichment factor	0.112	0.374	0.300	1007	3356	4020	13394
Female plaice blood plasma composition (Reference range)	41 (13-102)	307	137 (112-190)	0.024 (0.011-0.050)	0.129	0.712 (0.452-1.023)	3.287
Female enrichment factor	0.103	0.310	0.331	1468	4434	3737	11285

Table 10.2. contd....

Table 10.2. contd.

	Zn ppm	Zn/Ca mmol mol⁻¹	Se ppm	Se/Ca mmol mol⁻¹	Sr ppm	Sr/Ca mmol mol⁻¹	Ba ppm	Ba/Ca mmol mol⁻¹
Seawater	0.00033	0.00049	0.00013	0.00017	7.886	8.738	0.015	0.011
Marine fish blood plasma composition (range)	13 (9)	63	31 (7)	126	11	3.903		
Blood plasma enrichment factor	40618	130079	237648	761071	0.139	0.447		
All plaice blood plasma composition (RSD, %)	12 (21)	60	0.368 (23)	1.473	1.03 (16)	3.716	0.008 (70)	0.019
All enrichment factor	38232	124467	2742	8925	0.131	0.425	0.536	1.746
Male plaice blood plasma composition (Reference range)	13 (9-17)	65	0.375 (0.234-0.516)	1.535	0.998 (0.764-1.284)	3.685	0.008 (0.003-0.025)	0.019
Male enrichment factor	40067	133494	2791	9300	0.127	0.422	0.543	1.809
Female plaice blood plasma composition (Reference range)	12 (7-17)	52	0.361 (0.232-0.475)	1.341	1.078 (0.858-1.398)	3.607	0.007 (0.003-0.026)	0.016
Female enrichment factor	35296	106586	2691	8126	0.137	0.413	0.483	1.459

again more than 10 times higher than those in seawater, while the concentrations of conservative-type elements are not substantially different. This suggests that biological elements may be maintained at a certain level to fulfill a physiological requirement.

Since the changes in major elements/Ca ratios during incorporation of the elements from seawater to body are greater than the spatiotemporal change in seawater elemental composition, elemental partitioning during the incorporation step likely contributes to variation in the final otolith composition. Furthermore, because variations in these biological factors are species-dependent, analyzing elemental composition in tuna blood is essential for elucidating the mechanism of elemental partitioning into the otolith. Although elemental partitioning between the plasma and endolymph has been analyzed by Kalish (1991) and other investigators, the data appear to remain insufficient for determining or discussing the general mechanism of elemental partitioning in marine fish at this step. As for $\delta^{18}O$, isotopic fractionation appears to be ignorable at this step due to the lack of a selective transportation mechanism between seawater and endolymph.

In contrast, the $\delta^{13}C$ is expected to undergo substantial change between seawater and endolymph due to the addition of metabolic carbon. Both $\delta^{13}C$ in food and dissolved inorganic carbon (DIC) in environmental water have been identified as factors contributing to variations in otolith $\delta^{13}C$ (Kalish 1991; Schwarcz et al. 1998; Solomon et al. 2006; Nonogaki et al. 2007; Tohse and Mugiya 2008). Although some studies have shown that more carbon is incorporated from DIC into the otolith than from food (approximately 20% from feed and approximately 80% from DIC; Solomon et al. 2006; Tohse and Mugiya 2008), no sufficient discussion has been made in previous ecological studies due to difficulty in the interpretation of $\delta^{13}C$ data.

Otolith $\delta^{13}C$ values reported in previous studies range from -7 to $-10‰$ (VPDB) (Table 10.4), which are much lower than the corresponding values of DIC in seawater, which range from $+0.5$ to $+2‰$ (Kroopnick 1985; Tagliabue and Bopp 2008). Based on the fact that the $\delta^{13}C$ of organic matter from food ranges from -20 to $-25‰$, there is little doubt that respiration-derived carbons resulting from the decomposition of organic matter contribute to otolith composition. Because change in the proportion of respiration-derived metabolic carbons is expected to lead to substantial change in otolith $\delta^{13}C$, the proportion of DIC-derived and metabolism-derived carbons can be a key factor contributing to variation in otolith $\delta^{13}C$.

Even given the significant metabolic control, there is some evidence that the otolith $\delta^{13}C$ may record environmental and/or ecological information on bluefin tuna. The otolith $\delta^{13}C$ values of juvenile Pacific bluefin tuna have been found to drop >2‰ when the tuna were transferred and reared in the sea cage off the Amami Islands, Japan (J.C.S., unpubl. data). Isotope compositions are similar between right and left otoliths in bluefin tuna (Høie et al. 2004). Pacific bluefin tuna larvae grown in the same environment have similar values of otolith $\delta^{13}C$ (Kitagawa et al. 2013). These phenomena also suggest the potential of $\delta^{13}C$ as a parameter to be used for the elucidation of ecological characteristics of fish.

Biological factors in each life history phase

As the biological and physiological characteristics of fish change as the life history phase advances from egg, hatching, larval/juvenile, metamorphosis, and adult phases, the elemental composition of the otolith also varies from phase to phase through the life history. Ruttenberg et al. (2005) analyzed the elemental composition of the otolith in five fish species (*Galaxias maculates, Oxylebius pictus, Sebastes atrovirens, Dascyllus marginatus, Stegastes beebei,* and *Thalassoma bifasciatum*) and found higher ratios of Mg/Ca, Mn/Ca, and Ba/Ca around the core than along the margin of the otolith. Brophy et al. (2004) have reported a higher Mn/Ca ratio in the otolith core of Atlantic herring (*Clupea harengus*). de Pontual et al. (2003) observed a gradient of decreasing Sr/Ca ratio and K concentration from the core to the margin of the otolith of common sole (*Solea solea*). This appears to reflect change in otolith elemental composition through the life history phases of this fish species, starting from an early-life stage through metamorphosis to settlement. It is unclear whether this change occurs between seawater and endolymph or between endolymph and the otolith. It is also possible that the multiple fractionation of an element occurs in a different manner in each phase of the life history.

The line analysis profile of a tuna otolith along its growth axis has also shown variations in otolith elemental composition that may be due to varying biological factors according to the development and growth of the individual. Wang et al. (2009) demonstrated that Na/Ca, Mg/Ca, and Mn/Ca ratios in the otolith of southern bluefin tuna (*Thunnus maccoyii*) are relatively high in the core through to the first inflection point, and then decline along the growth axis. The first inflection point corresponds to 43 days of age, which is about the time of metamorphosis. On this basis, they concluded that these variations occurred through the ontogenetic process. Macdonald et al. (2013) found that Li/Ca and Mg/Ca ratios in the otolith of South Pacific albacore (*Thunnus alalunga*) show a core-to-edge decreasing trend whereas the Mn/Ca ratio peaks at a point 200–1000 μm from the core and then declines towards the edge in the otolith.

Although the elemental composition in the otolith core and its vicinity is likely to be strongly affected by biological factors, if the environmental effect (via biological factors) is large enough to cause difference in otolith composition among populations, the pattern of elemental composition along the otolith growth axis from the core to edge could be used for population discrimination. Between the Philippine Sea and the Sea of Japan, which have been identified as spawning grounds for Pacific bluefin tuna (Chen et al. 2006; Tanaka et al. 2007), differences of the elemental composition in the otolith may be detected by analysis of the same age class and same ontogenetic portion to minimize the effects of biological factors. However, it should be remembered that discriminating different populations based on differential elemental compositions around the otolith core is not equivalent to classifying each individual to the same population based on the absence of such differences. Because the elemental composition in the otolith core appears to be somewhat variable among individuals, if the effect of an environmental difference is smaller than that of individual difference, then different populations can have a similar otolith composition. Caution must therefore be exercised when interpreting data that are strongly influenced by biological factors.

Other effects

Impurities (vaterite and organic matter) in the otolith

Impurities in the otolith, such as organic matter and other crystaline structures (e.g., vaterite), may result in elemental and isotopic variations between the endolymph and otolith or in the otolith. The major component of the otolith is calcium carbonate, which forms aragonite crystals. In a precise sense, biogenic calcium carbonate is a hybrid of calcium carbonate and organic matter and also contains some elements as impurities. The use of otolith elemental and stable isotopic composition as a natural tag is based on the assumption that elements in seawater are incorporated into otolith aragonite. Elemental partitioning between seawater and organic matter is likely to differ substantially from that between seawater and aragonite; heterogeneous distribution of organic matter in the otolith may also affect the distribution of elements and isotopic composition in the otolith.

Although such variation mechanisms have rarely been discussed for the otolith, contamination of organic matter is a common problem in various organisms commonly used in geochemical and paleoenvironmental studies that contain biogenic calcium carbonate, such as corals, foraminifera, and bivalves, and its removal has been attempted in many studies (Mitsuguchi et al. 2001; Watanabe et al. 2001; Barker et al. 2003; Pena et al. 2005). Recent studies have suggested that the presence of organic matter in a solution during calcification affects the incorporation of elements into calcium carbonate (Stephenson et al. 2008; Wang et al. 2009; Shirai et al. 2014). Organic matter in the otolith may also cause instrumental analytical artifacts (Schöne et al. 2010; Shiao et al. 2014). The growth rings of the otolith consist of transparent and opaque zones, which are thought to result from differential compositions of organic matter (Jolivet et al. 2008). It should be carefully determined whether the variation in elemental composition is due to environmental factors or to the composition of organic matter.

Although a normal otolith is composed of aragonite, the occasional formation of vaterite is also known to occur. Given that there is a substantial difference in elemental composition between otolith vaterite and aragonite (Tomas and Geffen 2003; Melancon et al. 2005; Tzeng et al. 2007), otoliths should be carefully assessed for the presence of vaterite, although to our knowledge no study has shown the presence of vaterite in tuna otoliths.

Artifacts in sample and data analyses

As described earlier, the elemental composition of otolith may vary according to individual development and growth conditions. Thus, if the pattern of elemental variation within the otolith has not been identified, artifacts associated with an inappropriate handling of otolith data or an inappropriate selection of analyzed portions may cause apparent elemental variations.

Table 10.3 and Fig. 10.2 show a simulation of how a virtual otolith variation can change depending on the selection of analysis. The otolith variation patterns shown are modifications from the variation profile for South Pacific albacore reported by

Table 10.3. Virtual variation patterns of otolith used for the simulation of how analytical results can change depending on the selection of otolith portion presented in Fig. 10.2. See the caption of Fig. 10.2 for more details.

Otolith distance (μm)	Elemental ratio (mmol mol⁻¹)				Otolith distance (μm)	Elemental ratio (mmol mol⁻¹)			
	Pattern A	Pattern B	Pattern C	Pattern D		Pattern A	Pattern B	Pattern C	Pattern D
0	150	75	5	102	490	20	10	40	25
10	140	80	5	100	500	20	10	30	24.5
20	130	85	5	98	510	20	10	20	24
30	120	90	5	96	520	20	10	30	23.5
40	110	95	5	94	530	20	10	40	23
50	100	100	5	92	540	20	10	50	22.5
60	90	105	5	90	550	20	10	60	22
70	80	110	5	88	560	20	10	70	21.5
80	70	115	5	86	570	20	10	60	21
90	60	120	5	84	580	20	10	50	20.5
100	50	125	5	82	590	20	10	40	20
110	40	130	5	80	600	20	10	30	19.5
120	30	135	5	78	610	20	10	20	19
130	20	140	5	76	620	20	10	30	18.5
140	20	130	5	74	630	20	10	40	18
150	20	120	5	72	640	20	10	50	17.5
160	20	110	5	70	650	20	10	60	17
170	20	100	5	68	660	20	10	70	16.5
180	20	90	5	66	670	20	10	60	16
190	20	80	5	64	680	20	10	50	15.5
200	20	70	5	62	690	20	10	40	15
210	20	60	15	60	700	20	10	30	14.5
220	20	50	15	58					

Averaged ranges of otolith distance (μm)	Average of elemental ratios (mmol mol⁻¹)			
	Pattern A	Pattern B	Pattern C	Pattern D
0-100	100	100	5	92
0-200	63	105	5	82
0-300	49	79	8	72
0-400	43	64	10	63
0-500	38	52	17	56
0-600	35	45	21	50
0-700	33	40	25	45
0-40	130	85	5	98
30-60	105	98	5	93
50-90	85	108	5	89
70-110	65	118	5	85
610-640	20	10	35	18.3
630-660	20	10	55	17.3
650-680	20	10	60	16.3
670-700	20	10	45	15.3
0-10	145	78	5	101
10-30	130	85	5	98
10-50	120	90	5	96
10-70	110	95	5	94

The remaining left-hand table rows:

Otolith distance (μm)	Pattern A	Pattern B	Pattern C	Pattern D
230	20	40	15	56
240	20	30	15	54
250	20	20	15	52
260	20	10	15	50
270	20	10	15	48
280	20	10	15	46
290	20	10	15	44
300	20	10	15	42
310	20	10	15	40
320	20	10	15	38
330	20	10	15	36
340	20	10	15	34
350	20	10	15	32
360	20	10	15	31.5
370	20	10	15	31
380	20	10	15	30.5
390	20	10	15	30
400	20	10	15	29.5
410	20	10	20	29
420	20	10	30	28.5
430	20	10	40	28
440	20	10	50	27.5
450	20	10	60	27
460	20	10	70	26.5
470	20	10	60	26
480	20	10	50	25.5

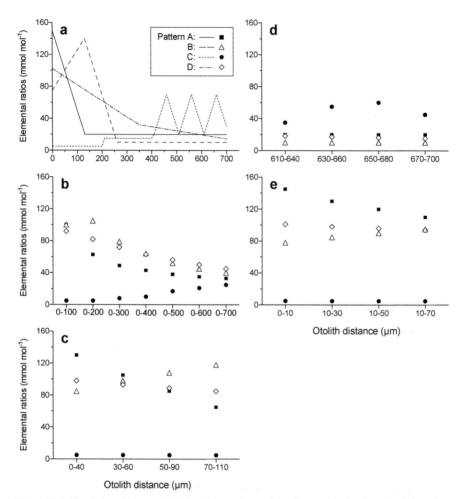

Figure 10.2. Simulation of how analytical results can change depending on the selection of otolith portion. (a) The otolith variation patterns modified from the elemental profile for South Pacific albacore reported by Macdonald et al. (2013), including pattern A (solid line, closed squares), characterized by enrichment in the otolith core; pattern B (dashed line, open triangles), where the concentration shows an increasing trend from the core, reaching a peak at the first inflection point and then declining; pattern C (dotted line, closed circles), where the concentration is constant in an early stage, then increasing with growth, and finally showing repeated, periodic fluctuations; and pattern D (long dashed dotted line, open diamonds), characterized by a gradual decrease in concentration. By changing the range for data re-sampling from the otolith virtual variation patterns, the following effects were simulated; (b) the effect of differential sampling dates on the whole otolith composition; (c) the effect of minor deviations of the analyzed spot from the core on analysis results; (d) the effect of minor deviations of analyzed spot at the otolith edge or of differential sampling dates on analysis results; and (e) the effect of the spot size used for analyzing the core on analysis results.

Macdonald et al. (2013), including pattern A, characterized by enrichment in the otolith core; pattern B, where the concentration shows an increasing trend from the core, reaching a peak at the first inflection point and then declining; pattern C, where the concentration is constant in an early stage, then increasing with growth, and finally showing repeated, periodic fluctuations; and pattern D, characterized by a gradual decrease in concentration (Fig. 10.2a). By changing the range for data re-sampling from the variation profile of otolith composition, the following effects were simulated: (1) the effect of differential sampling dates on the whole otolith composition (Fig. 10.2b); (2) the effect of minor deviations of the analyzed spot from the core on analysis results (Fig. 10.2c); (3) the effect of minor deviations in the analyzed spot at the otolith edge or of differential sampling dates on analysis results (Fig. 10.2d); and (4) the effect of the spot size used for analyzing the core on analysis results (Fig. 10.2e).

The results of this simulation demonstrate that different analysis results may be produced from otoliths with identical elemental variation patterns due to differential sampling dates, minor deviation within the analyzed portion and different diameters of the analyzed spot. It should be noted that this simulation adopts planar data resampling and that volume-based calculation will result in smaller effects. This simulation suggests that, for population discrimination and the estimation of migration routes based on otolith composition, it is essential to minimize or resolve the effect of biological factors by matching for days after hatching and for age estimated by daily and annual rings. If possible, an elemental variation profile along the growth axis should be obtained before determining the appropriate analysis portion and spot size.

The otolith is a three-dimensional structure with unequal growth rates along different axes. As shown in Shiao et al. (2009), calcium carbonate is deposited on the distal and proximal sides of the core area in the sub-adult to adult stages of southern bluefin tuna. This phenomenon is further evidenced in Fig. 2 of Wells et al. (2012). When the otolith powders are milled through the core area along a transverse plane, calcium carbonates from the real core deposited at the early life stage (<1 month) will be mixed with those from surrounding areas deposited at the sub-adult stage. Therefore, the analysis results of otolith powder from the core area of an adult tuna cannot truly reflect the real stable isotopic composition deposited during the early stage. This artifact has been previously demonstrated by Shiao et al. (2009).

Review of Studies Using Otolith Elemental Composition in Tunas

Although various assumptions and appropriate interpretation are required when using otolith elemental composition or stable isotopic composition as a natural tag, many reports demonstrate the usefulness of otolith elemental and stable isotopic compositions for the estimation of migration paths/routes and population structure of tunas. Table 10.4 summarizes the region, analytical instruments, and methods used and elements and stable isotopes analyzed in previous studies demonstrating the usefulness of otolith for ecological investigations of tunas.

Rooker et al. (2001; 2003) analyzed several elements (Li, Mg, Ca, Mn, Sr, and Ba) in otoliths of Pacific bluefin tuna and Atlantic bluefin tuna, and found that whole otolith composition is different between nursery grounds. The examined nursery grounds for Pacific bluefin tuna were the Sea of Japan and the east coast of the Japanese archipelago in the Northwest Pacific Ocean, while those for Atlantic bluefin tuna are the East Atlantic Ocean (the Atlantic coast of Europe and the Mediterranean Sea) and the West Atlantic Ocean (the east coast of continental North America). Because

Table 10.4. Summary of the location, analytical instruments and methods used, and elemental and stable isotopic compositions in previous studies of otolith for ecological investigations of tunas.

Species	Location	Treatment of Otolith	Analyzing device	Elements	Stable isotopes	Range of values	Reference
Thunnus thynnus	north Atlantic Ocean, Mediterranean Sea	Micro-drilling	Stable isotope mass spectrometer		$\delta^{13}C$, $\delta^{18}O$	-10~7 ‰, -1.5~-0.5 ‰	Rooker et al. (2014)
Thunnus alalunga	south Pacific Ocean	Transect ablation	LA-ICP-MS	Li/Ca, Mg/Ca, Mn/Ca, Cu/Ca, Sr/Ca, Ba/Ca, Pb/Ca		1~7 µmol/mol, 0.1~1.0 mmol/mol, 1~4 µmol/mol, 1~2 µmol/mol, 2~3 mmol/mol, 1~2 µmol/mol, 0.1~0.5 µmol/mol	Macdonald et al. (2013)
Thunnus albacares	western Pacific Ocean, central Pacific Ocean	Micro-drilling	Stable isotope mass spectrometer		$\delta^{13}C$, $\delta^{18}O$	-11.0~-8.5 ‰, -3.0~-1.5 ‰	Wells et al. (2012)
Thunnus alalunga	north Atlantic Ocean, Mediterranean Sea	Spot ablation	LA-ICP-MS	Mg, Mn, Sr, Ba		20 ppm, 0.5~5.0 ppm, 2000 ppm, 1.0~3.5 ppm	Davis et al. (2011)
Thunnus orientalis	north Pacific Ocean	Micro-drilling	Stable isotope mass spectrometer		$\delta^{13}C$, $\delta^{18}O$	-11~7 ‰, -2.7~-1.1 ‰	Shiao et al. (2010)
Thunnus thynnus	Gulf of St. Lawrence	Micro-drilling	Stable isotope mass spectrometer		$\delta^{13}C$, $\delta^{18}O$	-10~7 ‰, -2.5~-0.5 ‰	Schloesser et al. (2010)
Thunnus maccoyii	Indian Ocean	Transect ablation	LA-ICP-MS	Na/Ca, Mg/Ca, Mn/Ca, Sr/Ca, Ba/Ca		0.005~0.014 ppm/ppm, 0.0002~0.0015 ppm/ppm, 0.00002~0.00014 ppm/ppm, 0.002~0.006 ppm/ppm, 0.000001~0.000012 ppm/ppm	Wang et al. (2009)
Thunnus thynnus	north Atlantic Ocean, Mediterranean Sea	Whole and Micro-drilling	Stable isotope mass spectrometer		$\delta^{13}C$, $\delta^{18}O$	-10~7 ‰, -2.5~-0.5 ‰	Rooker et al. (2008a)
Thunnus thynnus	north Atlantic Ocean, Mediterranean Sea	Whole and Micro-drilling	Stable isotope mass spectrometer		$\delta^{13}C$, $\delta^{18}O$	-10~7 ‰, -2.5~-0.5 ‰	Rooker et al. (2008b)

Table 10.4. contd....

Table 10.4. contd.

Species	Location	Treatment of Otolith	Analyzing device	Elements	Stable isotopes	Range of values	Reference
Thunnus thynnus	north Atlantic Ocean Mediterranean Sea	Whole	Solution based ICP-MS	Li		0.1~0.5 ppm	Rooker et al. (2003)
				Mg		15~50 ppm	
				Ca		35~40 ppm	
				Mn		0.5~2.0 ppm	
				Sr		1000~1500 ppm	
				Ba		0.5~2.0 ppm	
Thunnus thynnus	western Atlantic Ocean Mediterranean Sea	Whole	ID ICP-MS and ICP-MS	Li		0.1~0.4 ppm	Secor et al. (2002)
				Na		2500~4500 ppm	
				Mg		15~50 ppm	
				K		500~1000 ppm	
				Ca		35~40 ppm	
				Mn		0.5~2.0 ppm	
				Sr		1000~1500 ppm	
				Ba		0.5~2.0 ppm	
Thunnus orientalis	north Pacific Ocean	Whole	Solution based ICP-MS	Li		0.5~3.5 ppm	Rooker et al. (2001)
				Mg		10~70 ppm	
				Ca		36~39 ppm	
				Mn		0.5~2.5 ppm	
				Sr		1150~1400 ppm	
				Ba		0.5~1.5 ppm	
Thunnus maccoyii	western Australia southern Australia south Africa	Transect scan	wavelength dispersive electron prove microanalysis and proton-induced X-ray emission	Na		2000~3500 ppm	Proctor et al. (1995)
				S		300~400 ppm	
				Cl		200~400 ppm	
				K		300~500 ppm	
				Ca		40~42 ppm	
				Sr		1200~2000 ppm	

the former nursery grounds were marginal sea areas while the latter were in the open ocean, respectively, environmental differences or differential early life stages in different marine areas might lead to differential otolith elemental compositions. Since time- and growth-related changes in otolith elemental composition have been identified as mentioned above, it is necessary to unify the scope of analysis of otoliths with matched age classes to determine whether the difference in whole otolith elemental composition reflects different nursery grounds, as discussed earlier.

Wang et al. (2009) and Macdonald et al. (2013) analyzed the elemental composition of tuna otoliths through the life history by a transect analysis from the otolith core to edge using laser ablation ICP-MS and found that Sr/Ca and Ba/Ca ratios change in line with annual growth rings. These changes are considered to reflect changes in environmental factors, such as salinity fronts and upwelling currents. However, as described above, the compositions of these elements change according to multiple factors, including change in habitat and individual physiological changes, making it difficult to precisely interpret any observed changes. Even if environmental factors are taken into account, considerations are also needed for the relationship between the rate of element incorporation into the otolith and the tuna's swimming speed in each marine area, as well as discrimination between vertical and horizontal migration. Both Atlantic and Pacific bluefin tunas have two spawning grounds, which poses a question if the southern bluefin tuna also have different spawning grounds in the Indian Ocean. Wang et al. (2009) investigated this question by comparing the elemental composition of the otolith core for the southern bluefin tuna adult between feeding and spawning grounds. All tuna analyzed by Wang et al. (2009) showed similar elemental composition at the larval stage, which does not support the existence of different spawning grounds in the Indian Ocean.

Stable isotopes (e.g., $\delta^{18}O$) are less susceptible to individual physiological changes and thus can be used for more accurate life-history estimation. Rooker et al. (2008a,b) demonstrated the effectiveness of otolith $\delta^{18}O$ in discriminating nursery grounds for Atlantic bluefin tuna between the Atlantic west coast (North American coast) and Atlantic east coast (the Mediterranean Sea or European coast). The difference in nursery grounds is most clearly reflected in $\delta^{18}O$, which shows about 1‰ difference between populations in the Atlantic west coast and those in the Atlantic east coast and is assumed by the authors to reflect the $\delta^{18}O$ of surface water in each marine area. Rooker et al. (2014) have also used this parameter and demonstrated that individual Atlantic bluefin tuna in the North Atlantic originate from various nursery grounds and most adults that spawn in the Mediterranean Sea are migratory individuals that originate from the nursery grounds of the Atlantic east coast. As mentioned above, the $\delta^{18}O$ of oceanic seawater is homogeneous at a depth of 500–2000 m, but is subject to change in the surface layer, which is the major swimming area for bluefin tuna, due to precipitation, concentration via evaporation, and dilution by river water and glacial melt. Therefore, in the natural environment, the $\delta^{18}O$ of environmental water, as well as its temperature, is likely to influence otolith $\delta^{18}O$. The surface water temperature also changes according to air temperature. A close interaction between these factors and otolith $\delta^{18}O$ of Atlantic bluefin tuna appears to enable discrimination between the two nursery grounds on both sides of the Atlantic Ocean.

The effectiveness of otolith $\delta^{18}O$ as a marker for estimating the spawning ground has been demonstrated for Pacific bluefin tuna as well, because the water temperature and salinity of surface waters differ between the areas off the eastern coast of Taiwan or the Okinawan Islands and the south Sea of Japan, two major spawning grounds in the West Pacific (Shiao et al. 2010). Otolith $\delta^{13}C$ and $\delta^{18}O$ have also been shown to be effective in estimating nursery grounds for Pacific individuals of yellowfin tuna (*Thunnus albacares*); young-of-the-year individuals found in offshore areas of the Hawaiian Islands have been found to originate from a nearby nursery ground in the coastal Hawaiian Islands (Wells et al. 2012).

Whole-life otolith $\delta^{18}O$ profiles also clearly reveal the migration of young-of-the-year southern bluefin tuna from the tropic spawning ground to the temperate nursing grounds in southern Australia (Shiao et al. 2009). Southern bluefin tuna experienced water temperatures decreased by 8–10°C inferred from the otolith $\delta^{18}O$ values increase by approximately 2‰ during the ages from one–12 months. However, the application of otolith $\delta^{18}O$ to reconstruct the water temperature experienced by adult southern bluefin tuna is not possible, due to the thermal conservation ability in tuna that maintains their brain and body temperature above the water temperature.

Although otolith $\delta^{13}C$ and $\delta^{18}O$ values are known to display inter-annual variations, these variations are not believed to affect the ability to discriminate between populations, because differences between nursery grounds are greater than observed inter-annual variations. Schoesser et al. (2010) used otolith $\delta^{13}C$ and $\delta^{18}O$ to identify nursery grounds for Atlantic bluefin tuna individuals captured in Gulf of St. Lawrence, Canada, and found that 99% of the individuals originated from a nursery ground in the Atlantic west coast (North American coast), regardless of capture year (1970s, 1980s, or 1990s) or capture location. Thus, otolith $\delta^{18}O$ can be used as a natural tag to identify nursery grounds more effectively than with other elemental compositions, as it substantially varies among tuna populations from different nursery grounds, has a well-established variation mechanism and is almost exclusively affected by environmental factors. Its inter-annual variations are smaller than the difference between nursery grounds and no need for age class matching is another important advantage for its application to ecological studies. Recent analytical techniques will enable the accurate determination of the $\delta^{18}O$ in sub-microgram quantities of sample. Kitagawa et al. (2013) have successfully elucidated the $\delta^{18}O$ in the small-size otoliths of larval bluefin tuna (pooled 100 otoliths, 1.3 µg in total) using the technique of Ishimura et al. (2004; 2008). Application of this technique to otolith life transects will provide a more detailed migratory ecology of tunas.

In contrast, otolith $\delta^{13}C$, which is another parameter commonly used with otolith $\delta^{18}O$, has not shown significant inter-site variations in any reports. This is probably because otolith $\delta^{13}C$ is subject to the combined effects from feed and DIC (dissolved inorganic carbon) and thus shows no clear difference among populations. Although otolith $\delta^{13}C$ may not be an effective marker for the discrimination of nursery grounds, it still appears to provide some information for stock discrimination (Schoesser et al. 2010; Rooker et al. 2014) as well as the developmental stage and metabolism of fish.

Future Prospects Regarding Studies of Tuna Otoliths

New natural migratory proxy

Pacific bluefin tuna are known to either remain in the Pacific Ocean near Japan or to migrate as far as the east Pacific Ocean off the coast of Mexico before first spawning. These two major migration patterns could not be estimated by conventional otolith natural tags. The development of new tags that can reveal the difference between these migration patterns will provide new insights into the migration ecology or stock dynamics of tunas. Two potential migratory proxies: radiocarbon and neodymium isotopes are introduced here.

Radiocarbon composition

Calcium carbonate, the major component of the otolith, contains radiocarbon with mass number 14. Radiocarbon has been used for age determination of long-lived fish, as it shows the peak from a previous nuclear test (i.e., bomb peak) (Campana and Jones 1998; Campana 2001; Kerr et al. 2004; Andrews et al. 2005). The distribution of radiocarbon in oceanic DIC shows a decreasing trend in the vertical direction and unique patterns of radiocarbon composition ($\Delta^{14}C$) in each marine area (Fig. 10.3). The $\Delta^{14}C$ of surface water DIC is lower in waters off the coast of Mexico than in waters near Japan. The $\Delta^{14}C$ of particulate organic carbon (POC) is strongly influenced by the $\Delta^{14}C$ of surface water DIC (McNicol and Aluwihare 2007). Food-derived radiocarbon, which is expected to be altered more or less through the food chain and undergo more complex changes than those of DIC or POC, is likely to be influenced by the composition of DIC in the marine area, the original carbon source. Thus, $\Delta^{14}C$ of any carbon sources likely show characteristic composition in each marine area. Therefore, it may be possible to estimate the radiocarbon characteristics of the water mass from the otolith $\Delta^{14}C$ and thereby narrow down the marine area inhabited by tuna.

As the otolith carbon isotopic ratio is strongly influenced by the contribution ratios of DIC and metabolic carbon, it should be determined which, of DIC or feed, is more strongly reflected in the value. It may also be helpful to estimate the $\Delta^{14}C$ of metabolic carbon from the $\Delta^{14}C$ of soft tissues, such as muscle for the application of the otolith $\Delta^{14}C$ to the ecological research.

The recent technological innovation in radiocarbon analysis has enabled the detection of radiocarbon in the order of 1–50 µg carbon (equivalent to several tens to several hundreds of µg of otolith) (Santos et al. 2007; Ruff et al. 2010; Salehpour et al. 2013). This sample amount is comparable to that required for the analysis of carbon/oxygen stable isotopic compositions. Thus, the composition of the combined $\delta^{13}C$, $\Delta^{14}C$ and $\delta^{18}O$ isotopes may be a new proxy for migration studies.

Neodymium isotopic ratio

Neodymium (Nd) isotopic ratio is likely to be a promising proxy for identifying the migratory route of fully marine fish such as tunas. This is similar to the use of the Sr isotope in diadromous and freshwater fish, which has also been shown to be highly

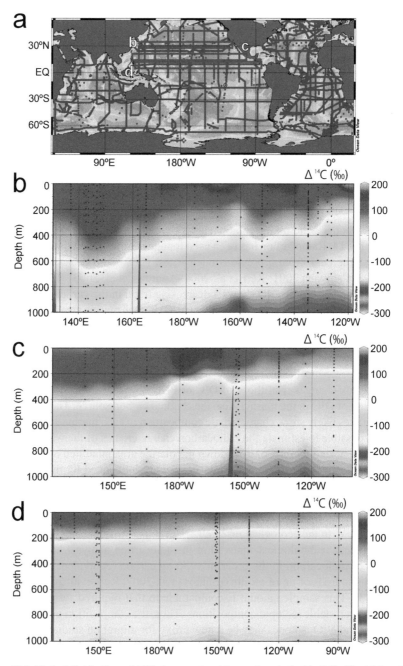

Figure 10.3. Vertical distributions of Δ^{14}C along west-east transections in the North Pacific. (a) Location of the sections shown in Figs. (b)-(d) are indicated by red rectangles. (b-d) Vertical distributions of Δ^{14}C. Color contour was produced from the data of black points. Color scale on right side of the panel. Figures were made from Δ^{14}C data compiled by Key et al. (2004) with software Ocean Data View.

effective in estimating the migration ecology of fish species. This is not only because the radiogenic isotopes ($^{87}Sr/^{86}Sr$) shows substantial regional difference, allowing for limiting the geographic area, but because mass-dependent isotopic fractionation associated with biological activities or analytical procedures can be corrected using the ratio of stable isotopes ($^{86}Sr/^{88}Sr$), so that the effect of the ratio of only the radiogenic signature is extracted. Nd is one of the oceanic elements with such characteristics (i.e., differential isotopic compositions in different marine areas and the effect of radiogenic signature can be extracted using the composition of stable isotopes).

The composition of Nd, in terms of concentration and isotopic ratio, substantially varies depending on the marine area, water mass, and water depth (Amakawa et al. 2004; Lacan et al. 2012). For example, its ε value ($=[((^{143}Nd/^{144}Nd)_{measured}/(^{143}Nd/^{144}Nd)_{\text{Chondritic Uniform Reservoir}})-1] \times 10^4$) in the Sea of Japan ranges from -8.9 to -7.2 while that in the Pacific Ocean ranges from -7 to -2. Nd isotopic ratio is usually determined using analytical techniques such as Thermal Ionization Mass Spectrometry (TIMS) and multi-collector inductively coupled plasma mass spectrometry (MC-ICP-MS). The mass-dependent fractionation can be corrected during analysis using non-radiogenic isotopic ratio (e.g., $^{146}Nd/^{144}Nd$), so that only the effect of radioactive isotopes, namely the $^{143}Nd/^{144}Nd$ ratio, can be determined.

With the recent advances in analytical technique that have enabled the determination of Nd isotopic ratio in a very small amount of sample, the reconstruction of change in water mass based on the Nd isotopic ratio of foraminiferal shell has been attempted in the paleoenvironmental field. The details of analysis of Nd isotopes in a foraminiferal shell are provided in a review by Tachikawa et al. (2014). With the state-of-the-art technology, ~1 ng of Nd is required for determining the isotopic ratio, with smaller amounts associated with larger error (H. Amakawa, pers. comm.). Given an otolith Nd concentration of 0.2–1 ng g^{-1}, as reported by Arslan et al. (2002), 1–5 g of otolith is required to retrieve 1 ng of Nd, which may not be practical at present. However, if the wide variation in Nd isotopic ratio among marine areas can be used to compensate for low measurement precision due to a small amount of sample, the difference in Nd isotopic ratio among marine areas may be detectable from otoliths. Future advances in analytical technique will further reduce the amount of samples needed for analysis, which may make this parameter an effective tool for the estimation of migratory routes of tunas. Because the values of the Nd isotopic ratio (and also $\Delta^{14}C$) show substantial change along a vertical gradient in the open ocean, it is necessary to estimate water depth from an estimated water temperature calculated from the $\delta^{18}O$ as well as the degree of body temperature adjustment (Kitagawa et al. 2006).

Perspectives

In conclusion, the use of otolith elemental and stable isotopic compositions as natural tags is highly promising for elucidating the migration ecology of tunas. For the otolith elemental and stable isotopic compositions of tunas to be used in population discrimination or migration route estimation, it is important to understand the process by which various elements are incorporated into the otolith. The major contribution to variations in the otolith elemental composition of tunas is thought to be via the combined effects of environmental and biological factors. Thus, rearing experiments under an artificially controlled environment could be the necessary complement to precisely identify the factors responsible for these variations. It may, however, be difficult to link an artificial environment with the wild environment, and it is therefore often difficult to assess the effects of each factor separately. When estimating population structures or migration routes from the otolith elemental and isotopic composition of marine fish, like tunas, it is at least necessary to minimize the effects of biological and physiological factors, such as life history stage, age, somatic growth rate, and maturity, to enable identification of as many relevant environmental factors as possible, such as temperature, salinity, and the elemental composition of environmental water (e.g., Sturrock et al. 2012). In addition, the use of otolith elemental and isotopic compositions in combination with conventional techniques, such as the mark-recapture method, bio-logging, and genetic population structure analysis is expected to provide a broader view of migration ecology of tunas.

The application of the latest geochemical technologies will most likely provide breakthroughs in otolith-based ecological migration studies of tunas. New natural tags such as the Nd isotopic ratio and Δ^{14}C could be useful tools for detecting the difference between migration patterns of individuals. The use of new proxies in combination with proven markers, such as the elemental composition and δ^{18}O, will further advance ecological studies of tunas. Analysis of stable isotopic ratios of trace carbonates has recently been performed for the detailed, high-resolution analysis of ecological trajectories. Ishimura et al. (2004; 2008) developed an analytical system to determine the δ^{13}C and δ^{18}O of submicrograms of calcium carbonate (less than 1/100th of that required by conventional analytical methods) and applied their analytical technique to the very small otoliths of larval bluefin tuna (Kitagawa et al., 2013). Recently, the determination of otolith δ^{18}O in a very small sample (on the order of several μm spot-size) using secondary ion mass spectrometry (SIMS) has also been reported (Shiao et al. 2014). In addition to these, other new proxies being established in the geochemical field include nitrogen isotopic ratio present in trace amount in calcium carbonate (Uchida et al. 2008; Yamazaki et al. 2011a; 2011b; 2013) and clumped isotope as a measure of water temperature virtually independent of seawater δ^{18}O (Eiler 2007; 2011). We strongly believe that the continued feedback of geochemical techniques using the latest high-precision, high-resolution analytical instruments will bring otolith-based migration ecology studies of tunas to a new stage of understanding.

References

Amakawa, H., D.S. Alibo and Y. Nozaki. 2004. Nd concentration and isotopic composition distributions in surface waters of Northwest Pacific Ocean and its adjacent seas. Geochem. J. 38: 493–504.

Andrews, A.H., E.J. Burton, L.A. Kerr, G.M. Cailliet, K.H. Coale, C.C. Lundstrom and T.A. Brown. 2005. Bomb radiocarbon and lead–radium disequilibria in otoliths of bocaccio rockfish (*Sebastes paucispinis*): a determination of age and longevity for a difficult-to-age fish. Mar. Freshwater Res. 56: 517–528.

Arslan, Z. and A. Paulson. 2002. Analysis of biogenic carbonates by inductively coupled plasma–mass spectrometry (ICP-MS). Flow injection on-line solid-phase preconcentration for trace element determination in fish otoliths. Anal. Bioanal. Chem. 372: 776–785.

Barker, S., M. Greaves and H. Elderfield. 2003. A study of cleaning procedures used for foraminiferal Mg/Ca paleothermometry. Geochem. Geophy. Geosy. 4.

Bath, G.E., S.R. Thorrold, C.M. Jones, S.E. Campana, J.W. McLaren and J.W. Lam. 2000. Strontium and barium uptake in aragonitic otoliths of marine fish. Geochim. Cosmochim. Acta. 64: 1705–1714.

Brand, W., T.B. Coplen, J. Vogl, M. Rosner and T. Prohaska. 2014. Assessment of International Reference Materials for Stable Isotope Ratio Analysis 2013. Pure. Appl. Chem. 86: 425–467.

Brophy, D., T.E. Jeffries and B.S. Danilowicz. 2004. Elevated manganese concentrations at the cores of clupeid otoliths: possible environmental, physiological, or structural origins. Mar. Biol. 144: 779–786.

Brown, R.J. and K.P. Severin. 2009. Otolith chemistry analyses indicate that water Sr:Ca is the primary factor influencing otolith Sr: Ca for freshwater and diadromous fish; but not for marine fish. Can. J. Fish. Aquat. Sci. 66: 1790–1808.

Bruland, K.W. and M.C. Lohan. 2003. Controls of trace metals in seawater. Ocean. Mar. Geochem. 5: 23–47.

Buckel, J.A., B.L. Sharack and V.S. Zdanowicz. 2004. Effect of diet on otolith composition in *Pomatomus saltatrix*, an estuarine piscivore. J. Fish. Biol. 64: 1469–1484.

Carey, F.G., J.W. Kanwisher and E.D. Stevens. 1984. Bluefin tuna warm their viscera during digestion. J. Exp. Biol. 109: 1–20.

Campana, S.E. and C.M. Jones. 1998. Radiocarbon from nuclear testing applied to age validation of black drum, *Pogonias cromis*. Fish. Bull. 96: 185–192.

Campana, S.E. 1999. Chemistry and composition of fish otoliths: pathways, mechanisms and applications. Mar. Ecol. Prog. Ser. 188: 263–297.

Campana, S.E. 2001. Accuracy, precision and quality control in age determination, including a review of the use and abuse of age validation methods. J. fish. biol. 59: 197–242.

Chen, K.S., P. Crone and C.C. Hsu. 2006. Reproductive biology of female Pacific bluefin tuna *Thunnus orientalis* from south-western North Pacific Ocean. Fish. Sci. 72: 985–994.

Coplen, T.B. 2011. Guidelines and recommended terms for expression of stable-isotope-ratio and gas-ratio measurement results. Rapid. Commun. Mass. Spectrom. 25: 2538–2560.

Curti, E. 1999. Coprecipitation of radionuclides with calcite: estimation of partition coefficients based on a review of laboratory investigations and geochemical data. Appl. Geochem. 14: 433–445.

Davis, C.A., D. Brophy, T. Jeffries and E. Gosling. 2011. Trace elements in the otoliths and dorsal spines of albacore tuna (*Thunnus alalunga*, Bonnaterre, 1788): An assessment of the effectiveness of cleaning procedures at removing postmortem contamination. J. Exp. Mar. Biol. Ecol. 396: 162–170.

de Pontual, H., F. Lagardère, R. Amara, M. Bohn and A. Ogor. 2003. Influence of ontogenetic and environmental changes in the otolith microchemistry of juvenile sole (*Solea solea*). J. Sea. Res. 50: 199–211.

de Villiers, S., G.T. Shen and B.K. Nelson. 1994. The SrCa-temperature relationship in coralline aragonite: Influence of variability in (Sr/Ca) seawater and skeletal growth parameters. Geochim. Cosmochim. Acta. 58: 197–208.

Dietzel, M., N. Gussone and A. Eisenhauer. 2004. Co-precipitation of Sr^{2+} and Ba^{2+} with aragonite by membrane diffusion of CO_2 between 10 and 50°C. Chem. Geol. 203: 139–151.

Eiler, J.M. 2007. "Clumped-isotope" geochemistry—The study of naturally-occurring, multiply-substituted isotopologues. Earth. Planet. Sci. Lett. 262: 309–327.

Eiler, J.M. 2011. Paleoclimate reconstruction using carbonate clumped isotope thermometry. Quat. Sci. Rev. 30: 3575–3588.

Elsdon, T.S. and B.M. Gillanders. 2003. Reconstructing migratory patterns of fish based on environmental influences on otolith chemistry. Rev. Fish. Biol. Fish. 13: 217–235.

Elsdon, T.S., B.K. Wells, S.E. Campana, B.M. Gillanders, C.M. Jones, K.E. Limburg, D.H. Secor, S.R. Thorrold and B.D. Walther. 2008. Otolith chemistry to describe movements and life-history parameters of fishes: hypotheses, assumptions, limitations and inferences. Oceanogr. Mar. Biol.: An Annual Review 46: 297–330.

Fowler, A.J., S.E. Campana, S.R. Thorrold and C.M. Jones. 1995. Experimental assessment of the effect of temperature and salinity on elemental composition of otoliths using laser ablation ICPMS. Can. J. Fish. Aquat. Sci. 52: 1431–1441.

Gaetani, G.A. and A.L. Cohen. 2006. Element partitioning during precipitation of aragonite from seawater: a framework for understanding paleoproxies. Geochim. Cosmochim. Acta. 70: 4617–4634.

Gaillardet, J., J. Viers and B. Dupré. 2003. Trace elements in river waters. Ocean. Mar. Geochem. 5: 225–272.

Gillanders, B.M. 2005. Otolith chemistry to determine movements of diadromous and freshwater fish. Aquat. Living. Resour. 18: 291–300.

Grossman, E.L. and T.L. Ku. 1986. Oxygen and carbon isotope fractionation in biogenic aragonite: temperature effects. Chem. Geol. 59: 59–74.

Hoppema, M., F. Dehairs, J. Navez, C. Monnin, C. Jeandel, E. Fahrbach and H.J.W. de Baar. 2010. Distribution of barium in the Weddell Gyre: Impact of circulation and biogeochemical processes. Mar. Chem. 122: 118–129.

Hwang, P.P. and T.H. Lee. 2007. New insights into fish ion regulation and mitochondrion-rich cells. Comp. Biochem. Physiol. A-Mol. Integr. Physiol. 148: 479–497.

Høie, H., E. Otterlei and A. Folkvord. 2004. Temperature-dependent fractionation of stable oxygen isotopes in otoliths of juvenile cod (*Gadus morhua* L.). ICES J. Mar. Sci. 61: 243–251.

Ishimura, T., U. Tsunogai and T. Gamo. 2004. Stable carbon and oxygen isotopic determination of sub-microgram quantities of $CaCO_3$ to analyze individual foraminiferal shells. Rapid. Commun. Mass. Spectrom. 18: 2883–2888.

Ishimura, T., U. Tsunogai and F. Nakagawa. 2008. Grain-scale heterogeneities in the stable carbon and oxygen isotopic compositions of the international standard calcite materials (NBS 19, NBS 18, IAEA-CO-1, and IAEA-CO-8). Rapid. Commun. Mass. Spectrom. 22: 1925–1932.

Jacquet, S.H.M., F. Dehairs, D. Cardinal, J. Navez and B. Delille. 2005. Barium distribution across the Southern Ocean frontal system in the Crozet–Kerguelen Basin. Mar. Chem. 95: 149–162.1

Jacquet, S.H.M., F. Dehairs, M. Elskens, N. Savoye and D. Cardinal. 2007. Barium cycling along WOCE SR3 line in the Southern Ocean. Mar. Chem. 106: 33–45.

Jolivet, A., J.F. Bardeau, R. Fablet, Y.M. Paulet and H. de Pontual. 2008. Understanding otolith biomineralization processes: new insights into microscale spatial distribution of organic and mineral fractions from Raman microspectrometry. Anal. Bioanal. Chem. 392: 551–560.

Kalish, J.M. 1991. ^{13}C and ^{18}O isotopic disequilibria in fish otoliths: Metabolic and kinetic effects. Mar. Ecol. Prog. Ser. 75: 191–203.

Kerr, L.A., A.H. Andrews, B.R. Frantz, K.H. Coale, T.A. Brown and G.M. Cailliet. 2004. Radiocarbon in otoliths of yelloweye rockfish (*Sebastes ruberrimus*): a reference time series for the coastal waters of southeast Alaska. Can. J. Fish. Aquat. Sci. 61: 443–451.

Kerr, L.A. and S.E. Campana. 2013. Chemical composition of fish hard parts as a natural marker of fish stocks. pp. 205–234. *In*: S.X. Cadrin, L.A. Kerr and S. Mariani (eds.). Stock Identification Methods: Applications in Fishery Science (Second Edition). Academic Press, London, Waltham, San Diego.

Key, R.M., A. Kozyr, C.L. Sabine, K. Lee, R. Wanninkhof, J.L. Bullister, R.A. Feely, F.J. Millero, C. Mordy and T.H. Peng. 2004. A global ocean carbon climatology: results from Global Data Analysis Project (GLODAP). Glob. Biogeochem. Cycle 18.

Kim, S.T. and J.R. O'Neil. 1997. Equilibrium and nonequilibrium oxygen isotope effects in synthetic carbonates. Geochim. Cosmochim. Acta. 61: 3461–3475.

Kim, S.T., J.R. O'Neil, C. Hillaire-Marcel and A. Mucci. 2007a. Oxygen isotope fractionation between synthetic aragonite and water: Influence of temperature and Mg^{2+} concentration. Geochim. Cosmochim. Acta. 71: 4704–4715.

Kim, S.T., A. Mucci and B.E. Taylor. 2007b. Phosphoric acid fractionation factors for calcite and aragonite between 25 and 75°C: revisited. Chem. Geol. 246: 135–146.

Kimura, S., Y. Kato, T. Kitagawa and N. Yamaoka. 2010. Impacts of environmental variability and global warming scenario on Pacific bluefin tuna (*Thunnus orientalis*): spawning grounds and recruitment habitat. Prog. Oceanogr. 86: 39–44.

Kinsman, D.J.J. and H.D. Holland. 1969. The co-precipitation of cations with $CaCO_3$—IV. The co-precipitation of Sr^{2+} with aragonite between 16° and 96°C. Geochim. Cosmochim. Acta. 33: 1–17.

Kitagawa, T., H. Nakata, S. Kimura, T. Itoh, S. Tsuji and A. Nitta. 2000. Effect of ambient temperature on the vertical distribution and movement of Pacific bluefin tuna *Thunnus thynnus orientalis*. Mar. Ecol. Prog. Ser. 206: 251–260.

Kitagawa, T., H. Nakata, S. Kimura and S. Tsuji. 2001. Thermoconservation mechanisms inferred from peritoneal cavity temperature in free-swimming Pacific bluefin tuna *Thunnus thynnus orientalis*. Mar. Ecol. Prog. Ser. 220: 253–263.

Kitagawa, T., S. Kimura, H. Nakata and H. Yamada. 2006. Thermal adaptation of Pacific bluefin tuna *Thunnus orientalis* to temperate waters. Fish. Sci. 72: 149–156.

Kitagawa, T., S. Kimura, H. Nakata and H. Yamada. 2007. Why do young Pacific bluefin tuna repeatedly dive to depths through the thermocline? Fish. Sci. 73: 98–106.

Kitagawa, T., T. Ishimura, R. Uozato, K. Shirai, Y. Amano, A. Shinoda, T. Otake, U. Tsunogai and S. Kimura. 2013. Otolith $\delta^{18}O$ of Pacific bluefin tuna *Thunnus orientalis* as an indicator of ambient water temperature. Mar. Ecol. Prog. Ser. 481: 199–209.

Kroopnick, P.M. 1985. The distribution of ^{13}C of ΣCO_2 in the world oceans. Deep-Sea. Res. Pt. I. 32: 57–84.

Lacan, F., K. Tachikawa and C. Jeandel. 2012. Neodymium isotopic composition of the oceans: A compilation of seawater data. Chem. Geol. 300: 177–184.

LeGrande, A.N. and G.A. Schmidt. 2006. Global gridded data set of the oxygen isotopic composition in seawater. Geophys. Res. Lett. 33.

Macdonald, J.I., J.H. Farley, N.P. Clear, A.J. Williams, T.I. Carter, C.R. Davies and S.J. Nicol. 2013. Insights into mixing and movement of South Pacific albacore *Thunnus alalunga* derived from trace elements in otoliths. Fish. Res. 148: 56–63.

Marohn, L., E. Prigge, K. Zumholz, A. Klügel, H. Anders and R. Hanel. 2009. Dietary effects on multi-element composition of European eel (*Anguilla anguilla*) otoliths. Mar. Biol. 156: 927–933.

McNichol, A.P. and L.I. Aluwihare. 2007. The power of radiocarbon in biogeochemical studies of the marine carbon cycle: Insights from studies of dissolved and particulate organic carbon (DOC and POC). Chem. Rev. 107: 443–466.

Melancon, S., B.J. Fryer, B.S.A. Ludsin, J.E. Gagnon and Z. Yang. 2005. Effects of crystal structure on the uptake of metals by lake trout (*Salvelinus namaycush*) otoliths. Can. J. Fish. Aquat. Sci. 62: 2609–2619.

Meybeck, M. 2003. Global occurrence of major elements in rivers. Ocean. Mar. Geochem. 5: 207–223.

Millero, F.J. 2003. Physicochemical controls on seawater. Ocean. Mar. Geochem. 6: 1–21.

Milton, D.A. and S.R. Chenery. 2001. Sources and uptake of trace metals in otoliths of juvenile barramundi (*Lates calcarifer*). J. Exp. Mar. Biol. Ecol. 264: 47–65.

Mitsuguchi, T., T. Uchida, E. Matsumoto, P.J. Isdale and T. Kawana. 2001. Variations in Mg/Ca, Na/Ca, and Sr/Ca ratios of coral skeletons with chemical treatments: Implications for carbonate geochemistry. Geochim. Cosmochim. Acta. 65: 2865–2874.

Nonogaki, H., J.A. Nelson and W.P. Patterson. 2007. Dietary histories of herbivorous loricariid catfishes: evidence from $\delta^{13}C$ values of otoliths. Environ. Biol. Fish. 78: 13–21.

Patterson, W., G. Smith and K. Lohmann. 1993. Continental paleothermometry and seasonality using the isotopic composition of aragonite otoliths of freshwater fishes. Clim. Change. Cont. Isot. Rec. 27: 199–202.

Pena, L.D., E. Calvo, I. Cacho, S. Eggins and C. Pelejero. 2005. Identification and removal of Mn-Mg-rich contaminant phases on foraminiferal tests: Implications for Mg/Ca past temperature reconstructions. Geochem. Geophys. Geosyst. 6.

Perrier, C., F. Daverat, G. Evanno, C. Pécheyran, J.L. Bagliniere and J.M. Roussel. 2011. Coupling genetic and otolith trace element analyses to identify river-born fish with hatchery pedigrees in stocked Atlantic salmon (*Salmosalar*) populations. Can. J. Fish. Aquat. Sci. 68: 977–987.

Radtke, R.L., P. Lenz, W. Showers and E. Moksness. 1996. Environmental information stored in otoliths: insights from stable isotopes. Mar. Biol. 127: 161–170.

Radtke, R.L., W. Showers. E. Moksness and P. Lenz. 1998. Environmental information stored in otoliths: insights from stable isotopes (vol. 127, pg 161, 1996). Mar. Biol. 132: 347–348.

Rimstidt, J.D., A. Balog and J. Webb. 1998. Distribution of trace elements between carbonate minerals and aqueous solutions. Geochim. Cosmochim. Acta. 62: 1851–1863.

Rohling, E.J. and S. Cooke. 1999. Stable oxygen and carbon isotope ratios in foraminiferal carbonate. pp. 239–258. *In*: B.K. Sen Gupta (ed.). Modern Foraminifera, Kluwer Academic Publishers. Dordrecht, The Netherlands.

Rooker, J.R., D.H. Secor, V.S. Zdanowicz and T. Itoh. 2001. Discrimination of northern bluefin tuna from nursery areas in the Pacific Ocean using otolith chemistry. Mar. Ecol. Prog. Ser. 218: 275–282.

Rooker, J.R., D.H. Secor, V.S. Zdanowicz, G. de Metrio and L.O. Relini. 2003. Identification of Atlantic bluefin tuna (*Thunnus thynnus*) stocks from putative nurseries using otolith chemistry. Fish. Oceanogr. 12: 75–84.

Rooker, J.R., J.R. Alvarado Bremer, B.A. Block, H. Dewar, G. De Metrio, A. Corriero, R.T. Kraus, E.D. Prince, E. Rodriguez-Marin and D.H. Secor. 2007. Life history and stock structure of Atlantic bluefin tuna (*Thunnus thynnus*). Rev. Fisher. Sci. 15: 265–310.

Rooker, J.R., D.H. Secor, G. de Metrio, A.J. Kaufman, A.B. Ríos and V. Ticina. 2008a. Evidence of trans-Atlantic movement and natal homing of bluefin tuna from stable isotopes in otoliths. Mar. Ecol. Prog. Ser. 368: 231–239.

Rooker, J.R., D.H. Secor, G. de Metrio, R. Schloesser, B.A. Block and J.D. Neilson. 2008b. Natal homing and connectivity in Atlantic bluefin tuna populations. Science. 322: 742–744.

Rooker, J.R., H. Arrizabalaga, I. Fraile, D.H. Secor, D.L. Dettman, N. Abid, P. Addis, S. Deguara, F.S. Karakulak, A. Kimoto, O. Sakai, D. Macias and M.N. Santos. 2014. Crossing the line: migratory and homing behaviors of Atlantic bluefin tuna. Mar. Ecol. Prog. Ser. 504: 265–276.

Ruff, M., S. Szidat, H.W. Gäggeler, M. Suter, H.A. Synal and L. Wacker. 2010. Gaseous radiocarbon measurements of small samples. Nucl. Instrum. Methods. Phys. Res. Sect. B-Beam. Interact. Mater. Atoms. 268: 790–794.

Ruttenberg, B.I., S.L. Hamilton, M.J. Hickford, G.L. Paradis, M.S. Sheehy, J.D. Standish, O. Ben-Tzvi and R.R. Warner. 2005. Elevated levels of trace elements in cores of otoliths and their potential for use as natural tags. Mar. Ecol. Prog. Ser. 297: 273–281.

Schloesser, R.W., J.D. Neilson, D.H. Secor and J.R. Rooker. 2010. Natal origin of Atlantic bluefin tuna (*Thunnus thynnus*) from Canadian waters based on otolith $\delta^{13}C$ and $\delta^{18}O$. Can. J. Fish. Aquat. Sci. 67: 563–569.

Salehpour, M., K. Håkansson and G. Possnert. 2013. Accelerator mass spectrometry of ultra-small samples with applications in the biosciences. Nucl. Instrum. Methods. Phys. Res. Sect. B-Beam. Interact. Mater. Atoms. 294: 97–103.

Santos, G.M., J.R. Southon, S. Griffin, S.R. Beaupre and E.R. Druffel. 2007. Ultra small-mass AMS ^{14}C sample preparation and analyses at KCCAMS/UCI Facility. Nucl. Instrum. Methods. Phys. Res. Sect. B-Beam. Interact. Mater. Atoms. 259: 293–302.

Schwarcz, H.P., Y. Gao, S. Campana, D. Browne, M. Knyf and U. Brand. 1998. Stable carbon isotope variations in otoliths of Atlantic cod (*Gadus morhua*). Can. J. Fish. Aquat. Sci. 55: 1798–1806.

Schöne, B.R., Z. Zhang, D. Jacob, D.P. Gillikin, T. Tütken, D. Garbe-Schönberg and A. Soldati. 2010. Effect of organic matrices on the determination of the trace element chemistry (Mg, Sr, Mg/Ca, Sr/Ca) of aragonitic bivalve shells (*Arctica islandica*)—Comparison of ICP-OES and LA-ICP-MS data. Geochem. J. 44: 23–37.

Shiao, J.C., T.F. Yui, H. Høie, U. Ninnemann and S.K. Chang. 2009. Otolith O and C stable isotope composition of southern bluefin tuna *Thunnus maccoyii* (Pisces: Scombridae) as possible environmental and physiological indicators. Zool. Stud. 48: 71–82.

Shiao, J.C., S.W. Wang, K. Yokawa, M. Ichinokawa, Y. Takeuchi, Y.G. Chen and C.C. Shen. 2010. Natal origin of Pacific bluefin tuna *Thunnus orientalis* inferred from otolith oxygen isotope composition. Mar. Ecol. Prog. Ser. 420: 207–219.

Shiao, J.C., S. Itoh, H. Yurimoto, Y. Iizuka and Y.C. Liao. 2014. Oxygen isotopic distribution along the otolith growth axis by secondary ion mass spectrometry: Applications for studying ontogenetic change in the depth inhabited by deep-sea fishes. Deep-Sea. Res. Pt. I. 84: 50–58.

Shirai, K., B.R. Schöne, T. Miyaji, P. Radarmacher, R.A. Krause, Jr. and K. Tanabe. 2014. Assessment of the mechanism of elemental incorporation into bivalve shells (*Arctica islandica*) based on elemental distribution at the microstructural scale. Geochim. Cosmochim. Acta. 126: 307–320.

Solomon, C.T., P.K. Weber, J.J. Cech, Jr., B.L. Ingram, M.E. Conrad, M.V. Machavaram and R.L. Frankli. 2006. Experimental determination of the sources of otolith carbon and associated isotopic fractionation. Can. J. Fish. Aquat. Sci. 63: 79–89.

Stephenson, A.E., J.J. DeYoreo, L. Wu, K.J. Wu, J. Hoyer and P.M. Dove. 2008. Peptides enhance magnesium signature in calcite: insights into origins of vital effects. Science 322: 724–727.

Sturrock, A.M., C.N. Trueman, A.M. Darnaude and E. Hunter. 2012. Can otolith elemental chemistry retrospectively track migrations in fully marine fishes? J. Fish. Biol. 81: 766–795.

Sturrock, A.M., E. Hunter, J.M. Miltona and C.N. Trueman. 2013. Analysis methods and reference concentrations of 12 minor and trace elements in fish blood plasma. J. Trace. Elem. Med. Biol. 27: 273–285.

Tachikawa, K., A.M. Piotrowski and G. Bayon. 2014. Neodymium associated with foraminiferal carbonate as a recorder of seawater isotopic signatures. Quat. Sci. Rev. 88: 1–13.

Tagliabue, A. and L. Bopp. 2008. Towards understanding global variability in ocean carbon-13. Glob. Biogeochem. Cycle. 22.

Tanaka, Y., M. Mohri and H. Yamada. 2007. Distribution, growth and hatch date of juvenile Pacific bluefin tuna *Thunnus orientalis* in the coastal area of the Sea of Japan. Fish. Sci. 73: 534–542.

Tarutani, T., R.N. Clayton and T.K. Mayeda. 1969. The effect of polymorphism and magnesium substitution on oxygen isotope fractionation between calcium carbonate and water. Geochim. Cosmochim. Acta. 33: 987–996.

Tesoriero, A.J. and J.F. Pankow. 1996. Solid solution partitioning of Sr^2, Ba^2, and Cd^2 to calcite. Geochim. Cosmochim. Acta. 60: 1053–1063.

Thorrold, S.R., S.E. Campana, C.M. Jones and P.K. Swart. 1997. Factors determining delta C-13 and delta O-18 fractionation in aragoniticotoliths of marine fish. Geochim. Cosmochim. Acta. 61: 2909–2919.

Tohse, H. and Y. Mugiya. 2008. Sources of otolith carbonate: experimental determination of carbon incorporation rates from water and metabolic CO_2, and their diel variations. Aquat. Biol. 1: 259–268.

Tomas, J. and A.J. Geffen. 2003. Morphometry and composition of aragonite and vaterite otoliths of deformed laboratory reared juvenile herring from two populations. J. Fish. Biol. 63: 1383–1401.

Townsend, D.W., R.L. Radtke, S. Corwin and D.A. Libby. 1992. Strontium:calcium ratios in juvenile Atlantic herring *Clupea harengus* L. otoliths as a function of water temperature. J. Exp. Mar. Biol. Ecol. 160: 131–140.

Tsukamoto, K. and I. Naka. 1998. Do all freshwater eels migrate? Nature 396: 635–636.

Tzeng, W.N., C.W. Chang, C.H. Wang, J.C. Shiao, Y. Iizuka, Y.J. Yang, C.F. You and L. Lozys. 2007. Misidentification of the migratory history of anguillid eels by Sr/Ca ratios of vaterite otoliths. Mar. Ecol. Prog. Ser. 348: 285–295.

Uchida, A., M. Nishizawa, K. Shirai, H. Iijima, H. Kayanne, N. Takahata and Y. Sano. 2008. High sensitivity measurements of nitrogen isotopic ratios in coral skeletons from Palau, western Pacific: Temporal resolution and seasonal variation of nitrogen sources. Geochem. J. 42: 255–262.

Walther, B.D. and S.R. Thorrold. 2006. Water, not food, contributes the majority of strontium and barium deposited in the otoliths of a marine fish. Mar. Ecol. Prog. Ser. 311: 125–130.

Walther, B.D. and K.E. Limburg. 2012. The use of otolith chemistry to characterize diadromous migrations. J. Fish. Biol. 81: 796–825.

Wang, C.H., Y.T. Lin, J.C. Shiao, C.F. You and W.N. Tzeng. 2009. Spatio-temporal variation in the elemental compositions of otoliths of southern bluefin tuna *Thunnus maccoyii* in the Indian Ocean and its ecological implication. J. Fish. Biol. 75: 1173–1193.

Watanabe, T., V. Kiron and S. Satoh. 1997. Trace minerals in fish nutrition. Aquaculture 151: 185–207.

Watanabe, T., M. Minagawa, T. Oba and A. Winter. 2001. Pretreatment of coral aragonite for Mg and Sr analysis: Implications for coral thermometers. Geochem. J. 35: 265–269.

Watson, E.B. 1996. Surface enrichment and trace-element uptake during crystal growth. Geochim. Cosmochim. Acta. 60: 5013–5020.

Watson, E.B. 2004. A conceptual model for near-surface kinetic controls on the trace-element and stable isotope composition of abiogenic calcite crystals. Geochim. Cosmochim. Acta. 68: 1473–1488.

Wells, R., J.R. Rooker and D.G. Itano. 2012. Nursery origin of yellowfin tuna in the Hawaiian Islands. Mar. Ecol. Prog. Ser. 461: 187–196.

Yamazaki, A., T. Watanabe, N.O. Ogawa, N. Ohkouchi, K. Shirai, M. Toratani and M. Uematsu. 2011a. Seasonal variations in the nitrogen isotope composition of Okinotori coral in the tropical western Pacific: A new proxy for marine nitrate dynamics. J. Geophys. Res.: Biogeosciences (2005–2012) 116.

Yamazaki, A., T. Watanabe and U. Tsunogai. 2011b. Nitrogen isotopes of organic nitrogen in reef coral skeletons as a proxy of tropical nutrient dynamics. Geophys. Res. Lett. 38.

Yamazaki, A., T. Watanabe, N. Takahata, Y. Sano and U. Tsunogai. 2013. Nitrogen isotopes in intra-crystal coralline aragonites. Chem. Geol. 351: 276–280.

Yokouchi, K., N. Fukuda, K. Shirai, J. Aoyama, F. Daverat and K. Tsukamoto. 2011. Time lag of the response on the otolith strontium/calcium ratios of the Japanese eel, *Anguilla japonica* to changes in strontium/calcium ratios of ambient water. Environ. Boil. Fish. 92: 469–478.

IV. Behavioral or Physiological Aspects

CHAPTER 11

Metabolic Limits and Energetics

Timothy D. Clark

Introduction

Tunas are impressive athletes that swim continuously and can undertake extensive vertical and horizontal migrations throughout the oceans of the world (Block et al. 2001; Graham and Dickson 2004; Block et al. 2005). Such athleticism demands a range of morphological, physiological and biochemical adaptations to ensure that energy and oxygen can be obtained and utilized most effectively. Adaptations for athleticism are obvious in the gross external morphology of tunas, including a fusiform body to reduce drag, fin grooves to increase streamlining, a high-aspect-ratio tail with a narrow caudal peduncle, and finlets across the trailing edges of the body (Dewar and Graham 1994; Graham and Dickson 2004). A closer examination reveals thin gill epithelia and a large gill surface area relative to body size, obligate ram ventilation, anterior-medial body position of slow-twitch oxidative (red) muscle, high aerobic capacity of fast-twitch (white) muscle, and countercurrent vascular heat exchangers (*retia mirabilia*) that function to retain metabolic heat in particular regions of the body and create a thermal excess relative to ambient water (commonly termed regional heterothermy or regional endothermy) (Carey and Teal 1969; Muir and Hughes 1969; Giovane et al. 1980; Carey et al. 1984; Bushnell and Brill 1992; Dewar et al. 1994; Mathieu-Costello et al. 1996; Stevens et al. 2000; Gunn et al. 2001; Marcinek et al. 2001; Graham and Dickson 2004). While these attributes have captured the interest of scientists for many decades, they have also contributed to the inherent difficulties in studying live and unstressed tunas.

Bluefin tunas stand out from all other tunas by attaining an extremely large adult body size (up to 680 kg in Atlantic bluefin; Collette and Nauen 1983), having more regions served by countercurrent heat exchangers, maintaining a greater environment-to-organism temperature differential, routinely undertaking trans-oceanic migrations, and roaming into deep and high-latitude waters that are too cold for other

---

AUniversity of Tasmania and CSIRO Agriculture Flagship, Hobart, Tasmania 7004, Australia.
  Email: timothy.clark.mail@gmail.com

tunas to tolerate (Brill et al. 1999; Block et al. 2001; Blank et al. 2004; Graham and Dickson 2004; Dagorn et al. 2006). Studies of live bluefins have necessarily relied on advancements in techniques and technologies to provide insights into the ecology and physiology of these remarkable animals. This chapter focuses on recently acquired knowledge concerning the metabolism and energetics of juvenile bluefin tunas (approximately 8–23 kg body mass (M_b) or two to four years old), including aspects of the circulatory oxygen transport system that support such an athletic lifestyle. Since much of the metabolic work on bluefin tunas is in its infancy, the chapter draws upon data from other tuna species for support where necessary.

Standard and Routine Metabolism

Standard Metabolic Rate (SMR) represents the basic cost of living and is defined as the minimum maintenance metabolic rate of a post-absorptive, resting ectotherm at a given temperature (Brett and Groves 1979; Priede 1985). Like all tunas, bluefins must swim continuously to maintain water flow across the gills and prevent sinking, which consequently prevents them from ever achieving the necessary state of complete rest that is required to fulfil the definition of SMR. Despite this, for comparative purposes, earlier studies estimated SMR in tunas either by measuring the fraction of oxygen removed from water passed over the gills of anaesthetized tunas (Stevens 1972), measuring whole-animal rates of oxygen consumption ($\dot{M}o_2$) from immobilized tunas (Brill 1979; Brill 1987), or extrapolating curves of swimming speed versus $\dot{M}o_2$ back to zero velocity (Gooding et al. 1981; Graham and Laurs 1982; Graham et al. 1989; Dewar and Graham 1994). Despite the fact that many of these early measurements were conducted under conditions that were likely to cause stress and elevated metabolism in the tunas, the resulting estimates of SMR set the precedent that tunas had metabolic rates that far exceeded other active teleosts (by two- to 10-fold; Korsmeyer and Dewar 2001). For logistical reasons, the vast majority of early studies used small-bodied tropical tunas (typically <4 kg), and no attempts were made to measure $\dot{M}o_2$ of any bluefin species.

The first decade of the 21st century saw the publication of the first measurements of $\dot{M}o_2$ in bluefin tunas. Blank et al. (2007a) housed individual Pacific bluefin (M_b = 8.3 kg) in a large swim tunnel respirometer for up to six days in order to quantify the effects of swimming speed on $\dot{M}o_2$ at a constant temperature of 20°C. The lowest Routine Metabolic Rate (RMR) of 222 ± 24 mg kg^{-1} h^{-1} was obtained when fish were swimming between 0.75 and 1.0 body lengths per second (BL s^{-1}). Extrapolating the speed versus $\dot{M}o_2$ relationship to zero velocity yielded a SMR of 120 ± 26 mg kg^{-1} h^{-1}, which was higher than the SMR estimated for yellowfin in the same study (91 ± 13 mg kg^{-1} h^{-1}) (Blank et al. 2007a). Nevertheless, these estimates of SMR were markedly lower than previously reported for yellowfin and other tuna species, even once differences in temperature and body size were accounted for (see Blank et al. 2007a). These new data questioned the dogma of exceptionally high SMR and RMR in tunas, and emphasized the need to provide extended recovery times if baseline levels of metabolism were desired.

A study by Fitzgibbon et al. (2008) used a large, sea-based, flexible polypropylene bag (12 m diameter, 2.5 m depth) as a respirometer in an attempt to estimate RMR of southern bluefin at 19°C (mean M_b = 19.6 kg). That study reported an extremely high RMR of 460 ± 35 mg kg^{-1} h^{-1} (2.1-fold higher than Pacific bluefin), again fuelling the dogma that tunas have exceptionally high rates of baseline metabolism. Standardizing for differences in body mass between the studies on Pacific and southern bluefin only accentuates the difference in RMR between the species (2.5-fold higher in southern bluefin when using a mass scaling exponent of 0.5; 3.2-fold higher in southern bluefin when using an exponent of 0.8). Could it be true that southern bluefin have two- to three-fold higher baseline energetic requirements than Pacific bluefin at the same temperature and body mass? Noting that the two closely-related species have similar haemoglobin concentration, heart rate, and relative heart mass (see Discussion of Clark et al. (2010a)), it seems most likely that the reported difference in RMR between species is an experimental artefact rather than a biological phenomenon. Indeed, the polypropylene bag respirometer suffered from significant issues associated with unreliable volume estimation (achieved by approximating the output of a water pump and the duration of its use), inconsistent oxygen mixing dynamics, and the requirement to have two or three individual fish in the bag in order to promote measurable declines in water oxygen content. Figure 11.1 illustrates personal observations of $\dot{M}o_2$ in the polypropylene respirometer when housing two southern bluefin (18.8 and 23.1 kg) and when filled with a well-quantified water volume of 300,000 ± 20,000 L (achieved

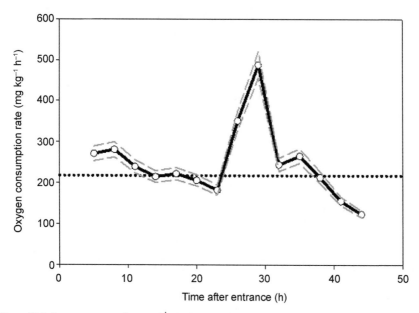

Figure 11.1. Oxygen consumption rate ($\dot{M}o_2$) of two southern bluefin tuna (*T. maccoyii*) swimming together in a sea-based polypropylene bag respirometer, as a function of time after entrance into the respirometer. White circles and bold line represent the 3-hour mean values calculated assuming water volume in the respirometer was 300,000 L, whereas lower and upper dashed lines represent values calculated assuming water volume was 280,000 L and 320,000 L, respectively (encompasses the potential error around the assumed volume of 300,000 L). Routine metabolic rate (RMR; 218 mg kg^{-1} h^{-1}) was calculated as the mean of all data points excluding the elevated points at 26 and 29 hours, and is represented by the horizontal dotted line. Note that absolute values should be taken with caution given the significant issues with this respirometer setup (see text).

by measuring the volume of subsamples of output water from the pump over timed periods, and accurately timing the duration of the pump's use during respirometer filling and water exchange periods). It is evident from these observations that RMR of southern bluefin is likely to be very similar to that of Pacific bluefin. Subsequent published studies using the polypropylene bag have reported progressively lower estimates of RMR for southern bluefin (down to 305 ± 15 mg kg^{-1} h^{-1} in Fitzgibbon and Seymour (2009)), providing further support that Pacific and southern bluefin are unlikely to have greatly contrasting metabolic rates (Table 11.1). The author is not aware of any similar work on Atlantic bluefin, yet there is little reason to believe that the metabolic physiology of this species would be markedly different from the other bluefin species, at least at the same water temperature and body mass.

Perhaps the most striking finding concerning the RMR of bluefin tunas came from a study by Blank et al. (2007b), in which juvenile Pacific bluefin (mean M_b = 8.8 kg) were found to significantly increase their metabolic rate as water temperature decreased below 15°C. Despite the fact that each tuna was maintained in a swim tunnel respirometer at a constant speed of 1 BL s^{-1}, the increase in RMR at lower

Table 11.1. Documented values for different levels of metabolic rate of bluefin tuna species with associated body size and water temperature.

Reference	Body mass (kg)	Body length (cm)	Water temp. (°C)	Routine $\dot{M}o_2$ (mg kg^{-1} h^{-1})	Active $\dot{M}o_2$ (mg kg^{-1} h^{-1})
Southern bluefin					
Fitzgibbon et al. (2007)	10.0 ± 0.4 (8.8–12.4)	84 ± 2 (76–91)	19.3 ± 0.4	366 ± 33 (300–456)	1,290
Fitzgibbon et al. (2008)	19.6 ± 1.9 (14.0–23.2)	106 ± 2 (93–117)	19.0 ± 0.3	460 ± 35 (346–539)	–
Fitzgibbon and Seymour (2009)	19.8 ± 0.5 (15.9–22.7)	105 ± 1 (101–112)	17.5 ± 0.7	305 ± 15 (272–357)	1,016
Fitzgibbon et al. (2010)	10.2 ± 0.3 (8.8–12.8)	84 ± 1 (76–91)	19.6 ± 0.2	384 ± 31 (300–456)	655
T.D. Clark (unpublished data)	20.9 ± 2.2 (18.8–23.1)	108 ± 5 (103–113)	16.7 ± 0.3	218 ± 14 (124–282)	488
Pacific bluefin					
Blank et al. (2007a)*	8.3 ± 0.8 (7.1–9.4)	74 ± 3 (70–78)	20.0 ± 0.5	222 ± 24 (195–250)	700
Blank et al. (2007b)*	8.8 ± 0.8 (7.4–9.9)	76 ± 4 (70–84)	20.0 ± 0.1	193 ± 10 (158–237)	450
Blank et al. (2007b)*	8.8 ± 0.8 (7.4–9.9)	76 ± 4 (70–84)	**8.0 ± 0.1**	331 ± 28 (250–419)	400
Blank et al. (2007b)*	8.8 ± 0.8 (7.4–9.9)	76 ± 4 (70–84)	**25.0 ± 0.1**	256 ± 9 (236–286)	450
Clark et al. (2010a)*	10.4 ± 0.2 (9.7–11.0)	81 ± 0 (80–83)	20.0 ± 0.5	174 ± 9 (139–206)	490
Clark et al. (2013)*	9.7	77	20.0 ± 0.1	177	–

Data are means (with ranges in parentheses). $\dot{M}o_2$ is oxygen consumption rate. 'Active $\dot{M}o_2$' refers to the highest $\dot{M}o_2$ recorded in the corresponding study, where relevant. Note that two additional temperatures are given for Blank et al. (2007b), indicated in bold type. The author is not aware of any existing data for Atlantic bluefin. *Fish held in swim tunnel at defined speed rather than voluntarily choosing a swimming speed.

temperatures was concomitant with an increase in tail beat frequency and movement of the tuna to the front of the swim tunnel. This finding was replicated in a follow-up study (Clark et al. 2013) and it was further discovered that heart rate essentially plateaued below 15°C, concomitant with the increase in tail beat frequency and $\dot{M}o_2$. This may represent the only example of a fish species increasing rather than decreasing $\dot{M}o_2$ with a decrease in water temperature. It is tempting to conclude that the increase in tail beat frequency and $\dot{M}o_2$ of juvenile Pacific bluefin below 15°C may act as a mechanism to increase metabolic heat production in the swimming muscles and thus stabilize muscle temperature and muscle power production during sojourns into cold water (Graham and Dickson 1981; Blank et al. 2007b). Future work comparing the locomotory and thermoregulatory responses of tuna species with different capacities for heterothermy will be valuable for deciphering the validity of this possibility (e.g., see Dizon et al. 1977; Dewar et al. 1994).

Maximum Metabolism and Aerobic Scope

Few data exist on the $\dot{M}o_2$ of bluefin tunas at elevated swimming speeds, and no study has successfully quantified Maximum Metabolic Rate (MMR). Blank et al. (2007a) documented a typical J-shaped increase in $\dot{M}o_2$ with increasing swimming speed in juvenile Pacific bluefin at 20°C (8.3 ± 0.8 kg), concluding that values are generally elevated above those for yellowfin at comparable swimming speeds. The highest swimming speed in that study (1.8 BL s^{-1}) elicited a mean $\dot{M}o_2$ of 498 ± 55 mg kg^{-1} h^{-1} in Pacific bluefin, although an examination of Fig. 1 in that study indicates that $\dot{M}o_2$ reached at least 700 mg kg^{-1} h^{-1} in some individuals at some point in the experimental protocol (Blank et al. 2007a). These elevated values of $\dot{M}o_2$ are quite high for a fish but not exceptional compared with other athletic species at the same temperature even when body mass scaling is taken into consideration (e.g., Clark et al. 2011). Values of $\dot{M}o_2$ approaching 1,300 mg kg^{-1} h^{-1} have been reported for southern bluefin at 19°C (Table 11.1; Fitzgibbon et al. 2007), but these values are likely to be erroneous considering the issues with the respirometry techniques detailed above, as well as the fact that the fish should not have been close to MMR at the time of the measurements (free-swimming at routine speeds and digesting a moderate-sized meal).

Despite the absence of empirical data on MMR, enough information exists for different species of tunas to make a first set of calculations on the metabolic limits of juvenile bluefins (M_b ~ 8–23 kg). Convective oxygen transport through the cardiovascular system is described by the Fick equation:

$$\dot{M}o_2 = f_H \cdot V_s \cdot (CaO_2 - CvO_2),$$

where f_H is heart rate, V_s is cardiac stroke volume, and $CaO_2 - CvO_2$ is the difference in arterial (CaO_2) and venous (CvO_2) oxygen content (a measure of tissue oxygen extraction). Cardiac output ($\dot{V}_b$) is the product of f_H and V_s. While this equation can be used to provide an estimate of circulatory oxygen transport limits, it has the potential to underestimate whole-animal $\dot{M}o_2$ because it does not account for any oxygen usage by tissues in direct contact with ambient water (e.g., gills, skin). It is not known how significant the latter could be in bluefins, but it is likely to be minor in comparison with the oxygen acquired via the circulatory oxygen transport cascade.

Table 11.2. Circulatory variables of various tuna species, and their estimated contributions to oxygen consumption rates at routine and maximum activity in juvenile bluefin tunas at 20°C.

Variable	Routine activity	Maximum activity	References
f_H, beats min^{-1}	25–40	120–130	(Clark et al. 2008b; Clark et al. 2013)
V_s, ml kg^{-1}	1.0–1.4	1.0–1.4	(Farrell 1996; Korsmeyer et al. 1997a; Korsmeyer et al. 1997b; Brill and Bushnell 2001; Blank et al. 2002)
$\dot{V}_b$, ml kg^{-1} min^{-1}	25–56	120–182	*
CaO_2, mg O_2 ml^{-1} (ml dl^{-1})	0.21 (15)	0.26 (18)	(Korsmeyer et al. 1997b; Clark et al. 2008a)
CvO_2, mg ml^{-1} (ml dl^{-1})	0.16 (11)	0.10 (7)	(Korsmeyer et al. 1997b; Clark et al. 2008a)
CaO_2–CvO_2, mg ml^{-1} (ml dl^{-1})	0.05 (4)	0.16 (11)	*
$\dot{M}o_2$, mg kg^{-1} h^{-1} (mg kg^{-1} min^{-1})	78–168 (1.3–2.8)	1,152–1,746 (19.2–29.1)	*

f_H is heart rate, V_s is cardiac stroke volume, $\dot{V}_b$ is cardiac output, CaO_2 is arterial oxygen content, CvO_2 is venous oxygen content, CaO_2–CvO_2 is tissue oxygen extraction, and $\dot{M}o_2$ is oxygen consumption rate.
* Calculated from given variables (see text).

Tunas are thought to have a limited ability to increase V_s, possibly due to their high proportion of compact myocardium in the ventricle (Agnisola and Tota 1994; Bushnell and Jones 1994; Tota and Gattuso 1996). Values of V_s are generally reported as 1.0–1.4 ml beat^{-1} kg^{-1} across tuna species (Korsmeyer et al. 1997a; Brill and Bushnell 2001; Blank et al. 2004), and so estimates of the scope in $\dot{V}_b$ can be made by multiplying f_H by 1.0 or 1.4. Thus, estimates of $\dot{V}_b$ for bluefins at 20°C are 25–56 ml kg^{-1} min^{-1} in completely resting and unfed fish (assuming f_H of 25–40 beats min^{-1}), and 120–182 ml kg^{-1} min^{-1} in maximally exercising fish (assuming f_H of 120–130 beats min^{-1}) (for heart rate data, see Clark et al. 2008b; Clark et al. 2013) (Fig. 11.3). Tuna blood respiratory measurements suggest that CaO_2 of bluefins may be around 0.21 mg O_2 ml^{-1} (15 ml dl^{-1}) in routinely swimming fish at 20°C (Farrell 1996; Korsmeyer et al. 1997b; Clark et al. 2008a), and this may increase to around 0.26 mg O_2 ml^{-1} (18 ml dl^{-1}) during exercise with splenic release of erythrocytes (Korsmeyer et al. 1997b; Clark et al. 2008a). Based on a venous partial pressure of oxygen (PvO_2) of 6 kPa, a routine value of CvO_2 is likely to be around 0.16 mg O_2 ml^{-1} (11 ml dl^{-1}) for bluefins (Korsmeyer et al. 1997b; Clark et al. 2008a), and this is likely to decrease during exercise to a minimum CvO_2 of about 0.10 mg O_2 ml^{-1} (7 ml dl^{-1}) at a PvO_2 of around 3 kPa (Korsmeyer et al. 1997b; Clark et al. 2008a). Using these values, CaO_2–CvO_2 would have a minimum of 0.05 mg O_2 ml^{-1} (4 ml dl^{-1}) in resting fish and a maximum of 0.16 mg O_2 ml^{-1} (11 ml dl^{-1}) during exercise (Fig. 11.3).

Thus, the circulatory adjustments of bluefin tunas at 20°C occurring at two extreme levels of exercise can be estimated to result in minimum $\dot{M}o_2$ values of 78–168 mg kg^{-1} h^{-1} and maximum $\dot{M}o_2$ values of 1,152–1,746 mg kg^{-1} h^{-1}. These calculations and estimates are presented in an illustrative form in Fig. 11.3 and they equate to an estimated mean aerobic scope of around 1,326 mg kg^{-1} h^{-1} (~12-times minimum $\dot{M}o_2$). These theoretical upper limits for MMR at 20°C are quite impressive, particularly for a fish in the 8–23 kg mass range if MMR does not scale isometrically.

Given the information noted above concerning the effect of cooling on swimming speed, RMR and heart rate, it seems likely that bluefins use greater proportions of their available aerobic scope and heart rate scope when they encounter cooler water temperatures (e.g., <15°C) (see Clark et al. 2013). This has the potential to limit other (non-locomotory) aerobic processes such as digestion, until such time that the bluefin returns to warmer waters. An interesting phenomenon reported in bluefin tunas is the presence of reversed (decreased affinity with decreasing temperature) or thermally-independent (no change in affinity with temperature) haemoglobin-oxygen binding (Rossi-Fanelli and Antonini 1960; Carey and Gibson 1983; Brill and Bushnell 2006; Clark et al. 2008a), which may play some role in ensuring oxygen delivery to working muscles (e.g., swimming muscles, ventricle) during sojourns into cold water (see Clark et al. 2010b).

Digestion

The energy utilized during the ingestion, digestion, absorption and assimilation of a meal is commonly referred to as Specific Dynamic Action (SDA) or Heat Increment of Feeding (HIF). In the majority of fishes, the SDA process generally accounts for $16 \pm 1\%$ of the gross energy ingested in a meal (termed 'SDA coefficient') (Secor 2009), highlighting the significant contribution of SDA to fish energy budgets. While measurements of SDA in bluefin tunas remained elusive for many years due to the inherent difficulties with studying such large and high performance fishes, it was assumed that SDA would be a large component of daily aerobic metabolic requirements because tunas have high rates of food consumption and digestion (Korsmeyer et al. 1996).

An initial attempt to measure SDA in juvenile southern bluefin (19.8 ± 0.5 kg) reported that peak $\dot{M}o_2$ reached as high as 1,290 mg kg^{-1} h^{-1} during digestion of large sardine meals at ~19°C and that the SDA coefficient was $35 \pm 2\%$, the latter implying relatively inefficient food processing and assimilation in bluefin tunas compared with other fishes (Fitzgibbon et al. 2007). A follow-up study by the same authors again reported that the SDA coefficient for southern bluefin was around 30–35%, yet made a further suggestion that a postprandial increase in swimming speed was responsible for artificially inflating estimates of the SDA coefficient (Fitzgibbon and Seymour 2009). Once the metabolic cost of higher swimming speed was subtracted from SDA estimates, the study concluded that the correct SDA coefficient was around 20% and therefore still at the high end of the range reported for other fishes (Fitzgibbon and Seymour 2009). These studies were conducted using the large, sea-based polypropylene bag outlined above, which suffered from the abovementioned issues that compromised $\dot{M}o_2$ measurements. Personal observations (unpubl. data) of southern bluefin confirmed major issues with trying to quantify any aspect of the SDA process using the polypropylene bag respirometer (Fig. 11.2; raw data from 'Trial 4' in Fitzgibbon and Seymour (2009)).

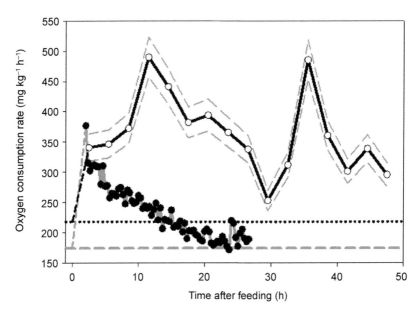

Figure 11.2. Comparison of the postprandial metabolic response for two southern bluefin after eating a combined total of 1.64 kg of high-lipid sardines (3.9% of combined tuna body mass; 1,509 kJ; 17°C) (black open circles and bold black line) versus one Pacific bluefin after eating 0.45 kg of a mixture of squid, sardine and vitamin-enriched gelatine (4.1% of tuna body mass; 1,680 kJ; 20°C) (black closed circles and bold grey line). Dashed grey lines around the bold black line represent the same information as indicated in Fig. 11.1. Data for southern bluefin are personal observations by the author, although the same data have been included in a modified form in Fitzgibbon and Seymour (2009). Data for Pacific bluefin are from Clark et al. (2010a). Horizontal lines represent the mean routine metabolic rate (RMR) calculated for southern bluefin (black dotted line; see Fig. 11.1) and Pacific bluefin (grey dashed line; see Clark et al. 2010a).

In light of the significant concerns surrounding the reported SDA measurements for southern bluefin, a subsequent research program using juvenile Pacific bluefin (10.4 ± 0.2 kg) employed different techniques to provide higher resolution measurements of SDA. Using a custom-designed swim tunnel respirometer and controlling for swimming speed during the postprandial period, clear and consistent SDA curves were established across a range of meal sizes for Pacific bluefin at ~20°C (Clark et al. 2010a). That study reported peak $\dot{M}o_2$ up to 440 mg kg^{-1} h^{-1} during digestion of large sardine meals, and a SDA coefficient of 9.2 ± 0.7% (range 7.5–11.6%) across meal sizes spanning 4.1 to 12.6% of tuna body mass (Clark et al. 2010a). Moreover, the duration of the SDA event never exceeded 33 hours in even the largest meals, highlighting relatively rapid digestive processes in comparison with other fishes at equivalent

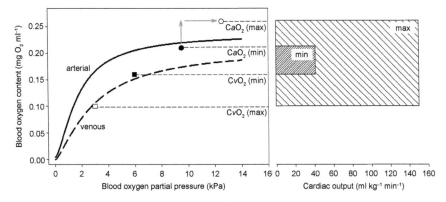

Figure 11.3. Estimated values for oxygen content of arterial (CaO_2) and venous (CvO_2) blood (left panel) and cardiac output (right panel) of routinely swimming (min) and maximally exercising (max) juvenile bluefin tunas in the size range of ~8–23 kg. Hyperbolic regressions in the left panel represent blood-oxygen equilibrium curves for arterial (solid line) and venous (dashed line) blood. The area of the boxes in the right panel represents the oxygen consumption rate (Mo_2) under the two extreme levels of activity, showing the relative contributions of tissue oxygen extraction (CaO_2–CvO_2) and cardiac output. The grey vertical arrow represents an increase in CaO_2 with exercise due to splenic release of erythrocytes, while the grey horizontal arrow represents an increase in arterial oxygen partial pressure due to increased gill ventilation. See text and Table 11.2 for more information.

temperatures (see Discussion of Clark et al. 2010a). Figure 11.2 compares the SDA traces obtained using the polypropylene bag respirometer (southern bluefin) versus the swim tunnel respirometer (Pacific bluefin) during the digestion of comparable meals. Thus, the most reliable data available for bluefins suggest that juveniles of these species have rapid and efficient food processing capacities in comparison with other fishes, corroborating reports of rapid gut clearance and high growth rates in tunas (Schaefer 1984; Olson and Boggs 1986; Polacheck et al. 2004). Despite this digestive efficiency, SDA is likely to be a significant component of daily energy budgets in bluefin tunas because of high and regular ingestion of prey (Bestley et al. 2008; Whitlock et al. 2013).

It has been established that species of the subgenus *Thunnus*, including the bluefins, possess the greatest capacity for regional heterothermy, including a visceral heat exchanger to retain heat produced in the viscera during digestion (Graham and Dickson 2004). A warm viscera is thought to speed digestion and gut evacuation for the next feeding opportunity because of the thermal enhancement of digestive enzyme activity (Stevens and McLeese 1984). The magnitude of the thermal increment in the viscera provides a good indication of meal size (Carey et al. 1984; Gunn et al. 2001; Bestley et al. 2008; Clark et al. 2008b; Whitlock et al. 2013), and it has been shown recently that the thermal increment in the viscera tracks the SDA response with a two to three hour lag in 10.4 ± 0.2 kg Pacific bluefin (Clark et al. 2010a). While very unusual across the animal kingdom, the pronounced visceral thermal increment of bluefin tunas can be used to understand foraging patterns and energy acquisition from the environment (Bestley et al. 2008; Whitlock et al. 2013).

Energetics

The recent advances in our knowledge of bluefin tunas, outlined in this and other chapters of this volume, provide insight into the daily energetics of these ecologically and economically important species. Estimates of the energy intake of wild, free-roaming bluefin tunas have been made based on archival tag measurements of visceral temperature profiles. Juvenile southern bluefin in the body length range of 93–111 cm have been estimated to consume about 0.75 kg of food per day when foraging to the southwest of the Australian coast (Bestley et al. 2008). This would equate to about 5.25 MJ d^{-1} (1,255 kcal d^{-1}) if the tunas were feeding on Australian sardines (*Sardinops sagax*), which are an important prey species in this region and have an energy content of around 7 MJ kg^{-1} (1,673 kcal kg^{-1}) (Clark et al. 2008b). This estimated daily energy intake for southern bluefin compares favourably with values estimated for similar-sized Pacific bluefin that were tagged and released off Baja California (mean ~1,000–1,500 kcal d^{-1}) (Whitlock et al. 2013), thus apparently providing a trustworthy indication of energy intake and confirming that the two geographically-separated species have similar foraging ecology and energetics. Estimates of the energy intake of larger bluefins can become more challenging due to less-pronounced visceral thermal increments after feeding, as well as confounding effects due to repeated and extreme vertical migrations into very cold water (e.g., Block et al. 2001).

While the energy intake of at least juvenile bluefins is becoming clearer as we learn to utilize the thermal increment in the viscera associated with digestion, our understanding of energy expenditure remains limited to laboratory-based studies with bluefins in respirometers. A lot of progress stems from new advances in tagging technologies in combination with rigorous controlled experiments. For example, heart rate can act as an excellent proxy for energy expenditure associated with activities such as locomotion and SDA (Clark et al. 2008b; Clark et al. 2010c), although the situation becomes more complex with changes in ambient temperature due to the unusual responses in tail beat frequency, $\dot{M}o_2$ and heart rate displayed by juvenile bluefins (see above; Clark et al. 2013). Great progress to understand the energetics of bluefins over broad temporal and spatial scales will be made with multi-sensor tagging technologies, such as those that record temperature, heart rate and three-dimensional acceleration (Clark et al. 2010c), as these will provide insight into the behaviour and physiology that underlie energy intake and expenditure.

Growth, gonadal development and reproduction are discussed thoroughly in other chapters of this volume and are obviously critical considerations in the lifetime energetics of bluefins. Indeed, energy investment into reproduction must be extreme; female southern bluefin spawn on average every 1.1 days during the breeding season which peaks in October and February, with the average spawning batch fecundity being 6.0 million oocytes or 57 oocytes per gram of body weight (Farley and Davis 1998).

Conclusions

There is no doubt that bluefin tunas are incredibly impressive and awe-inspiring animals that possess adaptations to set them apart from all other fishes. Nevertheless, the early precedent set for bluefins, and tunas in general, was perhaps a little too influenced by the desire to label every aspect of tuna physiology as being exceptional and unmatched in the fish world. Bluefins appear to have slightly elevated RMR compared with other active fishes, and their MMR also seems to be in the upper range for fishes once temperature and body size are taken into consideration. Bluefin heart rates are quite typical amongst the fishes, yet high cardiac stroke volume, blood oxygen carrying capacity and gill surface area may help to improve oxygen transport rates and support an athletic lifestyle. The unusual responses of bluefins to ambient temperature change are testament to their standing as impressive species that warrant our scientific attention. Nevertheless, while research on bluefin tunas has come a long way in recent times due to advancements in technologies and techniques, the challenges of studying these brilliant animals may worsen unless global populations of bluefins stabilize and improve. Further research into understanding the metabolic limits and energetic requirements of bluefins will help to forecast how these species will respond to anthropogenic stressors including changes in prey availability and ocean warming.

Acknowledgements

I have had many fruitful discussions over the years with people who share my interest and enthusiasm regarding tuna eco-physiology. I would like to especially thank Charles Farwell, Barbara Block, Anthony Farrell and Roger Seymour for their assistance, encouragement and guidance.

References

Agnisola, C. and B. Tota. 1994. Structure and function of the fish cardiac ventricle—flexibility and limitations. Cardioscience 5: 145–153.

Bestley, S., T.A. Patterson, M.A. Hindell and J.S. Gunn. 2008. Feeding ecology of wild migratory tunas revealed by archival tag records of visceral warming. J. Anim. Ecol. 77: 1223–1233.

Blank, J.M., C.J. Farwell, J.M. Morrissette, R.J. Schallert and B.A. Block. 2007a. Influence of swimming speed on metabolic rates of juvenile Pacific bluefin tuna and yellowfin tuna. Physiol. Biochem. Zool. 80: 167–177.

Blank, J.M., J.M. Morrissette, P.S. Davie and B.A. Block. 2002. Effects of temperature, epinephrine and Ca^{2+} on the hearts of yellowfin tuna (*Thunnus albacares*). J. Exp. Biol. 205: 1881–1888.

Blank, J.M., J.M. Morrissette, C.J. Farwell, M. Price, R.J. Schallert and B.A. Block. 2007b. Temperature effects on metabolic rate of juvenile Pacific bluefin tuna *Thunnus orientalis*. J. Exp. Biol. 210: 4254–4261.

Blank, J.M., J.M. Morrissette, A.M. Landeira-Fernandez, S.B. Blackwell, T.D. Williams and B.A. Block. 2004. *In situ* cardiac performance of Pacific bluefin tuna hearts in response to acute temperature change. J. Exp. Biol. 207: 881–890.

Block, B.A., H. Dewar, S.B. Blackwell, T.D. Williams, E.D. Prince, C.J. Farwell, A. Boustany, S.L.H. Teo, A. Seitz, A. Walli and D. Fudge. 2001. Migratory movements, depth preferences, and thermal biology of Atlantic bluefin tuna. Science 293: 1310–1314.

Block, B.A., S.L.H. Teo, A. Walli, A. Boustany, M.J.W. Stokesbury, C.J. Farwell, K.C. Weng, H. Dewar and T.D. Williams. 2005. Electronic tagging and population structure of Atlantic bluefin tuna. Nature 434: 1121–1127.

Brett, J.R. and T.D.D. Groves. 1979. Physiological energetics. pp. 279–352. *In*: W.S. Hoar, D.J. Randall and J.R. Brett (eds.). Fish Physiology, Bioenergetics and Growth, Vol. 8. Academic Press, New York.

Brill, R.W., B.A. Block, C.H. Boggs, K.A. Bigelow, E.V. Freund and D.J. Marcinek. 1999. Horizontal movements and depth distribution of large adult yellowfin tuna (*Thunnus albacares*) near the Hawaiian Islands, recorded using ultrasonic telemetry: implications for the physiological ecology of pelagic fishes. Mar. Biol. 133: 395–408.

Brill, R.W. and P.G. Bushnell. 2001. The cardiovascular system of tunas. pp. 79–120. *In*: B.A. Block and E.D. Stevens (eds.). Tuna: Physiology, Ecology, and Evolution, Vol. 19. Fish Physiology. Academic Press, San Diego.

Brill, R.W. and P.G. Bushnell. 2006. Effects of open- and closed-system temperature changes on blood O_2-binding characteristics of Atlantic bluefin tuna (*Thunnus thynnus*). Fish Physiol. Biochem. 32: 283–294.

Bushnell, P.G. and R.W. Brill. 1992. Oxygen transport and cardiovascular responses in skipjack tuna (*Katsuwonus pelamis*) and yellowfin tuna (*Thunnus albacares*) exposed to acute hypoxia. J. Comp. Physiol. B 162: 131–43.

Bushnell, P.G. and D.R. Jones. 1994. Cardiovascular and respiratory physiology of tuna—adaptations for support of exceptionally high metabolic rates. Env. Biol. Fishes 40: 303–318.

Carey, F.G. and Q.H. Gibson. 1983. Heat and oxygen exchange in the rete mirabile of the bluefin tuna, *Thunnus thynnus*. Comp. Biochem. Phys. A 74: 333–342.

Carey, F.G., J.W. Kanwisher and E.D. Stevens. 1984. Bluefin tuna warm their viscera during digestion. J. Exp. Biol. 109: 1–20.

Carey, F.G. and J.M. Teal. 1969. Regulation of body temperature by the bluefin tuna. Comp. Biochem. Phys. A 28: 205–213.

Clark, T.D., W.T. Brandt, J. Nogueira, L.E. Rodriguez, M. Price, C.J. Farwell and B.A. Block. 2010a. Postprandial metabolism of Pacific bluefin tuna (*Thunnus orientalis*). J. Exp. Biol. 213: 2379–2385.

Clark, T.D., C.J. Farwell, L.E. Rodriguez, W.T. Brandt and B.A. Block. 2013. Heart rate responses to temperature in free-swimming Pacific bluefin tuna (*Thunnus orientalis*). J. Exp. Biol. 216: 3208–3214.

Clark, T.D., K.M. Jeffries, S.G. Hinch and A.P. Farrell. 2011. Exceptional aerobic scope and cardiovascular performance of pink salmon (*Oncorhynchus gorbuscha*) may underlie resilience in a warming climate. J. Exp. Biol. 214: 3074–3081.

Clark, T.D., J.L. Rummer, C.A. Sepulveda, A.P. Farrell and C.J. Brauner. 2010b. Reduced and reversed temperature dependence of blood oxygenation in an ectothermic scombrid fish: implications for the evolution of regional heterothermy? J. Comp. Physiol. B 180: 73–82.

Clark, T.D., E. Sandblom, S.G. Hinch, D.A. Patterson, P.B. Frappell and A.P. Farrell. 2010c. Simultaneous biologging of heart rate and acceleration, and their relationships with energy expenditure in free-swimming sockeye salmon (*Oncorhynchus nerka*). J. Comp. Physiol. B 180: 673–684.

Clark, T.D., R.S. Seymour, R.M.G. Wells and P.B. Frappell. 2008a. Thermal effects on the blood respiratory properties of southern bluefin tuna, *Thunnus maccoyii*. Comp. Biochem. Phys. A 150: 239–246.

Clark, T.D., B.D. Taylor, R.S. Seymour, D. Ellis, J. Buchanan, Q.P. Fitzgibbon and P.B. Frappell. 2008b. Moving with the beat: heart rate and visceral temperature of free-swimming and feeding bluefin tuna. Proc. Roy. Soc. B. 275: 2841–2850.

Collette, B.B. and C.E. Nauen. 1983. FAO species catalogue. Vol. 2. Scombrids of the world. An annotated and illustrated catalogue of tunas, mackerels, bonitos and related species known to date. FAO Fisheries Synopsis 125: 1–137.

Dagorn, L., K.N. Holland, J.-P. Hallier, M. Taquet, G. Moreno, G. Sancho, D.G. Itano, R. Aumeeruddy, C. Girard, J. Million and A. Fonteneau. 2006. Deep diving behavior observed in yellowfin tuna (*Thunnus albacares*). Aquat. Living Resour. 19: 85–88.

Dewar, H. and J.B. Graham. 1994. Studies of tropical tuna swimming performance in a large water tunnel, 1—Energetics. J. Exp. Biol. 192: 13–31.

Dewar, H., J.B. Graham and R.W. Brill. 1994. Studies of tropical tuna swimming performance in a large water tunnel, 2—Thermoregulation. J. Exp. Biol. 192: 33–44.

Dizon, A.E., W.H. Neill and J.J. Magnuson. 1977. Rapid temperature compensation of volitional swimming speeds and lethal temperatures in tropical tunas (Scombridae). Env. Biol. Fishes 2: 83.

Farley, J.H. and T.L.O. Davis. 1998. Reproductive dynamics of southern bluefin tuna, *Thunnus maccoyii*. Fishery Bulletin 96: 223–236.

Farrell, A.P. 1996. Features heightening cardiovascular performance in fishes, with special reference to tunas. Comp. Biochem. Phys. A 113: 61–67.

Fitzgibbon, Q.P., R.V. Baudinette, R.J. Musgrove and R.S. Seymour. 2008. Routine metabolic rate of southern bluefin tuna (*Thunnus maccoyii*). Comp. Biochem. Phys. A 150: 231–238.

Fitzgibbon, Q.P. and R.S. Seymour. 2009. Postprandial metabolic increment of southern bluefin tuna *Thunnus maccoyii* ingesting high or low-lipid sardines *Sardinops sagax*. J. Fish Biol. 75: 1586–1600.

Fitzgibbon, Q.P., R.S. Seymour, J. Buchanan, R. Musgrove and J. Carragher. 2010. Effects of hypoxia on oxygen consumption, swimming velocity and gut evacuation in southern bluefin tuna (*Thunnus maccoyii*). Environ. Biol. Fishes 89: 59–69.

Fitzgibbon, Q.P., R.S. Seymour, D. Ellis and J. Buchanan. 2007. The energetic consequence of specific dynamic action in southern bluefin tuna *Thunnus maccoyii*. J. Exp. Biol. 210: 290–298.

Giovane, A., G. Greco, A. Maresca and B. Tota. 1980. Myoglobin in the heart ventricle of tuna and other fishes. Experientia 36: 219–20.

Gooding, R.M., W.H. Neill and A.E. Dizon. 1981. Respiration rates and low-oxygen tolerance limits in skipjack tuna, *Katsuwonus pelamis*. Fishery Bulletin 79: 31–48.

Graham, J.B. and K.A. Dickson. 1981. Physiological thermoregulation in the albacore *Thunnus alalunga*. Physiol. Zool. 54: 470–486.

Graham, J.B. and K.A. Dickson. 2004. Tuna comparative physiology. J. Exp. Biol. 207: 4015–4024.

Graham, J.B. and R.M. Laurs. 1982. Metabolic rate of the albacore tuna *Thunnus alalunga*. Mar. Biol. 72: 1–6.

Graham, J.B., W.R. Lowell, N.C. Lai and R.M. Laurs. 1989. O_2 tension, swimming-velocity, and thermal effects on the metabolic rate of the Pacific albacore *Thunnus alalunga*. Exp. Biol. 48: 89–94.

Gunn, J., J. Hartog and K. Rough. 2001. The relationship between food intake and visceral warming in southern bluefin tuna (*Thunnus maccoyii*). pp. 1009–1130. *In*: J.R. Sibert and J.L. Nielsen (eds.). Electronic Tagging and Tracking in Marine Fisheries, Vol. 1. Kluwer Academic Publishers, Dordrecht.

Korsmeyer, K.E. and H. Dewar. 2001. Tuna metabolism and energetics. pp. 35–78. *In*: B.A. Block and E.D. Stevens (eds.). Tuna—Physiology, Ecology, and Evolution. Academic Press, San Diego.

Korsmeyer, K.E., H. Dewar, N.C. Lai and J.B. Graham. 1996. The aerobic capacity of tunas—adaptation for multiple metabolic demands. Comp. Biochem. Phys. A 113: 17–24.

Korsmeyer, K.E., N.C. Lai, R.E. Shadwick and J.B. Graham. 1997a. Heart rate and stroke volume contributions to cardiac output in swimming yellowfin tuna—response to exercise and temperature. J. Exp. Biol. 200: 1975–1986.

Korsmeyer, K.E., N.C. Lai, R.E. Shadwick and J.B. Graham. 1997b. Oxygen transport and cardiovascular responses to exercise in the yellowfin tuna *Thunnus albacares*. J. Exp. Biol. 200: 1987–1997.

Marcinek, D.J., S.B. Blackwell, H. Dewar, E.V. Freund, C. Farwell, D. Dau, A.C. Seitz and B.A. Block. 2001. Depth and muscle temperature of Pacific bluefin tuna examined with acoustic and pop-up satellite archival tags. Mar. Biol. 138: 869–885.

Mathieu-Costello, O., R.W. Brill and P.W. Hochachka. 1996. Structural basis for oxygen delivery—muscle capillaries and manifolds in tuna red muscle. Physiology 113: 25–31.

Muir, B.S. and G.M. Hughes. 1969. Gill dimensions for three species of Tunny. J. Exp. Biol. 51: 271–285.

Olson, R.J. and C.H. Boggs. 1986. Apex predation by yellowfin tuna (*Thunnus albacares*): independent estimates from gastric evacuation and stomach contents, bioenergetics, and cesium concentrations. Can. J. Fish. Aquat. Sci. 43: 1760–1775.

Polacheck, T., J.P. Eveson and G.M. Laslett. 2004. Increase in growth rates of southern bluefin tuna (*Thunnus maccoyii*) over four decades: 1960 to 2000. Can. J. Fish. Aquat. Sci. 61: 307–322.

Priede, I.G. 1985. Metabolic scope in fishes. pp. 33–64. *In*: P. Tytler and P. Calow (eds.). Fish Energetics: New Perspectives. Croom Helm Ltd., London.

Rossi-Fanelli, A. and E. Antonini. 1960. Oxygen equilibrium of haemoglobin from *Thunnus thynnus*. Nature 186: 895–896.

Schaefer, K.M. 1984. Swimming performance, body temperatures and gastric evacuation times of the black skipjack, *Euthynnus lineatus*. Copeia 1984: 1000–1005.

Secor, S.M. 2009. Specific dynamic action: a review of the postprandial metabolic response. J. Comp. Physiol. B 179: 1–56.

Stevens, E.D., J.W. Kanwisher and F.G. Carey. 2000. Muscle temperature in free-swimming giant Atlantic bluefin tuna (*Thunnus thynnus* L.). J. Therm. Biol. 25: 419–423.

Stevens, E.D. and J.M. McLeese. 1984. Why bluefin tuna have warm tummies: temperature effect on trypsin and chymotrypsin. Am. J. Physiol. Reg. 1. 246: R487–494.

Tota, B. and A. Gattuso. 1996. Heart ventricle pumps in teleosts and elasmobranchs—a morphodynamic approach. J. Exp. Zool. 275: 162–171.

Whitlock, R.E., A. Walli, P. Cermeño, L.E. Rodriguez, C. Farwell and B.A. Block. 2013. Quantifying energy intake in Pacific bluefin tuna (*Thunnus orientalis*) using the heat increment of feeding. J. Exp. Biol. 216: 4109–4123.

CHAPTER 12

Swimming Performance of Pacific Bluefin Tuna

Tsutomu Takagi

Our experiences show that we bend our bodies and stay low to present a smaller area against strong wind to prevent being blown over. We know that migratory fish have characteristic morphologies such as a streamlined body shape and a relatively small cross-sectional area to reduce drag. What are the mechanisms of migratory fish for high-speed and long-distance travel? Functions used for migration and maneuvering are examined from the perspective of the hydrodynamic properties of Pacific bluefin tuna.

Hydrostatics of Bluefin Tuna: Gravity and Buoyancy

The gravitational force acts on all living creatures, including aquatic animals such as large Pacific bluefin tuna weighing over 100 kg. The gravitational force can be expressed by the following equation as the product of mass, m, and acceleration due to gravity, g.

$$Fg = mg.$$
Eq. 1

Generally, aquatic animals do not rapidly sink in water because a large buoyancy acts on them. This buoyancy acts vertically on them and its magnitude equals the weight of the displaced water. Thus, the larger the size, the greater the buoyant force. Greater buoyancy against gravity can be attained by a larger volume with less mass. Therefore, many fish species have swim bladders that provide sufficient buoyancy to prevent sinking to the sea bottom.

What forces act on fish during swimming, and how do these forces act? Figure 12.1 shows the forces acting on an individual fish. Usually, the buoyant and gravitational forces counteract each other vertically; therefore, if one is greater than the other, the

Faculty of Fisheries Sciences, Hokkaido University, Minato, Hakodate, Hokkaido 041-8611, Japan.
  Email: tutakagi@fish.hokudai.ac.jp

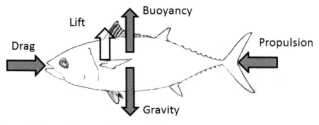

Figure 12.1. Schematic view of forces acting on a fish.

fish moves in the direction of the greater force vector. However, if an equilibrium state can be attained, the fish can remain at a constant depth. The gravitational force acting on the fish equals its weight, and this force cannot be rapidly changed without any intake of air, food, or water. Buoyancy, therefore, can be easily controlled by varying the volume of air using a swim bladder. Thus, this organ is important for depth control.

However, some scombrid fish species, such as the skipjack tuna and kawakawa, have no swim bladders, whereas other tuna species have swim-bladders that are not large enough to counteract the gravitational force. Thus, such tuna will sink without any moving action to prevent it. Magnuson (1978) reported that the body densities of some tuna were greater than that of seawater (1.025 g/cm^3) (Table 12.1). The body densities of skipjack tuna and kawakawa are greater than 1.09 g/cm^3 because they have no swim bladder. The body densities of both Albacore and yellowfin tuna are over 1.08 g/cm^3. Figure 12.2 shows the body densities of juvenile Pacific blue fin tuna (PBT) of different body sizes (Tamura and Takagi 2009). PBT have a swim bladder. As shown in Fig. 12.2, the larger the size, the greater the body density, and the difference between the maximal and minimal values of body density decreases on increasing the total length. The large differences in the small sizes were caused by the ontogeny of juveniles with uninflated swim bladders occurring in the early days after hatching, and their swim-bladder volume being relatively larger than that of other tuna. The body density of PBT converges to ~1.05–1.06 g/cm^3 in the adult stage.

As mentioned above, scombrid fish usually have greater body density than the density of sea water. How do they provide the upward force to support their submerged weight, and do they have some mechanism to produce this upward force? The hydrodynamics for producing this lift force to prevent sinking will be described next.

Table 12.1. The body density of scombrid fish and the absence of a swim bladder. The buoyant weight range was 2–4 kg for all species except the mackerel (<2 kg) (Magnuson 1978).

	Body density (g/cm³)	Swim bladder
Skipjack tuna	1.090	N
Kawakawa	1.088	N
Yellowfin tuna	1.087	Y
Albacore	1.082	Y
Bigeye tuna	1.047	Y
Chub mackerel	1.054	Y
Pacific Bluefin tuna	1.050–1.060	Y

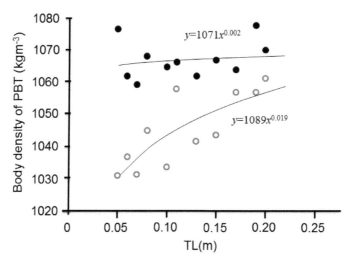

Figure 12.2. Relationship between the body density of a PBT and total length (TL).

Lift Force Compensating for Gravitational Force

Drag force acts on an object moving forward in a fluid, such as in the case of a swimming fish. The drag force acts as a resistance force in the current flow direction. Conversely, the lift force acts on the object perpendicularly to the drag force, and like the drag force, it can only be produced by motion in the flow as in the case of an aircraft (Fig. 12.1). Figure 12.3 shows the outline of a PBT with extended pectoral fins, and is larger than expected. This implies that the pectoral fins produce lift force like an aero foil.

An object that can produce a lift force must not necessarily have an aero foil shape. The body of the fish can produce a lift force depending on its angle of attack (this is the angle between the body axis and the current flow direction). We usually experience pushing against a wall and being moved in the opposite direction of the action. This is well known as the principle of action and reaction. This principle can be applied not only to solid objects but also to fluids. For instance, when a hand pushes water, the hand is pushed back by the water. If a fish pushes against the ambient water, the fish is

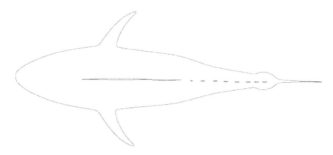

Figure 12.3. Profile line of a PBT from the top view.

pushed back by the water with a force equal in magnitude to the change in momentum of the water. Thus, if the body of a fish changes the momentum of the ambient water, the lift force acts on not only the pectoral fins but also the body.

When we observe the morphological feature of PBT, the caudal keel, which is a lateral ridge on the caudal peduncle, appears to provide stability for the pitching moment by producing lift on the aft section during swimming (Nauen and Lauder 2001a; Nauen and Lauder 2002). If lift force can be produced by the pectoral fins and the body to support the submerged weight, tuna can maintain their swimming depth. A higher lift force results in upward swimming, and they must swim continuously to avoid sinking. The following equation expresses the lift force:

$$FL = 0.5\rho_w C_L SU^2, \qquad\qquad \text{Eq. 2}$$

where F_L is the lift force, ρ_w is the sea water density, S is the wetted surface area, U is the swimming speed, and C_L is the coefficient of the lift force that depends on the body shape. The greater the speed and C_L, the greater the lift force produced. Figure 12.4 shows the C_Ls estimated by Computational Fluid Dynamics (CFD) analysis with different body sizes when the swimming speed was 1.5 TLs⁻¹. CFD is a numerical simulation technique for computing flow and pressure distribution around a body by solving the governing equation for the fluid dynamics. The author has applied CFD analysis to a PBT body during glide and swimming modes and estimated the drag and lift forces and their coefficients.

C_Ls were indicated when the angle of attack (AOA) was 0°. CFD models for both extended and retracted pectoral fins were developed, and the C_Ls were compared. As shown in Fig. 12.4, C_Ls for the model with the retracted pectoral fins were small, but in case of extended fins, C_Ls were more than double that of retracted fins for all body sizes. In particular, for 15–30 cm TL, the C_Ls with extended pectoral fins were much

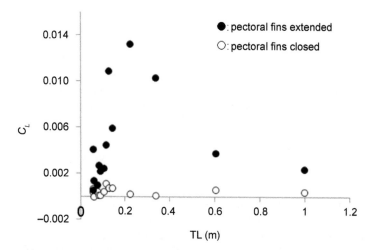

Figure 12.4. Coefficients of lift force for different PBT body sizes when the swimming speed was 1.5 TLs⁻¹ when the angle of attack was 0°. The PBT body did not make any undulatory motion or tail beating, so that the C_Ls here were estimated for glide mode.

greater. The possible reason was that the area of the pectoral fin was relatively larger in these body sizes. In the full lifecycle aquaculture of the PBT, there is a technical problem in that cultured juveniles 30–50 days after hatching often collide with the culturing tank walls because of burst swimming (Fukuda et al. 2010a; Fukuda et al. 2010b). This may relate the balance between the function of swimming control and propulsion.

For tuna to maintain swimming depth, the submerged weight and lift force must be in an equilibrium state so that the lift force can be easily estimated.

The buoyancy of the fish, B_f, is expressed by the following equation:

$$B_f = \rho_w V g,$$ Eq. 3

where V is the volume of the fish.

The weight of the fish is determined by the following equation:

$$W = \rho_f V g,$$ Eq. 4

where ρ_f is the body density.

Thus, Eqs. 3 and 4 can be used to derive the following equation:

$$B_f = W(\rho_w/\rho_f).$$ Eq. 5

If the lift force is equal to the difference between W and B_f, the lift force can be expressed by the following equation:

$$L = W - B_f = (\rho_f - \rho_w)/\rho_f \times W.$$ Eq. 6

If L is a positive value, the body weight is greater than the buoyant force. The greater the body density, the greater the sinking force. For instance, the body mass of a skipjack tuna is 4 kg, and its body density is 1.090 g/cm^3; so, the submerged weight is 0.239 gN. However, in case a PBT has the same body weight, the submerged weight is 0.100 gN, which is less than half that of a skipjack tuna. This means that tuna must produce lift forces using hydrofoils such as pectoral fins and caudal keels to support their submerged weights. Fish are surrounded by water, which has 800 times the density of air. Therefore, in water, 90% of the body weight can be supported by buoyancy, which is much greater than that in air. Thus, a hydrofoil does not need to be as large as the wing of a bird, but scombrid fish often extend their pectoral fins to produce lift during slow swimming.

Drag of a Fish: Friction and Pressure Drag

Resistance acts on a fish body in the direction opposite to the direction of motion during swimming. This resistance is roughly classified into two kinds of drag: friction drag and pressure drag. Friction drag is induced by the friction between fluid and the surface of the body. For fluid flow in the vicinity of a body surface, it is observed that the velocity of fluid flow decreases when approaching the body surface owing to friction (Fig. 12.5). One can see the velocity profile perpendicular to the body surface in the so-called boundary layer, the velocity gradient in which generates frictional drag.

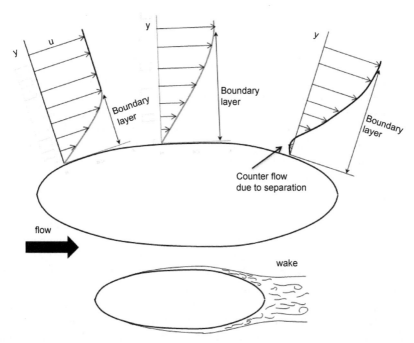

Figure 12.5. Varying velocity profile in the boundary layer along the surface in the downstream direction and the wake generated behind the body.

As shown in Fig. 12.5, the velocity flow in the boundary layer has a counter flow in the vicinity of the body surface. This is caused by a loss of the forward velocity due to friction, and a laminar flow profile is no longer maintained in the aft part of the body. The crush point where counter flow occurs is the so-called 'separation point'. Complex flow with eddies, the so-called 'wake' may develop behind the separation point. The pressure in the wake with eddies is lower than that in the upstream area in front of the body, so that a drag due to the pressure difference, the so-called 'pressure drag', acts on the body. The larger the wake, the greater the pressure drag, and the wake size depends on the body profile. Thus, the body shape of the aquatic animal must be developed to decrease the pressure drag as much as possible. The bodies of tuna are fusiform, which decreases wake size during swimming to reduce pressure drag.

Generally, this drag can be expressed by the following equation, which is nearly the same as that for the lift force:

$$FD = 0.5\rho_w C_D S U^2, \qquad \text{Eq. 7}$$

where C_D is the coefficient of drag.

Magnuson (1978) compared the theoretically estimated drag of the Atlantic mackerel with that found by experiment under the assumption that the total drag can be estimated as the friction drag from the Blasius boundary layer equation. As the result, the theoretical value was in good agreement with the experimental value when the current speed was $1 BLs^{-1}$. The estimated C_D was around 0.004 with S as the wetted surface area. Using CFD analysis (Tamura and Takagi 2009), estimated the C_D with extended and retracted pectoral fins of a PBT when the glide speeds were 1.5 TLs⁻¹ (Fig. 12.6). There were no large differences between the extended and closed pectoral fins, and C_Ds converged to ~0.010 with increasing body size. These values were greater than those estimated by Magnuson (1978) using the boundary layer equation.

There are two types of the boundary layers: one is the laminar flow boundary layer in which the flow pattern is laminar, and the other is the turbulent boundary layer in which the water particles are perpendicularly disturbed. Anderson et al. (2001) investigated the boundary layer on the body surface of a scup. As a result, they experimentally found that the boundary layer was laminar. In a laminar boundary layer, the friction drag is less than that in a turbulent boundary layer. Whether the boundary flow is laminar or turbulent depends on the Reynolds number (Re) as follows:

$$\text{Re} = Ul/v, \qquad\qquad\qquad \text{Eq. 8}$$

where l is the distance between the leading end of the body and the reference point of observation, and v is the dynamic viscosity. The larger the Reynolds number and the rougher the boundary substrate, the more turbulent the boundary layer. Considering the boundary layer on a smooth plate, a laminar boundary layer can be changed to turbulent when $\text{Re} = 4.0 \times 10^5$. This, therefore, causes the frictional drag increase.

The property of the boundary layer on the tuna's body surface has not been investigated to determine whether it is laminar or turbulent. Because the Reynolds number is large owing to the body size and the swimming speed is high, the boundary layer may be turbulent. However, a turbulent flow generally has less separation in the boundary layer; thus, the pressure drag can be less. As C_D and C_L in Figs. 12.4 and 12.6 were estimated for a swimming speed of 1.5 TLs⁻¹, Re was for the smallest individual and for the largest one. Therefore, if the total length is more than 0.6 m, a turbulent boundary layer is likely to develop on the body surface so that the friction drag is relatively high.

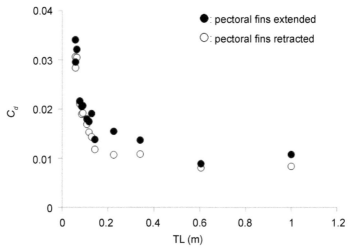

Figure 12.6. Drag coefficient of PBTs with different body sizes during gliding mode when the swimming speed was 1.5 TLs⁻¹ and the AOA was 0°.

Lift Force and Submerged Weight of a Bluefin Tuna

The submerged weight can be estimated using the body and water densities from Eq. 6, and the lift force is expressed by Eq. 2. The vertical movement of the individual is determined by the ratio of the submerged weight to the lift force. The ratio of the lift force to the submerged weigh $TLs^{-1}{}_w$ for different body sizes with extended pectoral fins and without tail beating when the current speed was 1.5 TLs^{-1} was calculated using the CFD analysis (Fig. 12.7). The lift force changes with the swimming speed, the AOA, and the hydrofoil effect such as from pectoral fins. When pectoral fins extend, a faster swimming speed and larger AOA increases the lift force.

The L/S_w of a PBT was in the range of 0.2–0.3 when the angle of attack (AOA) was 0° under the flow speed of 1.5 TLs^{-1}; therefore, this means that a PBT would sink under this condition. However, when the AOA was 4°, the L/S_w was 0.4, which is twice that for an AOA of 0°. Additionally, even if the AOA was 0°, the L/S_w was 0.8 when the flow speed was 3.0 TLs^{-1}. Therefore, juveniles and young adults must extend their pectoral fins and swim faster than 2.0 TLs^{-1} with a certain positive angle of attack to maintain their swimming depth. Tsuda (2009) reported that the cruising speed of young PBTs with fork lengths of 70–80 cm was around 2 BLs^{-1} by *in situ* experiments using archival tags with speed sensors. This partly supports the above estimation by CFD analysis. He and Wardle (1986) reported that large positive AOAs were measured for slow-swimming Atlantic mackerels. Thus, during slow swimming, PBT may fully extend their pectoral fins with relatively steep AOA.

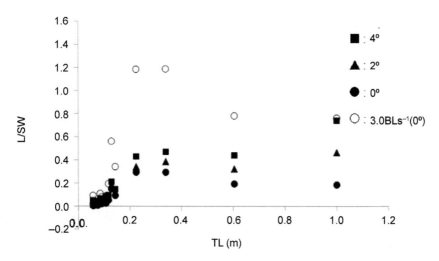

Figure 12.7. L/S_w of PBTs of different sizes with extended pectoral fins and without tail beating when the current speed was 1.5 TLs^{-1}. The densities of the fish body and the ambient water were 1050 and 1025 kg/m³, respectively.

Glide Swimming Mode for Negative Buoyant Fish

Weihs (1973) predicted that negative-buoyancy fish with densities greater than seawater, such as tuna, use the glide swimming mode to save kinetic energy for their migration (Fig. 12.8). The glide is a swimming mode in which fish move obliquely downward using gravitational force without tail beating, as in an air glider. The fish must move downward, but energy for propulsion is not expended during gilding. To recover the original swimming depth, fish must swim upward with tail beating, so that the distance covered during tail beating must be shorter than for continuous horizontal swimming. The shallower the glide angle, α, the longer the horizontal distance covered, as shown in Fig. 12.8. The glide angle α can be expressed by the coefficients of drag and lift force as follows:

$$\alpha = \tan^{-1}(C_D/C_L).$$ Eq. 9

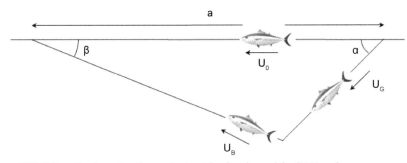

Figure 12.8. Schematic view of continuous horizontal swimming and the GAU mode.

The larger the coefficient of the lift force, the lesser the decrease in the angle of attack, so that the horizontal distance during gliding increases.

Figure 12.9 shows glide angles and glide speeds for different body sizes of PBT, obtained using CFD results. The glide speed can be expressed by the following equation:

$$U_G^2 = \frac{W}{0.5\rho_w S \sqrt{C_D^2 + C_L^2}},$$ Eq. 10

where W is the submerged weight and U_G is the glide speed.

With larger body size and smaller AOA, glide angles increased. However, the glide speeds were around 1.5 TLs^{-1} with little decreasing tendency over all body size ranges.

It is difficult to confirm whether free-ranging tuna have a preference for the glide and upward swimming (GAU) mode; however, Tsuda (2009) installed the archival tags that can measure swimming speed on wild PBTs, released them, and later retrieved the tags. As a result, PBTs often made GAU modes and the averaged glide speed was more than 1.0 FL/s. From this, as predicted by Weihs, it is possible that negative-buoyancy fish have the potential to save energy used for movement.

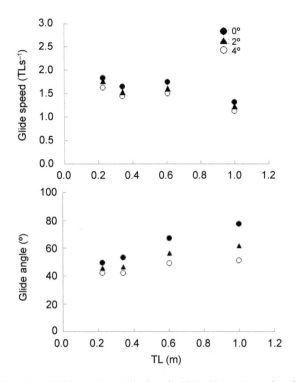

Figure 12.9. Glide angle and glide speed vs. body size of a PBT with varying angles of attack.

The Thrust Mechanisms of Scombrid Fish

It is well known that migratory fish have forms suitable for high-speed swimming. Without direct measurement, Wardle (1989) predicted that the maximum swimming speed of a PBT (the body size of which is around 2 m) is 80 km/hour by measuring its contracting muscle. Walters and Fierstine (1964) directly measured the swimming speed of the yellowfin tuna by putting it on a hook and measuring speed of the fishing line as it swam away. The maximum measured swimming speed was 75 km/hour.

Scombrid fish move forward through propulsive force produced by oscillating the caudal fin. How does the swimming speed change the diagnostic undulatory motion of such tail beating? Figure 12.10 shows the body axes of a juvenile PBT changing over one cycle of motion at a swimming speed of 0.53 ms^{-1}. The body axis in each time step started to bend at 0.4 TL from the snout to oscillate the caudal fin. The envelope curve of the body axes at different swimming speeds gives the amplitude of the body axes, and is shown in Fig. 12.11. This figure shows that the amplitudes for tail beating did not change with swimming speed.

Lindsey (1978) examined swimming forms in various fish and classified the swimming mode from the viewpoint of propulsion. According to the classification by Lindsey, the thunniform mode produces thrust force by laterally oscillating the caudal peduncle. The oscillating motion of PBT during swimming in our experiment

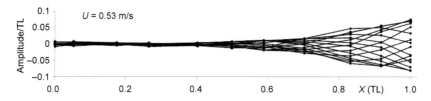

Figure 12.10. Body axes of a juvenile PBT (TL: 19 cm) for each time step over one cycle of oscillating motion at a swimming speed of 0.53 ms^{-1}.

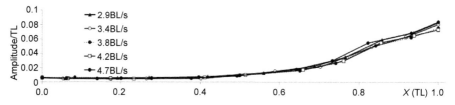

Figure 12.11. Envelope curve of the body axes of a juvenile PBT for the same individual shown in Fig. 12.10. The swimming speed was changed from 2.9 to 4.7 BLs^{-1}.

was different from the thunniform referred to by Lindsey because the bending point was located before that in the thunniform swimming mode. Figure 12.12 shows the relationship between the swimming speed and tail-beating frequency. The frequency increases linearly with increasing swimming speed. This implies that the swimming speed is controlled by the tail-beating frequency, rather than the amplitude. However, Webb (1971) reported that the swimming speed of the rainbow trout was inclined to be controlled by both the frequency and amplitude; thus, it was different from the PBT case.

The propulsive force is based on the principle of action and reaction. Nauen and Lauder (2001b) conducted experiments to visualize the ambient water flow around a swimming mackerel and proved that the jet produced after its caudal fin had oscillated,

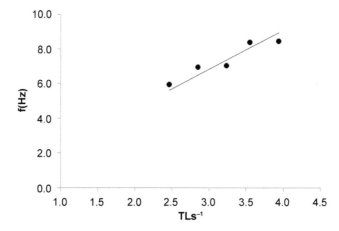

Figure 12.12. Relationship between the swimming speed and tail-beating frequency of a PBT.

resulting in thrust as a momentum reaction. In particular, high-speed migratory fish such as scombrid fish generate propulsive force by producing lift force using the caudal fin, which can function as a hydrofoil. We are generally inclined to think that the lift force is vertical. However, the lift force can work horizontally depending on the current direction, so that it produces a propulsive force by oscillations of the caudal fin. Tuna species have caudal fins, which can efficiently produce a lift force that acts as a propulsive force.

Figure 12.13 (A) shows a schematic view of a caudal fin producing lift force. As shown in the figure, a fish can be swimming at speed U, such that if there is no tail beating, the inlet flow speed into the caudal fin is the same as the swimming speed U. Hence, the lift force produced by the caudal fin is nearly perpendicular to the swimming direction, and the lift force does not act as a propulsive force. However, in tail beating, as the caudal fin moves laterally, the synthesized velocity, w, eventually directs the lift force obliquely such that the lift force acts as a propulsive force for its movement. This mechanism is the so-called theory of an oscillating wing. It can be applied not only to tuna species but also to marine mammals such as dolphins (Lighthill 1970; Nagai et al. 1996; Nakashima and Ono 1996).

Figure 12.13 (B) shows a schematic view of the lift force produced with faster tail beating. As the tail beating in the case of (B) is faster, the synchronized speed, w, is greater, and so the lift force is greater. This may be applied to tuna swimming, and quickly oscillating the caudal fin makes the speed of the inlet flow into the fin relatively high; therefore, greater lift force can be produced by the caudal fin, which acts as a highly functional hydrofoil with a high lift-to-drag ratio. As shown in Fig. 12.12, PBT swimming speed being controlled by the tail-beating frequency was based on the theory of the oscillating wing. Wardle et al. (1989) predicted that the tail-beating frequency of PBT maybe more than 10 Hz. Such a high frequency can produce a fast swimming speed of more than 10 BLs^{-1}. The author examined the speed of the juvenile PBT's burst swimming triggered by light stimulation and observed

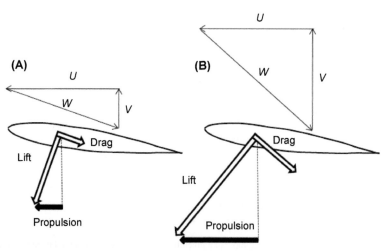

Figure 12.13. Schematic view of the lift force produced by the caudal fin of a PBT in the case of (A) a low oscillating frequency and (B) a high frequency.

a remarkably high swimming speed of more than 30 BLs^{-1}. This means that high burst swimming speeds can be attained by producing a high lift force and tail beating frequency of the caudal fin.

Thrust of a PBT During Swimming

The drag force acts on fish during forward motion, as shown in Fig. 12.1. Therefore, if fish cannot provide a thrust force equal to the drag, the fish cannot continuously swim forward at a constant speed. Thus, if the drag force during swimming is estimated, the thrust force can be derived. The drag on a fish during gliding can be easily measured because drag is the only force acting on it; however, if a thrust force is produced by an oscillating caudal fin, the thrust and drag forces cancel each other. Thus, in such a case, the force acting on an individual cannot be measured by experiments such as those involving a load sensor attached to an individual.

Nishio and Nakamura (2002) examined the hydrodynamics of the hydrofoil of a two-dimensional fish-like body with undulatory motion and measured the drag and thrust forces on it using a special procedure. They found that the stronger the anguilliform-like undulatory motion, the greater the drag force. It was considered that lateral movement strongly affected the drag force during forward movement owing to the apparent increase in the projected area.

The author estimated the coefficient of drag of a PBT during swimming using the previously mentioned CFD analysis (Takagi et al. 2013; Takagi et al. 2010; Takagi et al. 2006). Figure 12.14 shows the coefficients of drag for juvenile PBT during glide and tail-beating modes. K_d is the increase in the drag ratio of the coefficient during tail beating to that during gliding. The coefficient during tail beating was 1.8–2.0 times greater than that during gliding, as shown in Fig. 12.14. This may be caused by lateral oscillations, causing the projection area to the flow direction to increase. It was a hydrodynamic trade-off between thrust and drag during swimming behavior.

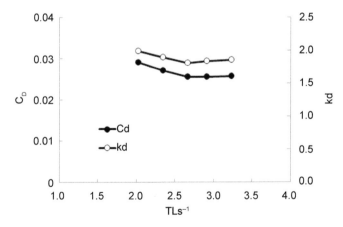

Figure 12.14. Coefficients of the drag force of a juvenile PBT (TL: 19 cm) during tail beating and gliding.

Glide and Upward Swimming Mode to Save Kinetic Energy

To reduce drag, less tail beating is needed during movement because tail beating for propulsion increases drag. How do tuna resolve this dilemma? The glide mode makes it possible to move horizontally; however, the fish must also sink vertically. Therefore, to recover the original swimming depth, upward swimming is needed. Takagi et al. (2013) estimated the kinetic energy of the GAU mode and compared that with the continuous horizontal swimming mode using CFD analysis. As shown in Fig. 12.8, the energy for the horizontal movement for the distance, a, can be expressed as shown in the following equations:

$$E_0 = D_0 a/\eta, \qquad\qquad\qquad \text{Eq. 11}$$

$$D_0 = 0.5\rho S U_0^2 \cdot k C_D, \qquad\qquad\qquad \text{Eq. 12}$$

where D_0 is the drag force acting on the fish during horizontal swimming; U_0, the horizontal swimming speed; C_D, the drag coefficient when the fish is straight without tail beating; η, the conversion efficiency of metabolic energy to kinetic energy ($0 < \eta < 1$); and k, the ratio of the drag, with tail beating (which we shall call active swimming), while producing thrust to the drag of straight gliding (Weihs 1973).

During the upward swimming phase, the propulsive force T with a constant upward swimming speed is equal to the sum of the path-direction component of the submerged weight W and drag force D_b; hence, T can be written as follows:

$$T = D_b + W\sin\beta. \qquad\qquad\qquad \text{Eq. 13}$$

Because the drag force during upward active swimming is k times that during gliding at the same speed, it is not different from active horizontal swimming. Eq. 13 can then be rewritten as follows:

$$T = 0.5\rho S U_B^2 k C_D + W\sin\beta, \qquad\qquad\qquad \text{Eq. 14}$$

Energy must be expended during upward active swimming, but kinetic energy is not consumed during downward gliding, because there is no tail beating. This can be expressed as follows (Weihs (1973)):

$$E = \frac{1}{\eta} T \frac{a\tan\alpha}{\tan\beta + \tan\alpha} \cdot \frac{1}{\cos\beta}. \qquad\qquad\qquad \text{Eq. 15}$$

To compare the energy required for horizontal swimming to GAU swimming modes, E/E_0 can be expressed by Eqs. 11–15 (recalling that $\tan\alpha = (\frac{C_D}{C_L})$), as follows (Takagi et al. 2013):

$$\frac{E}{E_0} = \frac{k C_D + l_1 C_L \tan\beta}{k l_1 (C_L \tan\beta + C_D)/l_0}, \qquad\qquad\qquad \text{Eq. 16}$$

where l_0 and l_1 are the rates of increase or decrease of C_{L0}/C_L and C_{L1}/C_L, respectively. C_{L0} and C_{L1} are the coefficients of the lift force during the horizontal and upward swimming, respectively.

For constant-speed downward gliding, the submerged weight should balance the vertical lift force component during the descent gliding phase; the following equation is then obtained (Weihs 1973):

$$W\cos\alpha = 0.5\rho SC_L U_G^2, \qquad \text{Eq. 17}$$

Here, U_G is the gliding speed as defined by Eq. 10, and C_L is the coefficient of the lift force acting on the fish.

Similarly, the equilibrium state for the lift force component during horizontal and upward swimming (Weihs 1973) can be written as follows:

$$W = 0.5\rho SC_{L0} U_0^2, \qquad \text{Eq. 18}$$

$$W\cos\beta = 0.5\rho SC_{L1} U_B^2, \qquad \text{Eq. 19}$$

Here, U_0 and U_B are the horizontal and upward swimming speeds, respectively.

The following equations can be derived by using Eqs. 18 and 19:

$$\cos\beta = l_1 \left(\frac{U_B}{U_G}\right)^2 \cos\alpha, \qquad \text{Eq. 20}$$

where E/E_0 can be defined by C_D, C_L, k, l_1, and l_0, and $m_1 = U_B/U_G$, as shown in Eq. 16.

The above mentioned formulation means that the parameter sets for the increasing lift/drag ratio during tail beating and the speed ratio U_B/U_G determine the kinetic energy saving by using the GAU mode.

Energy Saving Technique of Glide and Upward Swimming Mode

High lift force is necessary for long-distance travel and to maintain the same swimming depth under a low swimming speed. Thus, a high lift force enables less kinetic energy to be consumed. Therefore, E_0 for horizontal movement was estimated on the basis of the coefficient of the lift force with the pectoral fins fully extended. The maximal values of the lift coefficient during gliding were also used to evaluate the energy-saving ratio E/E_0. Figure 12.15 shows isoplethes maps of the kinetic energy ratio

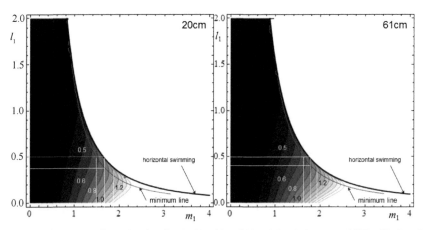

Figure 12.15. Energy saving ratio E/E_0 of PBTs (TL: 20 and 61 cm). Isoplethes map of E/E_0 with changing the speed ratio $m_1 (= U_B/U_G)$ and lift coefficient ratio $l_1 (= C_{L1}/C_L)$.

E/E_0 of PBTs (TL: 20 and 61 cm), with change in the swimming speed ratio, m_1, and the ratio l_1. The bold line in the figure indicates the combination of parameters m_1 and l_1, which allows PBT to swim horizontally. So if parameter (m_1, l_1) is below the line, the PBT cannot continuously swim horizontally at all. For instance, if l_1 is 0.5 in the case of TL = 61 cm (this means the coefficient of the lift during upward swimming is half that during gliding), the upward swimming speed should be less than 1.8 times that during gliding to achieve the GAU mode. If the upward swimming is done when m_1 = 1.8, the most efficient swimming is done when l_1 is around 0.4 because that is the crossing point between m_1 = 1.8 and the minimum line of the isoplethes map, as shown by the minimum line in the figure. The figure shows that increasing the upward swimming speed (m_1 is increasing) leads to an increase in the value of the isoline, so that there was no energy saving in the region where m_1 is greater than 2.0. However, when the upward swimming speed is decreased (m_1 is decreasing), E/E_0 also decreases. However, slow swimming speed requires a much longer time to travel the same distance; thus, the basal metabolic energy consumed during the traveling time and the total energy must be greater than the simple kinetic energy needed for movement. Hence, it is noted that Fig. 12.15 was estimated on the basis of the kinetic energy without considering basal metabolic energy.

The isopleth of E/E_0 can change depending on the Standard Metabolic Rate (SMR), so that before the SMR is taken into account, the SMR value should be normalized by dividing it by the kinetic energy during horizontal continuous swimming. Thus, Fig. 12.16 shows an E/E_0 isoplethes map taking into account the SMR during the time

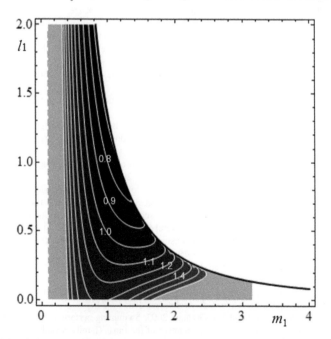

Figure 12.16. Isoplethes map taking into account the SMR for a PBT whose TL was 20 cm, which was assumed to be 0.3 times the kinetic energy during horizontal swimming.

consumed for movement for a PBT (TL: 20 cm). $SMR^* = SMR/E_0$ when $SMR^* = 0.3$. The figure shows a clear optimum point, which was the most efficient swimming mode in the range $1.0 < m_1 < 2.0$. Dewar and Graham (1994) and Sepulveda et al. (2003) reported that the ratio of SMR to the Gross Cost Of Transport (GCOT) of Bonitos and Yellowfin tuna was around 0.3; thus, Fig. 12.16 may provide useful results for the energy saving for the GAU mode of migration. If the movement is in a short time, Fig. 12.15 provides a more efficient swimming mode to save kinetic energy consumed during upward swimming, and its range should converge 1.0–2.0 times faster than the glide speed. However, if long-distance travel is needed, a PBT should provide for long periods and many cycles of the GAU mode.

The morphology of a PBT must determine its hydrodynamic properties; however, the swimming mode can be selected for the best performance for a particular purpose, such as energy saving, as shown in the present study. It is believed that behavior data acquired by many researchers using archival tags could help support and prove the above theoretical study, and it would help elucidate the swimming mode and behavior of aquatic animals.

References

Anderson, E.J., W.R. McGillis and M.A. Grosenbaugh. 2001. The boundary layer of swimming fish. J. Exp. Biol. 204: 81–102.

Dewar, H. and J. Graham. 1994. Studies of tropical tuna swimming performance in a large water tunnel - energetics. J. Exp. Biol. 192: 13–31.

Fukuda, H., S. Torisawa, Y. Sawada and T. Takagi. 2010a. Ontogenetic changes in schooling behaviour during larval and early juvenile stages of Pacific bluefin tuna *Thunnus orientalis*. J. Fish Biol. 76: 1841–1847.

Fukuda, H., S. Torisawa, Y. Sawada and T. Takagi. 2010b. Developmental changes in behavioral and retinomotor responses of Pacific bluefin tuna on exposure to sudden changes in illumination. Aquaculture 305: 73–78.

He, P. and C.S. Wardle. 1986. Tilting behaviour of the Atlantic mackerel, Scomber scombrus, at low swimming speeds. J. Fish Biol. 29: 223–232.

Lighthill, M.J. 1970. Aquatic animal propulsion of high hydro-mechanical efficiency. J. Fluid Mech. 44: 265–301.

Lindsey, C.C. 1978. Form, function, and locomotory habits in fish. Fish Physiology 7: 1–100.

Magnuson, J.J. 1978. Locomotion by Scombrid Fishes: Hydromechanics, Morphology, and Behavior. Academic Press, Inc., London.

Nagai, M., I. Teruya, U. Kazuhiro and T. Miyazato. 1996. Study on an oscillating wing propulsion mechanism (in Japanese). Trans. Jpn. Soc. Mech. Eng. B 62: 200–206.

Nakashima, M. and K. Ono. 1996. Dynamics of Two-Joint Dolphin-like Propulsion Mechanism: 1st Report, Analytical Model and Analysis Method. Trans. Jpn. Soc. Mech. Eng. B 62: 136–143.

Nauen, J.C. and G.V. Lauder. 2001a. Locomotion in scombrid fishes: visualization of flow around the caudal peduncle and finlets of the chub mackerel *Scomber japonicus*. J. Exp. Biol. 204: 2251–2263.

Nauen, J.C. and G.V. Lauder. 2001b. Three-dimensional analysis of finlet kinematics in the chub mackerel (*Scomber japonicus*). Biol. Bull. 200: 9–19.

Nauen, J.C. and G.V. Lauder. 2002. Hydrodynamics of caudal fin locomotion by chub mackerel, *Scomber japonicus* (Scombridae). J. Exp. Biol. 205: 1709–1724.

Nishio, S. and K. Nakamura. 2002. A study on the propulsive performance of fish-like motion using waving wing model (in Japanese). J. Soc. Nav. Archit. Jpn. 191: 17–24.

Sepulveda, C.A., K.A. Dickson and J.B. Graham. 2003. Swimming performance studies on the eastern Pacific bonito Sarda chiliensis, a close relative of the tunas (family Scombridae) I. Energetics. J. Exp. Biol. 206: 2739–2748.

Takagi, T., Y. Tamura and D. Weihs. 2013. Hydrodynamics and energy-saving swimming techniques of Pacific bluefin tuna. J. Theor. Biol. 336: 158–172.

Takagi, T., R. Kawabe, H. Yoshino and Y. Naito. 2010. Functional morphology of the flounder allows stable and efficient gliding: an integrated analysis of swimming behaviour. Aquat. Biol. 9: 149–153.

Takagi, T., Y. Tamura, H. Korte, M. Paschen, S. Okano, Y. Mitsunaga and W. Sakamoto. 2006. Functional morphology of swimming bluefin tuna based on CFD analysis: II. Efficiency of glide and tail-beat swimming modes, The Third International Symposium on Aero Aqua Bio-mechanisms ISABMEC 2006, Okinawa Convention Center, Ginowan, Okinawa, Japan.

Tamura, Y. and T. Takagi. 2009. Morphological features and functions of bluefin tuna change with growth. Fish. Sci. 75: 567–575.

Tsuda, Y. 2009. Swimming Mode of Pacific Bluefin Tuna (in Japanese). The Proceeding of The Japanease Society of Fisheries Science in Spring Meeting 2009 234.

Wardle, C.S., J.J. Videler, T. Arimoto, J.M. Franco and P. He. 1989. The muscle twitch and the maximum swimming speed of giant bluefin tuna, *Thunnus thynnus* L. J. Fish Biol. 35: 129–137.

Webb, P.W. 1971. The swimming energetics of trout: I. thrust and power output at cruising speeds. J. Exp. Biol. 55: 489–520.

Weihs, D. 1973. Mechanically efficient swimming techniques for fish with negative buoyancy. J. Mar. Res. 31: 194.

CHAPTER 13

Schooling Behavior of Pacific Bluefin Tuna

Hiromu Fukuda,[1,*] *Shinsuke Torisawa*[2] *and Tsutomu Takagi*[3]

General Introduction

"The tunny proper, the pelamys, and the bonito penetrate into the Euxine in summer and pass the summer there; as do also the greater part of such fish as swim in shoals with the currents, or congregate in shoals together. And most fish congregate in shoals, and shoal-fishes in all cases have leaders".

As Aristotle mentioned as above in 'The History of Animals' (Aristotle 384-322 BC), shoaling behavior is a common form of social aggregation among teleosts. Individuals in shoals are considered to receive ecological benefits, such as reduced predation risks, improved foraging, and a hydrodynamic advantage (Pitcher and Parrish 1993). On the other hand, fish in shoals also suffer under the increased food competition within the shoal members. Thus, shoaling behavior is considered to arise as a result of trade-offs made by individuals between cost and benefits.

Since the 1980's, humans have been consuming more than 60 million metric tons of marine fishes annually (FAO stat 2010), and most of those fishes form shoals. It can be said that shoal fishes have been targeted by the fishermen who developed their tools, fishing gears and fishing strategies to maximize the fishing efficiency. The shoal fishes also have been targeted by the scientists to evolve the behavioral ecology for the purpose of the management of fisheries resources. More than two thousand years after Aristotle described the migration and shoaling behavior of tuna-like species, the migration of shoals still attracts attention from both the fishing industries and the scientists.

---

[1] National Research Institute of Far Seas Fisheries, 5-7-1 Orido, Shimizu, Shizuoka 424-0902 Japan.
[2] Faculty of Agriculture, Kinki University, 3327-204 Naka-machi, Nara 631-8505 Japan.
[3] Faculty of Fisheries Sciences, Hokkaido University, Minato, Hakodate, Hokkaido, 041-8611 Japan.
* Email: fukudahiromu@affrc.go.jp

The term 'schooling' has an implication for the structure of 'shoaling'; the latter is just a group of fish that remain together, and the former shows directionally and temporally syntonic swimming (polarized swimming) (Pitcher and Parrish 1993). Schooling is therefore one of the behaviors exhibited by fish in shoals. Since bluefin tuna are particularly superior at both large-scale migration and high speed swimming, they exhibit typical schooling behavior throughout their life (Fig. 13.1). The traits of schooling in bluefin tuna change throughout their life according to their body size and behavior strategy in each growth stage. In this chapter, studies on schooling behavior in tuna and tuna-like species with emphasis on our works were reviewed, which addressed the schooling behavior of Pacific bluefin tuna *Thunnus orientalis* (here after PBT). We especially focused on the ontogenetic changes of schooling and the roll of the sensory organs for schooling in the PBT.

Figure 13.1. Schooling age-0 PBT in a sea net cage. This photo was taken by Dr. Ko Fujioka in Tosa-bay, Kochi, Japan.

Ontogenetic Change of Schooling Behavior

Introduction

The onset of schooling has been studied in many species as the key to understanding behavioral strategies in the early life history of pelagic fishes (e.g., Hunter and Coyne 1982; Gallego and Heath 1994; Nakayama et al. 2003; Masuda 2009). These studies have demonstrated the importance of the sense organs, the central nervous system, and development of swimming ability in the onset of schooling in pelagic/neritic fishes living in the sea.

Some field studies have elucidated the ecology of bluefin tuna during their early life stages (details are in Chapter 2. Early life history). PBT is known to spawn mainly in the Northwest Pacific Ocean. The duration between the spawning of eggs and the beginning of the juvenile stage, especially the growth until the onset of the juvenile stage, is considered to be critical for their survival (Tanaka et al. 2006; Satoh et al. 2013). During those stages, they distribute as high density larval populations (patches), in which the patches were advected together by the Kuroshio Current for several days post hatching (Satoh 2010).

However, little information is available about the behavior of individual fish in the open sea during early life stages. Developmental changes in the behavior of individuals may contribute to their survival: within a group, the fish that exhibit rapid growth may be at a competitive advantage when escaping from predators and/or preying on other fish and planktonic prey (Fuiman and Magurran 1994; Hunter 1972). A reason for this paucity of information is that until recently, larval and early juvenile tuna have been difficult to obtain and use for experimental purposes. Recently, a hatchery for PBT was developed in Japan in order to establish a stable supply of young fish for aquaculture and to enhance fish stocks (Sawada et al. 2005). The creation of this hatchery has enabled scientists to access their entire life cycle for experimental purposes under suitable conditions. Using fish from the hatchery, it became possible to observe the ontogenetic changes in the behavior of PBT during early life stages. Here, our study introduces the ontogenetic changes in their schooling behavior from the larval to the juvenile stages (Fukuda et al. 2010; 2011).

Materials and methods

We conducted a behavior observation for PBT larvae and juveniles obtained from the Oshima station of the Fisheries Laboratory, Kinki University, Japan. The behaviors of 10 fish of different ages (20, 23, 25, 27, 29, 31, 33, 36, 40, and 55 days after hatching (DAH)) were recorded by a digital video camera placed above experimental tanks (Fig. 13.2). The experiments were conducted on land in a circular water tank situated in a dark room. Illumination was achieved using a halogen light, which provided light (300 lx) conditions. Using the video images obtained, movements of the 10 individuals examined were digitized in order to obtain time series coordinate data at 0.1 second intervals for periods of 2 minutes. Behavior of each fish was evaluated using 2 indices: swimming speed, and separation swimming index (I_{ss}). I_{ss} is an index of parallel swimming in a fish group (Nakayama et al. 2003). In order to calculate I_{ss}, swimming vectors of an individual and its nearest neighbor were estimated over 1.0 seconds. Distances between paired swimming vectors were then divided by mean absolute values of the two vectors (Fig. 13.3). I_{ss} ranges from zero, when each individual is swimming exactly parallel to its neighbor, to two, when each individual is swimming in the direction opposite that of its neighbor. A mean I_{ss} value when the angle between swimming vectors is between 0 and 180° is used as an expected value

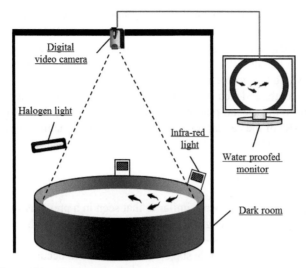

Figure 13.2. Schema of the experimental apparatus.

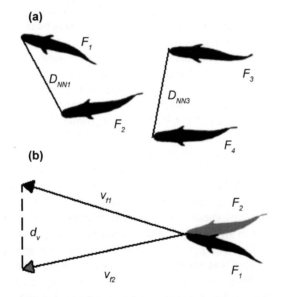

Figure 13.3. Measurement of I_{ss}. Movements of the F1 and F2 in 1 s were expressed as vectors (v_{f1} and v_{f2}). The distance between two vectors is dv. I_{ss} of F1 was defined as $2dv(v_{f1} + v_{f2})^{-1}$.

(1.27) when each individual swims in a random direction. The detailed information about the behavior experiment and the calculation of I_{ss} was described in Nakayama et al. (2003) and Fukuda et al. (2010), respectively.

We also conducted a morphological measurement for 14 different ages from 5 to 50 DAH to relate the morphological change and the development of behavior. For morphological measurements, 10 fish were randomly selected from the group and their body lengths measured. After measurement, each fish was fixed in Bouin's solution for 24 hours. After fixation, the fish were photographed under a light microscope and the Body Length (LB), Body Height (HB), upper jaw length (LJ), caudal peduncle height (HP), caudal fin height (HF), caudal fin length (LF) and area of the caudal fin (AF) were measured. To take into account any contraction induced by the fixation process, the measured lengths and area were converted into a proportion relative to body length (for length) or to the second power of body length (for area). We also estimated the aspect ratio (AR) of the caudal fin as an index of hydrodynamic characteristics due to caudal fin shape (Magnuson 1978).

Results and discussion

Larval and juvenile PBT grew as similar to that seen in a previous study (Miyashita et al. 2001); the mean total length (L_T) at 17 DAH was 8.3 mm; this length has been considered to indicate the post-flexion larval stage (Fig. 13.4A). The metamorphic change from the larval to juvenile stage was considered to occur between 17 and 25 DAH (8.3 and 26.2 mm L_T). Prior to this stage, the larvae swam in a straggly manner and did not exhibit parallel swimming, either in the experimental tank or the rearing tank. During this period, the larvae began to swim against the direction of flow of water in the rearing tank and momentarily exhibited chasing behavior in some couples of individuals. The swimming speed of PBT had not changed dramatically during this stage (8.3 to 26.2 mm L_T). The data distributions of the swimming speed in the larval and early-juvenile stages were distinctive, with some skewness at slow speeds and a minimum speed of zero (Fig. 13.4B). This implies that the larvae swam intermittently at speeds seen during sprinting, and that they had a specific sequence for stopping.

The decrease in I_{ss} from 25 DAH (26.2 mm L_T) onward ($P < 0.05$, n = 120, Steel-Dwass test) suggested the onset of schooling (Fig. 13.4C). The I_{ss} was significantly lower than the expected value for randomly swimming fish at 27 DAH (33.8 mm LT) ($P < 0.05$, n = 120, Steel test). The I_{ss} tended to decrease with growth and were stabilized at 36–40 DAH (77.1–80.1 mm L_T). The lowest values of I_{ss} (0.23), with small standard deviations, were found at 55 DAH and indicated tight schooling in this stage. The swimming speed of PBT increased rapidly after metamorphosis (Fig. 13.4B); the increase in swimming speed exceeded the rate of body growth from 25 to 40 DAH ($26 \cdot 2$–$80 \cdot 1$ mm L_T). The distribution of swimming speed was normal, and the minimum value was not zero from 31 DAH onward. This implies a change in swimming mode from intermittent high-speed swimming to continuous cruising. The swimming speed of fish at 40 and 55 DAH stabilized at approximately 3 L_T s^{-1} (27 and 47 cm s^{-1} respectively), and the distribution of swimming speed was normal.

Morphological analysis indicated some allometrical turning points during this period (Fig. 13.5; N = 140). At around 10 mm L_B, H_B, L_J, and H_F reached peak ratios in relation to L_B (Figs. 13.5a–13.5e). Thereafter, all these indices began to decrease except for H_F. H_B approached constant values around 30 mm L_B and were independent of further growth. Those morphological changes appeared with the metamorphosis from the larval to juvenile stage, and the metamorphosis considered to be completed around 20–30 mm L_B. L_J began to decline from 10 mm L_B, and gradually approached a constant value at 60 mm L_B.

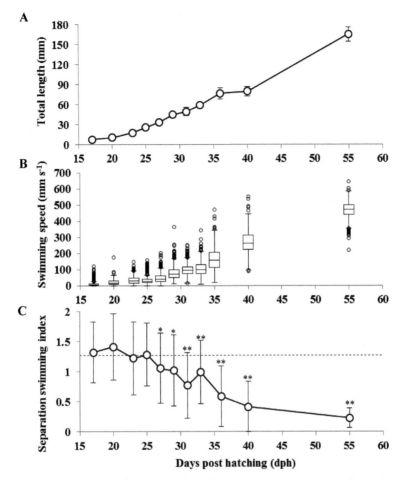

Figure 13.4. Changes in (A) total length, (B) swimming speed, and (C) separation swimming index of PBT during the experimental period. In (A) and (C), the mean values and the standards deviations are used for the open circles and the vertical bars. In (B), the thick line in the middle of the vertical box shows the median. The upper and lower boundaries of the vertical box show the top and bottom quartile lines, respectively. The bars extend up to 1·5 times the box height above and below the vertical box. Points beyond the bars are outliers. In (C), broken line indicates the expected value (1·27) for randomly swimming fish; *(P < 0·05) and **(P < 0·01), value is significantly smaller than 1·27 (steel multiple comparisons).

PBT larvae are known to distribute into patchy groups in the open sea (Satoh 2010; Tanaka et al. 2006). The first two weeks post hatching are suggested to be a critical period for their survival; fish that grow slowly in this period do not survive (Tanaka et al. 2006). Tanaka et al. (2008) also reported a low tolerance to starvation in larval PBT, claiming that these fish easily starve if they are not successful at foraging. Therefore, starvation itself and the consequently increased predation risk greatly contribute to the mortality of bluefin tuna larvae. Kato et al. (2008) suggested that the larvae, which has low swimming ability, might enhance their feeding success by the turbulence caused by sea surface winds.

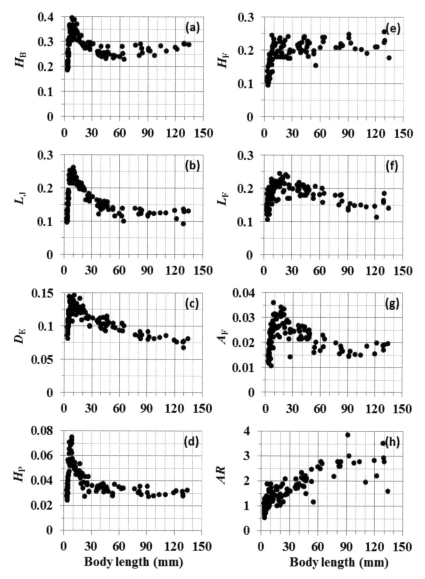

Figure 13.5. Changes in proportion of (a) body height, (b) upper jaw length, (c) eye diameter, (d) caudal peduncle height, (e) caudal fin height, (f) caudal fin length, and (g) area of caudal fin relative to body length (for length) or to the second power of body length (for area) and change in (h) aspect ratio during the experimental period.

Our observation of the morphological development of larval PBT suggested that foraging structures, such as parts of the head, develop at this stage to enable feeding on larger organisms. In contrast, the swimming organs such as the fin or body developed during the larval and early juvenile stages (20 mm L_B). Considering these facts, the relatively slow swimming in the larval stage might be the result of prioritizing the ability to forage in order to avoid starvation. Furthermore, the relatively wide separation of the larvae due to non-polarized swimming might serve to decrease competition for food and the risk of cannibalism (Sabate et al. 2010). For these larvae, however, momentary sprints are important to facilitate escape from predators or to strike planktonic prey (Domeniki and Blake 1997); we observed that these larvae engaged in intermittent high-speed swimming (Fig. 13.4B). Thus, the morphological and behavioral traits of larval PBT serve to enhance their foraging function, as mentioned above.

The swimming organs, especially the fin developed just before this stage. The ratio of the caudal fin area (AF) and L_F showed similar trends: peak values were reached at 15–20 mm L_B and decreased until around 100 mm L_B (Fig. 13.5g). The aspect ratio of the caudal fin (AR) tended to increase over the growth period, rapidly increasing in the period fish measuring 4 to 20 mm L_B and then only gradually increasing in the period that they grew after 20 mm L_B (Fig. 13.5h).

The above mentioned studies revealed the difference in the swimming modes of larvae and juveniles and shows that the onset of schooling occurred at the juvenile stage (Fig. 13.4C). Besides bluefin tuna, the Striped jack *Pseudocaranx dentex*, Chub mackerel *Scomber japonicus*, and Spanish mackerel *Scomberomorus niphonius* begin to form schools immediately after the completion of metamorphosis (Masuda and Tsukamoto 1998; Masuda et al. 2003; Nakayama et al. 2003; 2007; Sabate et al. 2010). Metamorphosis is generally important for the developmental changes in fish behavior.

During the larval stage, their intermittent sprinting swimming and the morphological traits seem to be suitable for striking behavior on prey organisms. They may have prioritized foraging over schooling in this period, in order to enhance growth and avoid starvation because of their high vulnerability to starvation (Tanaka et al. 2008). After the juvenile stage, PBT would start to form a school with a caudal fin adapted to the cruising swimming mode. The morphological development of the juvenile would not enhance only their cruising behavior, but also would affect their turning behavior. Our previous study showed the contribution of suitable morphological features to the development of maneuverability (Fukuda et al. 2011). Owing to the improved maneuverability, the time required for behavior transmission among individuals is shortened and the compactness and polarity of the fish schools progressively improve.

The onset of schooling occurs quite a while after this critical period in PBT (two weeks post hatching). After the critical period, the cruising swimming ability of juvenile bluefin tuna improves in order to facilitate migration. Then, the juvenile exhibits highly polarized swimming in schools to ensure that the fish remained within the group. Schooling may be advantageous for the survivors during this high mortality period, as they begin to migrate in fast-moving schools.

Effect of Light Intensity on Schooling Behavior

Introduction

In open waters, fish in a school synchronize with more than several hundred individuals. Swimming in a coordinated manner requires recognition of the swimming direction of neighbors and a decision on whether to follow these neighbors or to be followed by them. The basis of this leading/following is behavior transmission among neighboring individuals in a school (Nakayama et al. 2007; Fukuda et al. 2011), and it ensures the cohesion and plasticity of schools (Gerlotto et al. 2006). Since PBT had negative buoyancy, they swam even at night to produce the lift force. Accordingly, PBT has to perceive the motion of neighbors even in a dark condition to ensure that they remain within the group.

Vision and lateral lines have been indicated to be the most important part of sensory organs for school formation. This was addressed by Pitcher et al. (1976) and Partridge and Pitcher (1980), in that blinded saithe *Pollachius virens* can form schools. Beside the above studies, the importance of the vision and lateral lines for schooling was often studied (Cahn 1972; Glass et al. 1986; Higgs and Fuiman 1996; Miyazaki et al. 2000), respectively. About scombroid fishes, Cahn (1972) demonstrated that lateral lines were essential to adjust the distance and direction among individuals for Kawakawa *Euthynnus affinis*. Glass et al. (1986) studied a light intensity threshold for schooling in Atlantic mackerel *Scomber scombrus* and demonstrated that they could not form well-coordinated schools at light intensities below 1.8×10^{-7} µEs^{-1} m^{-2}. Suzuki et al. (2005) showed that Chub mackerel *S. japonicus* could form schools using only either vision or their lateral lines. These studies indicated that relative importance of vision and lateral lines for school formation might be different among the species. Our study about the effect of light intensity on schooling to reveal the role of vision for schooling are introduced here (Torisawa et al. 2011; Fukuda et al. 2014).

Materials and methods

The methods of observing schooling behavior of juvenile PBF were mentioned earlier. The behaviors of 10 individuals of different ages (20, 23, 25, 27, 29, 31, 33, 36, 40, and 55 DAH) were observed using a digital video camera under several light conditions. For age 20 to 36 DAH fish, behavior observations were conducted under both dark (<0.01 lx) and light (300 lx) conditions. For ages 40 to 55 DAH, six levels of light intensity condition were set: 0.01 lx, 0.05 lx, 0.5 lx, 5 lx, 30 lx and 300 lx. Under light conditions of less than 5 lx, infrared light-emitting diodes were used with a wavelength range >940 nm, and the digital video camera night function was used to record the fish. In these experiments, a shallow water depth was maintained to enable video recording under dark conditions using infrared light. Using the video images obtained, the movements of the 10 individuals examined were digitized in order to obtain time series coordinate data at 0.1 second intervals for periods of 2 minutes. The behavior of each fish was evaluated using two indices of swimming speed and I_{ss}.

Results and discussion

The lighting conditions had a strong influence on the swimming speed throughout the experiment period, at which they swam at a slower speed in dark conditions than in light conditions (Fig. 13.6). Especially under dark conditions of 20–29 DAH, they did not actively swim and mode values of swimming speed were approximately zero

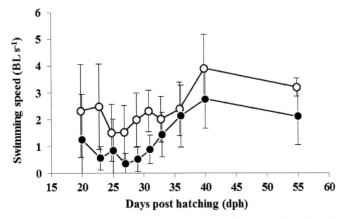

Figure 13.6. Changes in mean swimming speed during the experimental period. Vertical bars indicate standard deviations. White and black circles indicate swimming speed under light (300 lx) and dark (0.01 lx) conditions, respectively.

(Fig. 13.7a–e). This implies that they spent more than half of the recording time for floating without swimming in the experimental tank. Specimens examined under dark conditions at 31 DAH and onwards exhibited active swimming, and mode values of swimming speed tended to increase during this period (Fig. 13.7f–j). As described earlier, they showed intermittent high-speed sprinting under light conditions until 25 DAH (Fig. 13.7a–c), and thereafter, they showed a schooling behavior with a swimming mode of continuous cruising (Fig. 13.7d–j, Fig. 13.8).

Although PBT showed schooling behavior under light conditions after 27 DAH (Fig. 13.8), they did not show it under dark conditions (<0.01 lx) throughout the experimental period (until 55 DAH). The values of I_{ss} under dark conditions decreased from 1.54 to 1.09 (Fig. 13.8), indicating nearly random movement throughout the experiment period. Thus, the juveniles did not form polarized schools at dark conditions. Although they did not form schools under these dark conditions, they formed schools in a synchronized manner under dim light conditions such as 0.05–5 lx. In case of 40 DAH, I_{ss} marked the lowest value at 0.5 lx (Fig. 13.9a) and ranged from 0.32 to 0.64 at light intensities between 0.5 and 300 lx. This indicated that they keep its polarity of schooling in these conditions. At ≤0.5 lx, I_{ss} values significantly increased from 0.32 to 1.20 with decreasing light intensity. Values of I_{ss} for the 40 DAH tended to shift at a light intensity threshold of 0.5 lx. In case of 55 DAH, the values of I_{ss} ranged 0.27 to 0.43 at light intensities between 0.05 and 300 lx (Fig. 13.9b). Thus, juveniles formed well-polarized schools at these levels of light intensity. At this age, the I_{ss} marked high value (1.09) only in the condition of 0.01 lx. Values of I_{ss} for the 55 DAH fish tended to shift at a light intensity threshold of 0.05 lx.

The above mentioned results clearly indicate that light intensity strongly affects schooling behavior of PBT, which mainly depend on their vision to perceive the motion of neighbors and a situation to surround them. How the light intensity level affects their behavior was different by growth stage. For fish before the onset of schooling (25 DAH), dark conditions made PBT swim inactively (Fig. 13.7). Takashi et al.

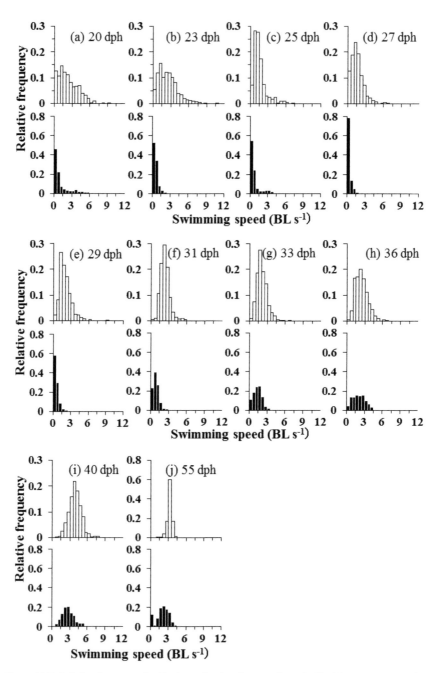

Figure 13.7. Relative frequency distributions of swimming speed standardized by body length of fish. White and black bars indicate light (300 lx) and dark (0.01 lx) conditions, respectively.

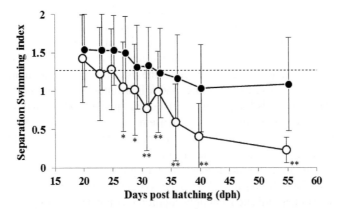

Figure 13.8. Changes in separation swimming index during the experimental period. Vertical bars indicate standard deviations. Empty and solid circles indicate SSI under light and dark conditions, respectively. Broken line indicates the expected value for randomly swimming fish (1.27), and asterisks below bars indicate values significantly smaller than that value (*p < 0.05, **p < 0.01; Steel multiple comparisons).

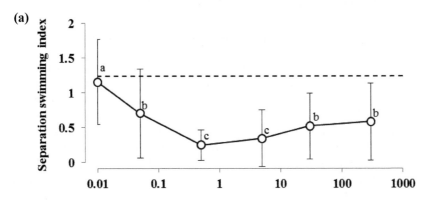

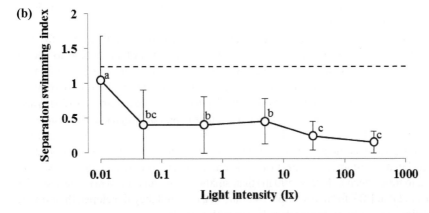

Figure 13.9. Separation swimming index under different light intensity (a: 40 DAH, b: 55 DAH). Significantly different groups are indicated by lower-case letters (p < 0.05, Steel-Dwass multiple comparisons). The dotted line indicates the expected value (1.27) for randomly swimming fish.

(2006) measured the volume of the swim bladder of the PBT larvae and reported that the swim bladder is functional at this stage and larger at night than during the day. At night, if the larvae and small juveniles floated without swimming by expanding the swim bladder, every individual should be advected together by ambient water current. This may present an advantage that reduces the risk of dispersion at night.

Fish after the onset of schooling, had a light intensity threshold for constant schooling due to their visibility (Fig. 13.9). Further, juveniles showed developmental changes in behavioral adaptation in response to low light intensity during the two weeks between 40 to 55 DAH. The juveniles of 55 DAH showed constant schooling under 0.05 lx where 40 DAH juveniles did not form it. The vision dependent schooling behavior was reported in other species such as gulf menhaden *Brevoortia patronus* and walleye pollock *Theragra chalcogramma* (Higgs and Fuiman 1996; Ryer and Olla 1998). Those studies indicated that light conditions could be a limiting factor for their diving behavior and/or habitat selection. The developmental change in a light intensity threshold for constant schooling, which was observed during the growth between 40 and 55 DAH, probably reflects the tendency of the fish to continue schooling 24 hours a day, or to alter their habitat from shallow to deeper ranges.

Schooling Behavior of PBT in the Open Ocean

PBT start to form schools right after the completion of metamorphosis from larvae to juveniles (around 25 DAH). Until then, they distribute as high density larval populations (patches), and the patches advect together by the current for at least a few weeks post hatching (Satoh 2010). Their swimming ability is not high at this stage, and that they do not swim aggressively by expanding their swim bladder to reduce their negative buoyancy at night instead of swimming with poor visibility.

At the same time as the onset of schooling (25 DAH), their swimming ability improved drastically (Fig. 13.4B). At around 40 DAH, their swimming speed reached 3.6 BL s^{-1} and further improvement was not confirmed after that. In these stages, they are not only advected but also start to migrate by themselves while maintaining a well-coordinated school.

Some of the field studies using recently developed small archival tags for tuna and tuna-like species showed a possibility that tuna expand their habitat to deeper seas with growth. Mitsunaga et al. (2013) reported the early juvenile Yellowfin tuna *Thunnus albacares* (20.5–24.0 cm fork length [L_F]) maintained well-coordinated schooling behavior at night in shallow water depth (<30 m). Okamoto et al. (2013) also reported coordinated vertical migration of a couple of age-0 Skipjack tuna *Katsuwonus pelamis* (42.5 and 44.0 cm L_F) during both daytime at the water depth of around 100 m and nighttime at the range of 20–90 m deep. Kitagawa et al. (2003) examined the diving behavior of immature PBT (47–144 cm L_F) and reported that larger individuals dived to much deeper depths (500 m) than smaller individuals (100 m). These studies indicated that PBT formed schools 24 hours in a day, and they developed their diving behavior to a depth of deeper than 100 m with growth.

Our experiments revealed the beginning of the behavioral adaptation in response to darker conditions (deeper conditions). Using the light intensity thresholds for school formation obtained by our experiment, the water-depth limits for forming well-organized schools could be calculated using the irradiance transmittance as follows;

$$I_z = I_0 \cdot e^{-Kz}$$ Eq. 1

Here, I_0 was set as 10,000 lx assuming fine weather in the daytime with a 4% reflection at the sea water surface and K was set as 0.118 assuming the optical water type of oceanic III (Jerlov 1976). The water-depth limits for forming well-organized schools were 84 m and 103 m for 40 and 55 DAH fish, respectively. In contrast, under full-moon conditions at a light intensity level of 0.2 lx, 40 DAH juveniles can form organized schools only at the sea surface, and 55 DAH juveniles can form well-organized schools at 11.4 m depth in oceanic III water. The current results indicate that early juvenile stage of PBT form better-organized schools during the night and become more adapted to deeper seas with growth, which coincide with their migration to the deep ocean.

After the early juvenile stages, PBT is known to migrate along the coastal area of Japan and some of the age-one fish leave the Western Pacific Ocean (WPO) for trans-Pacific migration to the Eastern Pacific Ocean (EPO) (Inagake et al. 2001). Recently, we observed schooling behavior of age-one fish using a fine-scale acoustic positioning and telemetry system (Fukuda, unpubl. data). PBT would migrate over the Pacific Ocean from WPO to EPO and vice versa within a school. Although there is no direct observation of schooling behavior of PBT older than age-two, fishery data have some information about their school. Since PBT are largely caught by the purse-seiners, good quality fishing data is accumulated for stock assessment purpose (ISC PBFWG 2012). Fukuda et al. (2012) examined the frequency distributions of the body length of PBT caught by purse-seiners that operated in the Sea of Japan for each set and reported that the sets, which had a mode at age-three, showed narrow distribution and were mostly occupied by a single age class. On the other hand the sets, which caught mainly age-six or older fish, did not have clear modes and multiple age class PBT contained within the same set. The relative abundance of various year-classes in the local fishing ground should affect the age structures of caught PBT by a purse seine though. The age-structure of PBT schools might alter at around age-three from a single age-structure to a multiple-age structure as seen in the above study.

According to Partridge et al. (1983), adult giant Atlantic Bluefin Tuna (ABT; around 2.5 m of fish length) formed schools and their school structure was suited for hunting. Newlands and Porcelli (2008) addressed the several school formation structures of ABT and reported that the trade-off between the hydro-dynamical and visual advantage, while affecting school internal structure, explains the shape of observed formations as a collective trade-off between school compactness and elongation. Although these studies did not make reference to PBT, they showed that the large adult bluefin tuna form schools.

Most recently, the International Scientific Committee for Tuna and Tuna-like Species in the North Pacific Ocean (ISC) conducted stock assessment for PBT to depict the population dynamics during 1952 to 2010 (ISC PBFWG 2012). It was estimated that Spawning Stock Biomass (SSB) in 2010 were at, or near, their lowest level and SSB has been declining for over a decade. Among the fisheries to catch PBT, the purse seiners historically have occupied the largest catch and the impact analysis suggests that a large footprint on PBT stock was marked by this gear (ISC PBFWG 2013). These purse seiners have been able to catch huge PBT schools by rapidly encircling them with advanced technology such as sophisticated sonar, which improves efficiency to catch as large a school as possible. It is certain that the schooling behavior, which developed as a major anti-predatory behavior, was vulnerable to human commercial fisheries. Currently, a number of studies have begun to address schooling behavior to improve sustainable usage of fish stocks, for example developing an abundance index (i.e., Willis 2008; Newlands and Porcelli 2008), revealing migration in an open ocean or utilizing Fish Aggregation Devices (i.e., Hilborn 2011; Mitsunaga et al. 2013; Schaefer et al. 2013). Also, they developed full-cycle aquaculture technique (i.e., Sabate et al. 2010; Fukuda et al. 2010). Those and further investigations for PBT schools will contribute to the reconstruction and sustainable use of this stock.

References

Aristotle, 384-322 BC. The History of Animals. Translated by D'Arcy Wentworth Thompson, eBooks@ Adelaide, The University of Adelaide Library.

Cahn, P.H. 1972. Sensory factors in the side-to-side spacing and positional orientation of the tuna, *Euthynnus affinis*, during schooling, Fishery Bulletin 70: 197–204.

Domenici, P. and R.W. Blake. 1997. The kinematics and performance of fish fast-start swimming. J. Exp. Biol. 200: 1165–1178.

FAO. 2010. Fishery and aquaculture statistics. FAO yearbook. ISSN 2070-6057.

Fuiman, L.A. and A. Magurran. 1994. Development of predator defences in fishes. Rev. Fish Biol. Fisher. 4: 145–183.

Fukuda, H., S. Torisawa, Y. Sawada and T. Takagi. 2010. Developmental changes in behavioral and retinomotor responses of Pacific bluefin tuna on exposure to sudden changes in illumination. Aquaculture 305: 73–78.

Fukuda, H., S. Torisawa, Y. Sawada and T. Takagi. 2010. Ontogenetic changes in schooling behaviour during larval and early juvenile stages of Pacific bluefin tuna *Thunnus orientalis*. J. Fish Biol. 76(7): 1841–1847.

Fukuda, H., Y. Sawada and T. Takagi. 2011. Ontogenetic changes in behavior transmission among individuals in the schooling of Pacific bluefin tuna *Thunnus orientalis*. Aquat. Living Resour. 24: 113–119.

Fukuda, H., M. Kanaiwa, I. Tsuruoka and Y. Takeuchi. 2012. A review of the fishery and size data for the purse seine fleet operating in the Japan Sea. Working paper submitted to the ISC PBFWG Meeting, November 2012, Honolulu, USA. ISC/12/PBFWG-3/03.

Fukuda, H., S. Torisawa and T. Takagi. 2013. Ontogenetic changes in schooling behavior and visual sensitivity during larval and juvenile stages in Pacific bluefin tuna, *Thunnus orientalis*. Bull. Fish. Res. Agen. 38: 135–139.

Gallego, A. and M.R. Heath. 1994. The development of schooling behaviour in Atlantic herring *Clupea harengus*. J. Fish Biol. 45: 569–588.

Gerlotto, F., S. Bertrand, N. Bez and M. Gutierrez. 2006. Waves of agitation inside anchovy schools observed with multibeam sonar: a way to transmit information in response to predation. ICES J. Mar. Sci. 63-8: 1405–1417.

Glass, C.S., C.S. Wardle and W.R. Mojsiewcz. 1986. A light intensity threshold for schooling in the Atlantic mackerel, *Scomber scombrus*. J. Fish Biol. 29 Supp. A: 71–81.

Hilborn, R. 2011. Modeling the Stability of Fish Schools: Exchange of individual Fish between School of Skipjack Tuna (*Katsuwonus pelamis*). Canadian Journal of Fisheries and Aquatic Sciences 48(6): 1081–1091.

Higgs, D.M. and L.A. Fuiman. 1996. Light intensity and schooling behavior in larval gulf menhaden. J. Fish Biol. 48: 979–991.

Hunter, J.R. 1972. Swimming and feeding behavior of larval Anchovy *Engraulis mordax*. Fishery Bulletin 70(3): 821–838.

Hunter, J.R. and K.M. Coyne. 1982. The onset of schooling in Northern anchovy larvae, *Engraulis mordax*. CalCOFI Report XXIII: 246–251.

Inagake, D., H. Yamada, K. Segawa, M. Okazaki, A. Nitta and T. Itoh. 2001. Migration of Young Bluefin Tuna, *Thunnus orientalis* Temminck et Schlegel, through archival tagging experiments and its relation with oceanographic conditions in the Western North Pacific. Bull. Nat. Res. Inst. Far Seas Fish. 38: 53–81.

Jerlov, N.G. and J. Piccard. 1959. Bathyscaph measurement of daylight penetration into the Mediterranean. Deep-Sea Res. 5: 201–204.

Kato, Y., T. Takebe, S. Masuma, T. Kitagawa and S. Kimura. 2008. Turbulence effect on survival and feeding of Pacific bluefin tuna *Thunnus orientalis* larvae, on the basis of a rearing experiment. Fish. Sci. 74: 48–53.

Kitagawa, T., S. Kimura, H. Nakata and H. Yamada. 2003. Diving patterns and performance of Pacific bluefin tuna *Thunnus thynnus orientalis* as recorded by archival tags. Otsuchi Mar. Sci. 28: 52–58.

Masuda, R. 2009. Behavioral ontogeny of marine pelagic fishes with the implications for the sustainable management of fisheries resources. Aqua-Bio Science Monographs 2: 1–56.

Masuda, R. and K. Tsukamoto. 1998. The ontogeny of schooling behaviour in the striped jack. J. Fish Biol. 52: 483–493.

Masuda, R., J. Shoji, S. Nakayama and M. Tanaka. 2003. Development of schooling behavior in Spanish mackerel *Scomber omorusniphonius* during early ontogeny. Fish. Sci. 69: 772–776.

Mitsunaga, Y., C. Endo and R. Babaran. Schooling behavior of juvenile yellowfin tuna *Thunnus albacores* around a fish aggregating device (FAD) in the Philippines. Aquat. Living Resour. 26: 79–84.

Miyashita, S., Y. Sawada, T. Okada, O. Murata and H. Kumai. 2001. Morphological development and growth of laboratory-reared larval and juvenile *Thunnus thynnus* (Pisces: Scombridae). Fish. B-NOAA. 99: 601–616.

Miyazaki, T., S. Shiozawa, T. Kogane, R. Masuda, K. Maruyama and K. Tsukamoto. 2000. Developmental changes of the light intensity threshold for school formation in the striped jack *Pseudocaranx dentex*. Mar. Ecol. Prog. Ser. 192: 267–275.

Nakayama, S., R. Masuda, J. Shoji, T. Takeuchi and M. Tanaka. 2003. Effect of prey items on the development of schooling behavior in chub mackerel *Scomber japonicus* in the laboratory. Fish. Sci. 69: 670–676.

Nakayama, S., R. Masuda and M. Tanaka. 2007. Onsets of schooling behavior and social transmission in chub mackerel *Scomber japonicus*. Behav. Ecol. Sociobiol. 61: 1383–1390.

Newland, N. and Porcelli. 2008. Measurement of the size, shape and structure of Atlantic bluefin tuna schools in the open ocean. Fish. Res. 91-1: 42–55.

Okamoto, S., H. Kiyofuji, M. Takei, H. Fukuda, Y. Ishikawa, H. Igarashi, S. Masuda and N. Sugiura. 2013. Vertical swimming behavior and habitat of age-0 skipjack tuna, *Katsuwonus pelamis*, based on archival tag data in the subtropical North Pacific during winter. Bull. Japan. Soc. Fish. Oceanogr. 77 (3): 155–163. In Japanese except abstract.

Partridge, B.L. and T.J. Pitcher. 1980. The sensory basis of fish schools: relative roles of lateral line and vision. J. Comp. Physiol. A 135: 315–325.

Partridge, B.L., J. Johansson and J. Kalish. 1983. The structure of schools of giant bluefin tuna in Cape Cod Bay. Environ. Biol. Fish. 9(3/4): 253–262.

ISC Pacific Bluefin Tuna Working Group. 2012. Stock assessment of Pacific Bluefin Tuna in 2012. Available at: http://isc.ac.affrc.go.jp/pdf/Stock_assessment/Stock%20Assessment%20of%20Pacific%20 Bluefin%20Assmt%20Report%20-%20May15.pdf.

ISC. 2013. Report of Pacific Bluefin Tuna Working Group Meeting 14–15 July, 2013, Busan, Korea. Plenary Report Annex 14 (In) Report of Pacific Bluefin Tuna Working Group Workshop Appendix 2: 25–72. Available at: http://isc.ac.affrc.go.jp/pdf/ISC13pdf/Annex%2014%20PB%20final%20version_0.pdf.

Pitcher, T.J., B.L. Partridge and C.S. Wardle. 1976. A blind fish can school. Science 194: 963–965.

Pitcher, T.J. and J.K. Parrish. 1993. Function of shoaling behaviour in teleosts. pp. 363–439. *In*: T.J. Pitcher (ed.). Behaviour of Teleost Fishes, 2 ed., Chapman & Hall, New York.

Ryer, C.H. and B.L. Olla. 1998. Effect of light on juvenile walleye pollock shoaling and their interaction with predators. Mar. Ecol. Prog. Ser. 167: 215–226.

Sabate, F.D., Y. Sakakura, Y. Tanaka, K. Kumon, H. Nikaido, T. Eba, A. Nishi, S. Shiozawa, A. Hagiwara and S. Masuma. 2010. Onset and development of cannibalistic and schooling behavior in the early life stages of Pacific bluefin tuna *Thunnus orientalis*. Aquaculture 301: 16–21.

Satoh, K. 2010. Horizontal and vertical distribution of larvae of Pacific bluefin tuna *Thunnus orientalis* in patches entrained in mesoscale eddies. Mar. Ecol. Prog. Ser. 404: 227–240.

Satoh, K., Y. Tanaka, M. Masujima, M. Okazaki, Y. Kato, H. Shono and K. Suzuki. 2013. Relationship between the growth and survival of larval Pacific bluefin tuna, *Thunnus orientalis*. Mar. Biol. 160: 691–702.

Sawada, Y., T. Okada, S. Miyashita, O. Murata and H. Kumai. 2005. Completion of the Pacific bluefin tuna *Thunnus orientalis* (Temminck et Schlegel) life cycle. Aquac. Res. 36: 413–421.

Schaefer, K.M. and D.W. Fuller. 2013. Simultaneous behavior of skipjack (*Katsuwonus pelamis*), bigeye (*Thunnus obsesus*), and yellowfin (*T. albacares*) tunas, within large multi-species aggregations associated with drifting fish aggregating devices (FADs) in the equatorial eastern Pacific Ocean. Mar. Biol. DOI: 10.1007/s00227-013-2290-9.

Suzuki, K., T. Takagi, S. Torisawa, H. Fukuda, O. Murata, S. Yamamoto and K. Miyashita. 2005. Math. Phys. Fish. Sci. 3: 15–20. ISSN 1348-6802 (In Japanese except Abstract).

Takashi, T., H. Kohno, W. Sakamoto, S. Miyashita, O. Murata and Y. Sawada. 2006. Diel and ontogenetic body density change in Pacific bluefin tuna, *Thunnus orientalis* (Temminck and Schlegel). larvae. Aquac. Res. 37: 1172–1179.

Tanaka, Y., K. Satoh, M. Iwahashi and H. Yamada. 2006. Growth-dependent recruitment of Pacific bluefin tuna *Thunnus orientalis* in the northwestern Pacific Ocean. Mar. Ecol. Prog. Ser. 319: 225–235.

Tanaka, Y., M. Mohri and H. Yamada. 2007. Distribution, growth and hatch date of juvenile Pacific bluefin tuna *Thunnus orientalis* in the coastal area of the Sea of Japan. Fish. Sci. 73: 534–542.

Tanaka, Y., K. Satoh, H. Yamada, T. Takebe, H. Nikaido and S. Shiozawa. 2008. Assessment of the nutritional status of field-caught larval Pacific bluefin tuna by RNA/DNA ratio based on a starvation experiment of hatchery-reared fish. J. Exp. Mar. Biol. Ecol. 354: 56–64.

Torisawa, S., H. Fukuda, K. Suzuki and T. Takagi. 2011. Schooling behaviour of juvenile Pacific bluefin tuna *Thunnus orientalis* depends on their vision development. J. Fish Biol. 79: 1291–1303.

Willis, J. 2008. Simulation model of universal law of school size distribution applied to southern bluefin tuna (*Thunnus maccoyii*) in the Great Australian Bight. Ecol. Model. 213(1): 33–44.

CHAPTER 14

Visual Physiology

Shinsuke Torisawa,[1,*] *Hiromu Fukuda*[2] and *Tsutomu Takagi*[3]

Vision is an important sensory system for most fish species. Fish eyes are similar to the eyes of terrestrial vertebrates such as birds and mammals, except that they have highly spherical lens. The retinas generally have both rod and cone cells (for scotopic and photopic vision, respectively), and most species have colour vision. Unlike humans, fish normally adjust their focus by moving the lens closer to, or further from, the retina.

Tuna (family Scombridae) and billfish (families Xiphiidae and Istiophoridae) have large eyes, which grow throughout their lives. The eyes of tuna and billfish are well developed (Tamura and Wisby 1963; Kawamura et al. 1981a; Fritsches et al. 2000; 2003a), suggesting that vision is an important sensory element for these diurnal hunters (Kawamura 1994). However, because of the difficulty of behavioural experiments and lack of knowledge of their visual capabilities, they were considered to have low visibility to dim light and colour blind until the end of the 20th century (Munz and McFarland 1973; Kawamura et al. 1981b).

Colour Vision

Do tuna and billfish distinguish colours? This question has occupied anglers and scientists for many years. Establishing that colour vision in this group of fish exists is particularly difficult, because behavioural colour discrimination experiments (the 'final proof' of colour vision ability in any animal) are impossible to undertake with such large, open ocean fish. Instead, scientists have focused on establishing whether billfish and tuna have the necessary hardware in their eyes to discriminate colours.

---

[1] Department of Fisheries, Faculty of Agriculture, Kinki University, 3327-204 Naka-machi, Nara 631-8505, Japan.
[2] National Research Institute of Far Seas Fisheries, 5-7-1 Orido, Shimizu, Shizuoka 424-0902, Japan.
[3] Faculty of Fisheries Sciences, Hokkaido University, Minato, Hakodate, Hokkaido 041-8611, Japan.
* Email: ns_torisawai@nara.kindai.ac.jp

Neither a study of extracted visual pigments (Munz and McFarland 1973), nor an electrophysiological (S-potential) investigation (Kawamura et al. 1981b), has found evidence of the presence of more than one visual pigment in the retina of tuna and billfish, leading to the conclusion that this group of fish is unlikely to distinguish colours. Fritsches et al. (2003a) measured the absorbance spectra of visual pigments within individual photoreceptors in different regions of the retina in the striped marlin *Tetrapturus audax* (Philippi), using microspectrophotometry. Three different cone photoreceptor visual retinal pigments were found in the area centralis region of the retina. This provided the first evidence of the basis of colour vision in the Istiophoridae. Furthermore, regional variations in photoreceptor density, type, and spatial arrangement indicated differing visual capabilities along different visual axes.

In juvenile and young Pacific bluefin tuna (PBT) *Thunnus orientalis*, two types of green cone cell and one type of blue cone cell have been found using molecular techniques for cone opsin genes (Miyazaki et al. 2008). Their spectral sensitivity ranged between 300–650 nm, with three similar peak wavelengths found using an electroretinogram (ERG) (Matsumoto et al. 2012). It was concluded that PBT have colour vision, and are photopic dichromats in the front and upward visual axes.

In addition, Nakamura et al. (2013) determined the genome sequence of the PBT, using next-generation sequencing technology to elucidate the genetic and evolutionary basis of optic adaptation of tuna. The authors identified five common fish visual pigment genes: red-sensitive (middle/long-wavelength sensitive; M/LWS), UV-sensitive (short-wavelength sensitive 1; SWS1), blue-sensitive (SWS2), rhodopsin (RH1, blue shift), and green-sensitive (RH2) opsin genes. Thus, the authors suggested that PBT has undergone evolutionary changes in three genes (RH1, RH2, and SWS2), which may have contributed to detecting blue-green contrast and measuring the distance to prey in the blue-pelagic ocean.

Visual Sensitivity

The light sensitivity of juvenile PBT has been electrophysiologically investigated using the ERG method. Matsumoto et al. (2009) compared the light sensitivities of two scombrid fish species, using the K value as an indicator of relative light sensitivity (Eguchi and Horikoshi 1984; Frank 2003). The K value is the irradiance (quanta·cm^{-2}·s^{-1}) required to generate half the intensity of the maximum ERG voltage of the fish's response to a light stimulation of 503 nm (green), which is close to the wavelengths of maximum sensitivities of the target species (Matsumoto et al. 2011). Matsumoto et al. (2009) used juvenile PBT (52–64 days post-hatch [dph], standard length [SL] 150–175 mm) and juvenile chub mackerel *Scomber japonicus* (107–109 dph, SL 122–150 mm). The authors found that the values of log K for PBT and chub mackerel were not significantly different from each other. The light intensities for K for PBT and chub mackerel were 1.38 and 1.86 lx, respectively. Therefore, visual sensitivity to dim light in juvenile PBT was not low relative to typical scombrid fish species, such as juvenile chub mackerel.

In general, visual light sensitivity in fish increases with eye size (the square of the eye's radius) (Higgs and Fuiman 1996; Miyazaki et al. 2000). Matsumoto et al.

(2011) reported that the visual light sensitivity of juvenile PBT has a tendency to increase slightly with growth between 28–64 dph, in individuals of between 29–175 mm SL; however, the visual light sensitivities of juveniles are still not sufficiently high for hunting in the deep ocean. Fritsches et al. (2003b) investigated the retinal specializations of the blue marlin (*Makairanigricans*), in relation to their eyes being adapted for high sensitivity in low light levels. From anatomically studying the density of ganglion cells in the retina, the authors found that blue marlin have Visual Acuities (VAs) of less than 10 cycles per degree, a surprisingly low visual resolution given the absolute size of the marlin eye. Cone photoreceptors, however, were present in high densities, resulting in a summation of cones and ganglion cells at a ratio of 40:1, even in the area of best vision. The optical sensitivity of the marlin eye is high, because of the large size of the cone photoreceptors. These results indicate that the marlin eye is specifically adapted to cope with the low light levels encountered during diving. Since the marlin is likely to use its vision at depth, this research approach could help estimate the limits of the marlin's vertical distribution, based on light levels.

Tuna and billfish have superior visual temporal resolution and optical sensitivity compared with their prey species because of their ability to maintain elevated body temperatures (endothermy), using highly specialized heating systems (Fritsches et al. 2005). This enhanced temporal resolution, and relatively high optical sensitivity, provides warm-blooded, and highly visual, oceanic predators, such as billfish and tuna, with a crucial advantage over their agile, cold-blooded prey.

Visual Acuity (VA)

Introduction

VA is an indicator of the spatial resolution of the visual processing system (i.e., vision); in humans, it is measured using so-called Landolt rings. To examine the spatial resolution of fish such as tuna, VA (min^{-1}) is a suitable parameter to determine and understand their visual spatial resolution.

In a behavioural experiment, Nakamura (1968) determined the VA of adult skipjack tuna *Katsuwonuspelamis*, and Kawamura et al. (1981a) histologically calculated the VA of adult PBT *Thunnus orientalis*. Although there are developmental changes in visual capability, and VA in many fish species increases with growth (Douglas and Djamgoz 1990; Shiobara et al. 1998; Shand et al. 2000; Miyagi 2001), juvenile PBT have not been investigated. Fritsches et al. (2003a) noted that the striped marlin has different visual capabilities along different visual axes. The position of the area with a high cell density, or the area centralis in the retina, is related to both habitat and the main visual axis of feeding behaviour (Kawamura et al. 1980; Collin and Pettigrew 1988; Shand et al. 2000). Knowledge of developmental changes in retinal topography gives important clues to behavioural changes that occur with growth. To measure the distribution of retinal cone density at each growth stage, it is necessary to investigate specimens at each stage. Recently, the full-cycle culture of PBT was achieved in the Fisheries Laboratory, Kinki University, Wakayama, Japan. By using the full-cycle-cultured specimens of this species, including juveniles, it was possible to investigate cone density distribution at each growth stage. We histologically investigated developmental changes in cone density distribution, the visual axis, and the minimum separable angle of cones.

Materials and methods

Six PBT at different growth stages, that had been full-cycle-cultured during 2003–2004 in the Fisheries Laboratory, Kinki University, were used (30 dph, total length [TL] 4.7 cm; 35 dph, TL6.2 cm; 41 dph, TL 6.5 cm; 46 dph, TL 11.1 cm; 80 dph, TL 31.8 cm; and one-year old, TL102.6 cm). The eyes of the specimens were enucleated, and fixed in Bouin's solution for 24 hours. Each retina was divided into nine regions in the 30- to 46-dph specimens, 25 regions in the 80-dph specimens, and 33 regions in the one-year old specimen, according to eye size. After paraffin embedding, the retinas were cut using a microtome into sections parallel to the retinal surface. The sections were then stained with haematoxylin–eosin. For quantitative analysis, the sections were examined under a light microscope, photographed, and the number of cones in 0.01 mm^2 of each region was counted.

Results and discussion

Photomicrographs of tangential sections of the retinal cones show that twin cones formed a regular mosaic of parallel rows, although there were occasional irregular single cones in each region from specimens at every stage. An example of a retinal tissue older than one-year post-hatch is shown in Fig. 14.1. The total number of twin and single cones in 0.01 mm^2 of each region was counted. The density distributions

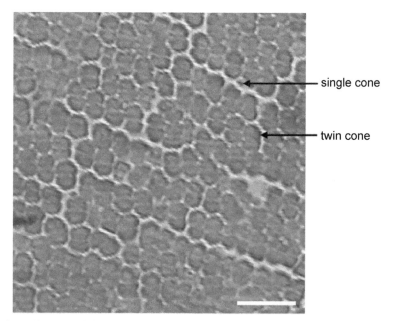

Figure 14.1. Photomicrograph of tangential section showing retinal cone arrangement of juvenile Pacific bluefin tuna. Scale bar = 20 μm.

of both cones from the right retinas of the specimens (30, 35, 41, 46, 80 dph, and one-year post-hatch) were then determined, graded, drawn, and displayed using contour lines at 25-cone intervals (Fig. 14.2). The specimens at even the youngest stage had already metamorphosed, and were feeding on fish prey (Miyashita et al. 1998).

The characteristics of the cone density distribution changed with growth stage. In the specimen aged 30 dph, the cone density in each peripheral area was five times (668/132 = 5.03) that of the bottom region. There was no definite direction of acute vision, such as a visual axis, since no specialized region of maximum cone density existed. It seems reasonable to assume that the distribution at this stage reflects the active differentiation of cones from stem cells in the ciliary marginal zone. From 35 to 46 dph, the areas with the greatest density were the ventral and ventro-temporal regions. Although the ratios of the maximum to minimum densities ranged from 1.55 to 2.03 (372/240 = 1.55, 544/268 = 2.03, and 400/208 = 1.92), which were lower than at other growth stages, the visual axis in each was upward and forward, as seen by the emerging wide areas of high cone density at these stages. In contrast, at 80 dph and one-year post-hatch, the areas of greatest density were limited to the ventro-temporal region, and the ratios in these specimens were essentially constant (584/172 = 3.40 and 240/72 = 3.33, respectively).

On the basis of the results, the region of maximum cone density, that is, the area centralis, gradually developed in the ventro-temporal area, with growth during the first 80 dph. Consequently, fish older than 80 dph have a visual axis that is clearly oriented upward and forward through the crystalline lens, indicating the direction and area of the most acute vision. The developmental changes that occur in the specialized retinal region probably reflect changes in fish habitat or feeding characteristics, such as prey species or striking direction. During culture, it was observed that juvenile tuna, from 35 to 46 dph, fed on fish prey at the water surface, in the direction of the visual axis.

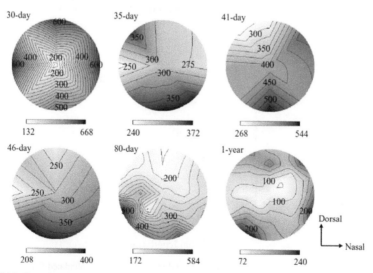

Figure 14.2. Cone density distributions of right eye retinas from six growth stages of juvenile bluefin tuna. The contour interval is 25 cells.

We postulate that this specialization in the retinal cone density distribution affects their feeding behaviour. This specialization also influences changes in the minimum separable angle of cone cells with growth. This angle was calculated in each specimen from the maximum density of cones, using the following equation (Tamura 1957):

$$\sin \alpha = 1/F \{2 \times 0.1 \times (1 + 0.25)/\sqrt{n}\} \qquad \text{Eq. 1}$$

where α is the minimum separable angle; F is the focal distance of the lens, which is 2.55 times the radius of the lens (Matthiessen's ratio); 0.25 is the degree of shrinkage; and n is the maximum number of cones per 0.01 mm^2.

Cone VA was also estimated from the minimum separable angle of cones, using the following equation:

$$VA = \pi/(\alpha \times 180 \times 60) \qquad \text{Eq. 2}$$

The cone VAs, calculated from the maximum cone densities in the retinas of specimens from 30, 35, 41, 46, 80 dph, and one-year post-hatch, were 0.048, 0.061, 0.084, 0.092, 0.240, and 0.347, respectively. From 80 dph to one-year post-hatch, maximum cone density decreased and lens radius increased, resulting in a slightly elevated cone VA. In contrast, from 30 to 80 dph, cone density did not decrease with increasing lens radius, so VA increased significantly. Therefore, whereas the cone VA in young adults increased slightly, in juveniles it increased considerably from 30 to 80 dph.

The relationship between TL and VA in PBT is shown in Fig. 14.3. The allometric function for this relationship, calculated using the least squares method, is VA = 0.0048 $L^{0.640}$ ($R^2 = 0.951$). Although the authors only examined a few individuals and divided the eyes into different sections because of the size of the eyes, the VA of cone cells tended to increase. The VA calculated using this function (0.262) was similar to the value obtained for PBT (0.27; Fork Length [FL] = 520 mm) by Kawamura et al. (1981a). In addition, this function is more similar to the function for a migratory

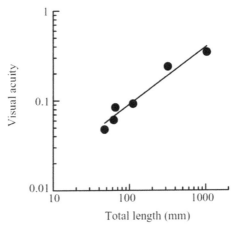

Figure 14.3. Relationship between log total length and log visual acuity, calculated from the cone cell density of Pacific bluefin tuna.

fish, the yellowtail *Seriolaquinqueradiata* (VA = 0.0051 $L^{0.622}$), than that for the red bream *Pagrus major* (VA = 0.0071 $L^{0.588}$) (Shiobara et al. 1998; Miyagi et al. 2001).

We examined developmental changes and specialization in the distribution of cone densities in juvenile PBT. However, Fritsches et al. (2003b) found that the convergence ratio of cone cells to ganglion cells was 40:1 in the area centralis of the retina in the blue marlin. To determine the visual capability of resolving objects by PBT, we should consider the convergence ratio of cone cells to ganglion cells, which indicates neural connections. Therefore, we compared the visual spatial resolutions calculated from cone and ganglion cell densities of PBT. Miyazaki (2013) reported that the visual spatial resolutions of juvenile PBT (TL 53–370 mm) calculated from ganglion cell densities, ranged from 15.6 to 4.7 minutes of arc. VA (min⁻¹) is the reciprocal of such a unit (min. of arc). Therefore, we recalculated the VAs of juvenile PBT; these ranged from 0.064 to 0.213, based on ganglion cell densities. These results are consistent with our results, and we calculated that the VAs of juvenile PBT of 6, 10, and 30 cm TL were approximately 0.06, 0.1, and 0.2, respectively.

Knowledge of the direction and degree of specialization with growth, based on the maximum cone density distribution and VA of juveniles, is important in understanding the behaviour of growing PBT.

Retinomotor Response and Behaviour

Introduction

Fish behaviour in relation to light intensity has been examined in several species using fish schooling indices. A light intensity threshold for school formation has been reported in several fish species (Hunter 1968; Glass et al. 1986; Higgs and Fuiman 1996; Ryer and Olla 1998; Miyazaki et al. 2000), and schooling behaviour in several fish species has been quantitatively analyzed using defined schooling variables (Masuda et al. 2003; Nakayama et al. 2003).

The migratory and diving behaviour of Atlantic bluefin tuna *T. thynnus* (ABT) and PBT have been investigated under natural conditions using archival tags (Block et al. 2001; Kitagawa et al. 2004). Kitagawa et al. (2004) suggested that the daily diving patterns of PBT occur in response to changes in ambient light at sunrise and sunset. The schooling behaviour of PBT, however, in relation to light intensity and visual characteristics, has not been examined. Tuna have well-developed vision (Tamura and Wisby 1963; Nakamura 1968; Kawamura et al. 1981a), which is considered the predominant sensory system in affecting behaviour (Kawamura 1994). Masuma et al. (2001) histologically examined the retinomotor responses of juvenile PBT under different light conditions. They reported that the transition from scotopic to photopic vision occurred at a light intensity of 7.52 lx, and that the transition took 15 minutes to complete. At dawn, the ambient light intensity rapidly increased from scotopic to photopic in 10 minutes. The researchers suggested that the time lag between changes in ambient light intensity and retinal adaptation could cause visual disorientation in juveniles (Masuma et al. 2001).

As stated previously, the Fisheries Laboratory at Kinki University has developed a full-cycle culture of PBT. Collisions with tank walls, however, have caused high mortality rates among juveniles and young adults (Miyashita et al. 2000). This behaviour may have been caused by environmental stimuli, such as lights and sounds. Juvenile PBT behaviour in relation to light intensity, using schooling variables and the retinomotor response (which indicates light or dark adaptation by juveniles), was examined. Schooling variables were compared with retinal adaptation ratios to investigate the relationship between behaviour and light intensity, in order to determine the role of the retinomotor response in the behaviour of PBT, especially in juvenile fish.

Materials and methods

Experimental fish

Full-cycle-cultured PBT, hatched on July 26, 2004 at the Kushimoto Oshima Station of the Kinki University Fisheries Laboratory, Wakayama, Japan, were used in this study. Approximately 100 juvenile PBT, 35–36 dph (mean ± S.D. TL 70 ± 8 mm; body mass 5.9 ± 1.8 g) were used in the experiments.

Analysis of schooling behaviour

Twenty juveniles were placed in a 2-m diameter water tank that was shielded from light by dark curtains. Video images of schooling behaviour were recorded using a digital video camera (DCR-TRV50; Sony, Tokyo, Japan) and halogen lamps (MHF-150L; Moritex, Tokyo, Japan) set above the tank. To limit swimming behaviour to two dimensions, water depth was maintained at 100–150 mm. By changing the number of white filter papers covering the halogen lamps, light intensities of 0.01, 0.5, 5, 50, and >300 lx were applied, as measured just above the water surface using an illuminometer (LM331; As One, Osaka, Japan). Under dark conditions of <5 lx, infrared light-emitting diodes (LEDs; K-00094; Akizuki, Tokyo, Japan) with a wavelength range >900 nm were used, and the digital video camera 'night' function was used to record the fish (see details in Fig. 13.2 of the previous chapter). After 60-minutes acclimatization to the experimental light conditions during daylight hours (Masuma et al. 2001), the schooling behaviour of 35–36-dph juveniles was recorded for 15-minute periods under the following light intensity levels: 0.01 lx (2.0×10^{-4} $\mu E \cdot m^{-2} \cdot s^{-1}$), 0.5 lx ($1.0 \times 10^{-2}$ $\mu E \cdot m^{-2} \cdot s^{-1}$), 5 lx ($1.0 \times 10^{-1}$ $\mu E \cdot m^{-2} \cdot s^{-1}$), 50 lx ($1.0$ $\mu E \cdot m^{-2} \cdot s^{-1}$), and 700 lx ($14$ $\mu E \cdot m^{-2} \cdot s^{-1}$).

The x and y positions of the fishs' snouts were recorded from the video images of schooling fish at 0.1 second intervals as time series data, using a personal computer and a digital video capture card. The procedure for recording the position of individuals and importing the data into a Microsoft Excel spreadsheet was automated, using a programme written in Visual Basic (Suzuki et al. 2003).

The schooling variable, the Separation Swimming Index (SSI), as defined by Nakayama et al. (2003), was calculated from the time series of the positions, under each condition. The SSI was analyzed using a particular video frame, together with additional frames from one second later. Vectors of one focal fish, and the nearest neighbouring fish, were measured as their movements over one second. To determine the SSI, the starting points of the neighbouring fish vectors were parallel translated to that of the focal fish, and the distance between the ends of the two vectors was divided by the average length of the two vectors (see details in Fig. 13.3 of the previous chapter).

Histological analysis of retinomotor response

At the end of the behavioural experiments, one experimental specimen from each experimental light condition was immediately fixed in Bouin's solution. Each retina was then enucleated and divided into nine regions. The retinas were dehydrated through an ethanol series, cleaned in xylene, embedded in paraffin, and sectioned into 6-mm thick slices. The sections were stained with haematoxylin and eosin. For quantitative analysis, the sections were examined under a light microscope, photographed, and the thickness from the outer membrane to the epithelium layer, the thickness of the pigment layer, and the migratory position of the cone ellipsoid, were measured to determine the retinomotor response (Fig. 14.4). The ratio of the light adaptation of the retina, the expansion of the pigment epithelium, and the contraction of the cone myoid were calculated relative to the thickness of the visual cell layer, under each experimental light intensity condition (Figs. 14.4a, 14.4b; Ali 1959; Masuma et al. 2001). The retinal indices indicating retinomotor pigment and cone position under different light intensities are expressed as the ratios p:v and m:v (i.e., the thickness of the pigment epithelium [p] relative to the visual cell thickness [pigment epithelium to the outer membrane, v] and the length of the cone myoid [m] relative to v).

(a)   (b)

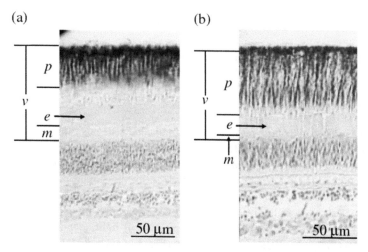

Figure 14.4 Photomicrographs of (a) dark-adapted, and (b) light-adapted retinal transverse sections, showing the thickness from the pigment epithelium to the outer membrane (*v*), the thickness of the pigment epithelium (*p*), the length of the cone myoid (*m*), and the ellipsoid of the cone (*e*). Scale bars = 50 μm.

Results

Schooling behaviour

The SSI of the 35–36 dph juveniles under various light intensities is shown in Fig. 14.5. The SSI was almost at a constant level under conditions of >5 lx. Under conditions of <5 lx, the SSI significantly increased with decreasing light intensity (Bonferroni test, each data point: n = 50, P < 0.01; Fig. 14.5). At 0.01 lx, the SSI was approximately 1.27, indicating close to random movement, as determined by averaging the SSI values integrated from 0° to 180°; therefore, fish swam randomly under low light conditions. In addition, the values of the SSI for the 35–36 dph juveniles were similar, and varied in relation to the light intensity threshold of 5 lx.

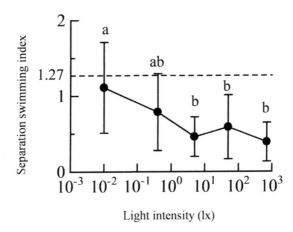

Figure 14.5. Separation swimming index (SSI) for 35–36 d post-hatch juveniles (●) under different light intensities (each data point, n = 50). Values are means ± S.D. Significantly different groups are indicated by lower-case letters (P < 0.01, Bonferroni multiple comparisons).

Retinomotor response

The retinal indices indicating retinomotor pigment and cone position in the 35–36 dph juveniles under different light intensities are shown in Fig. 14.6. The ratios m:v and p:v are expressed as the length of the cone myoid relative to the thickness from the pigment epithelium to the outer membrane, and the thickness of the pigment epithelium relative to the visual cell thickness. At >5 lx, the m:v values were low (0.05–0.06) and the p:v values were high (0.62–0.75) in the 35–36 dph juveniles, such that the retinal indices were nearly constant, and indicative of light adaptation. In contrast, p:v and m:v changed at light intensities <5 lx (Fig. 14.6). Under 0.01 lx, p:v and m:v were consistent with previously measured dark-adapted indices (Masuma et al. 2001).

Relationship between retinal indices and schooling indices

Schooling indices were compared to retinal light adaptation indices under each experimental light intensity condition. There was agreement between the ratios of

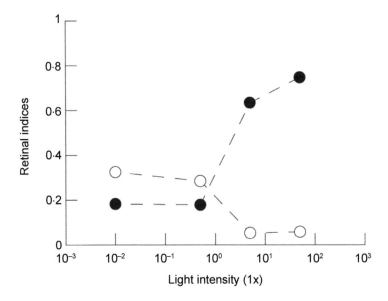

Figure 14.6. Retinal indices of 35–36 d post-hatch juveniles (n = 4) exposed to experimental light intensities. The retinal indices p:v (●) and m:v (○) indicate the expansion of the pigment epithelium and the contraction of the cone myoid, respectively. v, thickness from outer membrane to epithelium layer; p, thickness of pigment layer; m, length of myoid.

retinomotor movements and the schooling variables in relation to light intensity. The relationships between the retinal indices (m:v and p:v) and the SSI for 35–36 dph and 45–46 dph juveniles are shown in Fig. 14.7. Both variables of retinal indices (m:v and p:v) were correlated with the SSI (m:v, r = 0.96, n = 6, P < 0.01; p:v, r = 0.88, n = 6, P < 0.05) (Torisawa et al. 2007b).

Discussion

Schooling behaviour under different light intensities

Under lit conditions, the SSI of PBT (0.39–0.58) was high compared to that of chub mackerel (0.15–0.29; Nakayama et al. 2003), and closer to that of Atlantic Spanish mackerel *Scomberomorusmaculatus* (0.6–0.8; Masuda et al. 2003), which is a piscivorous species. Piscivorous species may form loose schools to minimize the risk of cannibalism.

Furthermore, the characteristics of juvenile PBT schooling behaviour varied with light intensity. At >5 lx, the SSIs were nearly constant, and fish could form well-organized schools. The SSI increased with decreasing light intensities <5 lx. With decreasing light intensities, fish schools gradually expanded, and parallel swimming changed to random movement. Juveniles swam almost randomly, and did not form schools at light conditions of 0.01 lx, based on their SSI values, which were close to random. The results indicated that the schooling behaviour characteristics of juveniles

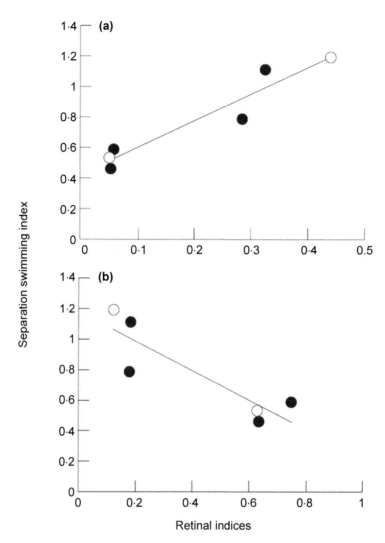

Figure 14.7. Relationship between the retinal indices (a) m:v, and (b) p:v, in juveniles 35–36 (•) and 45–46 (○) d post-hatch, and the separation swimming index (SSI). m, Length of myoid; v, thickness from outer membrane to epithelium layer; p, thickness of pigment layer (n = 6).

depend on light intensity. The light intensity threshold for schooling in juvenile PBT ranged from 0.5 to 0.01 lx.

   Light intensity thresholds for schooling in Pacific jack mackerel *Trachurussymmetricus* (Ayres) (105–141 mm TL) and striped jack *Pseudocaranxdentex* (Bloch and Schneider) (120 mm TL) have been reported to be 3.5 × 10^{-5} and 5 × 10^{-3} lx, respectively (Hunter 1968; Miyazaki et al. 2000). In addition, the threshold in scombrid species such as Atlantic mackerel *Scomberscombrus* L. (315–341 mm SL) is in the order of 10^{-5} lx (Glass et al. 1986). The light intensity threshold for schooling in juvenile PBT clearly differed from those of previously studied fish species, and was much higher than in other species.

Schooling behaviour in relation to retinomotor response

Comparing schooling indices with retinal light adaptation indices (m:v and p:v) at each experimental light intensity, both parameters of retinal indices were correlated with the SSI (m:v, r = 0.96, n = 6, P < 0.01; p:v, r = –0.88, n = 6, P < 0.05), suggesting that the retinomotor responses of PBT greatly affected schooling behaviour in 35–36 and 45–46 dph juveniles (Fig. 14.7). The effect of the lateral line system on schooling behaviour has been reported for several fish species (Pitcher et al. 1976; Partridge and Pitcher 1980). On the basis of the present results, however, the schooling behaviour of juvenile PBT mainly depends on vision. Furthermore, Higgs and Fuiman (1996) reported that changes in nearest neighbour angles and distances more closely paralleled movements of retinal pigment cells than cone cell migration in Gulf menhaden (*Brevoortiapatronus* [Goode]). In contrast, changes in the SSI of juvenile PBT were more closely correlated with cone cell migration than with retinal pigment cell movements, indicating that cones, which function at a high light intensity for photopic vision, play an important role in school formation in juveniles of this species. Therefore, we suggest that the light intensity threshold for schooling is relatively high in juvenile PBT, relative to other species.

In the present study, fish behaviour was quantified using schooling variables (SSI), and retinal indices were calculated, to investigate the relationship between vision characteristics and behaviour in juvenile PBT. The behaviour of juveniles was significantly affected by visual characteristics, suggesting that retinal indices of light adaptation reflect behavioural characteristics under different light intensities.

Juvenile PBT, which mainly swim at the sea surface and feed during the day, have a high light intensity threshold for schooling and a dependence on photopic vision, which indicate an adaptation to high ambient light intensity conditions. On the basis of the results, juvenile PBT (35–36 dph) should be kept at light intensities >5 lx in aquaria, to maintain the appropriate conditions and to avoid abnormal behaviour (e.g., collisions with tank walls). Therefore, the relationship between visual and behavioural characteristics is important in understanding the behaviour of juvenile PBT.

Schooling behaviour depends on vision development

The effects of vision development and light intensity on schooling behaviour during growth in juvenile PBT was investigated using both behavioural and histological approaches. The schooling behaviour of three juvenile age groups (25, 40, and 55 dph) was examined under various light intensities. The SSI was also measured under different light intensities. In addition, retinal indices of light adaptation in juvenile fish at each experimental light intensity, and their VA at six growth stages (25–55 dph), were histologically examined. During growth, the light intensity thresholds of the SSI decreased from 5 to 0.05 lx. The thresholds of light intensities for the light adaptation of retinas in juveniles (25–55 dph) similarly decreased from 5 to 0.05 lx.

In addition, the VA of juveniles increased from 0.04 to 0.17 with a decreasing SSI. These data clearly indicate that the characteristics of schooling behaviour strongly correspond to the degree of vision development. Juvenile PBT also appear to be more dependent on cone rather than rod cells under low light intensity conditions, resulting in a relatively high light intensity threshold for schooling. These results suggest that juveniles can adapt to darker conditions during growth by developing improved visual capabilities (Torisawa et al. 2011).

Behaviour in relation to light intensity

The thresholds of both indices of retinal adaptation, and schooling, shifted to lower light levels with growth, because older juvenile PBT could form schools using photopic vision, the sensitivity of which had greatly improved (Torisawa et al. 2011). Therefore, juveniles may detect and hunt prey at lower light levels, indicating that they possess the ability to adapt to dusk and dawn light conditions and deep, dark, seas. Kitagawa et al. (2003) examined the diving behaviour of immature PBT (FL 47–144 cm) using archival tags, and reported that larger individuals dived to much greater depths (500 m) than smaller individuals (100 m) did. The authors suggest that the diving performance of PBT is related to body size, and that a larger size enables the expansion of the vertical range of movement. A similar process of adaptation and specialization to deeper seas in juveniles was observed. The present experiments with juvenile PBT, adults of which can dive to depths of hundreds of metres, elucidate the developmental growth stages during which the fish obtain greater visual capabilities, which affect their behaviour in the deep sea.

These integrated analyses from the interdisciplinary fields of physiology and ethology help to elucidate the relationship between vision and behaviour in juvenile PBT. In addition, the comparative study of different growth stages revealed developmental shifts in, and relationships between, vision and behaviour. At 100,000 lx in sunlight, the water-depth limits for forming well-organized schools (calculated using the irradiance transmittance of the optical water type of oceanic IB: $c.$ 0.85 m^{-1}; Jerlov 1976) of 25 dph (TL 3.7 cm), 40 dph (TL 7.7 cm), and 55 dph (TL 14.6 cm) juvenile PBT were 60, 75, and 90 m, respectively. During the night, the 25-dph juveniles remain aggregated due to an inhibition for swimming. In contrast, under full-moon conditions at a light intensity level of 0.2 lx, 40 dph juveniles can form organized schools at the sea surface, and 55 dph juveniles can form well-organized schools at 10 m depth in oceanic IB water. Furthermore, juveniles may also be able to detect and hunt prey at similar water depths. These results indicate that juvenile PBT perceive and recognize schooling individuals and form better-organized schools during the night, and become better adapted to deeper seas with growth, which coincides with their migration to the deep ocean.

References

Ali, M.A. 1959. The ocular structure, retinomotor and photo-behavioural responses of juvenile Pacific salmon. Can. J. Zool. 37: 965–996.

Block, B.A., H. Dewar, S.B. Blackwell, T.D. Williams, E.D. Prince, C.J. Farwell, A. Boustany, S.L.H. Teo, A. Seitz, A. Walli and D. Fudge. 2001. Migratory movements, depth preferences, and thermal biology of Atlantic bluefin tuna. Science 293: 1310–1314.

Collin, S.P. and J.D. Pettigrew. 1988. Retinal topography in reef teleosts. I. Some species with well-developed area but poorly developed streaks. Brain Behav. Evol. 31: 269–282.

Douglas, R. and M. Djamgoz (eds.). 1990. The Visual System of Fish. Chapman & Hall, London.

Eguchi, E. and T. Horikoshi. 1984. Comparison of stimulus-response (V-log I) functions in five types of lepidopteran compound eyes (46 species). J. Comp. Physiol. A. 154: 3–12.

Frank, T.M. 2003. Effects of light adaptation on the temporal resolution of deep-sea crustaceans. Integr. Comp. Biol. 43: 559–570.

Fritsches, K.A., J.C. Partridge, J.D. Pettigrew and N.J. Marshall. 2000. Colour vision in billfish. Proc. R. Soc. B. 355: 1253–1256.

Fritsches, K.A., L. Litherland, N. Thomas and J. Shand. 2003a. Cone visual pigments and retinal mosaics in the striped marlin. J. Fish Biol. 63: 1347–1351.

Fritsches, K.A., N.J. Marshall and E.J. Warrant. 2003b. Retinal specializations in the blue marlin: eyes designed for sensitivity to low light levels. Mar. Freshw. Res. 54: 333–341.

Fritsches, K.A., R.W. Brill and E.J. Warrant. 2005. Warm eyes provide superior vision in Swordfishes. Curr. Biol. 15: 55–58.

Glass, C.W., C.S. Wardle and W.R. Mojsiewicz. 1986. A light intensity threshold for schooling in the Atlantic mackerel *Scomber scombrus*. J. Fish Biol. 29 (Suppl. A): 71–81.

Higgs, D.M. and L.A. Fuiman. 1996. Light intensity and schooling behaviour in larval gulf menhaden. J. Fish Biol. 48: 979–991.

Hunter, J.R. 1968. Effects of light on schooling and feeding of jack mackerel, *Trachurus symmetricus*. J. Fish. Res. Board Can. 25: 393–407.

Jerlov, N.G. 1976. Optical water types. pp. 132–137. *In*: N.G. Jerlov (ed.). Marine Optics. New York, NY: Elsevier Scientific.

Kawamura, G., W. Nishimura, S. Ueda and T. Nishi. 1981a. Vision in tunas and marlins. Mem. Kagoshima Univ. Res. Center South Pacific. 1: 3–47.

Kawamura, G., W. Nishimura, S. Ueda and T. Nishi. 1981b. Color vision and spectral sensitivity in tunas and marlins. Fisheries Sci. 47: 481–485.

Kawamura, G., R. Tsuda, H. Kumai and S. Ohashi. 1984. The visual cell morphology of *Pagrus major* and its adaptive changes with shift from pelagic to benthic habitat. Fisheries Sci. 50: 1975–1980.

Kawamura, G. 1994. Physiology of tunas and marlins. GekkanKaiyo. 26: 529–533.

Kawamura, G., S. Masuma, N. Tezuka, M. Koiso, T. Jinbo and K. Namba. 2003. Morphogenesis of sense organs in the bluefin tuna *Thunnus orientalis*. The Big Fish Bang. Institute of Marine Research, Bergen 123–135.

Kitagawa, T., S. Kimura, H. Nakata and H. Yamada. 2004. Diving behavior of immature, feeding Pacific bluefin tuna (*Thunnus thynnus orientalis*) in relation to season and area: the East China Sea and the Kuroshio–Oyashio transition region. Fish. Oceanogr. 13: 161–180.

Masuda, R., J. Shoji, S. Nakayama and M. Tanaka. 2003. Development of schooling behavior in Spanish mackerel *Scomberomorus niphonius* during early ontogeny. Fisheries Sci. 69: 772–776.

Masuma, S., G. Kawamura, N. Tezuka, M. Koiso and K. Namba. 2001. Retinomotor response of juvenile bluefin tuna *Thunnus thynnus*. Fisheries Sci. 67: 228–231.

Matsumoto, T., H. Ihara, Y. Ishida, T. Okada, M. Kurata, Y. Sawada and Y. Ishibashi. 2009. lectroretinographic Analysis of Night Vision in Juvenile Pacific Bluefin Tuna (*Thunnus orientalis*). Biol. Bull. 217: 142–150.

Matsumoto, T., T. Okada, Y. Sawada and Y. Ishibashi. 2011. Changes in the scotopic vision of juvenile Pacific bluefin tuna (*Thunnus orientalis*) with growth. Fish Physiol. Biochem. 37: 693–700.

Matsumoto, T., T. Okada, Y. Sawada and Y. Ishibashi. 2012. Visual spectral sensitivity of photopic juvenile Pacific bluefin tuna (*Thunnus orientalis*). Fish Physiol. Biochem. 38: 911–917.

Miyagi, M., S. Akiyama and T. Arimoto. 2001. The development of visual acuity in yellowtail *Seriola quinqueradiata*. Nippon Suisan Gakkaishi 67: 455–459.

Miyashita, S., K. Kato, Y. Sawada, O. Murata, Y. Ishitani, K. Shimizu, S. Yamamoto and H. Kumai. 1998. Development of digestive system and enzyme activities of larval and juvenile bluefin tuna, *Thunnus thynnus*, reared in the laboratory. *Suisanzoshoku* 46: 111–120.

Miyashita, S., Y. Sawada, N. Hattori, H. Nakatsukasa, T. Okada, O. Murata and H. Kumai. 2000. Mortality of northern bluefin tuna *Thunnus thynnus* due to trauma caused by collision during growout culture. J. World Aquac. Soc. 31: 632–639.

Miyazaki, T., S. Shiozawa, T. Kogane, R. Masuda, K. Maruyama and K. Tsukamoto. 2000. Developmental changes of the light intensity threshold for school formation in the striped jack *Pseudocaranx dentex*. Mar. Ecol. Prog. Ser. 192: 267–275.

Miyazaki, T., J. Kohbara, K. Takii, Y. Ishibashi and H. Kumai. 2008. Three cone opsin genes and cone cell arrangement in retina of juvenile Pacific bluefin tuna *Thunnus orientalis*. Fisheries Sci. 74: 314–321.

Miyazaki, T. Published on line: 18 June 2013. Retinal ganglion cell topography in juvenile Pacific bluefin tuna *Thunnus orientalis* (Temminck and Schlegel). Fish Physiol. Biochem.

Munz, F.W. and W.N. McFarland. 1973. The significance of spectral position in the rhodopsins of tropical marine fishes. Vision Res. 13: 1829–1864.

Nakamura, E.L. 1968. Visual acuity of two tunas, *Katsuwonus pelamis* and *Euthynnus affinis*. Copeia 1: 41–49.

Nakamura, Y., K. Mori, K. Saitoh, K. Oshima, M. Mekuchi, T. Sugaya, Y. Shigenobu, N. Ojima, S. Muta, A. Fujiwara, M. Yasuike, I. Oohara, H. Hirakawa, V.S. Chowdhury, T. Kobayashi, K. Nakajima, M. Sano, T. Wada, K. Tashiro, K. Ikeo, M. Hattori, S. Kuhara, T. Gojobori and K. Inoue. 2013. Evolutionary changes of multiple visual pigment genes in the complete genome of Pacific bluefin tuna. PNAS 110: 11061–11066.

Nakayama, S., R. Masuda, J. Shoji, T. Takeuchi and M. Tanaka. 2003. Effect of prey items on the development of schooling behavior in chub mackerel *Scomber japonicus* in the laboratory. Fisheries Sci. 69: 670–676.

Ryer, C.H. and B.L. Olla. 1998. Effect of light on juvenile walleye pollock shoaling and their interaction with predators. Mar. Ecol. Prog. Ser. 167: 215–226.

Shand, J., S.M. Chin, A.M. Harman, S. Moore and S.P. Collin. 2000. Variability in the location of the retinal ganglion cell area centralis is correlated with ontogenetic changes in feeding behavior in the black bream, *Acanthopagrus butcheri* (Sparidae, Teleostei). Brain Behav. Evol. 55: 176–190.

Shiobara, Y., S. Akiyama and T. Arimoto. 1998. Developmental changes in the visual acuity of red sea bream *Pagrus major*. Fisheries Sci. 64: 944–947.

Suzuki, K., T. Takagi and T. Hiraishi. 2003. Video analysis of fish schooling behavior in finite space using a mathematical model. Fish. Res. 60: 3–10.

Tamura, T. 1957. A study of visual perception in fish, especially on resolving power and accommodation. Fisheries Sci. 22: 536–557.

Tamura, T. and W.J. Wisby. 1963. The visual sense of pelagic fishes especially the visual axis and accommodation. Bull. Mar. Sci. Gulf Caribb. 13: 433–448.

Torisawa, S., T. Takagi, Y. Ishibashi, Y. Sawada and T. Yamane. 2007a. Changes in the retinal cone density distribution and the retinal resolution during growth of juvenile Pacific bluefin tuna *Thunnus orientalis*. Fisheries Sci. 73: 1202–1204.

Torisawa, S., T. Takagi, H. Fukuda, Y. Ishibashi, Y. Sawada, T. Okada, S. Miyashita, K. Suzuki and T. Yamane. 2007b. Schooling behaviour and retinomotor response of juvenile Pacific bluefin tuna *Thunnus orientalis* under different light intensities. J. Fish Biol. 71: 411–420.

Torisawa, S., H. Fukuda, K. Suzuki and T. Takagi. 2011. Schooling behaviour of juvenile Pacific bluefin tuna *Thunnus orientalis* depends on their vision development. J. Fish Biol. 79: 1291–1303.

CHAPTER 15

Physiology of Bluefin Tuna Reproduction

New Insights into Reproduction in Wild and Captive Bluefin Tuna Species

Koichiro Gen

Introduction

Bluefin tuna is one of the most important species in the fisheries industry, caught in large numbers and highly priced in the market. There are three species of bluefin tuna, which are the Pacific bluefin tuna (*Thunnus orientalis*), Atlantic bluefin tuna (*Thunnus thynnus*) and southern bluefin tuna (*Thunnus maccoyii*). Although Pacific bluefin tuna was classified as a subspecies of Atlantic bluefin tuna until the late 1990s, it has been found to belong to the distinct species based on morphological observations and genetic studies (Collette 1999). During the past decade, exploitation of their natural stocks has increased to dangerous levels through the pressure of commercial fishing. It is therefore important to develop captive breeding and seed production techniques for this species in order to reduce the negative impacts of fisheries on wild tuna stocks. Recently, research advances have enabled successful broodstock development, and natural or artificial spawning of bluefin tuna (Kumai 1997; Sawada et al. 2005; Mylonas et al. 2007). Based on our understanding and the current state of development of induced reproduction in bluefin tuna species, natural and artificial spawning are not expected to be stable in terms of periodicity, batch fecundity, and egg quality. Thus, improving reproductive efficiency is crucial for artificial propagation and seed production techniques, and as such advances in research on the reproductive biology of bluefin tuna are critical. This chapter outlines some areas of research needing further clarification.

Research Center for Tuna Aquaculture, Seikai National Fisheries Research Institute, Fisheries Research Agency, 1551-8 Taira-machi, Nagasaki 851-2213, Japan.
Email: kgen@affrc.go.jp

Beginning with the comprehensive study of Baglin published in 1982, the reproductive physiology of bluefin tuna has been reviewed over the years (Bayliff 1994; Schaefer 2001; Rooker et al. 2007; Schirripa 2011). This chapter provides an update on developments in this field, and also on assisted techniques for the control of sexual maturation and spawning for commercial aquaculture purposes.

Gonadal Differentiation

In teleost species in general, certain cells during early embryogenesis develop as progenitors of gametes, the Primordial Germ Cells (PGCs). The PGCs originate near the gut of the embryo and migrate to the developing gonads. Gene expression analysis of vas as a germ cell-specific marker revealed that vas-positive PGCs migrate and form two bilateral rows of cells close to the position of the future gonad (Yoon et al. 1997). Following migration of PGCs into the genital ridge, PGCs remain quiescent in the undifferentiated gonad before they start to proliferate and differentiate into oogonia and spermatogonia that will eventually develop into mature eggs or sperm. Despite the vas gene of Pacific bluefin tuna being identified in a previous study (Nagasawa et al. 2009), the proliferation and migration route of PGCs is still not clarified in this species.

Numerous studies have reported the process of sex differentiation in gonochoristic fishes, where individuals develop as females or males and remain the same sex throughout their life history. Yamamoto (1969) has summarized that fish are classified as either differentiated or undifferentiated species; the gonads in differentiated gonochoristic fishes such as salmonid species develop directly into ovaries or testes from undifferentiated gonads. By contrast, in undifferentiated gonochoristic species such as zebrafish (*Danio rerio*) and red seabream (*Pagrus major*), all individuals initially develop ovarian tissues, and subsequently differentiate into a testis in approximately half of the population. Furthermore, it has been shown that the timing and the process of gonadal differentiation varies among teleost species (Devlin and Nagahama 2002). Differentiation can occur very early as in medaka (*Oryzias latipes*) with increasing numbers of germ cells in presumptive females already at hatching, intermediate as in zebrafish (21–25 days post hatching), or late in juvenile stages as in red seabream (*Pagrus major*) (120–150 days post fertilization). Although it is still unclear in what type bluefin tuna species are classified and at what stage sex differentiation occurs, our recent study suggests that Pacific bluefin tuna belongs to differentiated gonochoristic species and also signs of morphological differentiation of either the ovary or testis are observed by 100 days post hatching (Goto, unpubl. observations). To date, with the exception of a few reports in the skipjack tuna (*Katsuwonus pelamis*) (Raju 1960), hermaphroditism has only been investigated in cultured Pacific bluefin tuna among bluefin tuna species. Although both ovarian and testicular tissues were investigated in the gonads of the same individual in the skipjack tuna, the distribution of ovaries and testes in gonads was completely mosaic in Pacific bluefin tuna (Sawada et al. 2006).

Ovarian Structure and Oogenesis

Ovarian structure

The ovaries of bluefin tuna are elongated and paired organs suspended within the abdominal cavity by the mesovarium. Adhesion occurs which isolates a part of the coelom, thereby forming the ovarian cavity into which ovulation occurs directly (Fig. 15.1). This ovarian cavity is connected only with its duct and is lined with mesothelium (Nagahama 1983a). Ovaries consist of ovigerous lamellae with ovarian follicles at different stages of development. Each follicle comprises of an oocyte and is surrounded by two layers of follicle cells, an outer theca cell layer and an inner granulosa cell layer. The granulosa cell layers are the sites of production of two pivotal steroid hormones (estrogen and maturation-inducing steroid), but their production depends on the provision of precursor steroids (androgens) by the thecal cell layer (Fig. 15.1). It is therefore considered that coordinated interaction between these cells is essential for the female endocrine function and fertility (Nagahama 1994).

In fish, the pattern of oocyte development has been divided into three types; synchronous, group-synchronous, and asynchronous types (Wallace and Selman 1981). The synchronous type is exhibited by those species spawning only once in their life history such as Japanese eel (*Anguilla japonica*) and coho salmon (*Oncorhynchus kisutch*). In this type of ovarian development, all oocytes advance in synchrony through

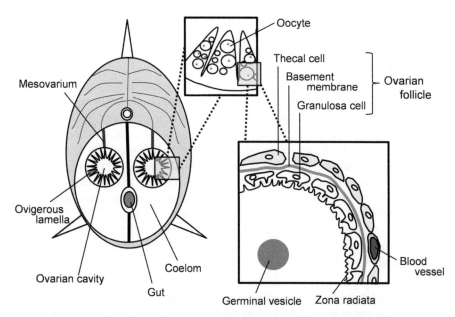

Figure 15.1. Schematic representation of the ovarian structure in bluefin tuna. Ovaries are suspended in the abdominal cavity by the peritoneal membrane called mesovarium. Ovaries consist of ovigerous lamellae with ovarian follicles at various stages of development. An ovarian follicle is composed of an oocyte surrounded by an outer thecal cell layer and an inner granulosa cell layer. The coordinated interaction between these cells is essential for the female endocrine function and fertility.

all phases of gametogenesis; thus, only one type of developing oocyte class is present in the ovary. The group-synchronous type is exhibited by seasonal spawners, which spawn one or more times during the annual reproductive season as in rainbow trout (*Oncorhynchus mykiss*) and European seabass (*Dicentrarchus labrax*). This ovarian type shows a group of yolked oocytes that is recruited and advances synchronously through further stages of development, whereas the rest of the oocyte population remains arrested. Asynchronous-type ovary is characteristic of multiple or batch spawner like bluefin tuna species and red seabream (*Pagrus major*); hence ovaries contain oocytes in all development stages and occur in species that spawn many times during the annual spawning season in daily or almost-daily intervals.

Ovarian development

Oocyte development has been extensively described in numerous teleosts with some differences depending on the teleost species. Oocytes of teleost species undergo a series of morphological changes during sexual maturation and can be classified into four phases (Nagahama 1983a); (i) proliferation phase, (ii) primary growth phase, (iii) secondary growth phase, and (iv) maturation or ovulatory phase (Fig. 15.2).

i) *Proliferation phase*: following the transformation PGCs into oogonia, oogonia proliferated through mitotic cell division forming oogonia nests in association with the pregranulosa cell. In the case of Atlantic bluefin tuna, each cell is spherical in outline measuring 8–14 μm in diameter with a prominent nucleolus (Sarasquete et al. 2002). After a prescribed time of mitotic cell division in teleost species, oogonia differentiated primary oocyte when chromosome development is stopped at the diplotene stage of the first meiotic prophase, referred to be the starting point for oocytes.

ii) *Primary growth phase*: the primary growth phase consists of two stages, chromatin nucleolus and perinucleolus stages. The oocyte of chromatin nucleolus (20–40 μm in Atlantic bluefin tuna) shows a strongly basophilic cytoplasm in histological sections, and is distinguished by a conspicuous nucleolus associated with chromatin threads (Sarasquete et al. 2002). It is generally found that this stage was relatively brief and it rarely observed, especially in maturing ovaries. During the perinucleolus stage, oocytes (45–100 μm in Atlantic bluefin tuna) have a slightly basophilic cytoplasm in histological sections, and contain multiple nucleoli adhering to the nuclear membrane (Fig. 15.2A). An electron microscopical study in Atlantic bluefin tuna demonstrated that the nuclear envelope exhibits numerous pores through which granular materials are exported from the nucleus to the cytoplasm (Abascal and Medina 2005).

iii) *Secondary growth phase*: this phase is divided into three different stages, yolk vesicle (also referred as cortical alveoli), oil droplet, and yolk globule stages (Sarasquete et al. 2002). A great deal of information is known regarding the process of secondary growth phase in fish, suggesting that the sequence of occurrence of the three stages varies with species; hence lipid droplets appear soon after yolk vesicle formation in the rainbow trout (*Oncorhynchus mykiss*), but they appear after both the yolk vesicle and yolk globule stages in the smelt

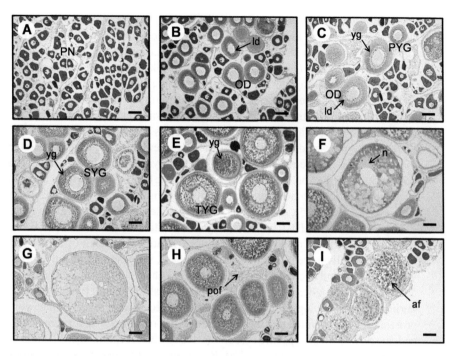

Figure 15.2. Light micrographs of histological sections from ovaries of Pacific bluefin tuna at various stages of ovarian development. (A) Perinucleolus stage; (B) Oil droplet stage; (C) Primary yolk globule stage; (D) Secondary yolk globule stage; (E) Tertiary yolk globule stage; (F) Migratory nucleus stage; (G) Mature stage; (H) Postovulatory follicle stage; (I) Atretic follicle stage. All sections were stained with hematoxylin and eosin. Scale bar 100 μm. PN: oocyte at perinucleolus stage; OD: oocyte at oil droplet stage; ld: lipid droplets; PYG: oocyte at primary yolk globule stage; SYG: oocyte at secondary yolk globule stage; TYG: oocyte at tertiary yolk globule stage; yg: yolk globule; n: germinal vesicle; pof: postovulatory follicle; af: atretic follicle.

(*Hypomesus japonicus*) (Nagahama 1983a). In Atlantic bluefin tuna and Pacific bluefin tuna, interestingly the yolk vesicle stage is very scarce, if not absent from, bluefin tuna oocytes as in other marine species; namely yolk globule appear after the commencement of the formation of lipid droplets in this species (Fig. 15.2B) (Abascal and Medina 2005; Gen et al. 2012). The yolk vesicles are eventually displaced to the periphery of the oocyte cytoplasm during the secondary growth phase, and are membrane-limited vesicles that stain with Periodic Acid-Schiff (PAS) for mucopolysaccharide and glycoprotein. In the case of the oil droplet stage, large amounts of neutral lipids are stored as lipid droplets in a more central position of the ooplasm of oocytes. Biochemical studies revealed that lipids are absorbed and accumulated in the oocyte from the plasma as very low-density lipoproteins, whereas the yolk vesicle is synthesized within the oocyte (Babin et al. 2007).

The start of the yolk globule stage is marked by the uptake of vitellogenin (Vtg). At the yolk globule stage, generally characteristic oocytes considerably increase in their volume as a results of a mainly Vtg incorporation; thus this process is referred to as vitellogenesis. The vitellogenesis can be divided into substages associated with the extent of yolk globule in ooplasm, the primary (Fig. 15.2C), the secondary (Fig. 15.2D), and tertiary yolk globule stages (Fig. 15.2E). The duration of vitellogenesis is extremely variable depending on the species. It can be a lengthy process, from several months to up to one year in species that spawn once a year with a mass of large eggs, like salmonids. In contrast, it lasts about one–two weeks in tilapias and tropical species that spawn almost every day. Fish Vtg is synthesized mainly in the liver under the control of estrogen, secreted into the blood, transferred through the vascular system, and incorporated into oocytes as described below. Growing oocytes have been shown to have unequal accumulation of different forms of Vtg by growing oocytes in several species (Sawaguchi et al. 2005). In the eastern Atlantic bluefin tuna, there are two types of Vtg (VtgA and VtgB), which have high similarities with those of other fishes (Pousis et al. 2011). Under captive conditions, relative levels of VtgA and VtgB mRNAs in the liver are low in the immature phase, increase significantly during vitellogenesis, and decline in the post-spawning phase. Furthermore, there was a trend towards higher mRNA levels in captive fish compared to wild females during oocyte development. Early studies in teleosts also revealed that yolk proteins are incorporated into the oocyte through a receptor-mediated process of endocytosis involving specific receptor (vitellogenin receptor: VtgR) (Babin et al. 2007). Recently, the cloning of VtgR sequences, and the expression profiles of its mRNA in ovaries have been identified in Atlantic bluefin tuna. Similar to other teleosts, the bluefin tuna VtgR has a number of very-low-density lipoprotein receptors, which has eight ligand binding repeats, and lacks a linked sugar domain (Pousis et al. 2012). Their expression profiles in the ovary have been characterized; transcripts of VtgR in captive females were less with spawning phase and increased in the post-spawning season. Moreover, a greater amount of VtgR mRNA was found in the wild than in captive Atlantic bluefin tuna (Pousis et al. 2012).

iv) *Maturation or ovulatory phase*: at the end of vitellogenesis, a maximum size range from 900 to 1200 mm for oocytes was recorded (900 μm Atlantic bluefin tuna; Sarasquete et al. 2002). At this stage, the nucleus lies at the center of the oocyte, which marks the end of the period of the secondary growth phase (Fig. 15.2F). Depending on the species, the maturation phase is often easier to observe, and has been verified by assessing volume increase, lipid droplet coalescence, yolk clarification, and germinal vesicle breakdown (GVBD) (Fig. 15.2G). In teleost species, unlike mammals, the GVBD corresponds to the first meiotic division immediately followed by preparation for the second division that will remain blocked at metaphase II until fertilization. Additionally, the oocyte volume and yolk clarification increase are more particular to fish (Atlantic bluefin tuna;

Sarasquete et al. 2002). These phenomena are due to the reorganization of lipoprotein yolk involving the action of proteolytic enzymes such as cathepsins. In marine pelagic eggs, like barfin flounder (*Verasper moseri*), this results in a notable rise of free amino acids and ions, leading to an osmotic disequilibrium, resulting in water entry and increased oocyte volume (Matsubara et al. 1999). Ovulation is the release of a mature oocyte from the follicle into the ovarian cavity. This process requires the separation of the oocyte from the granulosa layer, the rupture of follicle layers (also referred to as postovulatory follicles) and the expulsion of the oocyte.

Testicular Structure and Spermatogenesis

Testicular structure

In teleost species, testes are a pair of elongated structures composed of branching seminiferous tubules embedded in the stroma. A sperm duct (also referred to as the vas deferens) arises from the posterior mesodorsal surface of each elongated testis and leads to the urogenital papilla located between the rectum and the urinary ducts. Testicular structure in teleosts is variable from species to species, whereas two basic types, lobular and tubular types, can be identified according to the differentiation of germinal tissue. The testis of the lobular type, which is typical of most teleosts including bluefin tuna species, is composed of numerous lobules, which are separated from each other by a thin layer of fibrous connective tissues (Atlantic bluefin tuna; Abascal et al. 2004) (Fig. 15.3). Within the lobules, primary spermatogonia undergo numerous

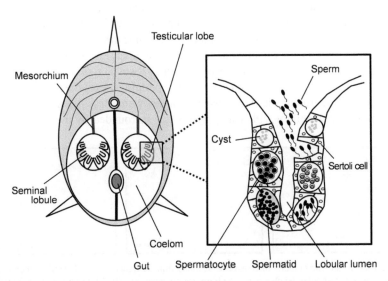

Figure 15.3. Schematic representation of the testicular structure in bluefin tuna. Testes are a pair of elongated structures composed of branching seminiferous tubules. The bluefin tuna has lobular unrestricted spermatogonial testes (referred to as lobular type), where spermatogonia undergo meiotic divisions to produce cysts containing multiple spermatogonial cells. In the testis, Leydig cells in the interstitial tissue and Sertoli cells inside the lobules are involved in steroidogenesis.

meiotic divisions to produce cyst containing several spermatogonial cells. During maturation, all of the germ cells within one cyst are at approximately the same stage of development. As spermatogenesis, and then spermiogenesis proceeds, the cysts expand and eventually rupture, liberating sperm into the lobular lumen which is continuous with the sperm duct. They divide by meiosis and give rise to spermatids from which spermatozoa are formed. The seminiferous tubules are packed with spermatozoa in the pre-spawning and spawning season (Schulz et al. 2010). In teleosts, as in other vertebrates, testicular interstitial cells (referred to as Leydig cells), which are distributed usually in the interstices of seminiferous tubules singly or in clusters are the main sites of production of androgens (Nagahama 1994). In addition, lobule boundary cells or Sertoli cells have also been implicated in steroidogenesis in some teleosts.

Testicular development

The differentiation pathway of germ cells into spermatozoa can be divided roughly into four phases (Schulz et al. 2010); (i) proliferation phase, (ii) meiotic division phase, (iii) spermiogenesis, and (iv) sperm maturation phase. In teleost species, the meiotic process in spermatogenesis is simpler than that in oogenesis, since male gametogenesis lacks both vitellogenesis and the long-term arrest in the first meiotic prophase.

(i) proliferation phase: spermatogenesis starts with the mitotic proliferation of type A spermatogonia (also referred as to primary spermatogonia), which are distributed all along the germinal epithelium. Type B spermatogonia (also referred as to secondary spermatogonia) resulting from successive mitoses of type A spermatogonia are found in small groups (Atlantic bluefin tuna; Abascal et al. 2004). After this proliferation phase, the germ cells enter meiosis (Fig.15.4A). (ii) meiotic division phase: primary spermatocytes are cells in prophase of the first meiotic division, and then proceed with the first meiotic division to produce secondary spermatocytes. These spermatocytes are smaller cells than the spermatogonia and the nucleolus is rarely visible. The second division of meiosis gives rise to the spermatids that are small, round and heterogeneous nuclei (Fig.15.4B). (iii) spermiogenesis: during spermiogenesis, the round spermatids transform into spermatozoa, that consist of an acrosome-less head, short midpiece and long single flagellum with scanty cytoplasm surrounded by the cell membrane

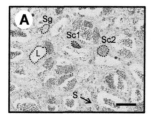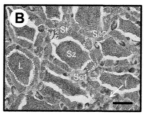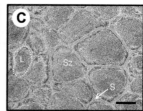

Figure 15.4. Light micrographs of transverse sections from testicular lobes of Pacific bluefin tuna at various stages of testicular development. (A) Early spermatogenesis; (B) Mid spermatogenesis; (C) Late spermatogenesis. All sections were stained with hematoxylin and eosin. Scale bar 50 µm. Sg: spermatogonia; Sc1: primary spermatocyte; Sc2: secondary spermatocyte; St: spermatid; Sz: spermatozoa; L: lobular lumen; S: Sertoli cell; I: interstitial tissue.

(Fig.15.4C). The aspect of the chromatin varies, being sometimes compact and homogeneous but usually granular with electron-lucent patches. (iv) sperm maturation phase: this phase is referred as the physiological change from nonfunctional gametes to mature spermatozoa fully capable of sperm motility and fertilization. In salmonids, acquisition of sperm motility is induced by the high pH of the seminal plasma in the sperm duct and involves the elevation of intrasperm cAMP levels (Miura et al. 1992).

The ultrastructure of Atlantic bluefin tuna sperm follows the general morphology known from perciform species. This type of sperm is referred to as the perciform sperm type (also refered as teleostean type II spermatozoon), which is characterized by the lateral insertion of the flagellum and the location of the centrioles outside the nuclear fossa (Abascal et al. 2004). The mean sperm concentration ($3.8 \pm 1.3 \times 10^{10}$ spermatozoa ml^{-1}) assessed in testis of captive Atlantic bluefin tuna (Suquet et al. 2010), which was lower than values observed in wild stock Atlantic bluefin tuna ($5.0–6.5 \times 10^{10}$ spermatozoa ml^{-1}) (Doi et al. 1982). The spermatozoa of most fish species are immotile in the testis, thereby, motility is induced after the spermatozoa are released into the aqueous environment. Numerous studies have reported that the duration of sperm motility in marine fish is generally higher than that in freshwater species including salmonids. This is also seen in Atlantic bluefin tuna that the spermatozoa being immotile at 140 seconds after activation as an energetic consequence of the high velocity ranging 215 to 230 micrometers s^{-1} (Cosson et al. 2008a). Additionally, it has been documented that several factors such as ion concentration, pH, osmotic pressure and dilution rate affect sperm motility (Cosson et al. 2008b).

Reproductive Cycle

In teleost species, ovaries and testes were roughly assigned to one of five developmental stages according to macroscopic and histological observations as follows (Murua et al. 2003). (i) immature stage: ovaries are thin, hollow tubes; nearly spherical, transparent oocytes (Fig. 15.5A). Presence of oocytes in oogonia, nuclear chromatin and perinucleolus stages. Testes translucent, filiform formed by spermatogonia, not organized in tubules (Fig. 15.6A). (ii) maturing or developing stage: ovaries are flaccid and the maturing process has started. The most advanced oocytes are in the early yolk stage. Testes are longer, wider, often of triangular or circular section, and the most predominant spermatogenic cells are spermatocytes. (iii) mature stage: this stage shows a considerable increase in gonadal weight resulting in an increase of gonadosomatic index (GSI: gonad weight/body weight x 100). In the case of asynchronous-type ovary like bluefin tuna, oocytes in several developmental stages exist together in an ovary; namely, the most advanced oocyte group has already reached the tertiary yolk globule stage and the other developmental oocyte groups are at the primary yolk globule stage or the oil droplet stage (Fig. 15.5B). The testis contains spermatogonia, spermatocytes and spermatids. Spermatozoa, which proliferate moderately within the lobule lumen initially, appear at this stage (Fig. 15.6B). (iv) spawning or ripe stage: most oocytes are at the nucleus migration stage or hydration stage. Postovulatory follicles may coexist with oocytes. The lobule lumens are full of spermatozoa with sparse peripheral spermatogonia and spermatocytes. (v) spent or resting stage: postovulatory follicles and atretic oocytes that failed to spawn are frequently observed. No oocyte develops

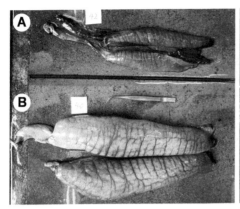

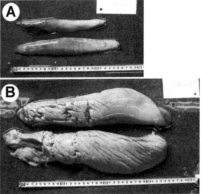

Figure 15.5. Macroscopic view of wild Pacific bluefin tuna ovaries. (A) Immature ovaries; (B) Mature ovaries. (Photos courtesy of Y. Ishihara, Tottori Prefectural Fisheries Research Center).

Figure 15.6. Macroscopic view of wild Pacific bluefin tuna testes. (A) Immature testes; (B) Mature testes. (Photos courtesy of Y. Ishihara, Tottori Prefectural Fisheries Research Center).

beyond the oil droplet stage and/or perinucleolus stage. Diffused residual spermatozoa combined with spermatogonia or spermatocytes are present in the empty space of the lobule lumen.

Based on the premise of two principal regions of spawning and juvenile production, the International Council for the Conservation of Atlantic Tuna (ICAAT) manages the Atlantic bluefin tuna under a two stock regime; one is the eastern Atlantic stock in the Mediterranean Sea, and one is the western Atlantic stock in the Gulf of Mexico (Carlsson et al. 2007; Boustany et al. 2008). Histological examinations revealed that the eastern Atlantic stock is in the immature phase from August to April, with only perinucleolus and oil droplet oocytes in ovaries and mainly spermatogonia and meiotic cells in the seminiferous epithelium. Mature bluefin tuna are observed in May, when ovaries were found in vitellogenic oocytes, and seminiferous lobules are progressively filled with spermatozoa. Hydrated oocytes and/or post-ovulatory follicles, signs of imminent and recent ovulation, have been found in actively spawning females captured in late June to early July. Fish are in the spent/resting phase from late July to September, with unyolked oocytes and late stages of atresia of vitellogenic oocytes and only residual spermatozoa in the testes (Baglin 1982; Medina et al. 2002; Corriero et al. 2005; Heinisch et al. 2008). In contrast, assessment of the reproductive condition of adult Atlantic bluefin tuna is limited in the western stock, for the reason, GSI values have been used to examine seasonal changes of maturation and spawning in this stock. Previous studies reported that GSI values observed well-developed ovaries in April and May, and were greater than 3.0% in both April and May. These results, together with hatching-data and back-calculated daily growth rate, indicate that Atlantic bluefin tuna begin spawning one month earlier in Gulf of Mexico than in the Mediterranean Sea (Karakulak et al. 2004b).

Pacific bluefin tuna is a highly migratory species found predominantly within the North Pacific Ocean. This species migrates from the Northern Pacific to the spawning grounds off the southern part of the Philippines from April to May, and move northward to off southwestern North Pacific Ocean in July (Nishikawa et al. 1985; Kitagawa et al. 2010). Mature Pacific bluefin tuna are also caught in a relatively small spawning area in the Sea of Japan in August, that is thought to have multiple spawning sites (Okiyama 1974). According to histological examinations of Pacific bluefin tuna in southern-western North Pacific Ocean, this stock spawns from late April through mid-June, when a high proportion of females have hydrated oocytes which coincide with the occurrence of postovulatory follicles as indicators of spawning events (Nishikawa et al. 1985; Chen et al. 2006). It is well accepted, however, that spawning in the Sea of Japan starts in the middle of June at the latest, and extends to the beginning of August (Tanaka 2006). At present, the detailed process of sexual maturation is largely unknown in both wild stocks, since almost all fish captured during sampling period possessed oocytes at the yolk globule or more developmental stage. With regards to captive-reared Pacific bluefin tuna, even now, much less research has been carried out on annual reproductive cycles (Hirota and Morita 1976; Miyashita et al. 2000). Previous observations reported that ovaries with unyolked oocytes were observed in cultured 4-year-old females during November to May. In the following months, ovarian development progressed with the gradual increase in the frequency of secondary and/ or tertiary yolk globule stages from May through July, and also postovulatory follicles were present in ovaries of females caught in July (Seoka et al. 2007). An additional indication that age-related alterations in the spawning period and sexual maturity occur in captive-reared females comes from our studies. In cultured 3-year-old females, ovarian development could be classified as immature and maturing on the basis of histological changes. For a female classified as immature, the ovaries had either only perinucleolus or oil droplet stage oocytes over the entire sampling period. In maturing females, the mean gonadosomatic index (GSI) gradually increased from June to reach a peak during the spawning season in July (Fig. 15.7A). Vitellogenic oocytes were first observed in ovaries of fish collected at the end of June. In the same period, yolk globule oocytes became apparent in 10% of the sampled fish, indicating that females are classified as maturing. Subsequently, ovarian development progressed with the gradual increase in the frequency of vitellogenic oocytes at the end of July (Fig. 15.7B). After the spawning period, the GSI decreased during the spent phase in August; the ovaries of all females were occupied only by the perinucleolus and oil droplet stages (Gen et al. 2012). By contrast, in the case of 4-year-old females, vitellogenic oocytes were already present in the ovaries of all fish collected in April. During the spawning period in the middle of May, oocytes in 100% of females proceeded to the tertiary yolk globule stage. Then, only perinucleolus stage oocytes were found in the ovary after the spawning season in November (Gen, unpubl. observations). This is completely different from the observations obtained in wild stock, in which the duration of spawning is confined to a very restricted period. In addition, most cultured female were not able to attain a GSI value >2 even in the age range from 5 to 12 years old in captivity (Miyashita et al. 2000), whereas mean GSI values >5 were reported for the wild population during the spawning season (Chen et al. 2006). In contrast to females, little information is available concerning testicular development of males during annual reproductive cycles. In the case of cultured 2- or 3-year-old males, testes with all stages of spermatogenesis were observed in June. Subsequently,

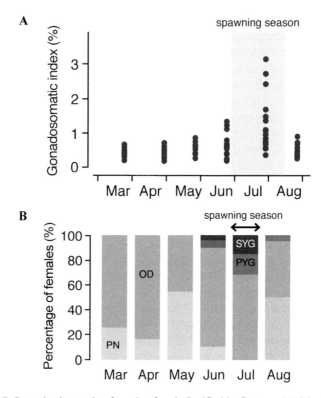

Figure 15.7. Reproductive cycle of captive female Pacific bluefin tuna. (A) Monthly changes in gonadosomatic index in 3-year-old females during sexual maturation reared at Amami Islands, Japan. The shaded area indicates the spawning season. Closed circles represent individual values for gonad somatic index. (B) Percentage of different oocyte stages in 3-year-old females during sexual maturation reared at Amami Islands, Japan. Ovaries were classified on the basis of the most advanced oocyte stage present in the ovary. PN: oocyte at perinucleolus stage; OD: oocyte at oil droplet stage; PYG: oocyte at primary yolk globule stage; SYG: oocyte at secondary yolk globule stage.

in the following months, the growth phase in both ages culminates in the formation of mature spermatozoa from May to July, demonstrating that a high proportion of males can reach first maturity at 2 years of age under intensive culture conditions (Seoka et al. 2007). Furthermore, although testicular development does not occur, the same individuals of 2 years old with functional mature testes were found in June corresponding to non-spawning season (Sawada et al. 2007).

Southern bluefin tuna are found in the southern extents of the Pacific, Indian, and Atlantic Oceans. Mature southern bluefin tuna are caught in a relatively small spawning area off the northwest coast of Australia that is thought to be their only spawning ground (Yukinawa 1987; Proctor et al. 1995). Despite the long history of the southern bluefin tuna fisheries, there is very little information available regarding the reproductive cycle of the wild stock. Thus far, annual fisheries catch data suggest that although its spawning is observed on the putative spawning ground during every month except July, most spawning occurs between September and April, with some low spawning levels seen in other months (Farley and Davis 1998).

Spawning

Spawning condition

Common to all bluefin tuna species is the strong relationship between spawning activity and sea surface temperature. Schaefer (2001) pointed out that surface temperature exceeding 24°C appears to stimulate spawning activity in all tuna species. Recent detailed analyses have also revealed that the most reproductively advanced female in Atlantic bluefin tuna was sampled when sea surface temperature was preferentially in the range from 22.6 to 27.5°C in the Gulf of Mexico and from 22.5 to 25.5°C in Mediterranean Sea, respectively (Karakulak et al. 2004a; 2004b; Garcia et al. 2005; Teo et al. 2007; Heinisch et al. 2008). Similar phenomena were observed in cultured Pacific bluefin tuna, which spawn in waters of surface temperatures of 21.6–29.2°C under natural conditions (Sawada et al. 2005). These results suggest that bluefin tuna are capable of spawning at temperatures below 24°C under natural conditions. A recent report based in wild Pacific bluefin tuna provided information to support this hypothesis; the spawning in the Sea of Japan stock occurred at temperatures as low as 20°C (ISC 2011). In contrast to Atlantic bluefin and Pacific bluefin tuna, little is known about the spawning traits of southern bluefin tuna. This species is observed to migrate to the spawning ground where the sea surface temperature is greater than 24°C and which are considered suitable for spawning (Yukinawa and Miyabe 1984; Bubner et al. 2012).

Spawning time

Reports from fisheries data suggest that tunas, like many pelagic marine fishes, are nocturnal spawners (McPherson et al. 1991; Schaefer 2001). Indeed, the spawning time has been observed for the eastern stock of Atlantic bluefin tuna within transport cages at night between 03:00 and 05:00 (Gordoa et al. 2009). However, studies of captive-reared bluefin tunas indicated that spawning also occurs during daylight and twilight hours. Captive-reared Pacific bluefin tuna have been reported to spawn between 17:00 and 23:30 hours (Masuma 2006a), 18:30 and 20:00 hours (Sawada et al. 2005) or between 17:00 and 19:00 hours (Kumai 1998; Masuma 2006a; Lioka et al. 2000). Additionally, there was a positive linear relationship between water temperature and spawning time in cultured Pacific bluefin tuna (e.g., spawning occurred later with increasing water temperature); suggesting the variation in spawning time may be related to temperature changes during the spawning season (Fig. 15.8) (Masuma et al. 2006b). An interest finding in artificially induced spawning fish by hormone treatment, females in adjacent cages appeared to shift their respective spawning times, although the hormonal therapy for the both cages were conducted during the same hours of the day (Rosenfeld et al. 2012).

Spawning behavior

Spawning is one of the most important behaviors of fish reproduction. The spawning behavior of captive-reared Pacific bluefin tuna exhibit courting and pairing behaviors

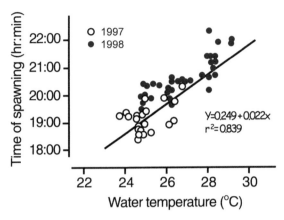

Figure 15.8. Relationship between water temperature and time of spawning for captive female Pacific bluefin tuna. The circles represent individual values from 1997 (open circles) and 1998 (closed circles), respectively. (Modified from Masuma et al. 2000).

in a sequence just before the simultaneous and adjacent release of gametes (Mimori et al. 2008). This behavior is apparently similar to that for captive yellowfin tuna *Thunnus albacares*, with more than one male commonly chasing after the same female, and lateral body markings and coloration body stripes and/or darker coloration are displayed by males during this period (Margulies et al. 2007). Interestingly, although tunas are considered to be surface spawners as noted previously, it has been observed that for cultured-Pacific bluefin tuna, spawning started at depths and exhibit a spiraling upward movement towards the surface from the bottom of floating net pens (Masuma 2006a). In contrast, information on wild bluefin tuna is limited though there are current unconfirmed reports of fisherman catching this species. More recently, researchers have employed physiological telemetry to monitor spawning activity in wild Atlantic bluefin tuna (Block et al. 2005; Teo et al. 2007). The results clearly indicate that a unique diving pattern is found on the majority of nights while the fish stayed at the spawning ground. This pattern consisted of frequent and brief oscillatory movements up and down through the mixed layer, which is believed to reflect recent courtship and spawning activity (Aranda et al. 2013a).

Spawning frequency and batch fecundity

As indicated above, the follicle tissues that contain hydrated oocytes collapse after ovulation, and remain in the ovary as an evacuated follicle. These structures are postovulatory follicles (POFs) (Fig. 15.2H) and retain their integrity for a day or more before being resorbed; thus POFs can be used to estimate spawning frequencies (Hunter and Macewicz 1985; Murua et al. 2003). The data obtained from a previous study indicated that in wild Atlantic bluefin tuna spawning frequency has been estimated at 1.2 days for Atlantic bluefin tuna in wild populations (Medina et al. 2002). This interval is similar to the observed frequencies of other members of tuna species except Pacific bluefin tuna in the south-western North Pacific Ocean (3.3 days; Chen et al. 2006); 1.62 days for Southern bluefin tuna (Farley and Davis 1998), 1.05 days for bigeye tuna (Sun et al. 2013), 1.27 to 1.99 days for yellowfin tuna (Itano 2000). These results suggested that bluefin tuna species capable of spawning almost daily, although the spawning frequency of individuals was not determined. In the case of cultured fish, based on sequencing analysis of mitochondrial DNA-D-loop region of fertilized eggs, eight haplotypes occurred at two to three consecutive days, suggesting that captive Pacific bluefin tuna females are capable to consecutively spawn multiple times during the spawning season (Nakadate et al. 2011). An additional indication is that a female spawned twice and three times a day at intervals for five days during the spawning period came from observations in the aquarium at Tokyo Sea Life Park (Mimori et al. 2008).

The batch fecundity is often estimated by counting the number of oocyte maturation follicles, migratory-nucleus (Fig. 15.2F) and hydrated oocytes (Fig. 15.2G) in the ovary (Murua et al. 2003). As described earlier, these follicles can be easily distinguished from other developmental stages in ovaries by their larger size (>750 μm). Most marine teleosts, like bluefin tuna species, have very high fecundity and it is not practical to count migratory-nucleus and hydrated oocytes in the ovary. For this reason, fish fecundity has often been determined using gravimetric methods, in which relative fecundity is the product of the number of oocytes per gram of body weight. Earlier studies in Atlantic bluefin tuna have shown that estimates of relative fecundity ranged from 97 to 137 oocytes/g Body Weight (BW) in the eastern Atlantic along the coast of Spain (Rodríguez-Roda 1967). Similar to the results obtained in the previous analysis, estimated from stereological quantifications was 93 oocytes/g BW (Medina et al. 2002). These fecundities of Atlantic bluefin tuna are greater than those estimated for other tuna species; 57 oocytes/g BW for southern bluefin tuna (Farley and Davis 1998), 31 oocytes/g BW for bigeye tuna *Thunnus obesus* (Nikaido et al. 1991), 67 oocytes/g BW for yellowfin tuna (Schaefer 1998).

Internal and External Determinants of Sexual Maturation

Growth, body size and age

It is widely accepted that size and/or age are important parameters of fish reproduction and usually represented by length at 50% maturity (L50; Triple and Harvey 1991). In the eastern Atlantic bluefin tuna, 50% of females in Mediterranean Sea were reproductively active at approximately 103 cm Curved Fork Length (CFL) at 3 years of age, and 100% maturity was reached between 115 and 121 cm CFL at age 4 or 5 years (Rodriguez-Roda 1967). Additionally, L50 value estimates have been further refined as described below (Corriero et al. 2005); 50% females of stock in the Mediterranean Sea reached sexual maturity at 104 cm Straight Fork Length (SFL) at 3 years of age and 100% at 130 cm SFL at age 4 or 5 years (Fig. 15.9). In contrast to the eastern stock, the size of females spawning in the Gulf of Mexico are estimated to be greater than 109 cm CFL, which corresponds to about 10 years old (Clay 1991). This disparity in age-at-maturity between eastern and western Atlantic bluefin tuna has been used as a major argument for separation into two stocks. In addition, the estimated age at 50% maturity was found to be 15.8 years old by reexamination of age at maturity for the western Atlantic using refined growth curve (Diaz 2011). Taken together, these finding

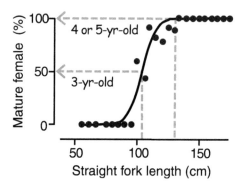

Figure 15.9. Size and age at maturity of female Atlantic bluefin tuna. Length at 50% maturity was estimated by fitting a logistic curve to the proportion of mature individuals at length (Modified from Corriero et al. 2005).

indicate that both female Atlantic bluefin tuna stocks display different reproductive schedules as reflected by the earlier age-at-maturity of the eastern stock, suggesting that it may result in a higher productivity of the eastern stock (Aranda et al. 2013b). Regarding Pacific bluefin tuna, sexual maturity of females in south-western North Pacific Ocean is estimated to occur at 5 years of age, and that the lengths of these fish are about 150 cm SFL (Harada 1980). Also, nearly all of the historical catch by the Taiwanese longline fleet is greater than 8 years old (190 cm), with a mode at ages 9 to 10 years (208 to 219 cm SFL). It is different from females in the Sea of Japan; maturity is estimated to occur at a younger age of 3 years or about 120 cm SFL (Tanaka 2006). In contrast to the wild population, it is uncertain and controversial at what age sexual maturity occurs in captive Pacific bluefin tuna. Since the first reported spawning in captivity was in 1979 (Kumai 1997), efforts at spawning have been met with some success demonstrating that spawning has mostly occurred for fish after 5 years in captivity (Lioka et al. 2000). Nevertheless, recent studies showed that females reared in captivity matured at 3 or 4 years of age (Seoka et al. 2007; Masuma et al. 2008; Gen et al. 2012). It is then clear that much more research is needed to acquire a better understanding of age-at-maturity in captive Pacific bluefin tuna. Southern bluefin tuna show the size at maturity for wild southern stock as 130 cm fork length, which was in effect an estimate of size at first maturity (Warashina and Hisada 1970; Shingu 1970). Further studies both on and off the spawning ground suggest that size at 50% maturity is around 150 to 160 cm fork length at 10 to 12 years of age (Davis 1995; Davis and Farley 2001). The youngest southern bluefin tuna caught by the Indonesia long-line fisheries on the spawning ground was 5 years old, although most are >8 years old and >150 cm (Farley et al. 2007; 2011). These observations suggest that captive rearing of female southern bluefin tuna may not shift age-specific reproductive maturity, despite some differences in indices of morphological features being apparent between captive and wild southern bluefin tuna.

Environmental factors

Photoperiod

Photoperiod has a major effect on the activity of the brain-pituitary axis and consequently on gonad maturation (Fig. 15.10). Unlike mammals, in teleost species, the pineal gland is directly photosensitive and secretes melatonin known as a photosensitive hormone. Thus, this organ conveys the photoperiodic information to the pituitary gland, which may translate photoperiodic information into neuroendocrine responses (Falcon et al. 2010). A great deal of information is known regarding the effects of photoperiod on fish reproduction (Migaud et al. 2010). In many salmonids, which spawn in autumn, gradually increasing photoperiods followed by decreasing shorter photoperiods play a dominant role in the regulation of reproductive cycles (Bromage et al. 2001; Davies and Bromage 2002). Moreover, photoperiod manipulation changes the incidence of sexual maturation, spawning time, fecundity and egg size in marine benthopelagic fish, Atlantic cod *Gadus morhua* (Hansen et al. 2001); hence, cod that were transferred from a natural photoperiod to a continuous light condition spawned earlier during the pre-breeding season, and had a low fecundity and smaller eggs than cod reared under natural conditions. Even though the photoperiodic response have so far been described in numerous teleost fish, however, no detailed information is available regarding the effects of photoperiod on reproductive processes of bluefin tuna species. As indicated earlier, breeding in Pacific bluefin tuna occurs from late May to early September under natural environmental conditions. This type of fish is well known as spring-to-summer spawning teleosts, in which the short daylenth in autumn is the predominant factor that causes the termination of the spawning period (Shimizu 2003). Consistent with this view, spawning of cultured 4-year-old Pacific

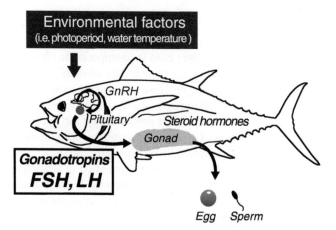

Figure 15.10. Schematic diagram of endocrine regulation of gametogenesis in fish. The process of gametogenesis and sexual maturation of gonads is regulated by the hypothalamic-pituitary-gonadal axis. Environmental cues (i.e., photoperiod, water temperature) are integrated by the brain to stimulate the GnRH. GnRH is transported to the pituitary and triggers the synthesis and/or release of FSH and LH. As a result, GtHs released from the pituitary stimulate the production of steroid hormones which in turn control gametogenesis.

bluefin tuna was observed almost daily from the end of May 2008 until the beginning of September 2008. Namely, the termination of the spawning occurs under the short daylenth condition but the range of water temperature remained unchanged during the spawning period (Gen et al. 2012), thereby lending additional support to the notion of the physiological role for photoperiod in the completion of sexual maturation of bluefin tuna (Fig. 15.11).

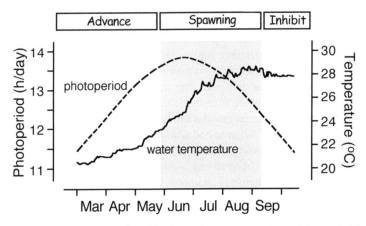

Figure 15.11. Diagrammatic representation of the changes in water temperature and photoperiod throughout sexual maturation in captive female Pacific bluefin tuna. The Solid line represents the daily average seawater temperature, and the dashed line represents the length of the photoperiod under natural conditions reared at the Amami Islands, Japan. The shaded are indicates the spawning season in captive 3-year-old females.

Water temperature

Bromage et al. (2001) demonstrated that photoperiod is the main factor entraining the endogenous processes or clocks which time reproduction, whereas other environmental factors are considered to be a secondary cues to photoperiod in phasing reproductive seasonality. Among these factors, water temperature obviously plays an important role in the control of all reproductive processes for gametogenesis (Fig. 15.10) (Van Der Kraak and Pankhurst 1997). Numerous studies have suggested that the effects of temperature are differentially expressed depending on fish species; for instance, elevated temperatures are required to cue maturation in spring-to-summer spawner such as mummichog *Fundulus heteroclitus* (Shimizu 2003), whereas increasing temperatures delays the onset of maturation and ovulation in autumn spawning fish such as salmonids (Pankhurst and King 2010). Currently, it remains unclear whether the water temperature influences sexual maturation in bluefin tuna species. However, it has been proposed that water temperature could play a dominant role in the regulation of vitellogenesis in bluefin tuna species known as spring-to-summer spawners, because ovarian development progressed in association with elevated water temperature in cultured fish (Gen et al. 2012), and it is thought that wild stock populations migrate from cold waters to warm tropical waters for spawning (Schaefer 2001). Furthermore, the optimal temperature during sexual maturation is important, and temperatures, which are too high and too low during final maturation of oocyte, may reduce fecundity and result in a high degree of atresia and poor egg quality (Pankhurst and King 2010; Okuzawa and Gen 2013). As already mentioned, there is little doubt that bluefin tuna spawn in waters of surface temperatures of around 24°C. In addition to the optimum temperature range, it has been postulated that the resulting changes in seawater temperature are also important for oocyte final maturation and spawning. This hypothesis is supported by the observation that a unique vertical movement which reflects spawning activity resulted in distinctive thermal profiles characterized by densely packed temperature oscillations that might drop between 2 to 3°C (Aranda et al. 2013a). Also, results obtained further support this hypothesis by demonstrating that a rapid rise of surface temperature from 23°C to 24°C, resulted in natural spawning to occur more efficiently in captive Pacific bluefin tuna (Masuma et al. 2008).

Other determinants

A range of studies have demonstrated the inhibitory and/or interruptive effects of stress on reproduction in teleost species (Schreck 2010). Although atresia is a common phenomenon in the mammalian ovary involving oocyte and the follicular wall degeneration, the presence of atretic follicles in teleost ovaries is frequently associated with environmental stress including population stress, reduced food abundance, excessive temperature and so on (Fig. 15.2I) (Wood and Van Der Kraak 2001). As in case of Atlantic bluefin tuna, females subjected to either severe acute stress or short-term starvation (up to 14 days) exhibited a decrease in gonad weight and had significantly higher population of atretic vitellogenic follicles than either wild or long-term caged individuals (Corriero et al. 2011). Its findings give considerable credence to the view that severe acute stress and/or starvation during the early spawning period is associated with a significant increase in the occurrence of atretic follicles in bluefin tuna species.

The vast majority of fish are iteroparous, which spawn several times during their life history. In long lived marine species, of interest is the finding that delayed maturation and non-annual spawning behavior exist in their lives, commonly referred to as skipped spawning. Rideout et al. (2005) has summarized that numerous marine and freshwater fish exhibit a range of 9 to 86% skipped spawning on an annual basis. The phenomenon is positively correlated with the reproductive life span, and is generally considered to be evidence of a trade-off between reproduction, growth and survival across the life span of the species (Rideout and Tomkiewicz 2011). In bluefin tuna species, catch data and several tagging studies have reported adult Atlantic and Pacific bluefin tuna in waters away from spawning areas during the spawning season (Zupa et al. 2009; Galuardi et al. 2010). Therefore, although skipped spawning has not been conclusively documented to date, it would be reasonable to assume that bluefin tuna are able to skip spawning in some years, or are using presently undocumented spawning areas (Secor 2008).

Also, lipid storage (Thorpe 2007), stock density (Pankhurst and Fitzgibbon 2006), and social communication (Stacey 2003) of control of reproduction in teleost species obviously exist, but they have been far less studied in bluefin tuna species (Mourente et al. 2001). It is therefore clear that further studies are needed to clarify the precise role and effect on their reproductive process.

Endocrine Control of Sexual Maturation

The regulation of gametogenesis in teleosts is largely dependent on the hypothalamic-pituitary-gonadal axis, where gonadotropin (GtH) is a critical modulator of gametogenesis and gonadal steroidogenesis (Gen et al. 2003; Swanson et al. 2003) (Fig. 15.10). In many teleost species, similar to tetrapod vertebrates, biochemical analysis and molecular cloning studies have demonstrated that there are two distinct GtHs, follicle-stimulating hormone (FSH; formerly known as GTH-I) and luteinizing hormone (LH; formerly known as GTH-II) (bigeye tuna: Okada et al. 1994; Atlantic bluefin tuna: Berkovich et al. 2013). Both GtHs are heterodimer glycoproteins each consisting of a common α-glycoprotein subunit and a unique β subunit (FSHβ and LHβ) that are non-covalently linked. Phylogenetic analyses indicate that teleost FSHβ sequences are more divergent than those of LHβ (Kato et al. 1993), suggesting that functional diversification of FSH has occurred in teleost species during the evolutionary process (Fig. 15.12). Immunocytochemical analyses demonstrated that both GtHs are synthesized in separate cells in the pituitary in bluefin tuna species; FSH-producing cells occupy a dorsal part of the proximal pars distalis (PPD), whereas LH-producing cells are present in the central part of the PPD and the external border of the pars intermedia (Kagawa et al. 1998; Rodriguez-Gomez et al. 2001). As in all vertebrates, GtHs are under the control of neuroendocrine factors released at hypothalamus. Of

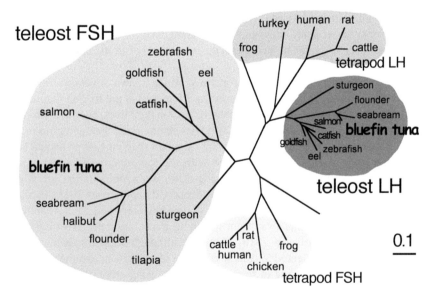

Figure 15.12. Phylogenetic tree of gonadotropin β-subunits in vertebrates. The unrooted phylogenetic tree was constructed using the neighbor-joining method after alignment of deduced amino acid sequences of the gonadotropin β−subunit proteins. The scale bar corresponds to estimated evolutionary distance units.

these factors, the hypothalamic decapeptide, gonadotropin-releasing hormone (GnRH) in fish plays a pivotal role in the synthesis and release of LH and possibly also FSH. However, the neuroendocrine mechanisms regulated by GnRH are not yet known in bluefin tuna species.

In teleost species, it is well accepted that GtHs perform separate functions and are found at different stages of the maturation process; namely, FSH is found in the pituitary and blood of fish undergoing initial gonadal growth and gametogenesis, whereas LH is predominant during the final stages of reproduction, i.e., final oocyte maturation and ovulation (Fig. 15.13). Mediation of these process of gametogenesis is carried out by steroid hormones, synthesized as a consequence of GtHs action, implying an indirectly impact of FSH and LH. In female fish in general, estradiol-17β (E_2) is the main estrogen that is essential for oocyte development and induction of the precursor of yolk protein (Vtg) in the liver. In the ovarian follicle, E_2 is produced in the granulosa cell layer by conversion of testosterone, which is produced in the theca cell layer (Nagahama 1994). In the case of Atlantic bluefin tuna, previous data obtained from the Mediterranean Sea stock has reported the correlation with plasma levels of E_2, Vtg, and the histological maturity stage of the ovaries. The plasma E_2 levels were low during the immature phase with only the perinucleolus stage in ovaries. Subsequently, their levels sharply increased during the mature stage in which the most advanced oocytes reached the secondary yolk globule stage. The levels of E_2 remained high in the pre-spawning phase, and rapidly decreased in the spent/resting stage. Furthermore, there was a good correlation between plasma E_2 levels and the appearance of vitellogenic oocytes and during the vitellogenic period, suggesting that like other teleost species, E_2 is implicated in the control of the oocyte growth by the synthesis of Vtg (Susca et al. 2001a).

The endocrinological contribution of GtHs, especially LH, to the control of the final oocyte maturation has been also demonstrated by teleost species. Numerous

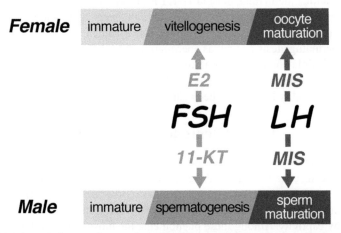

Figure 15.13. Schematic diagram of proposed gonadotropic regulation of ovarian and testicular development in teleost species. Gonadotropins perform separate functions and are found at different stages of the maturation process; data so far in numerous teleost species, indicate that FSH plays a significant role in vitellogenesis and spermatogenesis, whereas LH is important in processes involved in the final maturation and ovulation or spermiation.

studies have shown that this process, associated with progression to meiosis resumption, is an acute increase in plasma LH levels, and an LH-driven switch in the ovarian follicle steroidogenic pathway from the E_2 production during vitellogenesis, to the production of maturation-inducing steroids (MIS) (Fig. 15.13) (Nagahama and Yamashita 2008). The isolation of 17α, 20β-dihydroxy-4-pregnen-3-one (17α, 20β-DP) as MIS from amago salmon *Oncorhynchus masou* (Nagahama and Adachi 1985), and also 17α, 20β, 21-trihydroxy-4-pregnen-3-one (20β-S) is identified as the MIS of same other marine perciform species, like Atlantic crocker (*Micropogonias undulatus*) (Trant and Thomas 1989). Previous studies have demonstrated that plasma levels of 17α, 20β-DP were very low or nondetectable in vitellogenic females but were strikingly elevated in mature and ovulated salmonid species. Other studies also have shown that 17α, 20β-DP was the most effective inducer of GVBD in amago salmon postvitellogenic oocytes (Nagahama 1983b). Similar results were observed for 20β-S; it is effective in inducing oocyte maturation of Atlantic crocker and European sea bass (Suwa and Yamashita 2007). Therefore, 17α, 20β-DP and 20β-S have been recognized as potent MIS in many fish species. Although it is supposed that hormone-induced oocyte maturation is mediated by MIS in bluefin tuna species (Rosenfeld et al. 2012), no attention is given to a number of important subjects such as what type of MIS is primarily involved in final maturation of oocyte.

The most important androgen produced by the male gonad is considered to be 11-ketotestosterone (11-KT) in teleost fish, whereas testosterone is regarded as the pivotal hormone regulating spermatogenesis in tetrapods (Borg 1994). It was found in salmonids that rising 11-KT plasma levels mark the start of spermatogonial proliferation in the annual cycle, and increasing 11-KT concentrations were associated with the advancement of spermatogenesis towards spermation (Campbell et al. 2003). Furthermore, 11-KT has been shown to induce spermatogenesis by stimulating the differentiation of spermatogonia towards meiosis *in vitro* (Miura et al. 1991). Likewise, in male Atlantic bluefin tuna, plasma 11-KT is present in very low levels in immature phase, rises approaching spawning season to high levels, and then drops in the post-spawning phase (Susca et al. 2001b). Thus, it is assumed that 11-KT has important roles in spermatogenesis as well as spermiation in male bluefin tuna. At the present time, there is little doubt that bluefin tuna FSH and LH are involved in the control of ovarian and testicular steroidogenesis (Berkovich et al. 2013), although the functional gonadotropic duality has not yet been clarified.

Assisted Reproductive Techniques

Most aquacultured fish exhibit some degree of reproductive dysfunction, ranging from the complete absence of gonadal development observed in the freshwater eel to the absence of spawning observed in trout and salmon (Bromage et al. 1992). These are due to the absence of an appropriate environment for sexual maturation, as well as stresses imposed by captivity. Among these phenomena, the most common dysfunctions are the failure of females to undergo final oocyte maturation and ovulation, and those of males of reduced production of sperm and the reduced quality of the produced sperm (Zohar and Mylonas 2001). Therefore, the failure to undergo final maturation can be alleviated by treatment with various reproductive hormones (Gen et al. 2001; Zohar and Mylonas 2001; Mylonas et al. 2007). Of the hormones, gonadotropin-releasing hormone agonist (GnRHa) offer a great advantage due to their high potency and synthesis and stimulation of endogenous GtHs, especially LH, release (Mylonas and Zohar 2001). Recently, reproductive control of Atlantic bluefin tuna in captivity was examined for the first time using wild-caught broodstock that were administered with a control-release delivery system loaded with GnRHa. In females, post-treatment histological analysis of the ovaries demonstrated that the GnRHa-treatment of the females was effective in inducing oocyte maturation, ovulation and spawning (Corriero et al. 2007). An interesting observation was that a high proportion of GnRHa-treated females possessed oocyte maturation follicles, while captive untreated and wild tuna lacked oocyte maturation follicles. Additionally, hormonal treatment was found to reduce the rates of follicular atresia among treated Atlantic bluefin tuna female. In contrast to females, treatment with a GnRHa was not effective in inducing testicular maturation, although the percentage of spermiating tuna increased (Corriero et al. 2007; Mylonas et al. 2007). A subsequent study demonstrated that GnRHa treatment stimulated spermatogonial proliferation and reduced the presence of apototic germ cells. These results suggests that GnRHa administration is effective in enhancing germ cell proliferation and reducing apoptosis in captive males through the stimulation of luteinizing hormone release and testicular 11-KT production (Corriero et al. 2009).

Conclusion and Future Perspectives

Knowledge of the reproductive traits such as the maturational schedule and its regulation has contributed to reproductive control of bluefin tuna species, which is necessary for the sustainability of commercial tuna aquaculture. Also, fundamental investigations of the reproductive biology have enabled the rational exploitation and responsible management of wild populations (Schaefer 2001). Since the mid-2000s, there have been significant increases in our knowledge of the processes of ovarian and testicular development, the reproductive-related hormones including sex steroids, and the suitable conditions of spawning in wild and captive bluefin tuna species. Likewise, information has accrued on the hormonal induction of oocyte maturation and spermiation, as well as sperm cryopreservation. Nevertheless, to our knowledge, the study of bluefin tuna reproduction has been restricted almost exclusively to the Atlantic bluefin tuna. Furthermore, even for Atlantic bluefin tuna, there is not enough information concerning its reproduction as compared with other aquatic important species such as salmonids, seabream and cod. Because of a relatively long time to reach sexual maturity (>5 years old) and complexity and expense for its breeding, it quite difficult to clarify; however, advances in field studies should provide valuable information to resolve the problem with captive bluefin tuna. This is because most marine fish species generally exhibit some degree of reproductive dysfunction under intensive culture conditions, but this dysfunction does not occur in the natural setting. So, the ability to collect detailed information on wild tuna migration and thermal habitat is very important to understanding the processes that affect its reproduction at various stages in the near future. Furthermore, advent of transcriptomes and proteomes approaches, as well as genomes approaches (Chini et al. 2008; Gardner et al. 2012; Nakamura et al. 2013) will affect reproductive studies of this species in the future. These approaches will contribute to the identification of molecular pathways involved in ovarian and testicular development and will facilitate the comparative analyses of gonadal development between wild and captive fish. These researches will open up new understandings of bluefin tuna reproduction that can be used to answer some basic and applied questions.

Acknowledgements

The author wishes to thank Drs. Sukei Masuma, Takahiro Matsubara and Yukinori Kazeto for providing helpful criticisms and suggestions. The work conducted by the author was supported in part by Council for Science, Technology and Innovation (CSTI), Cross-ministerial Strategic Innovation Promotion Program (SIP), "Technologies for creating next-generation agriculture, forestry and fisheries" (funding agency: Bio-oriented Technology Research Advancement Institution, NARO); the Research Project for Utilizing Advanced Technologies in Agriculture, Forestry and Fisheries, Ministry of Agriculture, Forestry and Fisheries of Japan; a Grant-in-Aid for Scientific Research (C) (No-25450294) from Japan Society for the Promotion of Science.

References

Abascal, F.J., U. Megina and A. Medina. 2004. Testicular development in migrant and spawning bluefin tuna (*Thunnus thynnus* (L.)) from the eastern Atlantic and Mediterranean. Fish. Bull. 102: 407–417.

Abascal, F.J. and A. Medina. 2005. Ultrastructure of oogenesis in the bluefin tuna, *Thunnus thynnus*. J. Morphol. 264: 149–160.

Aranda, G., F.J. Abascal, J.L. Varela and A. Medina. 2013a. Spawning behaviour and post-spawning migration patterns of Atlantic bluefin tuna (*Thunnus thynnus*) ascertained from satellite archival tags. PLoS One 8: e76445.

Aranda, G., A. Medina, A. Santos, F.J. Abascal and T. Galaz. 2013b. Evaluation of Atlantic bluefin tuna reproductive potential in the western Mediterranean Sea. J. Sea Res. 76: 154–160.

Babin, P.J., O. Carnevali, E. Lubzens and W.J. Schnieder. 2007. Molecular aspects of oocyte vitellogenesis in fish. pp. 39–77. *In*: P. Babin, J. Cerda and E. Lubzens (eds.). The Fish Oocyte: From Basic Studies to Biotechnological Application. Springer, New York.

Baglin, R.E. 1982. Reproductive biology of western Atlantic bluefin tuna. Fish. Bull. 80: 121–134.

Bayliff, W.H. 1994. A review of the biology and fisheries for northern bluefin tuna, *Thunnus thynnus*, in the Pacific Ocean. FAO Fish. Tech. Paper. 336: 244–295.

Berkovich, N., A. Corriero, N. Santamaria, C.C. Mylonas, R. Vassallo-Aguis, F. de la Gandara, I. Meiri-Ashkenazi, V. Zlatnikov, H. Gordin, C.R. Bridges and H. Rosenfeld. 2013. Intra-pituitary relationship of follicle stimulating hormone and luteinizing hormone during pubertal development in Atlantic bluefin tuna (*Thunnus thynnus*). Gen. Comp. Endocrinol. 194: 10–23.

Block, B.A., S.L.H. Teo, A. Walli, A. Boustany, M.J.W. Stokesbury, C.J. Farwell, K.C. Weng, H. Dewar and T.D. Williams. 2005. Electronic tagging and population structure of Atlantic bluefin tuna. Nature 434: 1121–1127.

Borg, B. 1994. Androgens in teleost fishes. Comp. Biochem. Physiol. C-Toxicol. Pharmacol. 109: 219–245.

Boustany, A.M., C.A. Reeb and B.A. Block. 2008. Mitochondrial DNA and electronic tracking reveal population structure of Atlantic bluefin tuna (*Thunnus thynnus*). Mar. Biol. 156: 13–24.

Bromage, N., J. Jones, C. Randall, M. Thrush, B. Davies, J. Springate, J. Duston and G. Barker. 1992. Broodstock management, fecundity, egg quality and the timing of egg-production in the rainbow trout (*Oncorhynchus mykiss*). Aquaculture 100: 141–166.

Bromage, N., M. Porter and C. Randall. 2001. The environmental regulation of maturation in farmed finfish with special reference to the role of photoperiod and melatonin. Aquaculture 197: 63–98.

Bubner, E., J. Farley, P. Thomas, T. Bolton and A. Elizur. 2012. Assessment of reproductive maturation of southern bluefin tuna (*Thunnus maccoyii*) in captivity. Aquaculture 364: 82–95.

Campbell, B., J.T. Dickey and P. Swanson. 2003. Endocrine changes during onset of puberty in male spring chinook salmon, *Oncorhynchustshawytscha*. Biol. Reprod. 69: 2109–2117.

Carlsson, J., J.R. McDowell, J.E.L. Carlsson and J.E. Graves. 2007. Genetic identity of YOY bluefin tuna from the eastern and western Atlantic spawning areas. J. Hered. 98: 23–28.

Chen, K.S., P. Crone and C.C. Hsu. 2006. Reproductive biology of female Pacific bluefin tuna *Thunnus orientalis* from south-western North Pacific Ocean. Fish. Sci. 72: 985–994.

Chini, V., A.G. Cattaneo, F. Rossi, G. Bernardini, G. Terova, M. Saroglia and R. Gornati. 2008. Gene expressed in bluefin tuna (*Thunnus thynnus*) liver and gonads. Gene 410: 207–213.

Clay, D. 1991. Atlantic bluefin tuna (*Thunnus thynnus thynnus* (L.)): A review. Inter-Amer. Trop. Tuna Comm. Spec. Rep. 7: 89–180.

Collette, B.B. 1999. Mackerels, molecules, and morphology. 1999. pp. 149–164. *In*: B. Seret and J.Y. Sire (eds.). Proceedings of the 5th Indo-Pacific Fish Conference. Nouméa, New Caledonia; 3–8 November 1997.

Corriero, A., S. Karakulak, N. Santamaria, M. Deflorio, D. Spedicato, P. Addis, S. Desantis, F. Cirillo, A. Fenech-Farrugia, R. Vassallo-Agius, J.M. de la Serna, Y. Oray, A. Cau, P. Megalofonou and G. De Metrio. 2005. Size and age at sexual maturity of female bluefin tuna (*Thunnus thynnus* L. 1758) from the Mediterranean Sea. J. Appl. Ichthyol. 21: 483–486.

Corriero, A., A. Medina, C.C. Mylonas, F.J. Abascal, M. Deflorio, L. Aragón, C.R. Bridges, N. Santamaria, G. Heinisch, R. Vassallo-Agius, A. Belmonte Ríos, C. Fauvel, A. Garcia, H. Gordin and G. De Metrio. 2007. Histological study of the effects of treatment with gonadotropin-releasing hormone agonist (GnRHa) on the reproductive maturation of captive-reared Atlantic bluefin tuna (*Thunnus thynnus* L.). Aquaculture 272: 675–686.

Corriero, A., A. Medina, C.C. Mylonas, C.R. Bridges, N. Santamaria, M. Deflorio, M. Losurdo, R. Zupa, H. Gordin, F. de la Gandara, A. Belmonte Rìos, C. Pousis and G. De Metrio. 2009. Proliferation and apoptosis of male germ cells in captive Atlantic bluefin tuna (*Thunnus thynnus* L.) treated with gonadotropin-releasing hormone agonist (GnRHa). Anim. Reprod. Sci. 116: 346–357.

Corriero, A., R. Zupa, G. Bello, C.C. Mylonas, M. Deflorio, S. Genovese, G. Basilone, G. Buscaino, G. Buffa, C. Pousis, G. De Metrio and N. Santamaria. 2011. Evidence that severe acute stress and starvation induce rapid atresia of ovarian vitellogenic follicles in Atlantic bluefin tuna, *Thunnus thynnus* (L.) (Osteichthyes: Scombridae). J. Fish Dis. 34: 853–860.

Cosson, J., A.L. Groison, M. Suquet, C. Fauvel, C. Dreanno and R. Billard. 2008a. Studying sperm motility in marine fish: an overview on the state of the art. J. Appl. Ichthyol. 24: 460–486.

Cosson, J., A.L. Groison, M. Suquet, C. Fauvel, C. Dreanno and R. Billard. 2008b. Marine fish spermatozoa: racing ephemeral swimmers. Reproduction 136: 277–294.

Davies, B. and N. Bromage. 2002. The effects of fluctuating seasonal and constant water temperatures on the photoperiodic advancement of reproduction in female rainbow trout *Oncorhynchus mykiss*. Aquaculture 205: 183–200.

Davis, T.L.O. 1995. Size at first maturity of southern bluefin tuna. pp. 8. *In*: Report CCSBT/SC/95/9 prepared for the CCSBT Scientific Meeting.Simizu, Japan; 10–19 July 1995.

Davis, T.L.O. and J.H. Farley. 2001. Size distribution of southern bluefin tuna (*Thunnus maccoyii*) by depth on their spawning ground. Fish. Bull. 99: 381–386.

Devlin, R.H. and Y. Nagahama. 2002. Sex determination and sex differentiation in fish: an overview of genetic, physiological, and environmental influences. Aquaculture 208: 191–364.

Diaz, G.A. 2011. A revision of western Atlantic bluefin tuna age of maturity derived from size samples collected by the Japanese longline fleet in the Gulf of Mexico (1975–1980). Collect. Vol. Sci. Pap. ICCAT. 66: 1216–1226.

Doi, M.T. Hoshino, Y. Takiand and Y. Ogasawara. 1982. Activity of the sperm of the bluefin tuna *Thunnus thynnus* under fresh and preserved conditions. Bull. Japan. Soc. Sci. Fish. 48: 495–498.

Falcon, J., H. Migaud, J.A. Munoz-Cueto and M. Carrillo. 2010. Current knowledge on the melatonin system in teleost fish. Gen. Comp. Endocrinol. 165: 469–482.

Farley, J.H. and T.L.O. Davis. 1998. Reproductive dynamics of southern bluefin tuna, *Thunnus maccoyii*. Fish. Bull. 96: 223–236.

Farley, J.H., T.L.O. Davis, J.S. Gunn, N.P. Clear and A.L. Preece. 2007. Demographic patterns of southern bluefin tuna, *Thunnus maccoyii*, as inferred from direct aging data. Fish. Res. 83: 151–161.

Farley, J., P. Evenson and C. Proctor. 2011. Update on the length and age distribution of SBT in the Indonesian longline catch. pp. 10. *In*: Working paper CCSBT-ESC/1107/18 prepared for the CCSBT Extended Scientific Committee for the 16th meeting of the Scientific Committee. Bali, Indonesia; 19–28 July 2011.

Galuardi, B., F. Royer, W. Golet, J. Logan, J. Neilson and M. Lutcavage. 2010. Complex migration routes of Atlantic bluefin tuna (*Thunnus thynnus*) question current population structure paradigm. Can. J. Fish. Aquat. Sci. 67: 966–976.

Garcia, A., F. Alemany, J.M. de la Serna, I. Oray, S. Karakulak, L. Rollandi, A. Arigò and S. Mazzola. 2005. Preliminary results of the 2004 bluefin tuna larval surveys off different Mediterranean sites (Balearic Archipelago, Levantine Sea and the Sicilian Channel). Collect. Vol. Sci. Pap. ICCAT. 58: 1420–1428.

Gardner, L.D., N. Jayasundara, P.C. Castilho and B.A. Block. 2012. Microarray gene expression profiles from mature gonad tissues of Atlantic bluefin tuna, *Thunnus thynnus* in the Gulf of Mexico. BMC Genomics 13: 530.

Gen, K., K. Okuzawa, N. Kumakura, S. Yamaguchi and H. Kagawa. 2001. Correlation between messenger RNA expression of cytochrome p450 aromatase and its enzyme activity during oocyte development in the red seabream (*Pagrus major*). Biol. Reprod. 65: 1186–1194.

Gen, K., S. Yamaguchi, K. Okuzawa, N. Kumakura, H. Tanaka and H. Kagawa. 2003. Physiological roles of FSH and LH in red seabream, *Pagrus major*. Fish Physiol. Biochem. 28: 77–80.

Gen, K., H. Nikaido, Y. Kazeto, T. Matsubara, S. Sawaguchi and S. Masuma. 2012. Gonadal development and serum steroid levels during pubertal development in captive-reared Pacific bluefin tuna. pp. 11. *In*: Abstracts of the 7th International Symposium on Fish Endocrinology. Buenos Aires, Argentina; 3–6 September 2012.

Gordoa, A., M. PilarOlivar, R. Arevalo, J. Vinas, B. Moli and X. Illas. 2009. Determination of Atlantic bluefin tuna (*Thunnus thynnus*) spawning time within a transport cage in the western Mediterranean. ICESJ. Mar. Sci. 66: 2205–2210.

Hansen, T., O. Karlsen, G.L. Taranger, G.I. Hemre, J.C. Holm and O.S. Kjesbu. 2001. Growth, gonadal development and spawning time of Atlantic cod (*Gadusmorhua*) reared under different photoperiods. Aquaculture 203: 51–67.

Harada, T. 1980. Development and future outlook of studies on the aquaculture of tunas. pp. 50–58. *In*: Maguro Gyokyo Kyogikai Gijiroku, Suisancho-Enyo Suisan Kenkyusho (Proceedings of the Tuna Fishery Research Conference, Japan Fisheries Agency-Far Seas Fisheries Research Laboratory). Simizu, Japan; 6-7 February 1980.

Heinisch, G., A. Corriero, A. Medina, F.J. Abascal, J.M. de la Serna, R. Vassallo-Agius, A.B. Rios, A. Garcia, F. de la Gandara, C. Fauvel, C.R. Bridges, C.C. Mylonas, S.F. Karakulak, I. Oray, G. De Metrio, H. Rosenfeld and H. Gordin. 2008. Spatial-temporal pattern of bluefin tuna (*Thunnus thynnus* L. 1758) gonad maturation across the Mediterranean Sea. Mar. Biol. 154: 623–630.

Hirota, H. and M. Morita. 1976. An instance of the maturation of 3 full years old bluefin tuna cultured in the floating net. Bull. Japan. Soc. Sci. Fish. 42: 939 (in Japanese with English abstract).

Hunter, J.R. and B.J. Macewicz. 1985. Measurement of spawning frequency in multiple spawning fishes. pp. 79–94. *In*: R. Lasker (eds.). An Egg Production Method for Estimating Spawning Biomass of Pelagic Fish: Application to the Northern Anchovy, *Engraulis mordax*. U.S. Dep. Comm., NOAA Tech. Rep NMFS 36, Government Printing Office, Washington D.C.

ISC. 2011. Report of the Pacific Bluefin Tuna Working Group Workshop, 6–13 January, 2011, Shizuoka, Japan.

Itano, D.G. 2000. The reproductive biology of yellowfin tuna (*Thunnus albacares*) in Hawaiian waters and the western tropical Pacific Ocean. *In*: Project Summary. Pelagic Fisheries Research Program Report SOEST 00-01. JIMAR Contribution 00-328. Univ. of Hawaii.

Kagawa, H., I. Kawazoe, H. Tanaka and K. Okuzawa. 1998. Immunocytochemical identification of two distinct gonadotropic cells (GTH I and GTH II) in the pituitary of bluefin tuna, *Thunnus thynnus*. Gen. Comp. Endocrinol. 110: 11–18.

Karakulak, S., I. Oray, A. Corriero, D. Spedicato, D. Suban, N. Santamaria and G. De Metrio. 2004a. First information on the reproductive biology of the bluefin tuna (*Thunnus thynnus*) in the Eastern Mediterranean. Collect. Vol. Sci. Pap. ICCAT. 56: 1158–1162.

Karakulak, S., I. Oray, A. Corriero, M. Deflorio, N. Santamaria, S. Desantis and G. De Metrio. 2004b. Evidence of a spawning area for the bluefin tuna (*Thunnus thynnus* L.) in the eastern Mediterranean. J. Appl. Ichthyol. 20: 318–320.

Kato, Y., K. Gen, O. Maruyama, K. Tomizawa and T. Kato. 1993. Molecular cloning of cDNAs encoding two gonadotrophin β subunits (GTH-Iβ and -IIβ) from the masu salmon, *Oncorhynchus masou*: rapid divergence of the GTH-Iβ gene. J. Mol. Endocrinol. 11: 275–282.

Kitagawa, T., Y. Kato, M.J. Miller, Y. Sasai, H. Sasaki and S. Kimura. 2010. The restricted spawning area and season of Pacific bluefin tuna facilitate use of nursery areas: A modeling approach to larval and juvenile dispersal processes. J. Exp. Mar. Biol. Ecol. 393: 23–31.

Kumai, H. 1997. Present state of bluefin tuna aquaculture in Japan. Suisanzoshoku 45: 293–297 (in Japanese with English abstract).

Kumai, H. 1998. Studies on bluefin tuna artificial hatching, rearing and reproduction. Nippon Suisan Gakkaishi 64: 601–605 (in Japanese).

Lioka, C., K. Kani and H. Nhhala. 2000. Present status and prospects of technical development of tuna sea-farming. *In*: Recent advances in Mediterranean aquaculture finfish species diversification. Cah. Opt. Méditerr. 47: 275–285.

Margulies, D., J.M. Suter, S.L. Hunt, R.J. Olson, V.P. Scholey, J.B. Wexler and A. Nakazawa. 2007. Spawning and early development of captive yellowfin tuna (*Thunnus albacares*). Fish. Bull. 105: 249–265.

Masuma, S. 2006a. Maturation and spawning of bluefin tuna in captivity. pp. 15–19. *In*: Proceedings of the Kinki University International Symposium on Ecology and Aquaculture of Bluefin Tuna. Amami Oshima, Japan; 10–11 November.

Masuma, S., N. Tezuka, M. Koiso, T. Jinbo, T. Takebe, H. Yamazaki, H. Obana, K. Ide, H. Nikaido and H. Imaizumi. 2006b. Effects of water temperature on bluefin tuna spawning biology in captivity. Bull. Fish. Res. Agen. 4: 157–172 (in Japanese with English abstract).

Masuma, S., S. Miyashita, H. Yamamoto and H. Kumai. 2008. Status of bluefin tuna farming, broodstock management, breeding and fingerling production in Japan. Rev. Fish. Sci. 16: 385–390.

Matsubara, T., N. Ohkubo, T. Andoh, C.V. Sullivan and A. Hara. 1999. Two forms of vitellogenin, yielding two distinct lipovitellins, play different roles during oocyte maturation and early development of barfin flounder, Veraspermoseri, a marine teleost that spawns pelagic eggs. Dev. Biol. 213: 18–32.

McPherson, G.R. 1991. Reproductive biology of yellowfin tuna in the eastern Australian fishing zone, with special reference to the north-western coral sea. Aust. J. Mar. Freshwat. Res. 42: 465–477.

Medina, A., F.J. Abascal, C. Megina and A. Garcia. 2002. Stereological assessment of the reproductive status of female Atlantic northern bluefin tuna during migration to Mediterranean spawning grounds through the Strait of Gibraltar. J. Fish Biol. 60: 203–217.

Migaud, H., A. Davie and J.F. Taylor. 2010. Current knowledge on the photoneuroendocrine regulation of reproduction in temperate fish species. J. Fish Biol. 76: 27–68.

Mimori, R., S. Tada and H. Arai. 2008. Overview of husbandry and spawning of Bluefin tuna in the aquarium at Tokyo Sea Life Park. pp. 130–136. *In*: Proceedings of 7th International Aquarium Congress, Shanghai, China; 19–24 October 2008.

Miura, T., K. Yamauchi, H. Takahashi and Y. Nagahama. 1991. Hormonal induction of all stages of spermatogenesis *in vitro* in the male Japanese eel (*Anguilla japonica*). Proc. Natl. Acad. Sci. USA 88: 5774–5778.

Miura, T., K. Yamauchi, H. Takahashi and Y. Nagahama. 1992. The role of hormones in the acquisition of sperm motility in salmonid fish. J. Exp. Zool. 261: 359–363.

Miyashita, S., O. Murata, Y. Sawada, O. Okada, T. Kubo, Y. Ishitani, M. Seoka and H. Kumai. 2000. Maturation and spawning of cultured bluefin tuna, *Thunnus thynnus*. Suisanzoshoku 48: 475–488 (in Japanese with English abstract).

Mourente, G., C. Megina and E. Diaz-Salvago. 2001. Lipids in female northern bluefin tuna (*Thunnus thynnus thynnus* L.) during sexual maturation. Fish Physiol. Biochem. 24: 351–363.

Murua, H., G. Kraus, F. Saborido-Rey, P.R. Witthames, A. Thorsen and S. Junquera. 2003. Procedures to estimate fecundity of marine fish species in relation to their reproductive strategy. J. Northw. Atl. Fish. Sci. 33: 33–54.

Mylonas, C.C. and Y. Zohar. 2001. Use of GnRHa-delivery systems for the control of reproduction in fish. Rev. Fish Biol. and Fish. 10: 463–491.

Mylonas, C.C., C. Bridges, H. Gordin, A.B. Rios, A. Garcia, F. de la Gandara, C. Fauvel, M. Suquet, A. Medina, M. Papadaki, G. Heinisch, G. De Metrio, A. Corriero, R. Vassallo-Agius, J.M. Guzmán, E. Mañanos and Y. Zohar. 2007. Preparation and administration of gonadotropin-releasing hormone agonist (GnRHa) implants for the artificial control of reproductive maturation in captive-reared Atlantic bluefin tuna (*Thunnus thynnus thynnus*). Rev. Fish. Sci. 15: 183–210.

Nagahama, Y. 1983a. The functional morphology of teleost gonads. pp. 223–275. *In*: W.S. Hoar, D.J. Randall and E.M. Donaldson (eds.). Reproduction Endocrine Tissues and Hormones, Fish Physiology 9(A). Academic Press, New York.

Nagahama, Y., K. Hirose, G. Yong, S. Adachi, K. Suzuki and B. Tamaoki. 1983b. Relative *in vitro* effectiveness of 17α, 20β-dihydroxy-4-pregnen-3-one and other pregnene derivatives on germinal vesicle breakdown in oocytes of ayu (*Plecoglossusaltivelis*), amago salmon (*Oncorhychus rhodurus*), rainbow trout (*Salmogairdneri*), and goldfish (*Carassiusauratus*). Gen. Comp. Endocrinol. 51: 15–23.

Nagahama, Y. and S. Adachi. 1985. Identification of maturation-inducing steroid in a teleost, the amago salmon (*Oncorhynchus rhodurus*). Dev. Biol. 109: 428–35.

Nagahama, Y. 1994. Endocrine regulation of gametogenesis in fish. Int. J. Dev. Biol. 38: 217–229.

Nagahama, Y. and M. Yamashita. 2008. Regulation of oocyte maturation in fish. Dev. Growth Differ. 50: S195–S219.

Nagasawa, K., Y. Takeuchi, M. Miwa, K. Higuchi, T. Morita, T. Mitsuboshi, K. Miyaki, K. Kadomura and G. Yoshizaki. 2009. cDNA cloning and expression analysis of a vasa-like gene in Pacific bluefin tuna *Thunnus orientalis*. Fish. Sci. 75: 71–79.

Nakadate, M., T. Kusano, H. Fushimi, H. Kondo, I. Hirono and T. Aoki. 2011. Multiple spawning of captive Pacific bluefin tuna (*Thunnus orientalis*) as revealed by mitochondrial DNA analysis. Aquaculture 310: 325–328.

Nakamura, Y., K. Mori, K. Saitoh, K. Oshima, M. Mekuchi, T. Sugaya, Y. Shigenobu, N. Ojima, S. Muta, A. Fujiwara, M. Yasuike, I. Oohara, H. Hirakawa, V.S. Chowdhury, T. Kobayashi, K. Nakajima, M. Sano, T. Wada, K. Tashiro, K. Ikeo, M. Hattori, S. Kuhara, T. Gojobori and K. Inouye. 2013. Evolutionary changes of multiple visual pigment genes in the complete genome of Pacific bluefin tuna. Proc. Natl. Acad. Sci. USA 110: 11061–11066.

Nikaido, H., H. Miyabe and S. Ueyanagi. 1991. Spawning time and frequency of bigeye tuna, *Thunnus obesus*. Bull. Natl. Res. Ins. Far seas Fish. 28: 47–73.

Nishikawa, Y., M. Honma, S. Ueyanagi and S. Kikawa. 1985. Average distribution of larvae of oceanic species of scombroid fishes, 1956–1981. Bull. Far seas Fish. Res. Lab. 12: 1–99.

Okada, T., I. Kawazoe, S. Kimura, Y. Sasamoto, K. Aida and H. Kawauchi. 1994. Purification and characterization of gonadotropin-I and gonadotropin-II from pituitary glands of tuna (*Thunnus obesus*). Int. J. Peptide Protein Res. 43: 69–80.

Okiyama, M. 1974. Occurrence of the postlarvae of bluefin tuna, *Thunnus thynnus*, in the Japan Sea. Bull. Japan Sea Reg. Fish. Res. Lab. 25: 89–97.

Okuzawa, K., N. Kumakura, A. Mori, K. Gen, S. Yamaguchi and H. Kagawa. 2002. Regulation of GnRH and its receptor in a teleost, red seabream. pp. 95–110. *In*: I.S. Parhar (ed.). Gonadotropin-Releasing Hormone: Molecules and Receptors, Progress in Brain Research 141. Elsevier, Amsterdam.

Okuzawa, K. and K. Gen. 2013. High water temperature impairs ovarian activity and gene expression in the brain-pituitary-gonadal axis in female red seabream during the spawning season. Gen. Comp. Endocrinol. 194: 24–30.

Pankhurst, N.W. and Q.P. Fitzgibbon. 2006. Characteristics of spawning behaviour in cultured green back flounder Rhombosoleatapirina. Aquaculture 253: 279–289.

Pankhurst, N.W. and H.R. King. 2010. Temperature and salmonid reproduction: implications for aquaculture. J. Fish Biol. 76: 69–85.

Pousis, C., C. De Giorgi, C.C. Mylonas, C.R. Bridges, R. Zupa, R. Vassallo-Agius, F. de la Gandara, C. Dileo, G. De Metrio and A. Corriero. 2011. Comparative study of liver vitellogenin gene expression and oocyte yolk accumulation in wild and captive Atlantic bluefin tuna (*Thunnus thynnus* L.). Anim. Reprod. Sci. 123: 98–105.

Pousis, C., N. Santamaria, R. Zupa, C. De Giorgi, C.C. Mylonas, C.R. Bridges, F. de la Gandara, R. Vassallo-Agius, G. Bello and A. Corriero. 2012. Expression of vitellogenin receptor gene in the ovary of wild and captive Atlantic bluefin tuna (*Thunnus thynnus*). Anim. Reprod. Sci. 132: 101–110.

Proctor, C.H., R.E. Thresher, J.S. Gunn, D.J. Mills, I.R. Harrowfield and S.H. Sie. 1995. Stock structure of the southern bluefin tuna *Thunnus maccoyii*: an investigation based on probe microanalysis of otolith composition. Mar. Biol. 122: 511–526.

Raju, G. 1960. A case of hermaphroditism and some other gonadal abnormalities in the skipjack *Katsuwonus pelamis* (Linnaeus). J. Mar. Biol. Ass. India 2: 95–102.

Rideout, R.M., G.A. Rose and M.P.M. Burton. 2005. Skipped spawning in female iteroparousfishes. Fish Fish. 6: 50–72.

Rideout, R.M. and J. Tomkiewicz. 2011. Skipped Spawning in Fishes: More Common than You Might Think. Mar. Coast. Fish. 3: 176–189.

Rodriguez-Gomez, F.J., M.C. Rendon-Unceta, C. Pinuela, J.A. Munoz-Cueto, N. Jimenez-Tenorio and C. Sarasquete. 2001. Immunocytohistochemical characterization of pituitary cells of the bluefin tuna, *Thunnus thynnus* L. Histol. Histopath. 16: 443–451.

Rodríguez-Roda, J. 1967. Fecundidad del atún, *Thunnus thynnus* (L.), de la costa sudatlántica de España. Inv. Pesq. 31: 33–52.

Rooker, J.R., J.R.A. Bremer, B.A. Block, H. Dewar, G. De Metrio, A. Corriero, R.T. Kraus, E.D. Prince, E. Rodriguez-Marin and D.H. Secor. 2007. Life history and stock structure of Atlantic bluefin tuna (*Thunnus thynnus*). Rev. Fish. Sci. 15: 265–310.

Rosenfeld, H., C.C. Mylonas, C.R. Bridges, G. Heinisch, A. Corriero, R. Vassallo-Aguis, A. Medina, A. Belmonte, A. Garcia, F. De la Gandara, C. Fauvel, G. De Metrio, I. Meiri-Ashkenazi, H. Gordin and Y. Zohar. 2012. GnRHa-mediated stimulation of the reproductive endocrine axis in captive Atlantic bluefin tuna, *Thunnus thynnus*. Gen. Comp. Endocrinol. 175: 55–64.

Sarasquete, C., S. Cardenas, M.L.G. de Canales and E. Pascual. 2002. Oogenesis in the bluefin tuna, *Thunnus thynnus* L. A histological and histochemical study. Histol. Histopath. 17: 775–788.

Sawada, Y., T. Okada, S. Miyashita, O. Murata and H. Kumai. 2005. Completion of the Pacific bluefin tuna *Thunnus orientalis* (Temmincket Schlegel) life cycle. Aqua. Res. 36: 413–421.

Sawada, Y., M. Seoka, K. Kato, T. Tamura, M. Nakatani, S. Hayashi, T. Okada, K. Tose, S. Miyashita, O. Murata and H. Kumai. 2007. Testes maturation of reared Pacific bluefin tuna *Thunnus orientalis* at two plus years old. Fish. Sci. 73: 1070–1077.

Sawada, Y., M. Seoka, T. Okada, S. Miyashita, O. Murata and H. Kumai. 2006. Hermaphroditism in a captive-raised Pacific bluefin tuna. J. Fish Biol. 60: 263–265.

Sawaguchi, S., Y. Koya, N. Yoshizaki, N. Ohkubo, T. Andoh, N. Hiramatsu, C.V. Sullivan, A. Hara and T. Matsubara. 2005. Multiple vitellogenins (Vgs) in mosquitofish (*Gambusiaaffinis*): Identification and characterization of three functional Vg genes and their circulating and yolk protein products. Biol. Reprod. 72: 1045–1060.

Schreck, C.B. 2010. Stress and fish reproduction: The roles of allostasis and hormesis. Gen. Comp. Endocrinol. 165: 549–556.

Schaefer, K.M. 1998. Reproductive biology of yellowfin tuna (*Thunnus albacares*) in the eastern Pacific Ocean. Inter-Am. Trop. Tuna Comm., Bull. 21: 201–272.

Schaefer, K.M. 2001. Reproductive biology of tunas. pp. 225–270. *In*: B.A. Block and E.D. Stevens (eds.). Tuna. Physiology, Ecology, and Evolution. Academic Press, New York.

Schirripa, M.J. 2011. A literature review of Atlantic bluefin tuna. Age at maturity. Collect. Vol. Sci. Pap. ICCAT. 66: 898–914.

Schulz, R.W., L.R. de Franca, J.J. Lareyre, F. Le Gac, H. Chiarini-Garcia, R.H. Nobrega and T. Miura. 2010. Spermatogenesis in fish. Gen. Comp. Endocrinol. 165: 390–411.

Secor, D.H. 2008. Influence of skipped spawning and misspecified reproductive schedules on biological reference points in sustainable fisheries. Trans. Am. Fish. Soc. 137: 782–789.

Seoka, M., K. Kato, T. Kubo, Y. Mukai, W. Sakamoto, H. Kumai and O. Murata. 2007. Gonadal maturation of Pacific bluefin tuna *Thunnus orientalis* in captivity. Aquaculture Sci. 55: 289–292

Shimizu, A. 2003. Effect of photoperiod and temperature on gonadal activity and plasma steroid levels in a reared strain of the mummichog (*Fundulusheteroclitus*) during different phases of its annual reproductive cycle. Gen. Comp. Endocrinol. 131: 310–324.

Shingu, C. 1970. Studies relevant to distribution and migration of southern bluefin tuna. Bull. Far Seas Fish. Res. Lab. 3: 57–113.

Stacey, N. 2003. Hormones, pheromones and reproductive behavior. Fish Physiol. Biochem. 28: 229–235.

Sun, C.L., S.Z. Yeh, Y.J. Chang, H.Y. Chang and S.L. Chu. 2013. Reproductive biology of female bigeye tuna *Thunnus obesus* in the western Pacific Ocean. J. Fish Biol. 83: 250–271.

Suquet, M., J. Cosson, F. de la Gandara, C.C. Mylonas, M. Papadaki, S. Lallemant and C. Fauvel. 2010. Sperm features of captive Atlantic bluefin tuna (*Thunnus thynnus*). J. Appl. Ichthyol. 26: 775–778.

Susca, V., A. Corriero, C.R. Bridges and G. De Metrio. 2001a. Study of the sexual maturity of female bluefin tuna: purification and partial characterization of vitellogenin and its use in an enzyme-linked immunosorbent assay. J. Fish Biol. 58: 815–831.

Susca, V., A. Corriero, M. Deflorio, C.R. Bridges and G. De Metrio. 2001b. New results on the reproductive biology of the bluefin tuna (*Thunnus thynnus*) in the Mediterranean. Collect. Vol. Sci. Pap. ICCAT. 52: 745–751.

Suwa, K. and M. Yamashita. 2007. Regulatory mechanisms of oocyte maturation and ovulation. pp. 323–347. *In*: P. Babin, J. Cerda and E. Lubzens (eds.). The Fish Oocyte: From Basic Studies to Biotechnological Application, Springer, New York.

Swanson, P., J.T. Dickey and B. Campbell. 2003. Biochemistry and physiology of fish gonadotropins. Fish Physiol. Biochem. 28: 53–59.

Tanaka, S. 2006. Maturation of bluefin tuna in the Sea of Japan. pp. 7. *In*: Report ISC/06/PBFWG/09 prepared for the ISC Bluefin Tuna Working Group Meeting. Simizu, Japan; 16–20 January 2006.

Teo, S.L.H., A. Boustany, H. Dewar, M.J.W. Stokesbury, K.C. Weng, S. Beemer, A.C. Seitz, C.J. Farwell, E.D. Prince and B.A. Block. 2007. Annual migrations, diving behavior, and thermal biology of Atlantic bluefin tuna, *Thunnus thynnus*, on their Gulf of Mexico breeding grounds. Mar. Biol. 151: 1–18.

Thorpe, J.E. 2007. Maturation responses of salmonids to changing developmental opportunities. Mar. Ecol. Prog. Ser. 335: 285–288.

Trant, J.M. and P. Thomas. 1989. Isolation of a novel maturation-inducing steroid produced *in vitro* by ovaries of atlantic croaker. Gen. Comp. Endocrinol. 75: 397–404.

Triple, E.A. and H.H. Harvey. 1991. Comparison of methods used to estimate age and length of fishes at sexual maturity using populations of white sucker (*Catostomuscommersoni*). Can. J. Fish. Aquat. Sci. 48: 1446–1459.

Van Der Kraak, G. and N.W. Pankhurst. 1997. Temperature effects on the reproductive performance of fish. pp. 159–176. *In*: C.M. Wood and D.G. McDonald (eds.). Global Warming: Implications for Freshwater and Marine Fish, Society for Experimental Biology Seminar Series 61, Cambridge University Press, Cambridge.

Wallace, R.A. and K. Selman. 1981. Cellular and dynamic aspects of oocyte growth in teleosts. Am. Zool. 21: 325–343.

Warashina, L. and K. Hisada. 1970. Spawning activity and discoloration of meat and loss of weight in the southern bluefin tuna. Bull. Far Seas Fish. Res. Lab. 3: 147–165.

Wood, A.W. and G.J. Van Der Kraak. 2001. Apoptosis and ovarian function: Novel perspectives from the teleosts. Biol. Reprod. 64: 264–271.

Yamamoto, T. 1969. Sex differentiation. pp. 117–175. *In*: W. Hoar and D. Randall (eds.). Fish Physiology. Academic Press, New York.

Yoon, C., K. Kawakami and N. Hopkins. 1997. Zebrafish vasa homologue RNA is localized to the cleavage planes of 2- and 4-cell-stage embryos and is expressed in the primordial germ cells. Development 124: 3157–3165.

Yukinawa, M. and N. Miyabe. 1984. Report on 1983 research cruise of the R/V ShoyoMaru. Distribution of tuna and billfishes larvae and oceanographic observation in the eastern Indian Ocean October–December, 1983. Rep. Res. Div. Fish. Agency Jpn. 58: 1–103.

Yukinawa, M. 1987. Report on 1986 research cruise of the R/V Shoyo-Maru. Distribution of tunas and billfishes larvae and oceanographic observation in the eastern Indian Ocean January–March, Rep. Res. Div. Fish. Agency Jpn. 61: 1–100.

Zohar, Y. and C.C. Mylonas. 2001. Endocrine manipulations of spawning in cultured fish: from hormones to genes. Aquaculture 197: 99–136.

Zupa, R., A. Corriero, M. Deflorio, N. Santamaria, D. Spedicato, C. Marano, M. Losurdo, C.R. Bridges and G. De Metrio. 2009. A histological investigation of the occurrence of non-reproductive female bluefin tuna *Thunnus thynnus* in the Mediterranean Sea. J. Fish Biol. 75: 1221–1229.

V. New Insights: Mathematical model about Bluefin Tuna Ecology and Measuring Method for Swimming Behaviour of Bluefin Tuna

CHAPTER 16

Introduction to Modeling
Vertical Movement of Bluefin Tuna as a Stochastic Process

Minoru Kadota[1],* and *Eric White*[2]

Introduction

Mathematical models are an essential tool for conducting meaningful research in natural sciences. In almost all fields of science, mathematical models are used to formulate predictions and quantify experimental results in a precise and repeatable way. Models are constructed systematically, so that their functionality and predictive power can be rigorously verified.

The use of mathematical models has become an increasingly important tool with regard to the science of fisheries. Models are used in a variety of applications ranging from descriptions of microscopic objects—such as molecular structures of aquatic organisms—to the study of migration patterns of bluefin tuna, who dive to depths of hundreds of meters during their migration across the Pacific Ocean. Models are also used for studying the distribution of fishery resources in order to optimize resource management and maintenance of international fisheries. In this chapter, mathematics is presented as a tool for modeling physical phenomena. Even with no knowledge of advanced mathematics, we hope the reader will gain an understanding of the significance of these mathematical models and their applications.

[1] Temple University, Japan Campus, Tokyo, Japan.
[2] California Polytechnic State University, San Luis Obispo, CA.
* Email: tuf28758@tuj.temple.edu

Analysis and Synthesis

Before constructing a mathematical model, we first examine two fundamental concepts used for establishing physical theory. These two concepts can be traced back to René Descartes, the father of modern natural science. In his treatise *Rules for the Direction of the Mind*, we see early attempts to formulate what we now call the scientific method. Descartes proposes that we must:

> "reduce involved and obscure propositions step by step to those that are simpler, and then starting with the intuitive apprehension of all those that are absolutely simple, attempt to ascend to the knowledge of all others by precisely similar steps."

In other words, a complex phenomenon can be understood by first reducing it into component parts, which is a process called *analysis*. The reverse process, called *synthesis*, consists of combining many component parts in order to form a complex whole. Generally speaking, analysis is a process of deduction, by which complex concepts are reduced into simpler ideas in order to attain a better understanding of their entirety, while synthesis is a collection of simpler ideas that are used as a foundation for constructing new concepts. To paraphrase the mathematician Bernhard Riemann, every analysis requires a subsequent synthesis in order to confirm its accuracy with reference to our experience, and every synthesis, such as the universal laws of motion, rests upon a preceding analysis.

To illustrate the relation between analysis and synthesis, consider some examples from other fields of science and mathematics:

1. When factorizing prime numbers, integers are decomposed into prime numbers. In the treatment of polynomial factorization, these primes can be synthesized to form expanded polynomials.

2. In medical science, DNA analyses are performed to better understand the structure of genes and genomes. DNA synthesis is then used to create artificial gene sequences and DNA biosynthesis.

3. In chemistry, substances are analyzed in order to reduce their components into individual elements of matter. These elements can then be combined to synthesize complex compounds.

4. In signal processing, a Fourier analysis decomposes complicated functions into simple sine and cosine functions. Through synthesis, these simple trigonometric terms can be combined into more complicated signals, such as the digital simulation of tones produced by musical instruments.

Newton and Leibniz established the field of calculus through a similar mathematical systematization of the concepts of analysis and synthesis. Specifically, they applied analysis and synthesis to the study of continuous action and change. The process of differentiation consists of dividing a continuous action into infinitesimally small, or instantaneous, rates of change. By then summing, or integrating, each of these infinitesimal rates of change, the area bounded by a function can be calculated. Thus, differentiation and integration can be considered analogous to analysis and synthesis.

By applying these concepts to falling objects, Newton showed that the trajectories of objects could be described through differential equations involving position and time. By solving these equations through integration, Newton then showed that, given some initial conditions, the trajectory of an object could be predicted for all future times. We summarize these concepts as follows:

> Analysis plays an important role in the creation of a quantitative framework by which we can predict and describe a physical process. Such a framework can then be used to establish a foundation from which subsequent synthesis can be carried out. This synthesis, in turn, can be used to validate the quantitative model as an accurate descriptor of other physical phenomena. This cycle forms the cornerstone for the construction of a solid mathematical model.

Mathematical Models

Models are used as a means of expressing real-world phenomena in terms of abstract forms, and then extracting and validating essential physical processes from those abstractions. For instance, let us consider the everyday phenomenon of an apple falling from a tree. The essence of this phenomenon cannot be grasped if one focuses specifically on the apple. This is because objects do not have to be apples in order to fall from a tree! Our efforts should be focused on a more general property that is common to all objects, such as its mass, rather than on specific properties of the object itself. By constructing a physical model of point masses, whereby the mass of an object is taken to be localized at its center, it is possible to describe the collective motion of all objects as the sum of the motions of each individual point mass. Although modeling an object as a point mass disregards some attributes, such as its spatial extent and deformation, it can be a useful simplification for real-world problems in which the sizes of objects are much less than the distances that separate them.

While a physical model makes it possible to qualitatively interpret the motion of objects, meaningful and quantitative conclusions are attained only after the laws governing such physical models are converted into mathematical formulae. To see this, let us more closely examine the phenomenon of a falling object. The motion of a falling object can be considered the effect of a gravitational force on the object. From this perspective, one could establish the following hypothesis: because of gravity, all objects with mass are pulled toward the Earth. However, since the Earth is also an object with mass, it too will be pulled toward other objects that have mass. We would thus revise our theory so that it states: because of gravity, all objects with mass are pulled toward all other object having mass. Needless to say, this is the heart of Newton's law of universal gravitation, which is considered one of the most exemplary abstractions of real-world phenomena.

Having established a physical model, we are yet unable to make precise numerical predictions from the physical model alone. In order to quantify our model, we must first perform a large number of observations and experiments. From our observations, we could then establish numerical relations between the size of various masses, their relative distances, and the strength of their gravitational pull. From a similar observation, Newton stated that the gravitational force F between two objects is proportional to the product of their masses $m_1 \times m_2$ and is inversely proportional to the square of the distance between the two objects. This can be expressed in mathematical form as $F = Gm_1m_2/r_2$, where G is the gravitational constant and determined from experiment. By expressing the force F as a function of r, we create a mathematical quantity that can be integrated, differentiated, and manipulated in order to make precise numerical calculations concerning the motion of objects experiencing a gravitational force.

By quantifying our physical model in such a manner, we have created what is called a mathematical model. Armed with such a model, it is possible to analyze a far wider range of phenomena than what was initially targeted. Once a model has been validated for a range of observations, there is little need to repeat and record future observations associated with similar physical phenomena. Thus, additional labor and costs can be eliminated, and efforts can be focused solely on conditions pertaining to a specific outcome. Many incredible scientific achievements based on Newton's law of universal gravitation, such as landing on the moon or launching artificial satellites into the orbit, serve as testament to the power of a mathematical model.

In general, a mathematical model encapsulates the behaviors of a system that are allowed while excluding those that are not. By using differential equations to model his law of gravitation, Newton discovered that these equations were solved by elliptical trajectories. Planetary orbits follow elliptical paths through space, while non-elliptical solutions are considered unphysical. Thus, physical processes within a system are described by the subset of equations that are allowed by the corresponding mathematical model.

Consider a system that is described by some function f. Such a function maps a set of inputs X to a set of permissible outputs Y. As a mathematical model, this process converts the argument X to the value Y through the process $T : X \rightarrow Y$. Here, the operator T can be a real-valued function, such as f, or a more complicated operation such as differentiation or integration. In Fig. 16.1, we show examples of several operators that relate various types inputs and outputs. Note that inputs and outputs can be a number x, a vector $\mathbf{x}$, a function $f(x)$, or even something more complicated like a group or a set.

Classification of Systems

Depending on the system of interest, mathematical models can fall under several broad types of classifications. Generally, time-dependent systems are described by either continuous-time or discrete-time models. In continuous-time models, the time variable can span the entire real number line or some subset of it, such as the positive values $t > 0$. A function $f(t)$, which can have an infinite number of values between any two points in time, is typically found as the solution to a differential equation. An example of this is Newton's second law, often expressed as $F = ma$, which states that the net force on an object is equal to the mass times the acceleration. Since the acceleration is just the second derivate of the position x with respect to time, this equation can be expressed in terms of the differential equation

$$m\frac{d^2x(t)}{dt^2} - F(t) = 0. \qquad \text{Eq. 1}$$

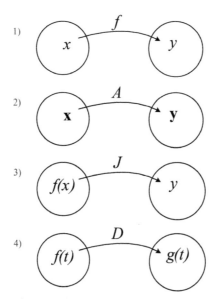

Figure 16.1. Examples of an input being transformed into an output through some operator T, such that $T : X \to Y$.
1) A function f transforms a number into a number, $y = f(x)$.
2) A matrix A transforms a vector into a vector, $y = Ax$.
3) An operator J transforms a function into a number, for example the definite integral $y = Jf(x) = \int_a^b f(x)dx$.
4) An operator D transforms a function into another function, for example the derivative $g(t) = Df(t) \, df(t)/dt$.

In discrete-time models, the time variable consists of a collection of separate and distinct points, and solutions such as $x(t_i)$ have finite values at each point in time, typically denoted x_i. Discrete-time systems are modeled using difference equations, for example $x_{i+1} = cx_i(1 - x_i)$. A collection of data xt recorded at sequential moments in time ($t = 1, 2, 3, ...$) is called a time series.

A deterministic system is one in which no randomness occurs as the solutions evolve in time. From a given set of initial conditions, a deterministic model will always produce the same output for all future times. In physics, all continuous-time systems that can be modeled by a differential equation are deterministic. When the same initial conditions result in different outcomes over time, the system is said to be stochastic. In order to describe such indeterminate systems, collections of random variables must be propagated through time using a stochastic model. A famous example of a stochastic process is the daily fluctuations of the stock market.

Consider a system that produces an output y for some input x. If the system is deterministic, a single value of x will always result in the same output value for y. If a source of randomness exists (such as noise), equal values of x can result in differing output values for y. In Fig. 16.2 we illustrate the behavior of various deterministic and stochastic systems with and without noise.

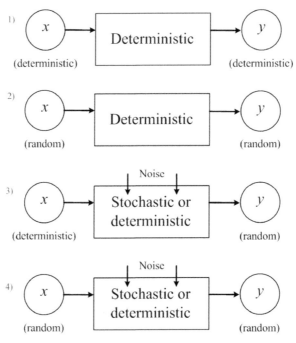

Figure 16.2. Examples of deterministic and stochastic systems. Top, a deterministic system, where a single value for *x* always produces the same output *y*. For both deterministic and stochastic systems, a random input (with or without noise) will always result in a random output.

Stochastic Analysis

A stochastic process is a collection of random variables having some form of correlation or distributional relationship between them. By finding the parameters of a stochastic process that are most likely to generate the observed collection of random variables, stochastic processes can be used to model future values of the variables. As previously mentioned, a collection of data Y_t recorded at sequential moments in time ($t = 1, 2, ...k$) is called a time series. The relationship between a time series and a stochastic process is best explained by analogy: a time series is to a stochastic process what a single number is to a random variable. While the outcome of a random variable, such as a dice roll, is a number, the outcome of a stochastic process is a discrete sequence of numbers (time series). Similarly, while an infinite amount of random trials would yield the exact probability for each outcome, an infinite collection of time series would yield the exact probability for each realization of the time series. A stochastic model allows global properties of a time series data to be described through statistics.

When a statistical analysis is performed, the sampled distribution is often considered to be an approximation of the 'true' distribution. For example, the mean and variance of a sample are assumed to be good approximations of the true mean and variance of the population being sampled. A similar concept applies to the analysis of a time series. That is, a single time series Y_k can be considered representative of a collection of *n* time series $\{Y_k^1, Y_k^2, ..., Y_k^n\}$. Note the distinction between the index

n, which corresponds to the total number of independently recorded time series, and k, which identifies a single point in time t_k. The following statistical measures are commonly used to describe a stochastic process:

$$\text{Ensemble average: } E[Y_k] = \mu_k = \frac{1}{n}\sum_{i=1}^{n} Y_k{}^i \qquad \text{Eq. 2}$$

$$\text{Variance:} \qquad \text{Var}[Y_k] = \sigma_k{}^2 = E[(Y_k - \mu_k)^2] \qquad \text{Eq. 3}$$

$$\text{Autocorrelation:} \quad \text{Cor}(Y_k, Y_i) = \frac{E[(Y_k - \mu_k)(Y_i - \mu_i)]}{\sigma_k \sigma_i} \qquad \text{Eq. 4}$$

The latter equation represents the autocorrelation for the time difference $\Delta t = t_k - t_i$. Note that the ensemble average and variance defined above correspond to n time series averaged at a single point in time t_k. In general, these averages may differ from the mean and variation of a single time series averaged over all points in time.

Stationary process

It possible that the statistical properties of a stochastic process remain unchanged over time. When the ensemble average and variance do not depend on the time t_k, so that $E[Y_k] = \mu$ and $\text{Var}[Y_k] = \sigma^2$ for all k, the distribution is said to be a stationary process. In other words, fluctuations of a stationary process will have the same statistical distributions no matter when one chooses to observe them. As an analogy, consider the average value obtained by simultaneously rolling 10 separate dice. If this average does not change in time, the process is stationary. Note that this does not say anything about the average value of a single die, or even whether the average of all 10 dice is equal to the expected value of 3.5. An example of a stochastic process often used to model time-series data is called white noise. White noise satisfies the following conditions:

$$E[Y_k] = 0, \qquad \text{Eq. 5}$$

$$\text{Var}[Y_k] = \sigma^2, \qquad \text{Eq. 6}$$

$$\text{Cor}(Y_k, Y_m) = 0. \qquad \text{Eq. 7}$$

White noise can be considered one of the simplest forms of a stationary stochastic process.

Ergodic process

In some cases, it is not possible to calculate the ensemble average at one particular point in time. However, when the mean value of any individual time series is equal to the ensemble average, the process is said to be ergodic (Breiman 1968)

$$\mu = \frac{1}{k}\sum_{i=1}^{k} Y_k{}^n \qquad \text{Eq. 8}$$

Returning to the dice analogy, the process is ergodic if the mean value of a single die, averaged over many rolls, is equal to the ensemble average of all 10 dice. Ergodic processes are important because an ensemble average can be obtained from the time average of a single series. Next we assume the time series of our recorded data can be modeled as an ergodic process.

Movement of Pacific Bluefin Tuna

Random variations within the movements of marine animals make it very challenging to accurately predict their motion. It is expected that any regularities within this uncertainty can be analyzed through the application of a stochastic model. In order to make sensible predictions, we must build a model by which we can precisely quantify these uncertainties. Our goal is to describe the characteristic trends of a collection of time series that, at first glance, appear to be completely random. For our time series, we consider the swimming depths of a Pacific bluefin tuna (*Thunnus orientalis*). Depths were recorded at discrete times using a data-logging package. Data for a single time series Y_t corresponds to depths $\{Y_1, Y_2, \ldots, Y_k\}$ recorded at times $t = t_1, t_2, \ldots, t_k$. At each moment in time t_k, the 'state' of a tuna is described by the stochastic variable X_k, defined as

$$X_k = \frac{Y_k}{Y_{k-1}}.$$

Eq. 9

The state variable X_k represents the fractional increase (or decrease) of the vertical position during the time interval $\Delta t = t_k - t_{k-1}$. A value of $X_k > 1$ corresponds to upward movement, $0 < X_k < 1$ corresponds to downward movement, and $X_k = 1$ corresponds to no change in depth. The stochastic variable X_k will be used to derive a formula that governs the state of the tuna for future times. The diagram in Fig. 16.3 shows X_t as a function of time for three Pacific bluefin tuna. The horizontal axis indicates time, while the vertical axis represents the stochastic variable X_k for each tuna. A traditional statistical analysis would project the values of X_k for a single time series into a single distribution. Such a distribution is projected onto the vertical axis (red) in Fig. 16.3 for a single tuna. For an ergodic process, this distribution would be equal for all time series. However, valuable time-dependent information is lost when time series are averaged in this way. Instead, we consider a novel method for viewing the evolution of the time series by plotting their ensemble distributions at different points in time. The distributions at $t_1, t_2, \ldots, t_k$ each represent an ensemble average of X_k at different points in time. For a stationary process, each of these distributions would be equal. Since we are interested in how these distributions evolve over time, we will consider a time-dependent equation for X_k in terms of the parameters μ and σ shown in Fig. 16.3. Next we show how this is done by constructing and solving a stochastic differential equation.

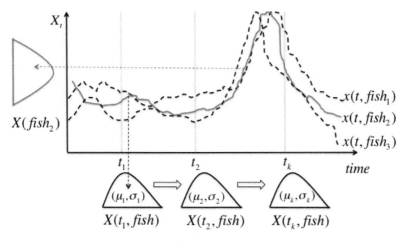

Figure 16.3. Time series for depth of Pacific bluefin tuna. The horizontal axis represents time, while the vertical axis represents the stochastic variable describing the state of the tuna fish. The average of a time series for a single tuna is projected onto the vertical axis (solid gray). Ensemble averages for all tuna fish at different times are projected downward onto the horizontal axis.

Stochastic Model for Swimming Depths

Suppose that the tuna depth at time $t = t_0$ is known to be Y_0. Our goal is to model the positions Y_t at all future times. In general, the position at time $t = t_k$ can be written as

$$Y_k = Y_{k-1}X_k = (Y_{k-2}X_{k-1})X_k = Y_0X_1X_2 \cdots X_k. \qquad \text{Eq. 10}$$

where X_k is defined in Eq. 9. Applying the natural logarithm to both sides of this equation gives

$$\ln(Y_k) = \ln(Y_0X_1X_2 \cdots X_k) = \ln(Y_0) + \sum_{i=1}^{k} \ln(X_i). \qquad \text{Eq. 11}$$

At this stage, two conditions must be satisfied: each $\ln(X_i)$ must be normally distributed with common mean μ and variance σ, and each $\ln(X_i)$ must be independent of the others.

We define a Wiener process as:

$$W_k = \sum_{i=1}^{k} \frac{\ln(X_i) - \mu}{\sigma} \sim N(0,k), \qquad \text{Eq. 12}$$

where $N(0, n)$ is a normal distribution with mean equal to zero and variance equal to k. Combining Eqs. 11 and 12 gives

$$\ln(Y_k) = \ln(Y_0) + k\mu + \sigma W_k \qquad \text{Eq. 13}$$

We now consider the change $\Delta \ln(Y_k) = \ln(Y_k) - \ln(Y_{k-1})$, which results from the change in depth between times t_{k-1} and t_k. Using the Taylor expansion for $\ln(z)$ about the point z_0:

$$\ln(z)|_{z_0} \simeq \ln(z_0) + \frac{z - z_0}{z_0} = \ln(z_0) + \frac{\Delta z}{z_0},$$

Eq. 14

we can express the difference $\Delta \ln(Y_k)$ as

$$\ln(Y_k) - \ln(Y_{k-1}) \simeq \ln(Y_{k-1}) + \frac{\Delta Y}{Y_{k-1}} - \ln(Y_{k-1}) = \frac{\Delta Y}{Y_{k-1}}.$$

Eq. 15

From Eq. 13, we find that

$$\Delta \ln(Y_k) = (k\mu + \sigma W_k) - ((k-1)\mu + \sigma W_{k-1}) = \mu \Delta t + \sigma \Delta W_k.$$

Eq. 16

In the limit that incremental changes Δ to each quantity become infinitesimally small, the discrete form can be changed into the continuous one. By equating the last terms in Eqs. 15 and 16, and substituting $t_k \to t$ in the continuum limit, we obtain the following stochastic differential equation:

$$\frac{dY}{Y_t} = \mu dt + \sigma dW_t,$$

Eq. 17

Where dW_t denotes the incremental change in a Wiener process, μ characterizes the mean rate of change of the depth Y_t, and σ measures the variation about this mean rate.

The solution to Eq. 17, called the geometric Brownian motion (Karatzas and Shreve 1991), has been applied to many financial problems such as the modeling of stock prices. When the initial condition Y_0 is known, the solution is given by

$$Y_t = Y_0 \exp\left[\left(\mu - \frac{\sigma^2}{2}\right)t + \sigma W_t\right].$$

Eq. 18

To solve this equation (Shreve 2004), the entire time series is first divided into shorter, non-overlapping intervals of time. Values for the mean growth rate μ and variation σ can then be estimated within each of these shorter segments of data. The resulting sequence of estimated parameters indicates behavioral changes over time. Length of data intervals is determined by the type of movement model and size of the observed time scales relative to behavioral changes of interest. In Fig. 16.4, the solution to the stochastic differential equation is used to simulate the swimming depths of a tuna, and the result is displayed along with recorded data.

The model we presented sufficiently captures the main aspects of a recorded time series, and the vertical dynamics of the tuna is accurately modeled. The next step would be to extend the model in order to describe a wider variety of behavioral patterns that depend on the model's parameters. While X_k correspond to changes in depth, it is important to realize that such a variable could represent any collection of states for each point t_k. For example, the state variable could be a multidimensional collection of the tuna's depth, size, location, the water's temperature, salinity level, etc.

The time series data can hold much information about different behavior types, such as when the tuna is foraging, feeding, traveling, or resting (Smouse et al. 2010;

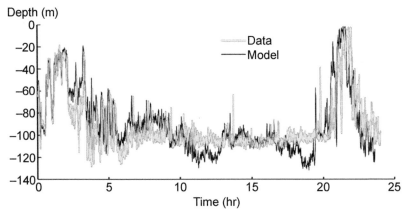

Figure 16.4. Time series for depths of Pacific bluefin tuna (*Thunnus orientalis*). The gray line represents recorded depths from data. The black line represents a simulation from the solution of the stochastic differential equation. Data was recorded in the waters offshore Kochi Prefecture, Japan. The datalogger recorded the tuna's depth at one-second intervals for 48 hours. The time series were binned by 30-second intervals, and for each bin ti the values of m and σ were computed. Good agreement can be found between the simulation and data.

Blackwell 1997; Preisler et al. 2004; Blackwell 1997; Kadota et al. 2011). With a mathematical model, these behavior patterns can be extracted and mapped onto meaningful biological descriptions. For instance, when a tuna encounters food, its pattern of movement may intensify. This results in larger values of μ and σ, which in turn indicates a greater change in the tuna's speed and depth. Observations of data suggest that tuna surface very quickly, tend to be inactive at night, and rest at deeper depths at night, the latter of which would correspond to smaller values of μ and σ. Since the parameter values are directly affected by these behavior patterns, the use of mathematical models is clearly of great importance to the accurate determination of these patterns from data.

Acknowledgements

The authors would like to thank Kinki University and the International Education and Research Center for Aquaculture Science of bluefin tuna and other cultured fish for support, as well as the Ministry of Education, Culture, Science, Sport and Technology, Japan, and the Japan Society for the Promotion of Science.

References

Blackwell, P.G. 1997. Random diffusion models for animal movement. Ecol. Model. 100(1-3): 87–102.
Breiman, L. 1968. Probability. Addison-Wesley, Reading, MA.
Kadota, M., E.J. White, S. Torisawa, K. Komeyama and T. Takagi. 2011. Employing relative entropy techniques for assessing modifications in animal behavior. Plos ONE 6(12): e28241. doi:10.1371/journal.pone.0028241.
Karatzas, I. and S.E. Shreve. 1991. Brownian Motion and Stochastic Calculus, Second Edition. Springer-Verlag, New York.

Preisler, H.K., A.A. Ager, B.K. Johnson and J.G. Kie. 2004. Modeling animal movements using stochastic differential equations. Environmetrics 15: 643–657. doi: 10.1002/env.636.

Shreve, S.E. 2004. Stochastic Calculus for Finance. Volume II: Continuous-Time Models. Springer-Verlag, New York.

Smouse, P.E., S. Focardi, P. Moorcroft, J.M. Morales, J. Kie and J. Reynolds. 2010. Stochastic modelling of animal movement. Phil. Trans. Roy. Soc. Lond. B 365: 2201–2211.

CHAPTER 17

Mathematical Modeling of Bluefin Tuna Growth, Maturation, and Reproduction Based on Physiological Energetics

Marko Jusup [1,*] and *Hiroyuki Matsuda* [2]

Introduction

Mathematical modeling of fish characteristics, such as growth or reproduction, aims at establishing relationships between two or more variables (e.g., body length as a function of age or fecundity as a function of body length) in a way that is as general as possible. In fisheries science, a very common approach used to infer the functional form that best describes the relationship between variables consists of fitting candidate functions to the data and performing goodness-of-fit analyses via statistical measures (e.g., Chen et al. 1992; Cailliet et al. 2006; Katsanevakis 2006; Quince et al. 2008). Often times, however, the best fitting function may turn out to be the one for which we have little theoretical justification, leaving us in the dark as to why a certain function fits the data very well. For this reason, mechanistic modeling from the first principles should be considered a preferable methodology whenever possible. Physiological energetics, by contrasting energy sources with sinks to determine the amount of energy available for growth and reproduction, represent a mechanistically-oriented approach in which mathematical models are built from the first principles of thermodynamics. As such, physiological energetics have long been used in fisheries science as a valuable means to

[1] Faculty of Sciences, Kyushu University, 6-10-1 Hakozaki, Higashi-ku, Fukuoka, 812-8581 Japan.
[2] Graduate school of Environment and Information Sciences, Yokohama National University, 79-7 Tokiwadai, Hodogaya-ku, Yokohama, 240-8501 Japan.
* Email: mjusup@gmail.com

probe the characteristics of fish metabolic rates and energy budgets. Recognizing this value, we overview the field of fish bioenergetics—a loose synonym for physiological energetics used here—as it pertains to tuna fishes and mathematical modeling thereof.

Our exposition of the subject is organized as follows. We begin by introducing 'traditional' bioenergetic approaches in the context of fisheries science, particularly emphasizing their empirical origin, consequent operational definitions of metabolic rates, and difficulties arising from the adoption of empirically sound concepts in modeling studies. We then summarize the main results of traditional bioenergetics pertaining to tuna fishes. Thereafter a novel bioenergetic approach—dubbed the formal metabolic theory of life (Sousa et al. 2008), but among practitioners better known as the Dynamic Energy Budget (DEB) theory (Nisbet et al. 2000; Sousa et al. 2010; Kooijman 2010)—is introduced in some detail. The application of DEB theory on modeling bluefin tuna growth, maturation, and reproduction is illustrated by a full life-cycle bioenergetic model for Pacific bluefin tuna (Jusup et al. 2011). We conclude this chapter by discussing several possible directions for future research of bluefin tuna physiological energetics.

Traditional Bioenergetic Approaches with a Focus on Fisheries Science

Traditional bioenergetic approaches aim at explaining how the energy contained in ingested food is used to power up various metabolic processes. Typically, particular interest is placed on growth, reproduction, respiration, excretion, and movement, though this list is by no means an exhaustive one. A recognizable feature of traditional approaches to bioenergetics concerns the way in which metabolic processes are defined. The definitions involved are often laid out in a practical or operational manner (Nisbet et al. 2012). For instance, to come up with an energy output that is representative of the reproduction rate, the number of eggs in a single batch may be counted and converted into energy by analyzing eggs for their chemical composition, and subsequently determining the energy equivalent of an egg. In other cases, changes in the respiration rate under multiple experimental conditions may be measured to deduce the fraction of energy required for the basal (or resting) metabolism as well as the energy requirements of movement. Note that metabolic rates are expressed in terms of energy per unit of time and (dry or wet) mass (e.g., $J \cdot d^{-1} \cdot kg^{-1}$).

The operational definitions of metabolic processes in traditional bioenergetic approaches reveal an empirical foundation whereby concepts from experimental studies became almost directly transferred to more theoretically-oriented modeling studies. As a consequence, a typical traditional bioenergetic model (see Kitchel et al. 1977 for a well-known example and Hansen et al. 1993 for an overview) contains one state variable and balances the energy budget of the form

$$C = G+R+F+U, \tag{1}$$

where the sole energy input is placed on the left hand-side of the equality, while various energy outputs are collected on the right hand-side. Let us interpret the terms appearing in Eq. (1).

Consumption (C) is the rate of feeding. It can be further decomposed into a product of the maximum consumption rate for a fish of a given size and an efficiency factor that ranges between zero and unity in order to account for suboptimal food availability. The maximum consumption rate increases with the size of the fish where an allometric functional relationship (a power law) is commonly assumed to hold.

Growth (G), is the rate of change (mainly increase) in body mass (B), which due to the chosen units must be written in the form $G = d_c B^{-1} dB/dt = d_c d\ln B/dt$. The parameter d_c is the average caloric density of the fish, expressed in $J \cdot kg^{-1}$. Depending on the main interest of the study, it is possible to decompose the growth term into contributions from somatic growth, gonad development, and the increase of (lipid) reserves.

Respiration (R), also dubbed metabolism, consists of three components. The standard (or basal) metabolic rate (M) refers to the amount of energy expended when the fish is at rest. Though seemingly a sound concept, the standard metabolic rate is somewhat ill-defined when considering tuna fishes because they never stop swimming. There are two reasons for such a behavior. First, having lost the ability to ventilate gills using the lower jaw and opercula, tunas are required to swim constantly with their mouth open, i.e., they are ram gill ventilators (Roberts 1975). The standard metabolic rate cannot be measured even in restrained tunas because a high gill perfusion rate is needed just to keep the fish alive (Stevens 1972). The second reason for constant swimming is the fact that scombrids, including tunas, are negatively buoyant and to counter negative buoyancy the fish rely on the lift produced by their pectoral fins (Magnuson 1978). To account for the energy expenditure caused by swimming in excess of the standard metabolic rate, a dimensionless factor called activity (A) is usually introduced. Finally, respiration increases in response to the costs of processing ingested food. Such an increase in respiration is known as Specific Dynamic Action (abbreviated to SDA in the text and S in equations). We can thus write $R = MA+S$. Similarly to the consumption rate, the standard metabolic rate is also assumed to change with the fish size according to an allometric relationship (i.e., a power law).

Egestion (F) is the discharge rate of undigested food from the digestive tract, whereas excretion (U) is the discharge rate of nitrogen waste products, primarily ammonia in fish (Smutna et al. 2002; Wilkie 2002). Both egestion and excretion are assumed to be related to the consumption rate.

Though conceptually simple, traditional bioenergetic models are not exactly straightforward to apply (Hansen et al. 1993; Chipps and Wahl 2008). Here, we single out some of the problems that arise in this context. Each term in Eq. (1) contains species-specific physiological parameters that require appropriate data for estimation. Even if we assume that such data is available, the parameters in the allometric relationships are notoriously difficult to estimate, at least in a statistically proper manner (Kaitaniemi 2004; Warton et al. 2006). Borrowing parameter values from other species is also not recommended because the theory does not offer any systematic way for performing cross-species comparisons. As a result, traditional bioenergetic modeling is exposed to criticism based on the model complexity (Hansen et al. 1993) as well as performance (Chipps and Wahl 2008).

Another difficulty arising in relation to the operational definitions of metabolic rates are slight differences in the way these definitions are operationalized among different studies (Nisbet et al. 2012). For instance, one way to estimate activity is to measure the increase in the respiration rate of fish swimming at progressively higher speeds and then fit an appropriate (non-linear) model from which the standard metabolic rate and the metabolic transport differential are estimated. The experimental procedure for measuring the respiration rate, aside from practical considerations (e.g., the design and the utilization of the swimming respirometer), is hampered by a number of factors that can affect respiration or swimming performance. These include stress (Sloman et al. 2000), gait transition (Korsmeyer et al. 2002), and simultaneous activation of red and white muscle fibers at higher speeds (Jayne and Lauder 1994), to name a few. Additional problems may appear in view of uncertainties as to what proportion of the increase in the respiration rate is actually directed towards the locomotory muscles (Kiceniuk and Jones 1977). Even if the experiments proceed smoothly, fitting of a non-linear model to the data is not entirely devoid of difficulties in the estimation and interpretation of the parameter values (Papadopoulos 2008; Papadopoulos 2009). Further problems emerge when trying to convert the respiration rate into energy consumption. The use of common oxy-calorific coefficients (Elliot and Davison 1975) is criticized on both experimental (Walsberg and Hoffman 2005) and theoretical (Sousa et al. 2006) grounds.

A different way to come up with the energetic requirements of movement is to consider the mechanical power necessary to produce the thrust for propelling the fish to a given speed, and then relate this power to the rate of metabolic (chemical) energy expenditure inside the muscle tissue. The state of affairs regarding the estimation of the mechanical power requirements of swimming is perhaps best summarized by Schultz and Webb (2002) in their remark on 'troubled swimming studies'. The authors note that the methods developed in the late 20th century still focus on the separation of drag and thrust, and that the drag estimates are as variable as in earlier studies. Consequently, much uncertainty remains also in the estimates of the actual power requirements of swimming. A particularly confusing issue is drag reduction which emerged in hope that fish possess the ability to reduce momentum losses beyond what is measured for rigid bodies. For instance, a study by Barrett et al. (1999) claims drag reductions of up to 70% in the case of a tuna-like autonomous mechanical model. Anderson et al. (2001), using digital particle tracking velocimetry to probe tangential and normal velocity profiles in the boundary layer of live swimming fish, dismiss the possibility of such drag reductions and state that comparisons between an undulating fish and a rigid body 'would not make sense'.

Considerable uncertainties concerning the power requirements of swimming are only matched with uncertainties involved in converting these requirements into the rate of metabolic energy expenditure in muscles. In fact, severe mismatches become apparent even when attempting to relate power requirements from hydromechanical considerations to the mechanical muscle power output measured *in vitro*. Note that the latter measurements are usually accompanied with *in vivo* electromyography to study—from a physiological perspective—patterns in muscle function during swimming (Altringham et al. 1993; Altringham and Ellerby 1999). It is also worth noting that, aside from the pioneering work of Webb (Webb 1971a; Webb 1971b; Ellerby 2010), the mechanics and physiology of fish swimming were mostly treated as separate lines of research, possibly contributing to the above-mentioned mismatches. In order to illustrate the magnitude of the mismatch, we provide an example. Rome et al. (2000) report, based on measurements of power generation by the red muscle of scup (*Stenotomus chrysops*) at 10°C, that the power output is between 0.7 and 7.7 W·kg⁻¹ when muscle bundles excised from various locations along the fish body are driven similarly to *in vivo* conditions in fish swimming at 30 cm·s⁻¹. Rome and Swank (1992) additionally observed that the maximum power output by scup red muscle more than doubles as the temperature increases from 10 to 20°C. By contrast, calculations made for a scup swimming at the same speed and at the temperature of 23°C indicate that the power output per unit red muscle mass needed to overcome friction drag is approximately 0.6 W·kg⁻¹ (Anderson et al. 2001). Though friction drag is not the total hydrodynamic drag, there still seems to be an order of magnitude difference between what the red muscle tissue can generate and the actual swimming requirements. Later by outlining a similar problem in tuna energetics, we confirm that the described situation is more than just a carefully selected exception. As a closing remark, we note that the studies based on respirometry also shed a little light on these issues. Part of the problem is that in order to make use of the respirometry data, we would need to account for propulsion, muscle, and metabolic (in)efficiencies, neither of which are precisely known (Schultz and Webb 2002).

What do Traditional Bioenergetic Approaches Tell us about Tuna Fishes?

Unique morphological and physiological adaptations of tunas, functional evolutionary convergence with lamnid sharks and cetaceans, and regional endothermy (for a succinct summary in the context of swimming performance see Blake 2004; see also Bernal et al. 2001; Graham and Dickson 2001; Motani 2002; Graham and Dickson 2004) are the topics that attracted substantial research interest, particularly in the context of metabolism and energetics (for a review see Korsmeyer and Dewar 2001). Though each of these topics is captivating in itself, there is a bigger picture to consider. If unique morphology and physiology result in greater swimming efficiency in comparison to other active fishes, then tunas may have an advantage in terms of the rate of net energy acquisition (i.e., energy gained through foraging minus the associated costs). Consequently, tunas may also have more energy to invest in growth and reproduction, resulting in a higher fitness and, ultimately, evolutionary success. With such a background, it should not come as a surprise that many tuna-related research efforts focused on discovering efficiencies in swimming and organ function relative to similarly-sized active fish species. Attempts to probe metabolism and energetics were, and still are, primarily documented in respirometry-based studies. These are often complemented with *in vitro* measurements of the mechanical muscle power output and *in vivo* electromyography. Alternative experimental approaches (e.g., Boggs and Kitchell 1991) are sporadic at best, while modeling studies (e.g., Essington et al. 2001; Chapman et al. 2011; Jusup et al. 2011) seem to be gaining some traction lately.

The most complete set of respirometry data and subsequent analyses known to us pertain to tropical tuna species, namely, yellowfin (*Thunnus albacares*), kawakawa (*Euthynnus affinis*), and skipjack (*Katsuwonus pelamis*) tunas (Dewar and Graham 1994; Korsmeyer et al. 1996; Korsmeyer and Dewar 2001). Of the numerous results yielded by respirometry-based studies, we believe that many are worth revisiting in order to form a basic understanding of tuna fishes in the context of physiological energetics. Starting from the standard metabolic rate, the values obtained by extrapolation to zero speed (because tunas never stop swimming) for a yellowfin of 51 cm fork length and a skipjack of 48 cm fork length at 24°C are 253 and 315 $mgO_2 \cdot kg^{-1} \cdot h^{-1}$, respectively. These values are roughly by a factor of three larger than the temperature-adjusted standard metabolic rates of similarly-sized salmonids. Because the standard metabolic rate is supposed to reflect the minimum aerobic energy expenditure on the maintenance processes, relatively large measured values in tunas are probably a consequence of higher maintenance costs incurred by hosting the specialized high-performance metabolic machinery. For instance, tunas are characterized by a gill surface area that is several times larger than in other similarly-sized bony fishes and a gill epithelial thickness that is up to an order of magnitude thinner. As a result, not only is the utilization of oxygen from the ventilation stream extremely high, but 50 to 60% of the total metabolic rate in restrained fish comes from gill respiration itself (Bushnell and Brill 1992). Tunas also have a comparatively large heart that beats at higher rates than in most bony fishes, producing a high cardiac output and high blood pressure. The resulting cardiac power generation relative to the body mass is thus several times to an order of magnitude larger in spinally-blocked or anesthetized tunas than in resting rainbow trout (Bushnell and Jones 1994). Additional tuna-specific adaptations may also be contributing to the maintenance costs. As an example, it is not fully understood to what extent does elevated mitochondrial capacity found in some tuna tissues (Moyes et al. 1992) affect the standard metabolic rate. Note that thermogenesis is unlikely to be a relevant factor here because the aerobic locomotor muscle should be producing minimal heat when fish are not swimming.

The results of respiration studies reveal how the swimming speed affects the respiration rate, and simultaneously provide some insight into the importance of activity as a sink of metabolic energy. After several hours of adaptation to swimming in a water tunnel, which is necessary for fish to recover from the handling stress, the correlation between the swimming speed and the respiration rate becomes closer. For a yellowfin tuna of 51 cm fork length at 24°C, the respiration rate raises exponentially from roughly 325 to 1200 $mgO_2 \cdot kg^{-1} \cdot h^{-1}$ as the swimming speed increases from 0.5 to 3 $L \cdot s^{-1}$ (Dewar and Graham 1994). When plotted on a semi-logarithmic plot with y-axis displayed using a logarithmic scale, the exponential relationship between the respiration rate and the swimming speed for yellowfin tuna exhibits a lower slope than the analogous relationships for salmonids. One is easily tempted to interpret the lower slope as a sign of an improved swimming efficiency (Dewar and Graham 1994; Korsmeyer et al. 1996). In reality, however, the lower slope is nothing more than an artifact of semi-logarithmic plotting and the fact that tunas have the higher standard metabolic rate (Korsmeyer and Dewar 2001). When the same data are plotted in a usual linear plot the difference between the slopes disappears, indicating that the respiration rate in both tunas and salmonids increases with the swimming speed in approximately the same manner. In other words, improvements in the swimming efficiency of tunas relative to salmonids are not observed in the plots of the respiration rate against the swimming speed obtained from laboratory water tunnels.

Respiration studies allow estimation of gross and net costs of transport (GCOT and NCOT, respectively) as another means of gaining insight into fish energetics, particularly the energetics of swimming. GCOT represents the total metabolic cost of traveling a predetermined distance at a given speed. NCOT is calculated from GCOT by subtracting the contribution from standard metabolism. For a yellowfin tuna of 51 cm fork length at 24°C, the minimum GCOT of 2.50 $J \cdot L^{-1}$ is reached at the swimming speed of approximately 2 $L \cdot s^{-1}$ (Dewar and Graham 1994). Comparably-sized ectothermic bony fishes at the same temperature should attain a considerably lower minimum GCOT of 1.83 $J \cdot L^{-1}$, presumably because the standard metabolic rate of yellowfin tuna is much higher. When adjusted for standard metabolism, the minimum NCOT of yellowfin tuna is 1.58 $J \cdot L^{-1}$, roughly the same as the minimum NCOT of ectothermic bony fishes (Dewar and Graham 1994). Though the speed at which the minimum NCOT is achieved seems to be higher in yellowfin tuna, comparable minimum NCOT values do not support the common conjecture of tunas being more efficient swimmers than ectothermic bony fishes. A similar conclusion follows from a comparative study on kawakawa tuna and chub mackerel in which no statistically significant differences in mean NCOT between the two species were found (Sepulveda and Dickson 2000).

Briefly turning our attention to the power requirements of swimming, from the minimum NCOT of 1.58 $J \cdot L^{-1}$ it follows that a yellowfin tuna of 51 cm fork length (2.1 kg), for the purpose of swimming at the optimal cruising speed of 2 $L \cdot s^{-1}$, generates a power output of around 3.16 W. Assuming this power output is entirely attributable to the red muscle, which in yellowfin tuna accounts for roughly 6.5% of body mass (Bernal et al. 2001), it would appear that approximately 23 $W \cdot (kg \ red \ muscle)^{-1}$ need to be generated for sustained swimming. However, the net muscle efficiency over a full contraction-relaxation cycle is usually below 30% (Smith et al. 2004), meaning that over 70% of metabolic energy expended by the muscle is turned into heat. We can, therefore, put the upper limit for the mechanical power output of yellowfin tuna red muscle during sustained swimming at 6.9 $W \cdot (kg \ red \ muscle)^{-1}$. An interesting comparison can be made at this point. Namely, a combination of theoretical studies involving fluid dynamics and experiments on a robotic model of bluefin tuna provide us with an independent estimate of mechanical power requirements for cruising in thunniform swimmers (Triantafyllou and Triantafyllou 1995; Barrett et al. 1999). The robotic model of a bluefin tuna of approximately 120 cm fork length achieved a speed 0.7 m s^{-1} (0.58 BL s^{-1}) with a power input to its motors of only 0.5283 W. Accompanying numerical modeling, depending on the value of kinematic parameters, estimated the mean mechanical power input at between 1.15 and 3.05 W at a cruising speed of 0.66 L s^{-1}. A living bluefin tuna similar in size to the robotic model (120 cm fork length, 35 kg body mass) would likely have somewhere between 4 and 13% red muscle mass as a percentage of body mass (Bernal et al. 2001). Hence mechanical power requirements of swimming relative to red muscle mass should be of the order of 1.0 $W \cdot (kg \ red \ muscle)^{-1}$ at common cruising speeds, with the highest plausible estimates yielding 2.2 $W \cdot (kg \ red \ muscle)^{-1}$. Given the differences in size and speed of the considered live yellowfin tuna and the robotic model, we surmise that the reported values compare rather favorably.

Additional comparisons, preferably with a recent respirometry data, should help instill more confidence into estimated mechanical power requirements of swimming. A good source of the data in this context is a study by Blank et al. (2007a), which reports the results of the measurements performed on both yellowfin and bluefin tunas. For bluefin tuna (average size 74 cm, 8.3 kg), the recorded respiration rate at the temperature of 20°C and at the swimming speed of 1.15 L s^{-1} is 267 mgO$_2$·kg^{-1}·h^{-1}. This respiration rate yields the minimum GCOT of 7.55 J·L^{-1}. From the standard metabolic rate of 120 mgO$_2$·kg^{-1}·h^{-1} estimated by the extrapolation to zero speed, the minimum NCOT of 4.16 J·L^{-1} is obtained. Accordingly, mechanical power requirements of swimming relative to red muscle mass should not exceed 4.3 W·(kg red muscle)$^{-1}$. For yellowfin tuna (average size 67 cm, 5.4 kg) at the same temperature and a similar swimming speed as in the case of bluefin tuna, the respiration rate is 216 mgO$_2$·kg^{-1}·h^{-1}. The corresponding minimum GCOT is 3.98 J·L^{-1}, whereas the minimum NCOT, given the standard metabolic rate of 91 mgO$_2$·kg^{-1}·h^{-1}, is 2.31 J·L^{-1}. From these values, mechanical power requirements of swimming relative to red muscle mass turn out to be 2.27 W·(kg red muscle)$^{-1}$. In summary, respirometry-based studies, a robotic model, and fluid dynamics-based modeling all produce similar estimates of the mechanical power output necessary for juvenile yellowfin and bluefin tunas to swim at their usual cruising speeds. Relative to red muscle mass this power output is of the order $O(1)$.

In comparison with mechanical power requirements of swimming estimated using the above-mentioned approaches, *in vitro* measurements of tuna red muscle power generation reveal an ability to deliver a puzzlingly high power output. Namely, peak mechanical power generation in red muscle of yellowfin tuna exceeds 60 W·(kg red muscle)$^{-1}$ and stays above 20 W·(kg red muscle)$^{-1}$ over a wide range of tail beat frequencies (Shadwick and Syme 2008). Even higher values, ranging between 44 and 75 W·(kg red muscle)$^{-1}$, are observable in skipjack tuna (Syme and Shadwick 2002). An inevitable conclusion is that the red muscle of tunas can generate enough mechanical power to exceed the requirements for sustained swimming by an order of magnitude. Such a large mismatch is reminiscent of the situation observed in the case of scup; muscle bundles excised from posterior locations along the fish body and stimulated similarly to *in vivo* conditions generated an order of magnitude more power than is required for sustained swimming. The difference between required and deliverable power outputs is perhaps best understood by considering that the latter may be more relevant for swimming near the maximum aerobic speed than at the typical sustained speed. Hence it is likely that fish activate only a small fraction of red muscle mass for cruising and keep a considerable spare capacity at their disposal for more vigorous activities.

While there is still some suspicion that tunas may be able to achieve higher swimming speeds than other fishes without recruiting the white muscle tissue (Katz et al. 2001), studies of swimming performance are yet to demonstrate an unquestionable advantage of the thunniform swimming mode. Hence, with the evidence of increased swimming efficiency in tunas remaining elusive, bioenergetic studies turned to explaining the high aerobic capacity of these fishes. Especially attractive was the fact that the respiration rate at the predicted maximum sustainable (i.e., aerobic) speed is well below the estimated maximum respiration rate—the former being around 1200 mgO$_2$·kg^{-1}·h^{-1} and the later above 2500 mgO$_2$·kg^{-1}·h^{-1} in a yellowfin tuna of 51 cm fork length at 24°C (Dewar and Graham 1994). As a result, a representation of tuna aerobic energetic costs emerged hypothesizing that the high aerobic scope is a specialization that allows simultaneous continuous swimming and the delivery of oxygen to other metabolic processes (Korsmeyer et al. 1996; Korsmeyer and Dewar

2001). These included standard metabolism, SDA, growth, and oxygen debt recovery. It was suggested that the oxygen debt incurred during episodes of burst swimming represents a considerable component of the tuna aerobic energy budget. The authors also identified a potential advantage of tunas over other fishes whereby the high aerobic capacity helps speed up lactate clearance in the white muscle tissue, thus allowing for a rapid recovery from exhaustive burst swimming. In the described representation of aerobic energetic costs, tunas were characterized as fish that expend large amounts of energy in a food-scarce pelagic environment in order secure even higher amounts of energy ingestion and assimilation vital not only for the survival, but also for growth, maturation, and reproduction. Tunas, therefore, became known as 'energy speculators'.

Though we have largely covered the main results in the context of traditional tuna bioenergetics, it is worth noting that recent studies attempted to quantify the energy expenditure attributable to SDA in two species of bluefin tuna (Fitzgibbon et al. 2007; Clark et al. 2010). The results are, unfortunately, somewhat conflicting. Measurements of SDA in southern bluefin tuna, *Thunnus maccoyii*, at temperatures between 18.2 and 21.3°C estimate the corresponding energy loss to be, on average, 35% of ingested energy (Fitzgibbon et al. 2007). Such a value is on the high end of recorded SDA in fishes (Secor 2009). On the other hand, measurements performed on juvenile Pacific bluefin tuna, *Thunnus orientalis*, at the temperature of 20°C indicate that SDA accounts for, on average, only 9.2% of ingested energy (Clark et al. 2010). The large difference between the two otherwise closely related species is surprising and quite possibly an artifact of the differences in the experimental procedure between the two studies. We can only hope that the future research will shed additional light on the subject and uncover the real energetic costs of SDA in bluefin tunas.

A Novel Bioenergetic Approach

The study of life is inevitably an interdisciplinary endeavor to which scientific fields like mathematics (Cohen 2004), chemistry (Wu and Schultz 2009), and physics (Mielczarek 2006) can contribute immensely. In recent years, an example of such an endeavor is embodied in the formal metabolic theory of life (Sousa et al. 2008), among practitioners better known as the Dynamic Energy Budget (DEB) theory (Kooijman 2010). To describe it in the shortest possible manner, perhaps it is best said that DEB theory builds a physics-like foundation for biological research. In order to accomplish this task, the theory identifies and makes use of universal mechanisms responsible for running metabolism of all life. One of the consequences is a remarkable level of generalization and formalism. Though biology takes the lead in identifying these universal mechanisms, important contributions from other sciences, particularly thermodynamics (Sousa et al. 2006), need to be acknowledged.

Applications of DEB theory are numerous and growing by the day, but a distinct success has been achieved in fisheries science. Some of the species for which a DEB-based model exists include zebra fish (Augustine et al. 2011), anchovy (Pecquerie et al. 2009), salmon (Pecquerie et al. 2011), bluefin tuna (Jusup et al. 2011; Jusup et al. 2014), eel (van der Meer 2011), flatfish (van der Veer et al. 2001; van der Veer et al. 2003; van der Veer et al. 2009; Freitas et al. 2012), sand goby (Freitas et al. 2011), and capelin (Einarsson et al. 2011). These and similar models are applied to a variety of issues such as optimizing aquaculture production (Serpa et al. 2012), interpreting fish otolith biomineralization (Fablet et al. 2011), predicting the effects of climate change (Teal et al. 2012), and analyzing the toxic effects of uranium (Augustine et al. 2012). Therefore, DEB theory represents a versatile and, even more importantly, integrative modeling framework through which the connections between a wide variety of datasets can be revealed. We firmly believe that the research of bluefin tuna, in its richness of experimental data, may greatly benefit from the development of DEB-based models for all bluefin species.

In what follows, we briefly describe the most important concepts of DEB theory. We start from a set of basic assumptions on which we build the so called standard DEB model in a step-by-step fashion. Aside from the basic assumptions that allow us to begin the model construction, several key assumptions are introduced along the way. We end the discussion by demonstrating the dynamics implied by the standard DEB model. Supplementary assumptions necessary to account for the specifics of bluefin tuna physiology are, together with the main results, left for later.

Basic assumptions

Biology offers a multitude of empirical evidence, often presentable in a stylized manner, upon which theoretical advances can be built (Sousa et al. 2008; Sousa et al. 2010). Basic assumptions of DEB theory largely follow from observations made on starving organisms or organisms in the embryonic stage. In both cases, growth occurs for a certain period of time without relying on an outside source of food. Starving adults are sometimes even able to reproduce. These observations suggest that the metabolizable organic compounds can be stored in anticipation of the periods of hardship or in preparation for certain life stages. Consequently, we make a conceptual division of a fish into reserve and structure compartments, where the status of the former (latter) is tracked by the state variable E (L) representing the energy stored in the reserve (structural length). The structural length is defined through its relationship with the volume of space, V, taken by the structure, i.e., $V = L^3$. An important idea here is that for an isomorphic individual—which is a good first approximation for a large part of the life of a fish—the ratio of any two length measurements remains constant in time. Hence, any well-defined physical length, L_w, can be related to the structural length via a proportionality constant called the shape factor, $\delta_M = L/L_w$. For a more intuitive grasp of the introduced concepts, it is useful to think of the reserve as being comprised of tissues that do not require maintenance and are metabolizable as a source of energy. The structure, on the other hand, requires continuous maintenance and is necessary for the survival of fish. Another fundamental observation is that fish mature and undergo stage transitions—the latter being brief episodes in life during which the behavior and energetics of an individual change considerably. For instance, embryos do not feed, juveniles feed but do not engage in reproduction, and adults do both. To keep track of fish maturation, we introduce the third state variable, E_H, called the level of maturity. Stage transitions are assumed to occur when the level of maturity crosses certain threshold values. A summary of the notation, equations, and the parameter values (to follow) is provided in Table 17.1.

Table 17.1. List of symbols, formulas, and parameter values (at the reference temperature of 20°C).

Description	Unit	Symbol, formula, value
Amount of energy in the reserve	J	E, Eq. (2)
Energy density	J·cm^{-3}	$[E] = E/L^3$, also see Eq. (3)
Structural volumetric length	cm	L, Eq. (4)
Level of maturity	J	E_H, Eq. (5)
Status of the reproduction buffer	J	$E_R = \int \dot{p}_R dt$ when $E_H \geq E_H^{\ p}$
Assimilation flow	J·d^{-1}	$\dot{p}_A = \{\dot{p}_{Am}\} f L^2$
Utilization (mobilization) flow	J·d^{-1}	$\dot{p}_C = E(\dot{v}[E_G]L^2 + \dot{p}_S)/(\kappa E + [E_G]L^3)$
Somatic maintenance flow	J·d^{-1}	$\dot{p}_S = \dot{p}_M + \dot{p}_T = [\dot{p}_M]L^3 + \{\dot{p}_T\}L^2$
Maturity maintenance flow	J·d^{-1}	$\dot{p}_J = k_J E_H$
Growth flow	J·d^{-1}	$\dot{p}_G$, Eq. (6a)
Maturation (reproduction) flow	J·d^{-1}	$\dot{p}_R$, Eq. (6b)
Shape correction function	–	$M_1(L, E_H)$, Eq. (10)
Shape factor	–	$\delta_M(E_H)$, Eq. (11)
Thermogenic efficiency	–	$M_2(E_H)$, Eq. (12)
Max. surf.-area-specific assimilation rate	J·cm^{-2}·d^{-1}	$\{\dot{p}_{Am}\}$, 160.53
Volume-specific cost of structure	J·cm^{-3}	$[E_G]$, 8828
Energy conductance	cm·d^{-1}	$\dot{v}$, 0.2386
Volume-specific somatic maintenance rate	J·cm^{-3}·d^{-1}	$[\dot{p}_M]$, 12.842
Surf.-area-specific somatic maintenance rate	J·cm^{-2}·d^{-1}	$\{\dot{p}_T\}$, 1635.6
Maturity maintenance rate coefficient	d^{-1}	k_J, 0.045166
Reserve allocation to soma	–	κ, 0.7807
Maturity-at-birth/-puberty	J	$E_H^{\ b}$, 0.7637/$E_H^{\ p}$, 2.548·10^7
Other maturities	J	$E_H^{\ j}$, 6.90·10^3 $E_H^{\ 2}$, 5.40·10^5 $E_H^{\ y}$, 9.70·10^5
Assimilation efficiency	–	$\kappa_X = 0.8$
Shape factor in the larval/adult stage	–	$\delta_M^{\ 1} = 0.2249/\delta_M^{\ 2} = 0.2704$

The dynamics of the three state variables is controlled by a total of six energy flows (Fig. 17.1). The assimilation flow ($\dot{p}_A$) controls the rate at which the energy from food enters the reserve. The utilization (mobilization or catabolic) flow ($\dot{p}_C$) is the rate at which the energy gets utilized (or mobilized) from the reserve. We assume the utilization flow to be divided in line with the κ-rule: the fraction $\kappa\dot{p}_C$ is used for somatic maintenance and growth, while the remainder $(1-\kappa)\dot{p}_C$ is allocated to maturity maintenance and maturation. Reproduction replaces maturation when the level of maturity reaches the threshold value that marks the end of the juvenile

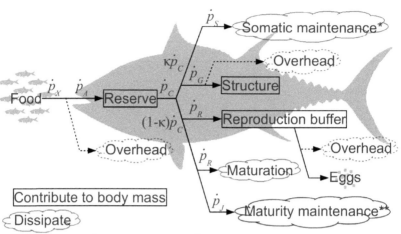

Figure 17.1. A schematic representation of the DEB-based model for bluefin tuna. Assimilation, growth, and reproduction overheads are paid directly from the corresponding energy flows. Somatic and maturity maintenance flows, the maturation flow, and overheads represent the dissipated energy, and therefore contribute to the respiration rate. Adapted with permission from Jusup et al. (2014).

and the beginning of the adult stage. The energy required to keep the fish alive and to maintain the current level of maturity in preparation for adulthood is determined by somatic and maturity maintenance flows ($\dot{p}_S$ and $\dot{p}_J$, respectively). The growth flow ($\dot{p}_G$) is responsible for delivering energy to the structure if $\kappa\dot{p}_C > \dot{p}_S$, resulting in the increase of the structural length. In much the same way, the increase in the level of maturity is driven by the maturation flow ($\dot{p}_R$) if $(1-\kappa)\dot{p}_C > \dot{p}_J$. At the onset of reproduction, i.e., when the level of maturity reaches the aforementioned threshold value, the reproduction flow ($\dot{p}_R$) takes over the maturation flow which is the reason why the same notation is used for both.

We are now in a position to turn the above considerations and the schematic representation in Fig. 17.1 into mathematical expressions. An equation describing the dynamics of the reserve compartment is

$$\frac{dE}{dt} = \dot{p}_A - \dot{p}_C. \tag{2}$$

The last equation can be cast into an equivalent form describing the dynamics of the reserve density, $[E]=E/L^3$, by writing

$$\frac{d[E]}{dt} = \frac{\dot{p}_A - \dot{p}_C}{L^3} - 3[E]\frac{d}{dt}\ln L. \tag{3}$$

For the structure compartment we have

$$\frac{dL}{dt} = \frac{\dot{p}_G}{3L^2[E_G]}, \tag{4}$$

where $[E_G]$ is a parameter called the volume-specific cost of structure. This parameter determines the amount of energy needed for every unit increase in the structural volume. The equation for maturity, when $E_H < E_H^p$, is

$$\frac{dE_H}{dt} = \dot{p}_R,$$ (5)

where E_H^p is the maturity-at-puberty, i.e., the threshold value that marks the end of the juvenile and the beginning of the adult stage. The connection between growth and maturation flows on the one hand and the utilization flow on the other is given by equations

$$\dot{p}_G = \kappa \dot{p}_C - \dot{p}_S,$$ (6a)

$$\dot{p}_R = (1 - \kappa)\dot{p}_C - \dot{p}_J,$$ (6b)

where $0 < \kappa < 1$ is a constant. Making equations (2–6) solvable (at least using numerical methods), requires specifying the dependence of assimilation, utilization, and both maintenance flows on the state variables.

Assimilation

The assimilation flow $(\dot{p}_A)$ can be interpreted as the rate at which the energy from ingested food is deposited in the reserve. Note, however, that the assimilation flow is not equivalent to the ingestion rate. The difference is largely a consequence of the fact that food and the reserve have different chemical compositions. Food, therefore, needs to be converted into the reserve and this conversion results in the release of energy in the form of heat and the metabolites. The ensuing energy losses are related to SDA (see Nisbet et al. 2012 for more detail). Another reason for differentiating the assimilation flow and the ingestion rate are the inefficiencies of the digestive system. In DEB theory, overall energy losses that follow the ingestion of food, but occur before assimilation into the reserve, are called the assimilation overhead. Fortunately enough, for a wide range of external conditions, the assimilation overhead is a constant fraction, $0 \leq C < 1$, of the ingestion rate, so that the assimilation flow equals $\dot{p}_A = \dot{p}_X - C\dot{p}_X = (1-C)\dot{p}_X$, where $\dot{p}_X$ denotes the ingestion rate. It is common to replace C with the assimilation efficiency of food into the reserve, $\kappa_X = 1 - C$. Now we can state the assumption that connects the assimilation flow to the structural length. Namely, the ingestion rate is assumed to be proportional to the squared structural length because ingestion can only happen at the interface between the fish and the environment. Again, the isomorphism approximation is crucial in order for the area of the interface separating the fish from the environment to be proportional to the squared structural length. The corresponding proportionality constant can be decomposed into two factors. The maximum surface-area-specific ingestion rate, $\{\dot{p}_{Xm}\}$, represents the physiological limitations of the digestive system, whereas food availability, $0 \leq f \leq 1$, corresponds to the environmental conditions. We can also define the maximum surface-area-specific assimilation rate by $\{\dot{p}_{Am}\} = \kappa_X \{\dot{p}_{Xm}\}$. Table 17.1 contains the final expression for the assimilation flow following from these considerations.

Utilization

Powering metabolic processes such as maintenance (both somatic and maturity), growth, maturation, and reproduction requires energy to be utilized (or mobilized) from the reserve. The rate at which this happens is determined by the utilization (or mobilization) energy flow ($\dot{p}_C$). Deriving the formula that describes how $\dot{p}_C$ depends on the state variables is perhaps one of the central problems of DEB theory and turns out to be quite technical (Sousa et al. 2008; Kooijman 2010). However, the problem can be simplified by assuming that the reserve density, appearing in Eq. (3), follows a first order dynamics (van der Meer 2006); during starvation, the reserve density decreases at a rate proportional to its current amount. From Eq. (3) it follows

$$\frac{\dot{p}_C}{L^3} + 3[E]\frac{d}{dt}\ln L = F(L)[E],\tag{7}$$

where $F = F(L)$ is, for the moment, an unknown function of L. Finding F involves inserting Eq. (7) back into Eq. (3) and observing that the equilibrium reserve density, $[E]^*$, satisfies the condition $F(L)[E]^* = \dot{p}_A/L^3$. The equilibrium reserve density should be at its maximum, $[E]^* = [E_m]$, when food in the environment is plentiful, i.e., $f = 1$. We thus obtain $F(L) = \{\dot{p}_{Am}\}/[E_m]L$. The ratio of the maximum surface-area-specific assimilation rate and the maximum reserve density is a constant that characterizes how fish of a given species utilize the energy from the reserve. Because of its importance, this ratio is treated as a fundamental parameter of DEB theory, usually denoted by $\dot{v}$ and called the energy conductance. As a consequence, we have

$$\dot{p}_C = E\left(\frac{\dot{v}}{L} - 3\frac{d}{dt}\ln L\right).\tag{8}$$

The last equation can be further transformed into the form found in Table 17.1 by using Eqs. (4) and (6a).

Maintenance

Keeping the fish alive and healthy, as well as maintaining its level of maturity, involves energy investments. In DEB theory, investment into the former is determined by the somatic maintenance flow ($\dot{p}_S$), whereas the investment into the latter is controlled by the maturity maintenance flow ($\dot{p}_J$). The physiological basis for somatic maintenance is primarily, but not exclusively, the protein turnover—continuous degradation and synthesis of proteins inside the cells of fish (Hawkins 1991; Houlihan 1991). To come up with a first order estimate of the energetic costs associated with the protein turnover, we simply note that these costs should increase in proportion to the number of cells. On the other hand, the number of cells should be approximately proportional to the structural volume or, when working with L, to the cubed structural length. In the case of tuna fishes, it may also be necessary to consider maintenance costs other than the protein turnover. These costs are primarily related to thermogenesis, where the term is used in the sense of fish possessing the ability to keep the temperature of some tissues above ambient seawater temperature (also called regional endothermy or heterothermy) present in all tuna species (Graham and Dickson 2001). In contrast to the costs of the protein turnover, it is more appropriate to model the energy expenditure for thermogenesis using a term proportional to the squared structural length because this expenditure needs to counteract the heat loss which occurs only through the outer surface. Hence a distinction between the volume-related ($\dot{p}_M$) and surface-area-related ($\dot{p}_T$) somatic maintenance costs is made, where $\dot{p}_S = \dot{p}_M + \dot{p}_T$, $\dot{p}_M \propto L^3$, and $\dot{p}_T \propto L^2$. The two corresponding proportionality constants, $[\dot{p}_M]$ and $\{\dot{p}_T\}$, are called the volume-specific and the surface-area-specific somatic maintenance costs, respectively. As before, mathematical expressions are summarized in Table 17.1.

To obtain a mathematical expression for the maturity maintenance flow ($\dot{p}_J$), we identify the level of maturity with the complexity of the structure. The second law of thermodynamics dictates that complexity would decrease without some form of maintenance (Sousa et al. 2010; Sousa et al. 2008). Analogously to the volume-related somatic maintenance costs, which can be rewritten in such a way that $\dot{p}_M$ is proportional to energy embedded into the structure, maturity maintenance flow is assumed to be proportional to the current level of maturity, i.e., $\dot{p}_J \propto E_H$. The proportionality constant is usually denoted by k_J and called the maturity maintenance rate coefficient. The resulting mathematical expression is presented in Table 17.1.

Temperature effects

The connection between metabolic rates in bony fishes and ambient temperature is well-captured by the Arrhenius equation (Clarke and Johnston 1999). According to this equation, if an energy flow, $\dot{p}_*$, is known at the reference temperature, T_0, then at another temperature, T, we have

$$\dot{p}_*(T) = \dot{p}_*(T_0)\exp\left(\frac{T_A}{T_0} - \frac{T_A}{T}\right), \tag{9}$$

where T_A is a constant characterizing the sensitivity of the fish to temperature variations. In the case of Pacific bluefin tuna, the constant T_A equals 5300 K. It is worth noting that DEB theory does not offer any mechanistic reasons for adopting Eq. (9). The Arrhenius equation apparently cannot be derived from the first principles at the present time, and thus represents a mere statistical formulation of an evolutionary end-result (Clarke 2004). Furthermore, some concerns may be raised as to what is the relevant temperature for regulating metabolic rates in regionally endothermic fish like bluefin tunas. The problem may not be overly serious if relevant body temperature is correlated with seawater temperature and we use the latter as a proxy. Experimental evidence is somewhat unclear in this context. Continuous temperature measurements inside the red muscle tissue of wild Pacific bluefin tuna show no correlation with seawater temperature (Marcinek et al. 2001). Even the average temperature of the peritoneal cavity recovered from archival tags deployed in Atlantic bluefin tuna seems to correlate only weakly with average seawater temperature ($R^2 = 0.08$, $p = 0.06$; our analysis of the data from Walli et al. 2005). On the other hand, the growth performance of Pacific bluefin tuna in captivity strongly depends on average temperature conditions at the rearing site (Masuma et al. 2008). This last piece of evidence provides important

384 Biology and Ecology of Bluefin Tuna

support for the idea of inserting a simply measurable proxy into Eq. (9) instead of searching for more complicated solutions.

The standard DEB model

Equations (2–6), (9), and Table 17.1 define the standard DEB model and provide all information needed to run numerical simulations. However, the details that account for the specifics of tuna physiology are still missing. Incorporating these specifics into the model becomes critically important when trying to reproduce the experimental data that pertains to the whole life cycle. The reason is that some parameter values, though valid during early development, become invalid for a more mature individual.

A picture of Bluefin Tuna Emerging from Bioenergetic Modeling

Before we could adapt the standard DEB model to our needs, we were faced with the selection of one target species among Atlantic (*Thunnus thynnus*), Pacific (*Thunnus Orientalis*; henceforth abbreviated PBT), and southern (*Thunnus maccoyii*) bluefin tunas. The choice fell on mid-sized PBT because of a huge amount of empirical information in existence. Readily available information included (i) a comprehensive coverage of embryonic development (Miyashita et al. 2000), (ii) detailed descriptions of larval morphology, physiology, and growth (Miyashita et al. 2001; Miyashita 2002; Sawada et al. 2005), (iii) the data on the growth of juveniles and adults in captivity and in the wild (Miyashita 2002; Masuma et al. 2008; Masuma 2009; Shimose et al. 2009), (iv) estimates of the reproductive output in captivity and in the wild (Chen et al. 2006; Masuma et al. 2008; Masuma 2009), (v) observations on the survival rates of juveniles and adults in captivity (Masuma et al. 2008), (vi) relationships between metabolic rates, swimming speed, and temperature in larvae and juveniles (Miyashita 2002; Blank et al. 2007a; Blank et al. 2007b), (vii) measurements of postprandial metabolic rates in juveniles (Clark et al. 2010), and (viii) a thorough analysis of the heat budget by means of the lumped capacitance model (Kitagawa et al. 2006; Kitagawa et al. 2007; Kubo et al. 2008). The above list was organized around the data comparable with the outputs of a DEB-based model, and hence could not be characterized as exhaustive. We do believe, however, that the richness of empirical information surrounding PBT is adequately illustrated.

Comparisons of the experimental data with the outputs of a theoretically sound model may reveal a mechanistic foundation for empirically observed patterns. For instance, we may hypothesize why PBT larvae have an accelerating growth rate up to one month after hatching, why early juveniles have a decelerating growth rate three to four months after hatching, and why the von Bertalanffy curve captures PBT growth remarkably well except during early development. If we, starting from the formulated hypotheses, construct a model whose outputs compare favorably with the data, then we also have a strong reason to believe that the mechanisms implemented into the model truly operate in nature. Because outputs of a DEB-based model are comparable to a wide variety of datasets, it may mechanistically underpin multiple empirical patterns at once. In this sense, DEB theory represents an integrative bioenergetic framework worthy of exploration.

A DEB-based model for Pacific bluefin tuna

The standard DEB model, though relied upon as a template, had to be supplemented with a limited number of assumptions on morphology and thermogenesis to be applicable to PBT. The first assumption concerns the near exponential growth that larval PBT undergo for around 30 days after hatching. In the context of DEB theory, such growth is indicative of the morphological (and possibly some physiological) changes that influence both energy assimilation and utilization. To capture the effect of the morphological changes on fish energetics, we assume that PBT develop as V1-morphs in the larval stage and, at least approximately, as isomorphs in all other stages (Kooijman et al. 2011; Kooijman 2010). In mathematical terms, all occurrences of the maximum surface-area-specific assimilation rate and the energy conductance must be multiplied by an auxiliary function of the form

$$M_1(L, E_H) = \begin{cases} 1, & E_H < E_H^b \\ L/L_b, & E_H^b \le E_H < E_H^j, \\ L_j/L_b, & E_H^j \le E_H \end{cases} \tag{10}$$

where L_b (L_j) is the structural length at which the level of maturity reaches the maturity-at-birth, E_H^b (maturity-at-metamorphosis, E_H^j). The function $M_1 = M_1(L, E_H)$ is often referred to as the shape correction function. The reason why morphological changes can affect the assimilation flow is intuitively understandable by considering that, for example, the digestive system may grow allometrically for a while causing the intestine to grow faster than the fish as a whole. Under such circumstances, the surface area through which food is assimilated is no longer proportional to the squared structural length, creating an advantage over a strict isomorph. It is a bit more difficult to see why morphological changes would affect energy utilization as well. Consider, however, that the energy conductance has the dimension length × time^{-1}. The dimension of length originates from a ratio of volume to surface-area that is proportional to length only in strict isomorphs. The consequence is that the utilization flow must improve similarly to the assimilation flow. Note that under this assumption the values of the surface-area-specific assimilation rate and the energy conductance in juveniles and adults depend on the feeding history in the larval stage, partially accounting for the natural variability in the parameter values of same-species individuals (Kooijman et al. 2011).

The second assumption specific to the DEB-based model for PBT is closely related to the first one and the fact that larval PBT undergo considerable morphological changes (Miyashita et al. 2001; Miyashita 2002). The isomorphism assumption is just a first order approximation even after larvae metamorphose into early juveniles. Miyashita et al. (2001), for example, report that only the preanal length reaches a constant proportion of the body length prior to metamorphosis. All other observed body parts, though the ratios of their size to the body length seem to be converging to a constant value, still exhibit allometric growth. Accordingly, we assume that the shape factor (δ_M) varies with the level of maturity in a way consistent with the substantial morphological changes during early development and an asymptotic approach to the final shape later. The corresponding mathematical expression is

$$\delta_M\left(E_H\right)=\frac{\delta_M^1\left(E_H^2-E_H^b\right)+\delta_M^2\left(E_H-E_H^b\right)}{E_H+E_H^2-2E_H^b},\; E_H^b\le E_H<E_H^p, \tag{11}$$

where $\delta_M^{\;l}$ ($\delta_M^{\;2}$) is the larval (adult) shape factor and $\delta_M(E_H^{\;2})=(\delta_M^{\;l}+\delta_M^{\;2})/2$.

One remaining assumption necessary to adapt the standard DEB model to PBT concerns the role of the excess red muscle tissue in continuous swimming and thermogenesis observable in all tuna species (Katz 2002; Graham and Dickson 2004). Because thermogenesis is not present in larval PBT, but becomes measurable in early juveniles some 80 to 120 days after hatching, the value of the surface-area-specific somatic maintenance rate must initially be zero and gradually increase as the fish mature. Therefore, we take into account the thermogenic energy dissipation by multiplying all occurrences of $\{\dot p_T\}$ by an auxiliary function of the form

$$M_2\left(E_H\right)=\begin{cases}0, & E_H<E_H^j \\ \left(E_H-E_H^j\right)/\left(E_H^y-E_H^j\right), & E_H^j\le E_H<E_H^y, \\ 1, & E_H^y\le E_H\end{cases} \tag{12}$$

where E_H^y marks the maturity of an individual with fully developed thermogenesis. The auxiliary function $M_2=M_2(E_H)$ is interpreted as the thermogenic efficiency.

Equations (2–6), (9), and Table 17.1 together with modifications (10–12) form a full life cycle DEB-based model for PBT—from an egg to an adult fish and its eggs. Having all definitions in place, we can move forward to explore how the model performs against the existing data, how our hypotheses link to physiological observations, and finally what are the implications for PBT ecology.

Modeling growth, maturation, and reproduction of Pacific bluefin tuna

A fully parameterized DEB-based model for PBT can reproduce many characteristics found in the existing datasets (Jusup et al. 2011; Jusup et al. 2014). Some of the more interesting examples are shown in Fig. 17.2. The focus is put on the ability of the model to capture the growth of the fish in all life stages.

Embryonic development begins in an egg of the estimated size of 0.972 mm (observed size 0.926–1.015 mm; Miyashita 2002), lasts between 19 and 43 hours depending on the temperature, and ends with newly hatched larvae of the estimated size of 3.07 mm total length (observed average size 3.08 mm TL; Sawada et al. 2005). From the first feeding, larvae grow in a near-exponential pattern and increase in size by approximately a factor of 13 some 30 to 40 days after hatching (Fig. 17.2a). In controlled hatchery conditions, we estimate that the larval stage lasts 36 days at the end of which assimilation and utilization of energy are supposedly as effective as in adult individuals. By contrast, the thermogenic efficiency is still at zero level and larvae, now turning into young juveniles, experience a period of an extraordinarily high, nearly constant growth rate.

The period of high growth in young juvenile PBT is illustrated in Fig. 17.2b. The growth rate is estimated at 0.45 cm·d⁻¹ and apparently maintained for around two months, allowing the fish to grow by approximately a factor of 7.5 during this period. Hence, over the first three months of life, PBT experience an astounding 13 × 7.5 ≈

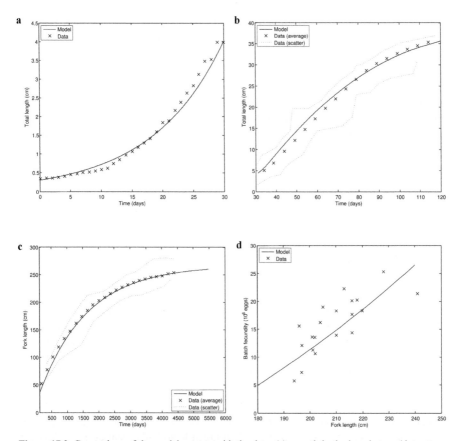

Figure 17.2. Comparison of the model outputs with the data: (a) growth in the larval stage (data source Sawada et al. 2005); (b) growth of early juveniles (data source Miyashita 2002); (c) growth of juveniles and adults (data source Masuma 2009); (d) batch fecundity in adults (data source Chen et al. 2006). Adapted with permission from Jusup et al. (2011).

100 times increase in body size. As the fish approach 30 cm fork length, some 90 to 120 days after hatching, the continuous improvement in the thermogenic efficiency finally results in considerable energy expenditure. A gradual drop in the growth rate ensues. Provided the food level is kept constant, the fish from this point on continue to grow in accordance with the von Bertalanffy curve.

Growth of juvenile and adult PBT consistent with the von Bertalanffy curve is illustrated in Fig. 17.2c. The growth rate is appreciably lower than in young juveniles and steadily decreases with the size of the fish. Ultimately, the fish approach an asymptotic body length dependent on the food level. For instance, captive PBT shown in Fig. 17.2 approach an estimated fork length of 267 cm, whereas theoretically, at $f = 1$, these fish could grow to 327 cm fork length. The von Bertalanffy growth rate as calculated from the estimated parameters is 0.249 y^{-1}. This value seems to be equally valid for juveniles older than 4 months after hatching and for reproductively active adults, contrary to the suggestion by Lester et al. (2004) that the von Bertalanffy curve is not a good description of somatic growth before maturation. We are also led to believe that the onset of reproduction in PBT does not require energy to be redirected from growth towards the reproductive organs, which is in line with the κ-rule, but contrary to some bioenergetic studies (e.g., Tsikliras et al. 2007; Ohnishi et al. 2012; for a review of resource allocation strategies between growth and reproduction in indeterminate growers see Heino and Kaitala 1999).

Turning the discussion to maturation and reproduction, we observe that the model is consistent with a variety of observations on PBT reproductive biology. For instance, age-at-puberty in captive fish is estimated at 3.2 years which corresponds to the fork length of close to 150 cm and compares favorably with the available reports (Hirota et al. 1977; Masuma et al. 2011). Batch fecundity of young adults is likely to be rather low, reaching around five million eggs as the fish approach 180 cm fork length. From there, the batch fecundity increases linearly to over 25 million eggs in PBT of 240 cm fork length (Fig. 17.2d). The estimate of the total number of batches per reproductive season is nine. Though consistent with observations in captivity (Masuma 2009), we have some reservations about such a low value. Namely, the model in its present form estimates dry mass of an egg to be 160 μg, which is three to four times higher than what measurements on eggs of yellowfin tuna indicate (Margulies et al. 2007). We believe that the cross-species comparison is appropriate in this case because other egg properties between PBT and yellowfin tuna are strikingly similar. The overestimation of egg dry mass may be caused by considerably higher water content of eggs in comparison to the mother's reserve, suggesting that the model allocates too much energy to the production of a single egg. As a consequence, the total number of batches per reproductive season may be higher than the current model estimate. We hold that the described issue represents an interesting topic for the future studies of bluefin tuna reproductive biology and energetics.

Having established that the model successfully reproduces a number of characteristics of captive PBT, a brief look at wild fish is in order. The comparison of growth between captive and wild PBT in juvenile and adult stages is displayed in Fig. 17.3. Individuals reared in captivity typically grow to a larger size than their wild counterparts of the same age. Even asymptotically the situation remains similar as wild fish attain the ultimate size of 252 cm fork length—approximately 15 cm less than captive PBT. Somewhat surprisingly, the difference in the same-age size can be largely explained by a 6.5°C higher average body temperature of captive fish. Food availability in the wild appears to be only marginally lower than in captivity, providing an explanation for the difference in ultimate sizes. These explanations strike us as reasonable given that the data for captive PBT come from a broodstock kept by the Fishery Research Agency of Japan at Amami station where the annual seawater temperature range at 10 m depth is 20 to 28°C due to a subtropical climate (Masuma 2009). By contrast, wild fish are known to spend much time at water temperatures below 20°C and venture into even colder waters below 15°C during migrations (Kitagawa et al. 2009; Boustany et al. 2010). An interesting side-effect of the lower average body temperature and lower food availability is that it takes 5.3 years for wild fish to sexually mature—over two years longer than captive PBT. This result is again in line with the observations as Chen et al. (2006) demonstrate that the age-at-puberty is, at most, six years in the wild.

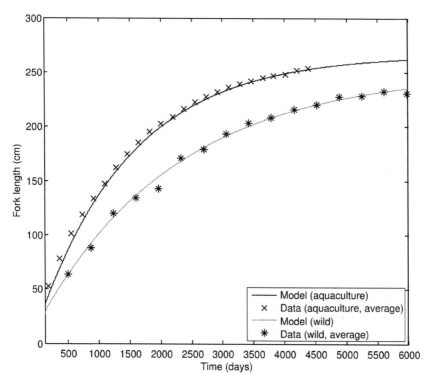

Figure 17.3. Comparison of growth between captive and wild Pacific bluefin tuna in juvenile and adult stages (data sources Masuma 2009 and Shimose et al. 2009). Adapted with permission from Jusup et al. (2011).

Physiological underpinning of the DEB-based model for Pacific bluefin tuna

Earlier we introduced three assumptions specific to the DEB-based model for PBT and then showed that the model outputs compare favorably with the existing data. As a result, we now have a good reason to believe that the stated assumptions are the stylized mathematical representations of concrete physiological characteristics of PBT. It is our next task to look into these characteristics more thoroughly and try to place them into a proper context given the model and the quality of its outputs.

One of the three PBT-specific assumptions made during model development stated that larval PBT behaved as a V1-morphic organism for a period of time after the onset of feeding. This assumption was largely motivated by the data on the near-exponential growth of PBT larvae (Sawada et al. 2005; Miyashita 2002; Miyashita et al. 2001) which in the context of DEB theory hinted at improvements in the way larval PBT assimilate and utilize energy. Focusing momentarily on improvements in energy assimilation, we identified multiple physiological mechanisms supportive of our assumption. For example, Kaji (2003) reported that the first appearance of gastric glands and pyloric caeca, as well as an increase in the specific activity of trypsine-like and pepsine-like digestive enzymes all took place while PBT were still in the larval stage. In addition, Miyashita (2002) reported that larval PBT exhibited allometric growth of the preanal length, the head length, the head height, the snout length, the upper jaw length, and the eye diameter, which likely boosted the ability of larvae to ingest ever larger food items. Evidence was also found to support the idea of improving energy utilization from the reserve. Highlighting an exceptional growth hormone activity, Kaji (2003) reported that in comparison with other previously examined fish species, larval PBT exhibited a very high and increasing ratio of the growth hormone immunoreactive cells volume to the pituitary volume. The RNA to DNA ratio as a proxy for protein synthesis followed a qualitatively similar increasing pattern towards the end of the larval stage and continued doing so even as the fish turned into young juveniles. Although other scenarios in DEB theory could account for the near-exponential growth (Kooijman et al. 2011), the physiological changes that take place in the larval stage of PBT offered a strong support for the V1-morph assumption used in the model construction.

Another assumption setting apart the DEB-based model for PBT from the standard DEB model stated that at the onset of thermogenesis the surface-area-related somatic maintenance costs gradually increased from zero to a value that contributed considerably to the overall energy budget of the fish. Similarly as before, the assumption was motivated by a pattern observed in fish growth. Namely, Miyashita (2002) presented the data that reveal a deceleration of the growth rate in young juvenile PBT some 90 to 120 days after hatching. In the context of DEB theory such a deceleration could have been explained by a previously non-existent sink of energy gradually coming on line. We found physiological evidence supportive of an extra energy sink in thermogenesis, its connection to the excess red muscle tissue, and continuous swimming. Kubo et al. (2008) reported that thermogenesis became detectable in PBT slightly smaller than 20 cm fork length, corresponding approximately to 80 days after hatching. Thermogenesis was readily measurable in fish of over 40 cm fork length, corresponding roughly to 120 days after hatching. Simultaneously, the volume ratio of red to white muscle increased non-linearly with the body length and attained a maximum value of around 12% when the fish approached 50 cm fork length. Note that our concern here was not the role of retia mirabilia nor conservation of heat (Graham and Dickson 2001; Graham and Dickson 2004). We were, in fact, oriented towards identifying the source of heat that in combination with retia mirabilia made thermogenesis in PBT possible. Excess red muscle emerged as the most likely explanation in this context because of the correspondence between red muscle development and the emergence of thermogenesis, as well as the involvement of red muscle in continuous swimming whereby metabolic energy is expended and heat created (Graham and Dickson 2001; Katz 2002). If excess red muscle were acting as a considerable sink of energy, the data should exhibit a surge in the respiration rate when comparing fish before and after the onset of thermogenesis. Unfortunately, at the time of writing we were not aware of any datasets of sufficient quality to test the above conjecture.

Implications for the ecology of Pacific bluefin tuna

To better understand the nature of the PBT energy budget, we can contrast the assimilation energy flow with the rate at which energy is dissipated or, in the adult stage, partly released into the environment in the form of eggs. The corresponding graph is shown in Fig. 17.4. Note that the non-dissipated energy ends up embedded either into the reserve or the structure, resulting in the increase of body mass. Even a superficial glance at Fig. 17.4 reveals that more than 80% of the assimilation energy flow is dissipated after PBT reach around 35 cm fork length. This percentage increases linearly with the body length, approaching 100% as the fish grow close to their asymptotic length. Another way to look at the same information is that the percentage of the assimilation energy flow available for the increase of body mass is less than 20% in young juveniles and decreases towards zero with age. It is important not to overlook that these results pertain to PBT in captivity where relatively high food availability is guaranteed. Though wild PBT also seem to experience relatively high food availability on average, favorable conditions may not prevail at all times. When food availability is suboptimal, given that much of the dissipation rate is attributable to maintenance costs (Fig. 17.4), PBT are quickly forced to rely on their reserve in order to stay alive. Yet, the maximum reserve capacity of $[E_m] = \{\dot{p}_{Am}\}/\dot{v} = 680 \text{ J·cm}^{-3}$ implied by our model is rather low. Such a reserve capacity provides enough energy to satisfy the somatic maintenance demands of juvenile PBT for less than a week. This result, although a bit surprising, is not so far-fetched when considering a starvation experiment by Boggs and Kitchell (1991) who observed that a number of skipjack, kawakawa, and yellowfin tuna showed signs of imminent death within the first week of fasting. In their experiment only one yellowfin tuna out of 63 experimental fish survived after being starved for 30 days. Therefore, empirical evidence strongly suggests that high maintenance costs pose a serious obstacle to survival of starving PBT even if, with the emergence of better data in the future, our estimate of the maximum reserve capacity turns out to be on the low end of the range. In this respect, our model is

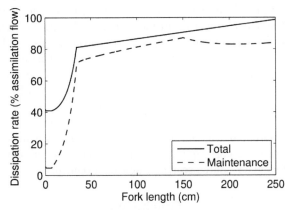

Figure 17.4. Rate at which energy is dissipated or, in the adult stage, partly released into the environment in the form of eggs as a percentage of the assimilation flow.

consistent with the notion of tuna fishes being energy speculators (Korsmeyer et al. 1996; Fitzgibbon et al. 2007).

Energy speculation in tunas, defined broadly as an adaptive mechanism whereby high rates of energy expenditure are speculated to bring even higher rates of energy ingestion and assimilation, can be put in a more general context by thinking about tuna fishes as an extreme case of a demand organism (Kooijman and Troost 2007; Jusup et al. 2011; Jusup et al. 2014; Lika et al. 2014). To substantiate this idea, we first note that individuals of all species belong somewhere on the spectrum ranging from supply organisms, which presumably evolved first, to demand organisms, which evolved later. Supply organisms have little 'behavioral flexibility', and hence developed considerable 'metabolic flexibility' to cope with ever-changing chemical and physical conditions of their environments. For instance, supply organisms tend to move little, depend on an active diffusion process for food, and have less sensitive sensory organs, but at the same time exhibit adjustable stoichiometric homeostasis, tolerance for a wide range of food availabilities, and the ability to shrink in starvation. A vast majority of species on Earth share the characteristics of supply organisms. By contrast, demand organisms are animals in which large energy expenditure must be met with a relatively stable and sizable food intake, thus resulting in diminished metabolic, but substantial behavioral flexibility. Demand organisms obey strict stoichiometric homeostasis, specialize on fewer types of food that satisfy their needs, and poorly tolerate starvation. However, they actively search for food, have highly developed sensory organs, exhibit a high peak metabolic rate supported by a closed circulatory systems, and function as endotherms to extend their ecological niches. Our DEB-based model suggests that the energy budget of PBT is dominated by high maintenance costs which, in turn, require the fish to maintain high food availability at all times and simultaneously reduce the set of possible growth paths. Low maximum reserve capacity further indicates that PBT cannot handle starvation very well which is likely to reflect on both the feeding and reproductive behavior of this species. Should feeding conditions become unfavorable, the survival of an individual PBT may depend on behavioral adaptations such as slower swimming (Boggs and Kitchell 1991), whereas adults may also spend energy that would otherwise be reserved for reproduction, causing a pause in spawning activities (Secor 2007). The above properties of the model emphasize the lack of metabolic flexibility in PBT, but this is not the whole picture. The model also confirms that high-performance metabolic machinery hosted by tuna fishes in general, works very well for wild PBT in the sense that they experience food availability that is only marginally lower than in captivity. In fact, there are no bioenergetic reasons why PBT should not reproduce throughout the year, pointing to a critical role of environmental factors in triggering egg production. We conclude that behavioral flexibility of tunas achieved through unique morphological and physiological adaptations is mostly sufficient to offset high maintenance costs and allows these remarkable animals to function as energy speculators in a resource-scarce pelagic environment.

Conclusion

We reviewed bioenergetic approaches in fisheries science with particular focus on the results that pertain to tuna fishes. We also outlined a novel bioenergetic approach called Dynamic Energy Budget theory and showed how it could be adapted to model growth, maturation, and reproduction of Pacific bluefin tuna. If the exposition was clear and persuasive enough to convince the reader of the usefulness of bioenergetic modeling and DEB theory in the context of Pacific bluefin tuna physiology and ecology, then the primary purpose of our effort was achieved. However, the presented results by no means exhausted the full potential of this approach. Reaching the full potential would require further generalizations, extensions, and coupling with complementary methods. In what follows, we sketched a few of the more obvious, yet promising, ideas.

Because DEB theory applies to all species in general and has been successfully applied to a number of fish species in particular, one way to extend the present model is the estimation of parameters values for other members of the genus *Thunnus* and subsequently for other genera in the tribe *Thunnini*. Then, having the same underlying model, we could make direct cross-species comparisons between the estimated parameters or functions thereof to find bioenergetic causes for the differences in physiological and ecological traits of tuna fishes. Among others, we could calculate the maximum attainable structural length, the minimum food availability needed for survival, the minimum time to maturation, the maximum reproductive rate, and the time till death by starvation. Even more could be achieved by looking if the estimated parameters and their functions vary systematically with the maximum structural length, thus exploring the kind of body-size scaling relationships that hold for tunas. Ideally, we would be led to a generic DEB-based model for tuna fishes, i.e., one that would capture bioenergetics of the whole *Thunnini* tribe reasonably well using a set of common parameter values in conjunction with body-size scaling relationships implied by DEB theory.

Another aspect of tuna bioenergetics where further advances are possible concerns the link between metabolic rates and aging. It has long been known that the caloric restriction significantly increases the average lifespan of organisms (Colman et al. 2009), suggesting a strong metabolic component to aging. To mathematically describe this component, DEB theory assumes that metabolic processes, particularly the utilization of energy from the reserve, cause the production of damage-inducing compounds (by means of activated mitochondria or changed genes). Such compounds have a twofold effect on the organism; they promote their own production and they generate damage that accumulates inside the organism, eventually affecting the survival probability and some other sub-lethal endpoints (e.g., assimilation and reproduction). When referring to damage, DEB theory takes the mainstream view of attributing the physiological decline in aged organisms to oxidative stress (van Leeuwen et al. 2010). The last statement actually reveals an inconsistency in the way aging is currently modeled in DEB theory—the accumulation of oxidative stress should be more closely related to the respiration rate than the utilization energy flow. Therefore, much work needs to be done to smooth out difficulties arising in the context of DEB-based modeling of aging (Jager 2013). In the case of tuna fishes, however, the prize may be worth the trouble because tunas are top predators and have the potential to live to an old age even in the wild, indicating that senescence may play an important ecological role.

A major step forward in exploiting the full potential of physiological energetics in the context of tuna research would be achieved by coupling DEB-based models with population dynamics approaches. Namely, the model we presented is capable of forecasting growth, maturation, and fertility of Pacific bluefin tuna in response to food availability and temperature in the environment. Fertility, however, can be presented in the form of a maternity function which is one of the two vital rates needed for simulating population dynamics. The other vital rate is the survival function which, as we mentioned previously, can be linked to metabolic rates defined in DEB-based models. In the wild, of course, individuals are affected by many other causes of death besides aging, but because tunas are top predators capable of surviving to an old age, survival function based solely on aging may provide insight into how tuna populations would look in an environment undisturbed by fishing. In particular, we could set the net reproductive output to unity and calculate food availability at which a given tuna population is in equilibrium with its environment. We could then assume that a certain fishing pressure is exerted causing now the fished population to adjust to a new equilibrium state. The properties of the fished population in the new equilibrium could be estimated together with the total catch. By testing a wide variety of fishing strategies, the strategy yielding the highest total catch could be identified. Finally, in more advanced applications, transient population dynamics taking place in-between two equilibrium states due to, for example, shifts in the fishing strategy could be modeled for the purpose of population management.

One last application of bioenergetic models for tuna fishes emphasized here concerns open research questions related to a particular life stage rather than the whole life-cycle. For instance, a recent study by Satoh et al. (2013) shows that in the wild higher food density and seawater temperature (if within the tolerance range) lead to better growth rates in Pacific bluefin tuna larvae, while survival rates positively correlate with the observed growth rates. Because the dependence of growth on the environmental conditions in the larval stage is captured by the presented DEB-based model, coupling it with biogeochemical ocean and Lagrangian dispersal models might help us better understand the role of environment in creating a suitable spawning habitat for tunas. The ensuing suite of models could be used to identify likely spawning grounds by simulating growth in natural conditions and registering the success rate at which larvae reach the size at metamorphosis. In this way, we might be able to better direct larval surveys, identify previously unknown spawning grounds, possibly reject some regions where spawning was suspected to occur, and ultimately obtain valuable information for the purpose of managing wild tuna stocks. Though we cannot be sure of the direction that bioenergetic research of tuna fishes will take in the future, we can safely assume that many exciting and important findings are yet to be uncovered. The importance of future findings lies in much more than just intellectual curiosity. What we are truly hoping for are the means by which tuna populations around the globe will be exploited in a responsible manner for the long-term benefit of humanity.

References

Altringham, J.D. and D.J. Ellerby. 1999. Fish swimming: patterns in muscle function. J. Exp. Biol. 202(23): 3397–3403.
Altringham, J.D., C.S. Wardle and C.I. Smith. 1993. Myotomal muscle function at different locations in the body of a swimming fish. J. Exp. Biol. 182(1): 191–206.

Anderson, E.J., W.R. McGillis and M.A. Grosenbaugh. 2001. The boundary layer of swimming fish. J. Exp. Biol. 204(1): 81–102.

Augustine, S., B. Gagnaire, C. Adam-Guillermin and S.A.L.M. Kooijman. 2012. Effects of uranium on the metabolism of zebrafish, *Danio rerio*. Aquat. Toxicol. 118: 9–26.

Augustine, S., B. Gagnaire, M. Floriani, C. Adam-Guillermin and S.A.L.M. Kooijman. 2011. Developmental energetics of zebrafish, *Danio rerio*. Comp. Biochem. Physiol. A Mol. Integr. Physiol. 159(3): 275–283.

Barrett, D.S., M.S. Triantafyllou, D.K.P. Yue, M.A. Grosenbaugh and M.J. Wolfgang. 1999. Drag reduction in fish-like locomotion. J. Fluid Mech. 392(1): 183–212.

Bernal, D., K.A. Dickson, R.E. Shadwick and J.B. Graham. 2001. Review: analysis of the evolutionary convergence for high performance swimming in lamnid sharks and tunas. Comp. Biochem. Physiol. A Mol. Integr. Physiol. 129(2): 695–726.

Blake, R.W. 2004. Fish functional design and swimming performance. J. Fish Biol. 65(5): 1193–1222.

Blank, J.M., C.J. Farwell, J.M. Morrissette, R.J. Schallert and B.A. Block. 2007a. Influence of swimming speed on metabolic rates of juvenile Pacific bluefin tuna and yellowfin tuna. Physiol. Biochem. Zool. 80(2): 167–177.

Blank, J.M., J.M. Morrissette, C.J. Farwell, M. Price, R.J. Schallert and B.A. Block. 2007. Temperature effects on metabolic rate of juvenile Pacific bluefin tuna *Thunnus orientalis*. J. Exp. Biol. 210(23): 4254–4261.

Boggs, C.H. and J.F. Kitchell. 1991. Tuna metabolic rates estimated from energy losses during starvation. Physiol. Zool. 64(2): 502–524.

Boustany, A.M., R. Matteson, M. Castleton, C. Farwell and B.A. Block. 2010. Movements of pacific bluefin tuna (*Thunnus orientalis*) in the Eastern North Pacific revealed with archival tags. Prog. Oceanogr. 86(1): 94–104.

Bushnell, P.G. and R.W. Brill. 1992. Oxygen transport and cardiovascular responses in skipjack tuna (*Katsuwonus pelamis*) and yellowfin tuna (*Thunnus albacares*) exposed to acute hypoxia. J. Comp. Physiol. B 162(2): 131–143.

Bushnell, P.G. and D.R. Jones. 1994. Cardiovascular and respiratory physiology of tuna: adaptations for support of exceptionally high metabolic rates. Env. Biol. Fish. 40(3): 303–318.

Cailliet, G.M., W.D. Smith, H.F. Mollet and K.J. Goldman. 2006. Age and growth studies of chondrichthyan fishes: the need for consistency in terminology, verification, validation, and growth function fitting. Env. Biol. Fish. 77(3-4): 211–228.

Chapman, E.W., C. Jørgensen and M.E. Lutcavage. 2011. Atlantic bluefin tuna (*Thunnus thynnus*): a state-dependent energy allocation model for growth, maturation, and reproductive investment. Can. J. Fish. Aquat. Sci. 68(11): 1934–1951.

Chen, K.-S., P. Crone and C.-C. Hsu. 2006. Reproductive biology of female Pacific bluefin tuna *Thunnus orientalis* from south-western North Pacific Ocean. Fisheries Sci. 72(5): 985–994.

Chen, Y., D.A. Jackson and H.H. Harvey. 1992. A comparison of von Bertalanffy and polynomial functions in modelling fish growth data. Can. J. Fish. Aquat. Sci. 49(6): 1228–1235.

Chipps, S.R. and D.H. Wahl. 2008. Bioenergetics modeling in the 21st century: reviewing new insights and revisiting old constraints. T. Am. Fish. Soc. 137(1): 298–313.

Clark, T.D., W.T. Brandt, J. Nogueira, L.E. Rodriguez, M. Price, C.J. Farwell and B.A. Block. 2010. Postprandial metabolism of Pacific bluefin tuna (*Thunnus orientalis*). J. Exp. Biol. 213(14): 2379–2385.

Clarke, A. 2004. Is there a Universal Temperature Dependence of metabolism? Funct. Ecol. 18(2): 252–256.

Clarke, A. and N.M. Johnston. 1999. Scaling of metabolic rate with body mass and temperature in teleost fish. J. Anim. Ecol. 68(5): 893–905.

Cohen, J.E. 2004. Mathematics is biology's next microscope, only better; biology is mathematics' next physics, only better. PLoS Biol. 2(12): e439.

Colman, R.J., R.M. Anderson, S.C. Johnson, E.K. Kastman, K.J. Kosmatka, T.M. Beasley, D.B. Allison, C. Cruzen, H.A. Simmons, J.W. Kemnitz and R. Weindruch. 2009. Caloric restriction delays disease onset and mortality in rhesus monkeys. Science 325(5937): 201–204.

Dewar, H. and J. Graham. 1994. Studies of tropical tuna swimming performance in a large water tunnel-energetics. J. Exp. Biol. 192(1): 13–31.

Einarsson, B., B. Birnir and S. Sigurðsson. 2011. A dynamic energy budget (DEB) model for the energy usage and reproduction of the Icelandic capelin (*Mallotus villosus*). J. Theor. Biol. 281(1): 1–8.

Ellerby, D.J. 2010. How efficient is a fish? J. Exp. Biol. 213(22): 3765–3767.

Elliott, J.M. and W. Davison. 1975. Energy equivalents of oxygen consumption in animal energetics. Oecologia 19(3): 195–201.

Essington, T.E., J.F. Kitchell and C.J. Walters. 2001. The von Bertalanffy growth function, bioenergetics, and the consumption rates of fish. Can. J. Fish. Aquat. Sci. 58(11): 2129–2138.

Fablet, R., L. Pecquerie, H. de Pontual, H. Høie, R. Millner, H. Mosegaard and S.A.L.M. Kooijman. 2011. Shedding light on fish otolith biomineralization using a bioenergetic approach. PLoS ONE 6(11): e27055.

Fitzgibbon, Q.P., R.S. Seymour, D. Ellis and J. Buchanan. 2007. The energetic consequence of specific dynamic action in southern bluefin tuna *Thunnus maccoyii*. J. Exp. Biol. 210(2): 290–298.

Freitas, V., S.A.L.M. Kooijman and H.W. van der Veer. 2012. Latitudinal trends in habitat quality of shallow-water flatfish nurseries. Mar. Ecol. Prog. Ser. 471: 203–214.

Freitas, V., K. Lika, J.I. Witte and H.W. van der Veer. 2011. Food conditions of the sand goby *Pomatoschistus minutus* in shallow waters: an analysis in the context of Dynamic Energy Budget theory. J. Sea Res. 66(4): 440–446.

Graham, J.B. and K.A. Dickson. 2001. Anatomical and physiological specializations for endothermy. Fish Physiology Series 19: 121–166.

Graham, J.B. and K.A. Dickson. 2004. Tuna comparative physiology. J. Exp. Biol. 207(23): 4015–4024.

Hansen, M.J., D. Boisclair, S.B. Brandt, S.W. Hewett, J.F. Kitchell, M.C. Lucas and J.J. Ney. 1993. Applications of bioenergetics models to fish ecology and management: where do we go from here? T. Am. Fish. Soc. 122(5): 1019–1030.

Hawkins, A.J.S. 1991. Protein turnover: a functional appraisal. Funct. Ecol. 5(2): 222–233.

Heino, M. and V. Kaitala. 1999. Evolution of resource allocation between growth and reproduction in animals with indeterminate growth. J. Evol. Biol. 12(3): 423–429.

Hirota, H., M. Morita and N. Taniguchi. 1976. An instance of the maturation of 3 full years old bluefin tuna cultured in the floating net. B. Jpn. Soc. Sci. Fish. 42: 939.

Houlihan, D.F. 1991. Protein turnover in ectotherms and its relationships to energetics. Adv. Comp. Environ. Physiol. 7: 1–43.

Jager, T. 2013. Making Sense of Chemical Stress. VU University, 107 p. PDF.

Jayne, B.C. and G.V. Lauder. 1994. How swimming fish use slow and fast muscle fibers: implications for models of vertebrate muscle recruitment. J. Comp. Physiol. A 175(1): 123–131.

Jusup, M., T. Klanjscek, H. Matsuda and S.A.L.M. Kooijman. 2011. A Full Lifecycle Bioenergetic Model for Bluefin Tuna. PLoS ONE 6(7): e21903.

Jusup, M., T. Klanjscek and H. Matsuda. 2014. Simple measurements reveal the feeding history, the onset of reproduction, and energy conversion efficiencies in captive bluefin tuna. J. Sea Res. 94: 144–155.

Kaitaniemi, P. 2004. Testing the allometric scaling laws. J. Theor. Biol. 228(2): 149–153.

Kaji, T. 2003. Bluefin tuna larval rearing and development—State of the art. Cahiers Options Méditerranéennes 60: 85–89.

Katsanevakis, S. 2006. Modelling fish growth: model selection, multi-model inference and model selection uncertainty. Fish. Res. 81(2): 229–235.

Katz, S.L. 2002. Design of heterothermic muscle in fish. J. Exp. Biol. 205(15): 2251–2266.

Katz, S.L., D.A. Syme and R.E. Shadwick. 2001. High-speed swimming: Enhanced power in yellowfin tuna. Nature 410(6830): 770–771.

Kiceniuk, J.W. and D.R. Jones. 1977. The oxygen transport system in trout (*Salmo gairdneri*) during sustained exercise. J. Exp. Biol. 69(1): 247–260.

Kitagawa, T., S. Kimura, H. Nakata and H. Yamada. 2006. Thermal adaptation of Pacific bluefin tuna *Thunnus orientalis* to temperate waters. Fisheries Sci. 72(1): 149–156.

Kitagawa, T., S. Kimura, H. Nakata and H. Yamada. 2007. Why do young Pacific bluefin tuna repeatedly dive to depths through the thermocline? Fisheries Sci. 73(1): 98–106.

Kitagawa, T., S. Kimura, H. Nakata, H. Yamada, A. Nitta, Y. Sasai and H. Sasaki. 2009. Immature Pacific bluefin tuna, *Thunnus orientalis*, utilizes cold waters in the Subarctic Frontal Zone for trans-Pacific migration. Env. Biol. Fish. 84(2): 193–196.

Kitchell, J.F., D.J. Stewart and D. Weininger. 1977. Applications of a bioenergetics model to yellow perch (*Perca flavescens*) and walleye (*Stizostedion vitreum vitreum*). J. Fish. Res. Board Can. 34(10): 1910–1921.

Kooijman, S.A.L.M. 2010. Dynamic energy budget theory for metabolic organisation (3rd edition). Cambridge University Press, 532 p.

Kooijman, S.A.L.M., L. Pecquerie, S. Augustine and M. Jusup. 2011. Scenarios for acceleration in fish development and the role of metamorphosis. J. Sea Res. 66(4): 419–423.

Kooijman, S.A.L.M. and T.A. Troost. 2007. Quantitative steps in the evolution of metabolic organisation as specified by the dynamic energy budget theory. Biol. Rev. Camb. Philos. Soc. 82(1): 113–142.

Korsmeyer, K.E. and H. Dewar. 2001. Tuna metabolism and energetics. Fish Physiology Series 19: 35–78.

Korsmeyer, K.E., H. Dewar, N.C. Lai and J.B. Graham. 1996. The aerobic capacity of tunas: adaptation for multiple metabolic demands. Comp. Biochem. Physiol. A: Comp. Physiol. 113(1): 17–24.

Korsmeyer, K.E., J.F. Steffensen and J. Herskin. 2002. Energetics of median and paired fin swimming, body and caudal fin swimming, and gait transition in parrotfish (*Scarus schlegeli*) and triggerfish (*Rhinecanthus aculeatus*). J. Exp. Biol. 205(9): 1253–1263.

Kubo, T., W. Sakamoto, O. Murata and H. Kumai. 2008. Whole-body heat transfer coefficient and body temperature change of juvenile Pacific bluefin tuna *Thunnus orientalis* according to growth. Fisheries Sci. 74(5): 995–1004.

Lester, N.P., B.J. Shuter and P.A. Abrams. 2004. Interpreting the von Bertalanffy model of somatic growth in fishes: the cost of reproduction. Proc. R. Soc. B 271(1548): 1625–1631.

Lika, K., S. Augustine, L. Pecquerie and S.A.L.M. Kooijman. 2014. The bijection from data to parameter space with the standard DEB model quantifies the supply–demand spectrum. J. Theor. Biol. 354: 35–47.

Magnuson, J.J. 1973. Comparative study of adaptations for continuous swimming and hydrostatic equilibrium of scombroid and xiphoid fishes. Fish. Bull. 71(2): 337–356.

Marcinek, D.J., S.B. Blackwell, H. Dewar, E.V. Freund, C. Farwell, D. Dau, A.C. Seitz and B.A. Block. 2001. Depth and muscle temperature of Pacific bluefin tuna examined with acoustic and pop-up satellite archival tags. Mar. Biol. 138(4): 869–885.

Margulies, D., J.M. Sutter, S.L. Hunt, R.J. Olson, V.P. Scholey, J.B. Wexler and A. Nakazawa. 2007. Spawning and early development of captive yellowfin tuna (*Thunnus albacares*). Fish. Bull. 105(2): 249–265.

Masuma, S. 2009. Biology of Pacific bluefin tuna inferred from approaches in captivity. Collect. Vol. Sci. Pap. ICCAT 63: 207–229.

Masuma, S., S. Miyashita, H. Yamamoto and H. Kumai. 2008. Status of bluefin tuna farming, broodstock management, breeding and fingerling production in Japan. Rev. Fish. Sci. 16(1-3): 385–390.

Masuma, S., T. Takebe and Y. Sakakura. 2011. A review of the broodstock management and larviculture of the Pacific northern bluefin tuna in Japan. Aquaculture 315(1-2): 2–8.

Mielczarek, E.V. 2006. Resource letter PFBi-1: physical frontiers in biology. Am. J. Phys. 74: 375.

Miyashita, S. 2002. Studies on the seedling production of the Pacific bluefin tuna, *Thunnus thynnus orientalis*. Bulletin of the Fisheries Laboratory of Kinki University 8: 171 (in Japanese).

Miyashita, S., Y. Sawada, T. Okada, O. Murata and H. Kumai. 2001. Morphological development and growth of laboratory-reared larval and juvenile *Thunnus thynnus* (Pisces: Scombridae). Fish. Bull. 99(4): 601–616.

Miyashita, S., Y. Tanaka, Y. Sawada, O. Murata, N. Hattori, K. Takii and H. Kumai. 2000. Embryonic development and effects of water temperature on hatching of the bluefin tuna, *Thunnus thynnus*. Suisan Zoshoku 48(2): 199–208 (in Japanese).

Motani, R. 2002. Scaling effects in caudal fin propulsion and the speed of ichthyosaurs. Nature 415(6869): 309–312.

Moyes, C.D., O.A. Mathieu-Costello, R.W. Brill and P.W. Hochachka. 1992. Mitochondrial metabolism of cardiac and skeletal muscles from a fast (*Katsuwonus pelamis*) and a slow (*Cyprinus carpio*) fish. Can. J. Zool. 70(6): 1246–1253.

Nisbet, R.M., M. Jusup, T. Klanjscek and L. Pecquerie. 2012. Integrating Dynamic Energy Budget (DEB) theory with traditional bioenergetic models. J. Exp. Biol. 215: 892–902.

Nisbet, R.M., E.B. Muller, K. Lika and S.A.L.M. Kooijman. 2000. From molecules to ecosystems through dynamic energy budget models. J. Anim. Ecol. 69(6): 913–926.

Ohnishi, S., T. Yamakawa, H. Okamura and T. Akamine. 2012. A note on the von Bertalanffy growth function concerning the allocation of surplus energy to reproduction. Fish. Bull. 110(2): 223–229.

Papadopoulos, A. 2008. On the hydrodynamics-based power-law function and its application in fish swimming energetics. T. Am. Fish. Soc. 137(4): 997–1006.

Papadopoulos, A. 2009. Hydrodynamics-Based Functional Forms of Activity Metabolism: A Case for the Power-Law Polynomial Function in Animal Swimming Energetics. PLoS ONE 4(3): e4852.

Pecquerie, L., L.R. Johnson, S.A.L.M. Kooijman and R.M. Nisbet. 2011. Analyzing variations in life-history traits of Pacific salmon in the context of Dynamic Energy Budget (DEB) theory. J. Sea Res. 66(4): 424–433.

Pecquerie, L., P. Petitgas and S.A.L.M. Kooijman. 2009. Modeling fish growth and reproduction in the context of the Dynamic Energy Budget theory to predict environmental impact on anchovy spawning duration. J. Sea Res. 62(2): 93–105.

Quince, C., B.J. Shuter, P.A. Abrams and N.P. Lester. 2008. Biphasic growth in fish II: empirical assessment. J. Theor. Biol. 254(2): 207–214.

Roberts, J.L. 1975. Active branchial and ram gill ventilation in fishes. Biol. Bull. 148(1): 85–105.

Rome, L.C. and D. Swank. 1992. The influence of temperature on power output of scup red muscle during cyclical length changes. J. Exp. Biol. 171(1): 261–281.

Rome, L.C., D.M. Swank and D.J. Coughlin. 2000. The influence of temperature on power production during swimming. II. Mechanics of red muscle fibres *in vivo*. J. Exp. Biol. 203(2): 333–345.

Satoh, K., Y. Tanaka, M. Masujima, M. Okazaki, Y. Kato, H. Shono and K. Suzuki. 2013. Relationship between the growth and survival of larval Pacific bluefin tuna, *Thunnus orientalis*. Mar. Biol.: 1–12.

Sawada, Y., T. Okada, S. Miyashita, O. Murata and H. Kumai. 2005. Completion of the Pacific bluefin tuna *Thunnus orientalis* (Temminck et Schlegel) life cycle. Aquac. Res. 36(5): 413–421.

Schultz, W.W. and P.W. Webb. 2002. Power requirements of swimming: do new methods resolve old questions? ICB 42(5): 1018–1025.

Secor, D.H. 2007. Do some Atlantic bluefin tuna skip spawning? Col. Vol. Sci. Pap. ICCAT 60(4): 1141–1153.

Secor, S.M. 2009. Specific dynamic action: a review of the postprandial metabolic response. J. Comp. Physiol. B 179(1): 1–56.

Sepulveda, C. and K.A. Dickson. 2000. Maximum sustainable speeds and cost of swimming in juvenile kawakawa tuna (*Euthynnus affinis*) and chub mackerel (*Scomber japonicus*). J. Exp. Biol. 203(20): 3089–3101.

Serpa, D., P. Pousão Ferreira, H. Ferreira, L.C. da Fonseca, M.T. Dinis and P. Duarte. 2012. Modelling the growth of white seabream (*Diplodus sargus*) and gilthead seabream (*Sparus aurata*) in semi-intensive earth production ponds using the Dynamic Energy Budget approach. J. Sea Res. 76: 135–145.

Shadwick, R.E. and D.A. Syme. 2008. Thunniform swimming: muscle dynamics and mechanical power production of aerobic fibres in yellowfin tuna (*Thunnus albacares*). J. Exp. Biol. 211(10): 1603–1611.

Shimose, T., T. Tanabe, K.-S. Chen and C.-C. Hsu. 2009. Age determination and growth of Pacific bluefin tuna, *Thunnus orientalis*, off Japan and Taiwan. Fish. Res. 100: 134–139.

Sloman, K.A., G. Motherwell, K.I. O'Connor and A.C. Taylor. 2000. The effect of social stress on the standard metabolic rate (SMR) of brown trout, *Salmo trutta*. Fish Physiol. Biochem. 23(1): 49–53.

Smith, N.P., C.J. Barclay and D.S. Loiselle. 2005. The efficiency of muscle contraction. Prog. Biophys. Mol. Biol. 88(1): 1–58.

Smutna, M., L. Vorlova and Z. Svobodova. 2002. Pathobiochemistry of Ammonia in the Internal Environment of Fish (a Review). Acta Vet. Brno 71(2): 169–181.

Sousa, T., T. Domingos and S.A.L.M. Kooijman. 2008. From empirical patterns to theory: a formal metabolic theory of life. Phil. Trans. R. Soc. B 363(1502): 2453–2464.

Sousa, T., T. Domingos, J.-C. Poggiale and S.A.L.M. Kooijman. 2010. Dynamic Energy Budget theory restores coherence in biology. Phil. Trans. R Soc. B 365(1557): 3413–3428.

Sousa, T., R. Mota, T. Domingos and S.A.L.M. Kooijman. 2006. Thermodynamics of organisms in the context of dynamic energy budget theory. Phys. Rev. E 74(5): 051901.

Stevens, E.D. 1972. Some aspects of gas exchange in tuna. J. Exp. Biol. 56(3): 809–823.

Syme, D.A. and R.E. Shadwick. 2002. Effects of longitudinal body position and swimming speed on mechanical power of deep red muscle from skipjack tuna (*Katsuwonus pelamis*). J. Exp. Biol. 205(2): 189–200.

Teal, L.R., R. Hal, T. Kooten, P. Ruardij and A.D. Rijnsdorp. 2012. Bio-energetics underpins the spatial response of North Sea plaice (*Pleuronectes platessa* L.) and sole (*Solea solea* L.) to climate change. Glob. Change Biol. 18(11): 3291–3305.

Triantafyllou, M.S. and G.S. Triantafyllou. 1995. An efficient swimming machine. Sci. Am. 272(3): 64–71.

Tsikliras, A.C., E. Antonopoulou and K.I. Stergiou. 2007. A phenotypic trade-off between previous growth and present fecundity in round sardinella Sardinella aurita. Popul. Ecol. 49(3): 221–227.

van der Meer, J. 2006. An introduction to Dynamic Energy Budget (DEB) models with special emphasis on parameter estimation. J. Sea Res. 56: 85–102.

van der Meer, J., H.W. van der Veer and J.I. Witte. 2011. The disappearance of the European eel from the western Wadden Sea. J. Sea Res. 66(4): 434–439.

van der Veer, H.W., J.F. Cardoso, M.A. Peck and S.A.L.M. Kooijman. 2009. Physiological performance of plaice *Pleuronectes platessa* (L.): A comparison of static and dynamic energy budgets. J. Sea Res. 62(2): 83–92.

van der Veer, H.W., S.A.L.M. Kooijman and J. van der Meer. 2001. Intra-and interspecies comparison of energy flow in North Atlantic flatfish species by means of dynamic energy budgets. J. Sea Res. 45(3): 303–320.

van der Veer, H.W., S.A.L.M. Kooijman and J. van der Meer. 2003. Body size scaling relationships in flatfish as predicted by Dynamic Energy Budgets (DEB theory): implications for recruitment. J. Sea Res. 50(2): 257–272.

van Leeuwen, I.M.M., J. Vera and O. Wolkenhauer. 2010. Dynamic energy budget approaches for modelling organismal ageing. Phil. Trans. R. Soc. B 365(1557): 3443–3454.

Walli, A., S.L. Teo, A. Boustany, C.J. Farwell, T. Williams, H. Dewar, E. Prince and B.A. Block. 2009. Seasonal movements, aggregations and diving behavior of Atlantic bluefin tuna (*Thunnus thynnus*) revealed with archival tags. PLoS ONE 4(7): e6151.

Walsberg, G.E. and T.C. Hoffman. 2005. Direct calorimetry reveals large errors in respirometric estimates of energy expenditure. J. Exp. Biol. 208(6): 1035–1043.

Warton, D.I., I.J. Wright, D.S. Falster and M. Westoby. 2006. Bivariate line-fitting methods for allometry. Biol. Rev. Camb. Philos. Soc. 81(2): 259–291.

Webb, P.W. 1971a. The swimming energetics of trout I. Thrust and power output at cruising speeds. J. Exp. Biol. 55(2): 489–520.

Webb, P.W. 1971b. The swimming energetics of trout II. Oxygen consumption and swimming efficiency. J. Exp. Biol. 55(2): 521–540.

Wilkie, M.P. 2002. Ammonia excretion and urea handling by fish gills: present understanding and future research challenges. J. Exp. Biol. 293(3): 284–301.

Wu, X. and P.G. Schultz. 2009. Synthesis at the interface of chemistry and biology. J. Am. Chem. Soc. 131(35): 12497–12515.

CHAPTER 18

A Method for Measuring the Swimming Behaviour of Pacific Bluefin Tuna

Kazuyoshi Komeyama[1],* and *Tsutomu Takagi*[2]

Introduction

Knowledge about the underwater behaviour of fish remains limited because technology sensitive enough to elucidate detailed behaviour (and behavioural transitions) through indirect observation has yet to be developed. Although techniques that can be used to track fish positions underwater have been introduced, deficiencies remain that require improvement.

One example of a tracking system used to track marine wildlife that surface is VHF (Very High Frequency) telemetry (Hooker et al. 2002), where animals are equipped with transmitters that emit signals at certain radio frequencies. Animal tracking has relied on VHF technology for many years, but other systems have been developed recently, such as satellite tracking telemetry (Jouventin and Weimerskirch 1990; Ferraroli et al. 2004; Hays et al. 2004; Sims et al. 2005), GPS (Global Positioning System) technology (Weimerskirch et al. 2005), and geolocation or global location sensing (Wilson et al. 1994; Block et al. 2001). These systems have allowed researchers to monitor and record the sequential positions of animals remotely. However, all of these telemetry methods send electrical signals to satellites, and so they only work when the animal is emerging from or moving towards the sea surface, because the transmission of electrical signals is generally reduced or inhibited under water (salt and fresh). Therefore, information on animal paths obtained by telemetry has been limited to just two dimensions.

[1] Faculty of Fisheries, Kagoshima University, 4-50-20, Shimoarata, Kagoshima 890-0056, Japan.
[2] Faculty of Fisheries Sciences, Hokkaido University 3-1-1, Minato-cho, Hakodate, Hokkaido 041-8611, Japan.
* Email: komeyama@fish.kagoshima-u.ac.jp

Two technologies are currently available for indirectly investigating animal movement underwater: acoustic telemetry (e.g., Hindell et al. 2002) and dead reckoning (Wilson and Wilson 1988). Acoustic telemetry uses ultrasonic transmitters (Harcourt et al. 2000; Klimley et al. 2001; Hindell et al. 2002; Mitamura et al. 2012). Thus, a receiver must be within a few hundred metres of the object, making it difficult to obtain data on wide-ranging species (Wilson et al. 2007). This measurement range problem needs to be addressed to extend its applicability to a variety of species. Acoustic telemetry is also not recommended for studying detailed fish movement patterns over a short timescale because natural noise often interrupts the signals from the acoustic transmitter. Thus, micro data loggers represent a more viable option combined with dead-reckoning. The dead-reckoning method overcomes the problem of measurement range, and its principle is simple. First, swimming vectors (such as the animal's speed, heading angle, and depth change) are measured for each interval using micro data loggers. Next, the velocity vectors are calculated and estimated indirectly based on the dataset. Finally, by integrating these velocity vectors, 3-dimensional trajectories may be constructed.

The disadvantage of micro data loggers is that they must be retrieved to access the stored data. Consequently, these loggers are usually used to study animals that exhibit homing behaviour or that inhabit a limited area to ensure a high probability of logger retrieval, such as marine turtles (Yasuda and Arai 2009), salmon (Tanaka et al. 2005; Tsuda et al. 2006), and marine mammals (Mitani et al. 2003). Furthermore, micro data loggers are useful for measuring swimming paths in difficult-to-access offshore areas, whereas complete datasets from such locations cannot always be collected by acoustic telemetry receivers. Furthermore, loggers are easily collected from fish placed in an enclosure, such as a cage. Dead-reckoning produces temporally finitely resolved, regular, sequential positional data with no gaps. The disadvantage of this method is that the obtained trajectories may, for various reasons, include accumulative error. One of this error is associated with the lack of reference points underwater. However, the greatest source of inaccuracy in dead-reckoning is drift from tidal currents and observational error. Since a trajectory is estimated by integrating velocity vectors, the inaccuracy increases with time. Therefore, when designing a research project using dead-reckoning, it is important to evaluate the potential effect of errors on analyses within the context of the ecological question being asked.

Examining the behaviour of cultivated fish in net cages is logistically problematic. Some studies have already described the behaviour of cultivated tuna in aquaculture net cages (Kubo et al. 2004; Okano et al. 2006; Kadota et al. 2011). However, these studies only measured the time series of 1-dimensional behavioural data (swimming depth, acceleration, swimming speed, and feeding behaviour), whereas 3-dimensional behavioural data should be measured to examine space use properly.

In this chapter, we assess the potential sources of drift error, due to ocean currents, and propose a new technique to remove drift error effectively. We present a case study for the visualization of bluefin tuna movement patterns in an aquaculture facility using dead-reckoning improved by the new method of removing current drift from the measurement data.

Specifically, we measured the swimming behaviour of cultivated Pacific bluefin tuna, *Thunnus orientalis*, in an aquaculture net cage using two types of micro data

402 *Biology and Ecology of Bluefin Tuna*

loggers to understand how the available living space in the cage is used. The main goal was to demonstrate a method for determining the swimming path of a fish and to clarify the ways in which reared bluefin tuna swim in a submerged net cage. The visualization technique used to produce 3-dimensional trajectories of Pacific bluefin tuna is described next .

Movement patterns must also be measured on a fine time scale to fully understand the turning behaviour of fish in response to obstacles, such as wall structures or other individuals. It is possible to measure fine-scale fish movement patterns in a controlled experiment within an enclosed space, such as a fish tank (Bégout Anras and Lagardère 2004), whereas such research would be more difficult in a field experiment. Movement data on fish grown in submerged net cages is expected to provide useful information for developing cages that cause less stress to reared fish.

Data Collection for Understanding Fish Behaviour in a Net Cage

A single Pacific bluefin tuna (PBT, fork length [FL], 0.51 m; estimated weight, 2.6 kg, taken from a regression curve fit to tuna farm records) was captured from a net cage (30 m diameter × 22 m depth) by angling. Two micro data loggers (PD3GT, Little Leonardo, Japan, dia. 21 × 117 mm long, 75 g in air; DST Comp-Tilt, Star-Oddi, Iceland, dia. 15 × 46 mm long 19 g in air) were inserted into a floating cellular material plate (90 mm long × 35 mm high; buoyancy adjusted to slightly more than its underwater weight) and were externally attached to the body of the tuna near the dorsal fin. The fish was then released back into the net cage at 0901 on March 6, 2010.

The PD3GT data logger recorded swimming speed and depth at one-second intervals on a flash memory drive from 0930 on March 6, until 1730 on March 7, 2010. A propeller was attached to the PD3GT to record the speed that the fish moved through the water. We confirmed the relationship between velocity and the number of propeller revolutions, as well as the stall speed, of the device (0.13–0.18 m·s^{-1}) in a preliminary experiment. In the main study, the tagged fish swam at speeds of over 0.28 m·s^{-1} without the propeller stopping. The device also measures swimming depth, 3-axis acceleration, and ambient water temperature. However, 3-axis acceleration and ambient temperature data were not used here, because these parameters are not needed to estimate trajectories.

The DST Comp-Tilt data logger recorded the heading of the fish, which was calculated using 2-dimensional geomagnetism at one-second intervals. Because of the limited memory capacity of the DST Comp-Tilt, data were divided into four phases of daily activity: dusk, 1700–1900; night-time, 2330–2430; dawn, 0500–0700; and daytime, 1130–1230. We set a longer measurement time at dusk and dawn, than at day or night, as a major change in illumination was expected. The data logger packages were detached from the fish using a timing device and were collected by a diver after three days.

To monitor the current profile near the net cage, an electromagnetic current meter (Infinity-EM, Alec Electronics) was fixed close to the net cage at a depth of 12 m. The Infinity-EM recorded the current speed and direction using 2-dimensional velocity. To estimate current velocities, 30 samples were obtained at one-second intervals every five minutes, and then values were averaged.

How do Pacific bluefin tuna swim in a cage?

A record of swimming behaviour was obtained from data retrieved over the 33-hours study period (Fig. 18.1). The fish moved between depths of 5 and 22 m, swimming near the cage bottom (15–22 m) during the day, and at a depth of around 10–15 m at night. There was a significant difference in swimming depth between day (mean ± SD, 17.0 ± 2.6 m; 0600–1800) and night (mean ± SD, 14.7 ± 2.9 m; 1800–0600) (*t*-test, $p < 0.001$; *F*-test, $p > 0.05$).

The fish moved at speeds of 0.7–0.8 m·s^{-1} (approximately 1.4 FL·s^{-1}), with a maximum speed of 3.6 m·s^{-1} (7.0 FL·s^{-1}). After release, the fish swam over 1 m·s^{-1} faster than the cruising speed from 1300 to the end of the measurement. This fast swimming rate is thought to indicate trauma. Kadota et al. (2011) observed that trauma behaviour lasted for four–six hours after release in behavioural measurements of cultivated bluefin tuna. In the current study, burst swims exceeding 2 m·s^{-1} were observed only after dark or close to dark (i.e., dusk), in addition to after release (Fig. 18.1), with a significant difference in swimming speed being recorded between day (mean ± SD, 0.73 ± 0.10 m·s^{-1}; 0600–1800) and night (mean ± SD, 0.77 ± 0.10 m·s^{-1}; 1800–0600) (*t*-test, $p < 0.001$; *F*-test, $p > 0.05$).

Previous research indicates that the gross cost of transport for bluefin tuna is minimum at speeds of 1.15–1.3 body lengths per second (BL·s^{-1}) (Blank et al. 2007), with bluefin tuna swimming most efficiently within this speed range (termed cruising). The cruising swimming speed of reared tuna has been documented to be around 1.1 BL·s^{-1} (Okano et al. 2006). Marcinek et al. (2001) reported that the mean travel speed of wild Pacific bluefin tuna is 1.1–1.4 FL·s^{-1}. Although we measured FL rather than BL, our results showed that the tagged tuna swam at 1.46 ± 0.18 FL·s^{-1} while cruising, which was calculated from its speed through the water. The tagged tuna in our study cruised at a slightly faster speed than that reported previously (Marcinek et al. 2001; Okano et al. 2006). The FL of individuals in previous studies was approximately 1 m, whereas it was 0.5 m in the current study. Thus, the optimum swimming speed might be influenced by the different sizes of the study individuals.

However, this difference may impact the measurement method used. It is possible that the actual swimming distance is underestimated because the measurement interval of acoustic telemetry is intermittent in time. In addition, Okano et al. (2006) estimated the swimming speed of reared bluefin tuna from the tail beat frequency using the formula $U = 0.65f$ (refer to Wardle et al. (1989)), where U is the swim speed and f is the tail beat frequency. However, Pacific bluefin tuna exhibit glide swimming, during which they stop beating their tails (Okano et al. 2006); thus, it is not possible to estimate swim speed using beat frequency. In our study, even if the tagged tuna glided while in the cage, we were able to measure its swim speed by using the propeller. This issue might also explain the slow speed observed in the previous studies.

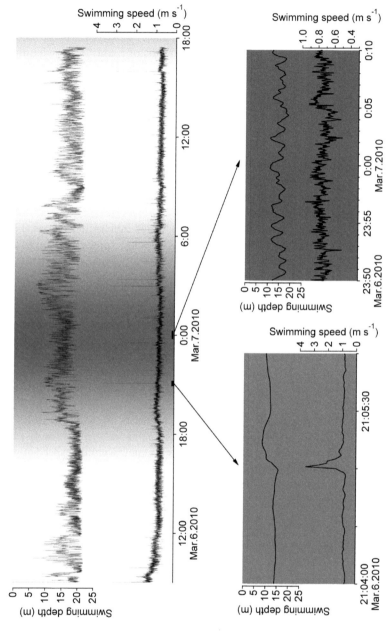

Figure 18.1. Top: Time series data for the swimming behaviour of the tagged fish showing the swimming depth and speed for 33 hours. White and black background colours represent day and night, respectively. The ceiling net of a submerged net cage is set at around 2 m under the surface, while and the bottom net is set at around 22 m deep. Bottom: The two slides show close-up sections of the top figure: left, burst swimming of the tagged tuna; right, typical vertical movement of the tagged tuna at midnight.

Visualization of the 3-Dimensional Trajectories of Tuna

Formulation of the dead-reckoning technique

The dead-reckoning technique was used to visualize the swimming paths of a single tuna in an aquaculture net cage. Figure 18.2 shows the swimming vectors. Swimming vectors were estimated based on speed $v(n)$, heading angle $\theta(n)$ calculated using azimuth $\theta_a(n)$, and depth $V_z(n)$ per measurement interval ($n\Delta t$) using the following formulae:

$$v_x(n\Delta t) = \sqrt{v(n\Delta t)^2 - v_z(n\Delta t)^2} \cdot \cos\big(\theta(n\Delta t)\big) \qquad (n = 1,2,\text{--}) \qquad (1)$$

$$v_y(n\Delta t) = \sqrt{v(n\Delta t)^2 - v_z(n\Delta t)^2} \cdot \sin\big(\theta(n\Delta t)\big) \qquad (n = 1,2,\text{--}) \qquad (2)$$

$$v_z(n\Delta t) = \frac{d\big((n+1)\Delta t\big) - d(n\Delta t)}{\Delta t} \qquad\qquad (n = 1,2,\text{--}) \qquad (3)$$

where $n\Delta t = 1$ corresponds to the measurement interval and the vector ($v_x(n\Delta t)$, $v_y(n\Delta t)$, $v_z(n\Delta t)$) is a locomotion vector at measurement time Δt. The 3-dimensional swimming trajectories could then be reconstructed from the observational dataset by integrating the locomotion vector with respect to time (i.e., summing equations (1)–(3)) to give the following:

$$x(n\Delta t) = \sum_{n=0}^{N} \sqrt{v(n\Delta t)^2 - v_z(n\Delta t)^2} \cdot \cos\big(\theta(n\Delta t)\big) \cdot \Delta t \qquad (4)$$

$$y(n\Delta t) = \sum_{n=0}^{N} \sqrt{v(n\Delta t)^2 - v_z(n\Delta t)^2} \cdot \sin\big(\theta(n\Delta t)\big) \cdot \Delta t \qquad (5)$$

$$z(n\Delta t) = \sum_{n=0}^{N} \frac{d\big((n+1)\Delta t\big) - d(n\Delta t)}{\Delta t} \qquad (6)$$

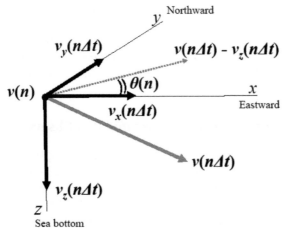

Figure 18.2. Schematic of the swim vector ($v_x(n\Delta t)$, $v_y(n\Delta t)$, $v_z(n\Delta t)$) of tagged fish. The coordinate system expressed a right handed coordinate system. In this figure, $v(n\Delta t)$ is the swimming speed, $\theta(n)$ is the horizontal heading of tagged tuna on n. $v_x(n\Delta t)$ and $v_y(n\Delta t)$ were calculated from $v(n\Delta t) - v_z(n\Delta t)$ with $h(n)$. $v_z(n\Delta t)$ was calculated from the change in swimming depth with $n\Delta t$. The researcher measured the swimming speed $v(n\Delta t)$, the horizontal heading $\theta(n)$, and swimming depth $z(n)$. Using these parameters as an estimation, the swimming vector $v(n\Delta t)$ deconstructed the fish position ($v_x(n\Delta t)$, $v_y(n\Delta t)$, $v_z(n\Delta t)$) after $n\Delta t$. We then calculated θ to estimate swimming vectors with $90 - \theta_a$. Where θ_a expressed the azimuth angle.

However, the horizontal swimming paths obtained using equations (4) and (5) include large cumulative errors (Shiomi et al. 2008; Komeyama et al. 2011), partly because of an observational error, but mostly because of the horizontal ocean current (Komeyama et al. 2013). Moreover, currents could also change in three dimensions. Using a measurement system similar to the PD3GT used in this study, Shiomi et al. (2008) suggested that the accumulated error was caused by the ocean current and, similarly, Mitani et al. (2003) emphasized the influence of current flow.

In contrast, the vertical swimming trajectories given by equation (6) contain less cumulative errors because the vertical positions of the fish were measured using the pressure sensor of PD3GT. Thus, the depth measurement is not likely to be influenced by the tidal current. Therefore, equations (4) and (5) were separated into three terms (tuna-location, current-drift, and accumulation of error), and then solved for tuna location. Thus, tuna location at measurement time Δt is written as follows:

$$x_{tuna} (n\Delta t) = \sum_{n=0}^{N} \sqrt{v(n\Delta t)^2 - v_z(n\Delta t)^2} \cdot cos\left(\theta(n\Delta t)\right) \cdot \Delta t + x_{cd}(n\Delta t) + x_\varepsilon (n\Delta t)$$

(7)

$$y_{tuna} (n\Delta t) = \sum_{n=0}^{N} \sqrt{v(n\Delta t)^2 - v_z(n\Delta t)^2} \cdot sin\left(\theta(n\Delta t)\right) \cdot \Delta t + y_{cd}(n\Delta t) + y_\varepsilon (n\Delta t)$$

(8)

where x_ε and y_ε represent observational errors and x_u and y_u represent errors associated with horizontal current velocities. Thus, the net errors are expressed as follows:

$$x_{cd}(n\Delta t) = \sum_{n=1}^{N} x_u(n\Delta t)$$

(9)

$$y_{cd}(n\Delta t) = \sum_{n=1}^{N} y_u (n\Delta t)$$

(10)

To correct the estimated positional data, the net errors of equations (9) and (10) were then removed by assuming constant linear drift. However, positions estimated in this way accumulate error with time and become cumulatively inaccurate, simply because the ocean current is not constant. Therefore, an alternative technique was required that did not assume constant current velocities to remove errors associated with ocean currents.

Removing Current Drift and Accumulated Observational Error

The summed observational data calculated using equations (4) and (5) include low-frequency drift, usually arising as a result of physical noise, which is mainly caused by the ocean current (corresponding to the second terms in equations (7) and (8)). If the biases caused by physical noise are not accounted for, they invalidate events related to the biological signals of interest and substantially decrease the power of the statistical analysis. Therefore, the removal of low-frequency drift is one of the most important steps in reconstructing 3-dimensional fish trajectories. However, this pre-processing step is, unfortunately, also one of the most precarious steps because the biological signal of interest may easily be removed if incorrect filters are applied.

In this study, we carefully analyzed the time series of ocean current data that were collected in an experimental net cage using a high-pass filter to remove the drift signal (which decreases slowly), as shown next.

Reconstructing Swimming Trajectories in a Net Cage

Spectra analysis of the circling of fish

We conducted a spectra analysis of the sine component of fish heading. In general, the fish swam clockwise in the net cage during each period (Fig. 18.3). The peaks in frequency were from 0.01 to 0.02 Hz at night and from 0.01 to 0.1 Hz during the day. Daytime peaks varied compared to night-time peaks. If the fish swam horizontally

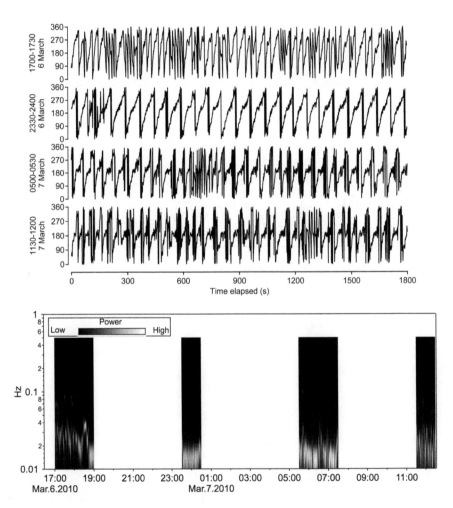

Figure 18.3. Top: The time series data showing the heading of the circling fish during each period. Values of degrees show the azimuth angle θ_a. For instance, 0 and 360 degrees mean North direction, 90 degrees means East, 180 degrees means South, 270 degrees means West. Bottom: The power spectra of the sine component of the heading $\theta(n)$ with time, which were obtained for each measuring period. The grey-scale expresses the power of frequency. Significant peaks appear in the frequency spectrum from 0.02 to 0.01 Hz per period. We found that the peak frequencies and the peaks were located from 0.01 Hz to 0.1 Hz during the daytime. At night, the peaks were located around 0.01 Hz.

at 1.46 FL s⁻¹ on an orbit of 30 m diameter, the lap time is estimated at 127 seconds. These significant, higher-frequency peaks equalled the time needed for a tuna to swim one circumference of the net cage within 10 to 100 seconds.

Spectra analysis of the tidal current

The results of the current profile measurements are shown in Fig. 18.4. Current speeds from 0.2 to 11 cm·s⁻¹ were observed (mean current velocity, 6.2 cm·s⁻¹ 1700–1900 March 6, 3.0 cm·s⁻¹ 2330–0030 March 6, 2.6 cm·s⁻¹ 0500–0700 March 7, 2.0 cm·s⁻¹

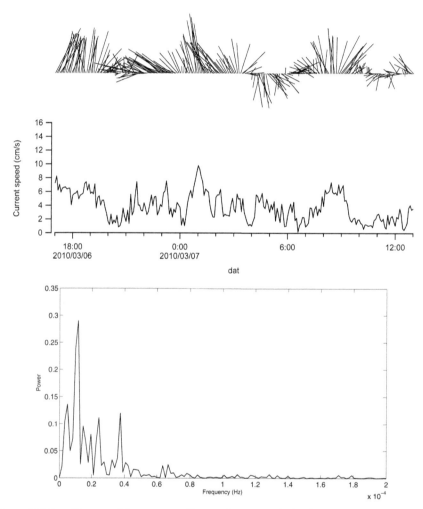

Figure 18.4. Top: Close-up of the current profile near the net cage during the measurement period. The stick diagram indicates the current velocities. An upward-oriented stick represents current flowing from south to north, while the stick length represents the velocity. The time series data of the black line expresses the current speed contemporary of the stick diagram. Bottom: The power spectra of the ocean current at a depth of 12 minutes in the net cage from March 6 to March 12. Significant peaks were observed at around 24.4 hours and 11.5 hours, with smaller significant peaks at 7.5 hours for one week.

1130–1230 March 7). No strong ocean currents were measured during the study period. For instance, ocean currents exceeding 20 cm·s⁻¹ could reduce the volume of the cage (Suzuki et al. 2009). Current data were not otherwise used in this study.

We conducted a spectra analysis on the time series of ocean current data to estimate which frequency contributed to drift. The power spectra of the ocean current at a depth of 12 m within the net cage were calculated over six days, including the day on which the experiment was conducted (Fig. 18.4). Although the dataset was not large, our analysis resolved the tidal signatures, with significant peaks at around 24.4 and 11.5 hours and a smaller significant peak at 7.5 hours. These spectral peaks are consistent with those reported by Stockwell et al. (2004), who used data from more than 19,000 monthly time series taken from 262 data buoy sites and discovered peaks at 24 hours, 12 hours, and 8 hours along the frequency spectrum. Thus, our analysis indicates that most of the ocean current power spectrum at a depth of 12 m was associated with the tidal signal, allowing us to remove the frequency that contributed a major portion of the drift by selecting cut-off values of less than 8 hours. In addition, although ocean currents fluctuate over short periods, this high-frequency fluctuation was disregarded as noise for reconstructing the 3-dimensional tuna trajectories in the present study, as we were only concerned with a time scale of a few minutes to a few hours. Thus, we calculated the energy that was concentrated at a frequency greater than 0.00021 Hz (fourth peak in Fig. 18.4, bottom), and found that 31% of the ocean current energy occurred at these higher frequencies. This fluctuation in ocean current energy needed to be removed from our observational data. For example, the reconstructed trajectories drift 180 m·h⁻¹ at 5 cm·s⁻¹ current velocity if the drift only affects the ocean current. Thus, the drift consists of both the tidal current and some observational error. This finding is consistent with that observed in a flume tank experiment described by Komeyama et al. (2013).

Reconstructed 3-dimensional trajectories

The swimming path reconstructed by dead reckoning was visualized, and the error bias in the southwest direction was confirmed (Fig. 18.5). When reconstructing the 3-dimensional swimming paths of bluefin tuna, Komeyama et al. (2011) assumed constant linear drift over time and, thus, subtracted drift error by means of the least-squares fit. However, external effects, such as ocean currents, vary over time, invalidating the linear drift assumption and leading to swimming paths that include a large error. Therefore, we compared the linear drift and high-pass filter methods, and evaluated the nonlinear components of the drift error from a frequency perspective.

Given the highly linear relationship between the *x*-component of the locomotion vector and time (Fig. 18.6, top), a linear trend was fitted to the data and the accumulated error removed to give a corrected swimming path (red line in Fig. 18.6, top). The blue line is the *x*-component of the locomotion vector that was removed by the high-pass filter with a cut-off frequency of 0.00021 (Fig. 18.6, bottom). Neither of these methods produced trajectories that fitted within the diameter of the cage (maximum 30 m), as the maximum and minimum values for both methods were greater than 15 m and less than −15 m, respectively. This result indicated that a nonlinear effect due to high frequency range (>0.00021 Hz) needed to be removed to reduce the amplitude of the

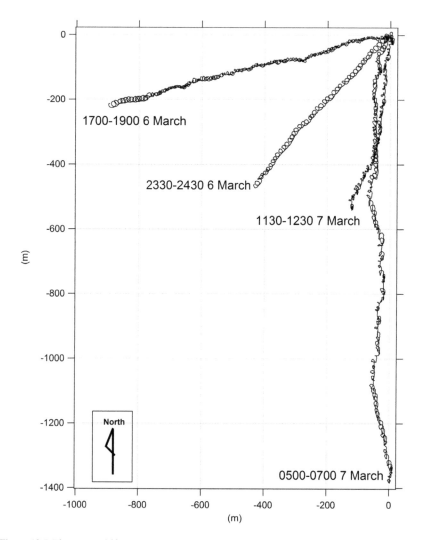

Figure 18.5. The current drift and observational error when reconstructing the trajectories. The swimming trajectories reconstructed for each period using the dead-reckoning technique showing the 2-dimensional coordinate system. The x coordinate shows eastward and the y coordinate shows northward. The starting point of each trajectory is the origin (0, 0). The centre of the aquaculture net cage is represented by the origin on the figure. In the figure, the upwards direction represents north.

trajectories. Therefore, we estimated appropriate cut-off values based on the above analysis, and used a high-pass filter to remove all frequencies contributing to drift in the observational data.

The black line in Fig. 18.6 reflects the removal of all frequencies less than 0.0015 Hz. The resultant data were then back-transformed into the time domain. The low-frequency part (<0.0015 Hz) of the x-component of the locomotion vector was removed by the high-pass filter and successfully removed drift from the data (Fig. 18.6). Therefore, we applied the same filter to all of the observational data, removing

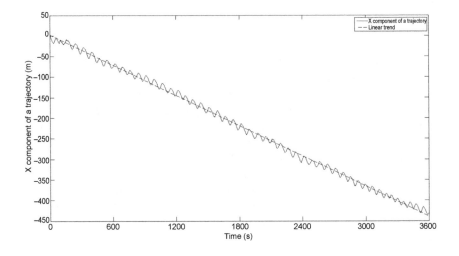

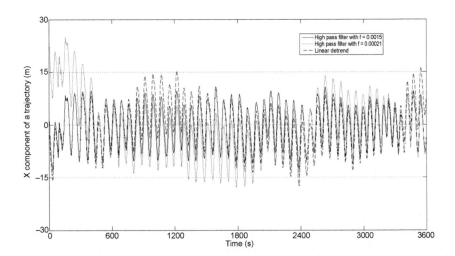

Figure 18.6. Top: The linear trend in the x-component of the locomotion vector through time. The dotted lines represent the linear trend of the locomotion vector. Bottom: Corrected trajectories calculated by subtracting a linear trend or using the designed high-pass filter. The maximum and minimum values are denoted by the dotted line (linear square regression method) and grey line (the high-pass filter with a cut-off frequency of 0.00021 Hz), respectively. The black line is the low-frequency (<0.0015 Hz) signal in the x-component of the locomotion vector that was removed by the high-pass filter. The maximum diameter of the net cage was 30 m. A radius of over ∓15 m meant that the trajectory fell outside of the net cage.

any drift due to low-frequency movement. The amplitude of the *x*-component of the 3-dimensional trajectory derived using this method did not exceed the radius of the net cage (Fig. 18.6).

These high-pass filters were designed using the MATLAB Signal Processing Toolbox with a third-order Butterworth filter type. To validate the swimming trajectories calculated in this study, we examined whether the trajectories fit into the aquaculture net cage. We chose the origin (0, 0, *d*(0)) as the starting point of a trajectory. It was difficult to obtain data on the diameter of the cage at each depth because the net cage had a bowl-like shape with a maximum diameter of 30 m near the frame on top of the cage, but a gradually decreasing diameter with increasing depth because of pull by the sinker. Therefore, data were estimated using Net Shape and Loading Analysis (NaLA) software (Takagi et al. 2002; Suzuki et al. 2009), which numerically estimated the geometry and internal forces acting on the net and rigging.

We reconstructed trajectories for all time periods (Fig. 18.7). We chose the origin as the starting point of each trajectory, but any point could have been selected without affecting the shape of the trajectory, swimming speed, or heading of the fish (although the distance between the fish and the cage wall may have changed). The calculated trajectories fit within the diameter (30 m) of the net cage. Almost all fish positions were located within each layer of the net cage. We then calculated the trajectories for the other time periods (dawn, daytime, dusk) in the same manner as the reconstructed trajectories and found that they also fitted within the maximum diameter of each layer of the net cage (Fig. 18.7). Partial trajectories that extended beyond the cage were observed fairly infrequently and could be explained by observational error in equations (7) and (8). Using this method, accuracy was confirmed at approximately 0.2 m, as described in Komeyama et al. (2013).

Applying the Dead-Reckoning Technique to Cultivated Bluefin Tuna

Examination of the obtained swimming path shows how a fish swims in an aquaculture net cage (Fig. 18.7). This study provides a useful initial evaluation of whether offshore submerged net cages represent suitable living spaces for bluefin tuna.

Swimming speed measurements taken by the PD3GT data logger represent speed through the water rather than relative to the ground, making them subject to tidal effects. Furthermore, the drift trajectory results shown in Fig. 18.5 imply that the dead-reckoning technique cannot accurately estimate swimming trajectories where there are ocean/tidal currents. Mitani et al. (2003) applied a dead-reckoning technique similar to the PD3GT used in this study to reconstruct the 3-dimensional trajectories of Weddell seals (*Leptonychotes weddellii*), and suggested that the accumulated error was the result of ocean currents. Shiomi et al. (2008) also recognized the influence of current flow when calculating trajectories. However, although these studies provided clear start and end points for calculating animal trajectories, the reconstructed end points were not consistent with the actual end points. In this study, the fish was in a net cage throughout the observational period. However, the fish appeared to move outside the net cage despite calm current conditions, making it difficult to determine the exact path, because of the influence of tidal currents. Previous studies (Komeyama et al. 2011) attempted to remove such effects by assuming a linear accumulation of errors. However, although this technique corrected a portion of the trajectory error, it did not completely remove temporal changes in the tidal component. Moreover, the current changes in three dimensions, making it difficult to measure current velocities.

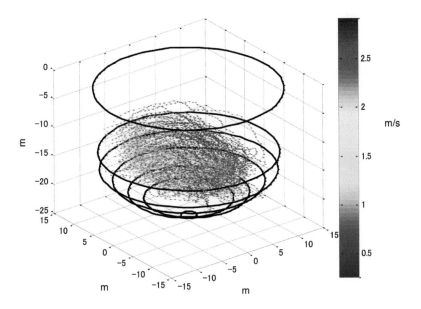

5:00

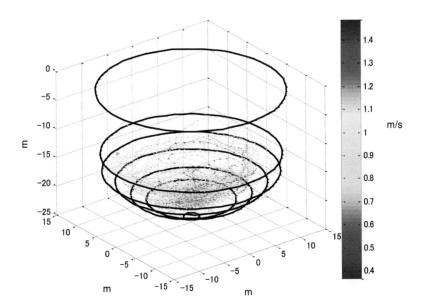

11:30

Figure 18.7. contd....

414   *Biology and Ecology of Bluefin Tuna*

Figure 18.7. contd.

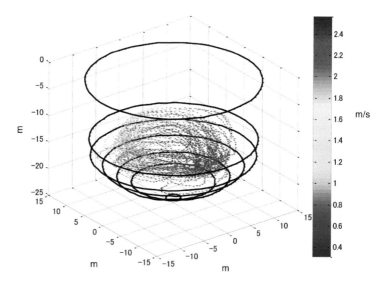

17:00

11:30

Figure 18.7. Reconstructed trajectories for each phase. Different coloured trajectories indicate different swimming speeds. Thick black circular lines in each layer represent the maximum diameter of the submerged net cage estimated using the Net Shape and Load Analysis system (NaLA; Takagi et al. 2003). The fish trajectories started at the origin. Therefore, researchers should exclude several trajectory periods from the start point when analyzing certain trajectory parameters, such a fish turning and the space use of aquaculture fish in net cages.

In contrast, the high-pass filter method used in the present study was effective at removing accumulated error, even when there were 3-dimensional changes in tidal flow. Even if current flow changes vertically, this method is likely to remove component periodically in frequency. In addition, the direction of the horizontal current in Fig. 18.4 and the direction of drift in Fig. 18.5 did not match (e.g., eastward current was observed at 0500–0700, but the observed drift was southward). The low-frequency component, which accumulated observational error, and periodically changing tidal current, could be removed by a high-pass filter, as demonstrated by this method. However, the response of the filter was extremely low near the starting point because of the characteristics of the filter. Therefore, further research is required to determine how to best overcome this problem (e.g., by excluding the first part of the trajectory from the analysis). The methods presented here generated sufficiently accurate trajectories for analyzing the turning performance and space use by fish cultivated in net cages.

By analyzing the frequency of the circling fish in combination with time series data for current velocity, we determined the 3-dimensional trajectories of a bluefin tuna circling within a submerged aquaculture net cage more accurately than has been accomplished in previous studies. The corrected paths help to determine swimming speed, inclination of the circle, and swimming depth, providing useful information about the space use of cultivated fish in aquaculture net cages. Given that a monitoring technique does not currently exist for cultivated tuna, researchers have not been able to determine whether a causal relationship exists between fish mortality and reduced living space. However, we believe that visualization of the trajectories of circling fish using the methods developed here will help increase the efficiency of bluefin tuna cultivation by contributing to future behavioural studies of how fish turn when they meet obstacles (e.g., cage walls and other fish) and how fish schools move in outdoor cages.

The distance at which tagged fish swim from the wall of a net cage remains poorly understood. However, existing knowledge could be improved by determining the location of a fish within the cage or by assessing the distance of fish trajectories from the wall. To use the method proposed in this study effectively, several trajectory pass points should be measured to ensure that they are correctly located and reflect the absolute coordinates.

Application of the proposed method may be limited to circling fish under aquaculture conditions, as additional challenges will arise when measuring trajectories of free-ranging aquatic animals in the open sea. However, exact trajectories in the open sea could potentially be estimated by removing any drift components that fluctuate periodically, such as tides. Alternatively, if noise sources cannot be identified, acoustic telemetry could be used to detect fish positions within estimated trajectories in the open sea.

Acknowledgements

We express our sincere gratitude to Dr. Minoru Kadota (Temple University, Japan), Dr. Shinsuke Torisawa (Kinki University), Dr. Yuichi Tsuda (Fisheries Laboratory, Kinki University), Dr. Katsuya Suzuki (Nittoseimo Co., Ltd.), Mr. Eita Ogata, Mr.

Kazunari Tanaka (Faculty of Fisheries, Kagoshima University), and the staff of Kinki University for providing the specimens required for our study.

We sincerely thank Mr. Masamchi Kanechiku (Kinki University) for his help and support with the NaLA system. We also thank Mr. Shigeru Asaumi (Furuno Electric Co., Ltd.), Mr. Tsugihiko Kobayashi (Taiyo A&F Co., Ltd.), and the Association of Marino-Forum 21 for their help and support. This study was financially supported by a Grant-in-Aid for Young Scientists (B) (23780200) from the Ministry of Education, Culture, Sports, Science and Technology and was supported in part by a Grants-in-Aid for Scientific Research (C) (22580214), and the Global COE program 'Centre of Aquaculture Science and Technology for Bluefin Tuna and Other Cultivated Fish' of Kinki University from the Japan Society for the Promotion of Science.

References

Bégout Anras, M.-L. and J.P. Lagardère. 2004. Measuring cultured fish swimming behaviour: first results on rainbow trout using acoustic telemetry in tanks. Aquaculture 240: 175–186.

Blank, J.M., C.J. Farwell, J.M. Morrissette, R.J. Schallert and B.A. Block. 2007. Influence of swimming speed on metabolic rates of juvenile Pacific bluefin tuna and yellowfin tuna. Physiol. Biochem. Zool. 80: 167–177.

Block, B.A., H. Dewar, S.B. Blackwell, T.D. Williams, E.D. Prince, C.J. Farwell, A. Boustany, S.L.H. Teo, A. Seitz, A. Walli and D. Fudge. 2001. Migratory movements, depth preferences, and thermal biology of Atlantic bluefin tuna. Science 293: 1310–1314.

Ferraroli, S., J.Y. Georges, P. Gaspar and Y. Le Maho. 2004. Endangered species—where leatherback turtles meet fisheries. Nature 429: 521–522.

Harcourt, R.G., M.A. Hindell, D.G. Bell and J.R. Waas. 2000. Three-dimensional dive profiles of free-ranging Weddell seals. Polar Biol. 23: 479–487.

Hays, G.C., J.D.R. Houghton and A.E. Myers. 2004. Endanqered species—Pan-Atlantic leatherback turtle movements. Nature 429: 522–522.

Hindell, M.A., R. Harcourt, J.R. Waas and D. Thompson. 2002. Fine-scale three-dimensional spatial use by diving, lactating female Weddell seals Leptonychotes weddellii. Marine Ecology Progress Series 242: 275–284.

Hooker, S.K., H. Whitehead, S. Gowans and R.W. Baird. 2002. Fluctuations in distribution and patterns of individual range use of northern bottlenose whales. Marine Ecology Progress Series 225: 287–297.

Jouventin, P. and H. Weimerskirch. 1990. Satellite tracking of Wandering Albatrosses. Nature 343: 746–748.

Kadota, M., E.J. White, S. Torisawa, K. Komeyama and T. Takagi. 2011. Employing relative entropy techniques for assessing modifications in animal behavior. PloS one 6: e28241.

Klimley, A.P., B.J. Le Boeuf, K.M. Cantara, J.E. Richert, S.F. Davis and S. Van Sommeran. 2001. Radio-acoustic positioning as a tool for studying site-specific behavior of the white shark and other large marine species. Mar. Biol. 138: 429–446.

Komeyama, K., M. Kadota, S. Torisawa, K. Suzuki, Y. Tsuda and T. Takagi. 2011. Measuring the swimming behaviour of a reared Pacific bluefin tuna in a submerged aquaculture net cage. Aquat. Living Resour. 24: 99–105.

Komeyama, K., M. Kadota, S. Torisawa and T. Takagi. 2013. Three-dimensional trajectories of cultivated Pacific bluefin tuna Thunnus orientalis in an aquaculture net cage. Aquacult. Environ. Interact. 4: 81–90.

Kubo, T., W. Sakamoto and H. Kumai. 2004. Correlation between oceanic environmental fluctuation and bluefin tuna behavior in the aquaculture pen. Proceedings of the International Symposium on SEASTAR2000 and Bio-logging Science (The 5th SEASTAR2000 Workshop).

Marcinek, D.J., S.B. Blackwell, H. Dewar, E.V. Freund, C. Farwell, D. Dau, A.C. Seitz and B.A. Block. 2001. Depth and muscle temperature of Pacific bluefin tuna examined with acoustic and pop-up satellite archival tags. Mar. Biol. 138: 869–885.

Mitamura, H., K. Uchida, Y. Miyamoto, T. Kakihara, A. Miyagi, Y. Kawabata, K. Ichikawa and N. Arai. 2012. Short-range homing in a site-specific fish: search and directed movements. J. Exp. Biol. 215: 2751–2759.

Mitani, Y., K. Sato, S. Ito, M.F. Cameron, D.B. Siniff and Y. Naito. 2003. A method for reconstructing three-dimensional dive profiles of marine mammals using geomagnetic intensity data: results from two lactating Weddell seals. Polar Biol. 26: 311–317.

Okano, S., Y. Mitsunaga, W. Sakamoto and H. Kumai. 2006. Study on swimming behavior of cultured Pacific bluefin tuna using biotelemetry. Memoirs of the Faculty of Agriculture of Kinki University 39: 79–82.

Shiomi, K., K. Sato, H. Mitamura, N. Arai, Y. Naito and P.J. Ponganis. 2008. Effect of ocean current on the dead-reckoning estimation of 3-D dive paths of emperor penguins. Aquat. Biol. 3: 265–270.

Sims, D.W., E.J. Southall, G.A. Tarling and J.D. Metcalfe. 2005. Habitat-specific normal and reverse diel vertical migration in the plankton-feeding basking shark. J. Anim. Ecol. 74: 755–761.

Stockwell, R.G., W.G. Large and R.F. Milliff. 2004. Resonant inertial oscillations in moored buoy ocean surface winds. Tellus A 56: 536–547.

Suzuki, K., S. Torisawa, T. Takagi. 2009. Numerical analysis of net cage dynamic behavior due to concurrent waves and current. Proceedings of the ASME (28th International Conference on Ocean, Offshore and Arctic Engineering) 1513–1520.

Takagi, T., K. Suzuki and T. Hiraishi. 2002. Development of the numerical simulation method of dynamic fishing net shape. Nippon Suisan Gakk 68: 320–326.

Tanaka, H., Y. Naito, N.D. Davis, S. Urawa, H. Ueda and M.A. Fukuwaka. 2005. First record of the at-sea swimming speed of a Pacific salmon during its oceanic migration. Marine Ecology Progress Series 291: 307–312.

Tsuda, Y., R. Kawabe, H. Tanaka, Y. Mitsunaga, T. Hiraishi, K. Yamamoto and K. Nashimoto. 2006. Monitoring the spawning behaviour of chum salmon with an acceleration data logger. Ecol. Freshw. Fish. 15: 264–274.

Wardle, C.S., J.J. Videler, T. Arimoto, J.M. Franco and P. He. 1989. The muscle twitch and the maximum swimming speed of giant bluefin tuna, *Thunnus Thynnus* L. J. Fish Biol. 35: 129–137.

Weimerskirch, H., A. Gault and Y. Cherel. 2005. Prey distribution and patchiness: Factors in foraging success and efficiency of wandering albatrosses. Ecology 86: 2611–2622.

Wilson, R.P. and M.P. Wilson. 1988. Dead reckoning - a new technique for determining penguin movements at sea. Meeresforschung 32: 155–158.

Wilson, R.P., B.M. Culik, R. Bannasch and J. Lage. 1994. Monitoring Antarctic environmental variables using penguins. Marine Ecology Progress Series 106: 199–202.

Wilson, R.P., N. Liebsch, I.M. Davies, F. Quintana, H. Weimerskirch, S. Storch, K. Lucke, U. Siebert, S. Zankl, G. Müller, I. Zimmer, A. Scolaro, C. Campagna, J. Plötz, H. Bornemann, J. Teilmann and C.R. McMahon. 2007. All at sea with animal tracks; methodological and analytical solutions for the resolution of movement. Deep Sea Research Part II: Topical Studies in Oceanography 54: 193–210.

Yasuda, T. and N. Arai. 2009. Changes in flipper beat frequency, body angle and swimming speed of female green turtles Chelonia mydas. Marine Ecology Progress Series 386: 275–286.

Index

For Product Safety Concerns and Information please contact our EU representative GPSR@taylorandfrancis.com Taylor & Francis Verlag GmbH, Kaufingerstraße 24, 80331 München, Germany

T - #0196 - 160425 - C39 - 234/156/19 - PB - 9780367737993 - Gloss Lamination